Automatic
Control
Systems

Automatic Control Systems

fifth edition

Benjamin C. Kuo

Professor of Electrical and Computer Engineering
University of Illinois at Urbana-Champaign

PRENTICE-HALL, INC., Englewood Cliffs, New Jersey 07632

Library of Congress Cataloging-in-Publication Data

KUO, BENJAMIN C., (date)
   Automatic control systems.

   Includes bibliographical references and index.
   1. Automatic control. 2. Control theory. I. Title.
TJ213.K8354 1987   629.8'3   86-17047
ISBN   0-13-054842-1

Editorial/production supervision
   and interior design: Elena Le Pera
Cover: Original design by Benjamin C. Kuo, adapted by Karen Stephens
Manufacturing buyer: Rhett Conklin

Printed in the United States of America

10 9 8 7 6 5 4 3 2

ISBN 0-13-054842-1   025

Prentice-Hall International (UK) Limited, *London*
Prentice-Hall of Australia Pty. Limited, *Sydney*
Prentice-Hall Canada Inc., *Toronto*
Prentice-Hall Hispanoamericana, S. A., *Mexico*
Prentice-Hall of India Private Limited, *New Delhi*
Prentice-Hall of Japan, Inc., *Tokyo*
Prentice-Hall of Southeast Asia Pte. Ltd., *Singapore*
Editora Prentice-Hall do Brasil, Ltda., *Rio de Janeiro*

Contents

9   FREQUENCY-DOMAIN ANALYSIS OF CONTROL SYSTEMS   *552*

10   FREQUENCY-DOMAIN DESIGN OF CONTROL SYSTEMS   *647*

APPENDICES

Preface

The main motivation for preparing the fifth edition of *Automatic Control Systems* is to demonstrate the importance of computer-aided learning in the field of control systems. Control systems represent one of the areas in which computers are heavily involved not only for analysis and design, but also as controllers. Most of the design methods in classical control theory rely heavily on trial-and-error and do not lead to a unique solution to a given design problem. In optimal control, the design algorithms are so complex that numerical iterations are often necessary. Thus, except for simple systems, it would be very tedious to carry out the analysis or design of a control system without the aid of a computer. While large-scale scientific computers are in abundance and easily accessible, the reader is still faced with the responsibility of acquiring or preparing programs that can be used on the computer to which he or she can access. These are all time-consuming tasks. The author believes that whenever possible, time should be spent on understanding and executing control problems, rather than on the drudgery of carrying out the calculations.

In the last few years, personal computers have become so powerful and available, that it is now possible for most people to use them to solve complex control system problems easily and efficiently. This development should make the study of control systems easier, as the student can go through large numbers of analysis and design runs in order to learn its theories.

The fifth edition is accompanied by two Computer Disks and a Software Manual. The disks contain programs that can be executed by an IBM PC or any compatible personal computer. These programs can be used for the solution of most of the problems presented in the book. The Computer Disks and/or Software Manual can be ordered from the Publisher using the reply card in the back of this book. The disks can be duplicated for multiple users. For more information, see Preface to Computer Disks and Software Manual.

The book is intended as an undergraduate introductory text on linear control systems. Although a substantial number of new developments have been made in modern control theory during the past twenty years, preparing suitable and meaningful material for a modern introductory course on control systems remains a difficult task. The problem is complicated because of the complexity of real-life control systems which are truly nonlinear, have time-varying parameters, and often demonstrate uncertain properties. It is generally recognized that a sharp division exists between optimal control theory and classical control techniques. The fact is that it is difficult to teach optimal control theory at the undergraduate introductory level; furthermore, many practical control system problems are still being solved in industry by classical methods. Although some of the techniques used in modern control theory are much more powerful and can solve more complex problems than can the classical methods, there are often more restrictions when it comes to the practical applications of these methods. The author believes that a modern control engineer should have a broad knowledge of both classical and optimal control techniques. A knowledge of modern control theory should enhance one's analytical perspective in solving complex control problems. Therefore, it is important to introduce the subject of state variables and state equations at the introductory level, even though the treatment may be less than rigorous from the theoretical standpoint. The exposure should allow the reader to prepare for more advanced studies in control systems, especially, modern control theory. It has been said that the study of control systems is essentially an attempt to learn a variety of techniques and solutions—all with the intention of solving only *one* problem; that is, to come up with a system that will satisfy all design specifications. Although the statement may sound a bit oversimplified, the various chapters of the text do represent alternative analysis and design methods such as the time-domain and the frequency-domain techniques.

The development of microcomputers has been explosive in the past few years. Today, many control systems feature digital controllers which contain microprocessors. This means that the beginning student needs to be exposed to digital control theory, so that he or she will realize the importance of seeking more advanced studies of the subject. The reader should find it useful and enjoyable to study digital control together with analog control, since a majority of the analysis and design methods of digital control are extensions of their analog counterparts.

The reader should keep in mind that although we are trying to promote the use of computers to take out the drudgery of control systems design and analysis, we should not be totally dependent on the computer. This is why we still have to rely heavily on the theory and analytical development presented in the book and use the computer only as a tool. It should be emphasized that the well-learned engineer should not only know how to use the computer, but also know how to interpret the results and to judge their accuracy. Quite often, the engineer should also know how to prepare the computer program in order to solve a given problem. Without the complete understanding of basic principles, the designer would not be able to do so.

The material assembled in this book is an outgrowth of a senior-level control system course taught by the author at the University of Illinois at Urbana-Champaign for many years. The book is also suitable for self-study and professional reference. A complete solutions manual is available from the publisher for qualified instructors.

The fifth edition has a total of ten chapters and two appendices. Chapter 1 presents the basic concepts of control systems. The definition and effects of feedback are given. Various types of control systems are defined and illustrated. Chapter 2 presents the mathematical foundation and preliminaries. The subjects covered are Laplace transform, z-transform, differential and difference equations, matrix algebra, and the applications of transform methods. Transfer functions and signal-flow-graph methods of modeling linear systems are discussed in Chapter 3. Chapter 4 covers the mathematical modeling of physical systems. Typical transducers and prime movers used in control systems are illustrated. The treatment of systems modeling cannot be exhaustive because in practice there are numerous types of control devices. Chapter 5 introduces the state-variable method of analyzing dynamical systems. Definitions of controllability and observability are given. Chapter 6 describes time-domain analysis of control systems. The performance of a control system is evaluated in the real-time domain and using pole-zero interpretation in the s-plane. Chapter 7 discusses the root locus method. In Chapter 8, various design methods using time-domain techniques are presented. One feature of this revised edition is that a class of design techniques is discussed immediately following the analysis topics. This way, the reader will still be fresh on the basic concepts so that the analysis and design techniques can be closely related and cross-referenced. Chapter 9 deals with frequency-domain analysis and the Nyquist stability criterion, and Chapter 10 discusses the design of control systems in the frequency domain.

The book does contain more material than can be covered in one semester. When used as a college text, some of the review material, such as the contents of Chapter 2, can be assigned as review reading. For courses that are intended to cover only the classical methods of analysis and design, Chapter 5 can be bypassed, and some sections in the design chapters that are related to state variables can be eliminated without loss of continuity. The coverage of digital control systems, although integrated with that of continuous-data systems, can easily be omitted if desired. The author feels that the material presented in the text can be arranged for either a one-semester or a two-quarter course.

Department of Electrical
 and Computer Engineering
 University of Illinois
 at Urbana-Champaign
 Urbana, Illinois 61801

B. C. KUO

PREFACE TO COMPUTER DISKS AND SOFTWARE MANUAL

The analysis and design of control systems has been sharply affected by the widespread use of personal computers. With the mainframe computer, the user may not have the convenience of getting access to the computer at any time or at any place. There may also be no convenient software package to conduct systems analysis and design; this is particularly true for engineers in small companies and for some engineering students.

Today, most schools and companies are equipped with personal computers that are easily accessible and with minimum initial costs.

The two Computer Disks contain programs that can be run on the IBM PC, XT, AT®, or any compatible personal computer. A family of these programs are devised for general linear continuous-data and discrete-data systems analysis and design. Some programs are devoted to reviews and exercises to accompany the text, *Automatic Control Systems*, fifth edition, by B. C. Kuo, published by Prentice-Hall, Inc., 1987.

The primary objective of the programs in the Computer Disks and Software Manual is not to solve highly complex systems but to educate. Thus, most of the programs are capable of handling up to an eighth-order system, which seems adequate for average day-to-day use. Some limitations do exist in terms of magnitudes of the system parameters that can be treated effectively. The main advantage is that the programs can be run on a moderately equipped PC with two floppy disk drives, 256K of memory, and a monochrome monitor.

The author hopes that the computer programs will be useful to engineers and students in their day-to-day activities of solving linear control systems problems. For those who are studying from the text, *Automatic Control Systems*, fifth edition, the Computer Disks and Software Manual should make the task more rewarding and enjoyable.

B. C. KUO

chapter one

Introduction

1.1 CONTROL SYSTEMS

In this introductory chapter we attempt to familiarize the reader with the following subjects:

1. What a control system is.
2. Why control systems are important.
3. What the basic components of a control system are.
4. Why feedback is incorporated into most control systems.
5. Types of control systems.

With regard to the first two items, we cite the example of the human being as perhaps the most sophisticated and the most complex control system in existence. An average human being is capable of performing a wide range of tasks, including decision making. Some of these tasks, such as picking up objects, or walking from one point to another, are normally carried out in a routine fashion. Under certain conditions, some of these tasks are to be performed in the best possible way. For instance, an athlete running a 100-yard dash has the objective of running that distance in the shortest possible time. A marathon runner, on the other hand, not only must run the distance as quickly as possible, but in doing so, he or she must control the consumption of energy so that the best result can be achieved. Therefore, we can state in general that in life there are numerous "objectives" that need to be accomplished, and the means of achieving the objectives usually involve the need for control systems.

In recent years control systems have assumed an increasingly important role in the development and advancement of modern civilization and technology. Practi-

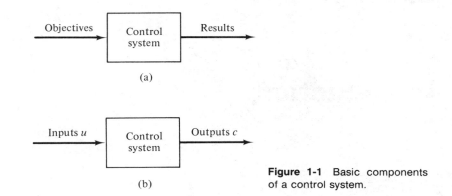

Figure 1-1 Basic components of a control system.

cally every aspect of our day-to-day activities is affected by some type of control system. For example, in the domestic domain, automatic controls in heating and air-conditioning systems regulate the temperature and humidity of homes and buildings for comfortable living. To achieve maximum efficiency in energy consumption, many modern heating and air-conditioning systems in large office and factory buildings are computer controlled.

Control systems are found in abundance in all sectors of industry, such as quality control of manufactured products, automatic assembly line, machine-tool control, space technology and weapon systems, computer control, transportation systems, power systems, robotics, and many others. Even such problems as inventory control, and social and economic systems control, may be approached from the theory of automatic controls.

Regardless of what type of control system we have, the basic ingredients of the system can be described by

1. Objectives of control.
2. Control system components.
3. Results.

In block diagram form, the basic relationship between these three basic ingredients is illustrated in Fig. 1-1(a).

In more scientific terms, these three basic ingredients can be identified with inputs, system components, and outputs, respectively, as shown in Fig. 1-1(b).

In general, the objective of the control system is to control the outputs c in some prescribed manner by the inputs u through the elements of the control system. The inputs of the system are also called the *actuating signals*, and the outputs are known as the *controlled variables*.

As a simple example of the control system fashioned in Fig. 1-1, consider the steering control of an automobile. The direction of the two front wheels may be regarded as the controlled variable c, or the output; the direction of the steering wheel is the actuating signal u, or the input. The control system or process in this case is composed of the steering mechanisms and the dynamics of the entire automobile. However, if the objective is to control the speed of the automobile, then

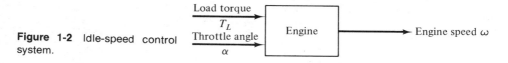

Figure 1-2 Idle-speed control system.

the amount of pressure exerted on the accelerator is the actuating signal, and the vehicle speed is the controlled variable. As a whole, we may regard the automobile control system as one with two inputs (steering and accelerator) and two outputs (heading and speed). In this case, the two controls and outputs are independent of each other; but in general, there are systems for which the controls are coupled. Systems with more than one input and one output are called *multivariable systems*.

As another example of a control system, we consider the idle-speed control of an automobile engine. The objective of such a control system is to maintain the engine idle speed at a relatively low value (for fuel economy) regardless of the applied engine loads (e.g., transmission, power steering, air conditioning, etc.). Without the idle-speed control, any sudden engine load application would cause a drop in engine speed which might cause the engine to stall. Thus, the main objectives of the idle-speed control system are (1) to eliminate or minimize the speed droop when engine loading is applied, and (2) to maintain the engine idle speed at a desired value. Figure 1-2 shows the block diagram of the idle-speed control system from the standpoint of inputs–system–outputs. In this case, the throttle angle α and the load torque T_L (due to the application of air conditioning, power steering, transmission, or brakes, etc.) are the inputs, and the engine speed ω is the output. The engine is the controlled process or system.

Open-Loop Control Systems (Nonfeedback Systems)

The idle-speed control system illustrated in Fig. 1-2 is rather unsophisticated and is called an *open-loop control system*. It is not difficult to see that the system as it is shown would not satisfactorily fulfill the desired performance requirements. For instance, if the throttle angle α is set at a certain initial value, which corresponds to a certain engine speed, when a load torque T_L is then applied, there is no way to prevent a drop in the engine speed. The only way to make the system work is to have means of adjusting α in response to a change in the load torque, in order to maintain ω at the desired level.

Because of the simplicity and economy of open-loop control systems, we may find this type of system in practical use in numerous situations. In fact, practically all automobiles manufactured prior to 1981 did not have an idle-speed control system.

An electric washing machine is another example of an open-loop system because, typically, the amount of machine wash time is entirely determined by the judgment and estimation of the human operator. A true automatic electric washing machine should have the means of checking the cleanliness of the clothes being washed continuously and turn itself off when the desired degree of cleanliness is reached.

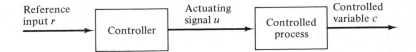

Figure 1-3 Elements of an open-loop control system.

The elements of an open-loop control system can usually be divided into two parts: the controller and the controlled process, as shown by the block diagram in Fig. 1-3. An input signal or command r is applied to the controller, whose output acts as the actuating signal u; the actuating signal then controls the controlled process so that the controlled variable c will perform according to some prescribed standards.

In simple cases, the controller can be an amplifier, mechanical linkages, or other control means, depending on the nature of the system. In more sophisticated electronics control, the controller can be an electronic computer, such as a microprocessor.

Closed-Loop Control Systems (Feedback Control Systems)

What is missing in the open-loop control system for more accurate and more adaptive control is a link or feedback from the output to the input of the system. To obtain more accurate control, the controlled signal $c(t)$ should be fed back and compared with the reference input, and an actuating signal proportional to the difference of the input and the output must be sent through the system to correct the error. A system with one or more feedback paths such as that just described is called a *closed-loop system*.

The block diagram of a closed-loop idle-speed control system is shown in Fig. 1-4. The reference input ω_r sets the desired idling speed. Ordinarily, when the load torque is zero, the engine speed at idle should agree with the reference value ω_r, and any difference between the actual speed and the desired speed caused by any disturbance such as the load torque T_L is sensed by the speed transducer and the

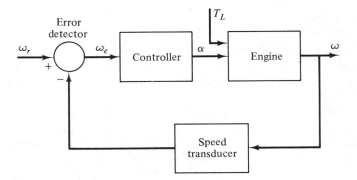

Figure 1-4 Closed-loop idle-speed control system.

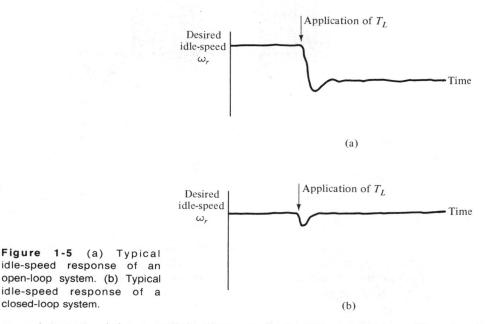

(a)

Figure 1-5 (a) Typical idle-speed response of an open-loop system. (b) Typical idle-speed response of a closed-loop system.

(b)

error detector, and the controller will operate on the difference and provide a signal to adjust the throttle angle α to correct the error.

Figure 1-5 illustrates a comparison of the typical performances of the open-loop and closed-loop idle-speed control systems. In Fig. 1-5(a), the idle speed of the open-loop system will drop and settle at a lower value after a load torque is applied. In Fig. 1-5(b) the ideal speed of the closed-loop system is shown to recover quickly to the preset value after the application of T_L.

The idle-speed control system illustrated above is also known as a *regulator system* whose objective is to maintain the system output at some prescribed level.

As another illustrative example of a closed-loop control system, Fig. 1-6 shows the block diagram of the printwheel control system of a word processor or electronic typewriter. The printwheel, which typically has 96 or 100 characters, is to be rotated to position the desired character in front of the hammer for printing. The character selection is done in the usual manner from a keyboard. Once a certain key on the keyboard is depressed, a command for the printwheel to rotate from the present position to the next position is initiated. The microprocessor computes the direction

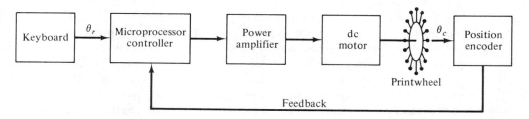

Figure 1-6 Printwheel control system.

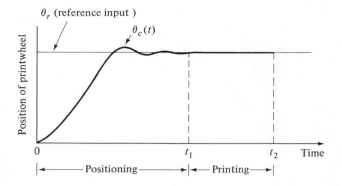

Figure 1-7 Typical input and output of the printwheel control system.

and the distance to be traveled, and sends out a control logic signal to the power amplifier, which in turn controls the motor that drives the printwheel. The position of the printwheel is detected by a position sensor whose output is compared with the desired position in the microprocessor. The motor is thus controlled in such a way as to drive the printwheel to the desired position. In practice, the control signals generated by the microprocessor controller should be able to drive the printwheel from one position to another sufficiently fast so that the printing can be done accurately within the specified time frame.

Figure 1-7 shows a typical set of input and output of the system. When a reference command input is given, the signal is represented as a step function. Since the electric circuit of the motor has inductance and the mechanical load has inertia, the printwheel cannot move to the desired position instantaneously. Typically, it will follow the response as shown, and settle at the new position after some time t_1. Printing cannot begin until the printwheel has come to a stop; otherwise, the character will be smeared. Figure 1-7 shows that after the printwheel has settled, the period from t_1 to t_2 is reserved for printing, so that after $t = t_2$, the system is ready to receive a new command.

1.2 WHAT IS FEEDBACK AND WHAT ARE ITS EFFECTS?

The motivation for using feedback illustrated by the examples in Section 1.1 is somewhat oversimplified. In these examples the use of feedback is shown to be for the purpose of reducing the error between the reference input and the system output. However, the significance of the effects of feedback in control systems is much more profound than is demonstrated by these examples. The reduction of system error is merely one of the many important effects that feedback may have upon a system. We show in the following sections that feedback also has effects on such system performance characteristics as stability, bandwidth, overall gain, impedance, and sensitivity.

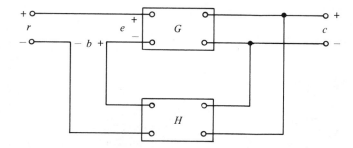

Figure 1-8 Feedback system.

To understand the effects of feedback on a control system, it is essential that we examine this phenomenon with a broad mind. When feedback is deliberately introduced for the purpose of control, its existence is easily identified. However, there are numerous situations wherein a physical system that we normally recognize as an inherently nonfeedback system may turn out to have feedback when it is observed in a certain manner. In general we can state that whenever a closed sequence of *cause-and-effect relationships* exists among the variables of a system, feedback is said to exist. This viewpoint will inevitably admit feedback in a large number of systems that ordinarily would be identified as nonfeedback systems. However, with the availability of the feedback and control system theory, this general definition of feedback enables numerous systems, with or without physical feedback, to be studied in a systematic way once the existence of feedback in the above-mentioned sense is established.

We shall now investigate the effects of feedback on the various aspects of system performance. Without the necessary background and mathematical foundation of linear system theory, at this point we can only rely on simple static system notation for our discussion. Let us consider the simple feedback system configuration shown in Fig. 1-8, where r is the input signal, c the output signal, e the error, and b the feedback signal. The parameters G and H may be considered as constant gains. By simple algebraic manipulations it is simple to show that the input–output relation of the system is

$$M = \frac{c}{r} = \frac{G}{1 + GH} \tag{1-1}$$

Using this basic relationship of the feedback system structure, we can uncover some of the significant effects of feedback.

Effect of Feedback on Overall Gain

As seen from Eq. (1-1), feedback affects the gain G of a nonfeedback system by a factor of $1 + GH$. The reference of the feedback in the system of Fig. 1-8 is negative, since a minus sign is assigned to the feedback signal. The quantity GH may itself include a minus sign, so the general effect of feedback is that it may increase or decrease the gain. In a practical control system, G and H are functions of frequency, so the magnitude of $1 + GH$ may be greater than 1 in one frequency range but less than 1 in another. Therefore, feedback could increase the gain of the system in one frequency range but decrease it in another.

Effect of Feedback on Stability

Stability is a notion that describes whether the system will be able to follow the input command. In a nonrigorous manner, a system is said to be unstable if its output is out of control or increases without bound.

To investigate the effect of feedback on stability, we can again refer to the expression in Eq. (1-1). If $GH = -1$, the output of the system is infinite for any finite input. Therefore, we may state that feedback can cause a system that is originally stable to become unstable. Certainly, feedback is a two-edged sword; when it is improperly used, it can be harmful. It should be pointed out, however, that we are only dealing with the static case here, and, in general $GH = -1$ is not the only condition for instability.

It can be demonstrated that one of the advantages of incorporating feedback is that it can stabilize an unstable system. Let us assume that the feedback system in Fig. 1-8 is unstable because $GH = -1$. If we introduce another feedback loop through a negative feedback of F, as shown in Fig. 1-9, the input–output relation of the overall system is

$$\frac{c}{r} = \frac{G}{1 + GH + GF} \tag{1-2}$$

It is apparent that although the properties of G and H are such that the inner-loop feedback system is unstable, because $GH = -1$, the overall system can be stable by properly selecting the outer-loop feedback gain F.

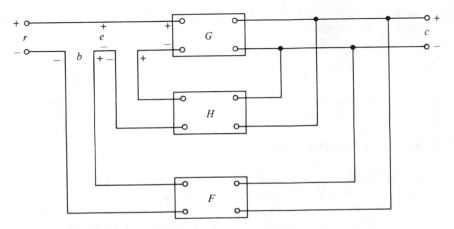

Figure 1-9 Feedback system with two feedback loops.

Effect of Feedback on Sensitivity

Sensitivity considerations often play an important role in the design of control systems. Since all physical elements have properties that change with environment and age, we cannot always consider the parameters of a control system to be completely stationary over the entire operating life of the system. For instance, the winding resistance of an electric motor changes as the temperature of the motor rises during operation. In general, a good control system should be very insensitive to these parameter variations while still able to follow the command responsively. We shall investigate what effect feedback has on the sensitivity to parameter variations.

Referring to the system in Fig. 1-8, we consider G as a parameter that may vary. The sensitivity of the gain of the overall system M to the variation in G is defined as

$$S_G^M = \frac{\partial M / M}{\partial G / G} \tag{1-3}$$

where ∂M denotes the incremental change in M due to the incremental change in G; $\partial M / M$ and $\partial G / G$ denote the percentage change in M and G, respectively. The expression of the sensitivity function S_G^M can be derived by using Eq. (1-1). We have

$$S_G^M = \frac{\partial M}{\partial G} \frac{G}{M} = \frac{1}{1 + GH} \tag{1-4}$$

This relation shows that the sensitivity function can be made arbitrarily small by increasing GH, provided that the system remains stable. It is apparent that in an open-loop system the gain of the system will respond in a one-to-one fashion to the variation in G.

In general, the sensitivity of the system gain of a feedback system to parameter variations depends on where the parameter is located. The reader may derive the sensitivity of the system in Fig. 1-8 due to the variation of H.

Effect of Feedback on External Disturbance or Noise

All physical control systems are subject to some types of extraneous signals or noise during operation. Examples of these signals are thermal noise voltage in electronic amplifiers and brush or commutator noise in electric motors.

The effect of feedback on noise depends greatly on where the noise is introduced into the system; no general conclusions can be made. However, in many situations, feedback can reduce the effect of noise on system performance.

Let us refer to the system shown in Fig. 1-10, in which r denotes the command signal and n is the noise signal. In the absence of feedback, $H = 0$, the output c is

$$c = G_1 G_2 e + G_2 n \tag{1-5}$$

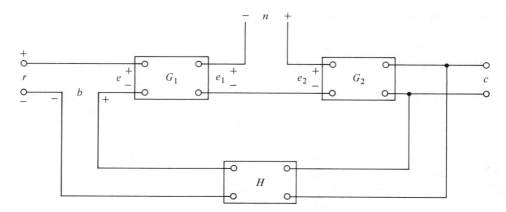

Figure 1-10 Feedback system with a noise signal.

where $e = r$. The signal-to-noise ratio of the output is defined as

$$\frac{\text{output due to signal}}{\text{output due to noise}} = \frac{G_1 G_2 e}{G_2 n} = G_1 \frac{e}{n} \tag{1-6}$$

To increase the signal-to-noise ratio, evidently we should either increase the magnitude of G_1 or e relative to n. Varying the magnitude of G_2 would have no effect whatsoever on the ratio.

With the presence of feedback, the system output due to r and n acting simultaneously is

$$c = \frac{G_1 G_2}{1 + G_1 G_2 H} r + \frac{G_2}{1 + G_1 G_2 H} n \tag{1-7}$$

Simply comparing Eq. (1-7) with Eq. (1-5) shows that the noise component in the output of Eq. (1-7) is reduced by the factor $1 + G_1 G_2 H$, but the signal component is also reduced by the same amount. The signal-to-noise ratio is

$$\frac{\text{output due to signal}}{\text{output due to noise}} = \frac{G_1 G_2 r / (1 + G_1 G_2 H)}{G_2 n / (1 + G_1 G_2 H)} = G_1 \frac{r}{n} \tag{1-8}$$

and is the same as that without feedback. In this case feedback is shown to have no direct effect on the output signal-to-noise ratio of the system in Fig. 1-10. However, the application of feedback suggests a possibility of improving the signal-to-noise ratio under certain conditions. Let us assume that in the system of Fig. 1-10, if the magnitude of G_1 is increased to G_1' and that of the input r to r', with all other parameters unchanged, the output due to the input signal acting alone is at the same level as that when feedback is absent. In other words, we let

$$c|_{n=0} = \frac{G_1' G_2 r'}{1 + G_1' G_2 H} = G_1 G_2 r \tag{1-9}$$

With the increased G_1, G'_1, the output due to noise acting alone becomes

$$c|_{r=0} = \frac{G_2 n}{1 + G'_1 G_2 H} \tag{1-10}$$

which is smaller than the output due to n when G_1 is not increased. The signal-to-noise ratio is now

$$\frac{G_1 G_2 r}{G_2 n / (1 + G'_1 G_2 H)} = \frac{G_1 r}{n} (1 + G'_1 G_2 H) \tag{1-11}$$

which is greater than that of the system without feedback by a factor of $(1 + G'_1 G_2 H)$.

In general, feedback also has effects on such performance characteristics as bandwidth, impedance, transient response, and frequency response. These effects will become known as one progresses into the ensuing material of this text.

1.3 TYPES OF FEEDBACK CONTROL SYSTEMS

Feedback control systems may be classified in a number of ways, depending upon the purpose of the classification. For instance, according to the method of analysis and design, feedback control systems are classified as linear and nonlinear, time varying or time invariant. According to the types of signal found in the system, reference is often made to continuous-data and discrete-data systems, or modulated and unmodulated systems. Also, with reference to the type of system components, we often come across descriptions such as electromechanical control system, hydraulic control systems, pneumatic systems, and biological control systems. Control systems are often classified according to the main purpose of the system. A positional control system and a velocity control system control the output variables according to the way the names imply. In general, there are many other ways of identifying control systems according to some special features of the system. It is important that some of these more common ways of classifying control systems are known so that proper perspective is gained before embarking on the analysis and design of these systems.

Linear versus Nonlinear Control Systems

This classification is made according to the methods of analysis and design. Strictly speaking, linear systems do not exist in practice, since all physical systems are nonlinear to some extent. Linear feedback control systems are idealized models that are fabricated by the analyst purely for the simplicity of analysis and design. When the magnitudes of the signals in a control system are limited to a range in which system components exhibit linear characteristics (i.e., the principle of superposition applies), the system is essentially linear. But when the magnitudes of the signals are extended outside the range of the linear operation, depending upon the severity of the nonlinearity, the system should no longer be considered linear. For instance, amplifiers used in control systems often exhibit saturation effect when their input signals become large; the magnetic field of a motor usually has saturation properties. Other common nonlinear effects found in control systems are the backlash or dead play between coupled gear members, nonlinear characteristics in springs, nonlinear frictional force or torque between moving members, and so on. Quite often, nonlinear characteristics are intentionally introduced in a control system to improve its performance or provide more effective control. For instance, to achieve minimum-time control, an on–off (bang-bang or relay) type of controller is used. This type of control is found in many missile or spacecraft control systems. For instance, in the attitude control of missiles and spacecraft, jets are mounted on the sides of the vehicle to provide reaction torque for attitude control. These jets are often controlled in a full-on or full-off fashion, so a fixed amount of air is applied from a given jet for a certain time duration to control the attitude of the space vehicle.

For linear systems there exists a wealth of analytical and graphical techniques for design and analysis purposes. However, nonlinear systems are very difficult to treat mathematically, and there are no general methods that may be used to solve a wide class of nonlinear systems.

Time-Invariant versus Time-Varying Systems

When the parameters of a control system are stationary with respect to time during the operation of the system, the system is called a time-invariant system. In practice, most physical systems contain elements that drift or vary with time. For example, the winding resistance of an electric motor will vary when the motor is being first excited and its temperature is rising. Another example of a time-varying system is a guided-missile control system in which the mass of the missile decreases as the fuel on board is being consumed during flight. Although a time-varying system without nonlinearity is still a linear system, the analysis and design of this class of systems are usually much more complex than that of the linear time-invariant systems.

Continuous-Data Control Systems

A continuous-data system is one in which the signals at various parts of the system are all functions of the continuous time variable t. Among all continuous-data control systems, the signals may be further classified as ac or dc. Unlike the general definitions of ac and dc signals used in electrical engineering, ac and dc control systems carry special significances. When one refers to an ac control system it usually means that the signals in the system are modulated by some kind of modulation scheme. On the other hand, when a dc control system is referred to, it does not mean that all the signals in the system are of the direct-current type; then there would be no control movement. A dc control system simply implies that the signals are unmodulated, but they are still ac signals according to the conventional definition. The schematic diagram of a closed-loop dc control system is shown in

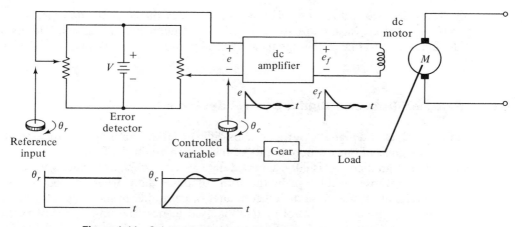

Figure 1-11 Schematic diagram of a typical dc closed-loop control system.

Fig. 1-11. Typical waveforms of the system in response to a step function input are shown in the figure. Typical components of a dc control system are potentiometers, dc amplifiers, dc motors, and dc tachometers.

The schematic diagram of a typical ac control system is shown in Fig. 1-12. In this case the signals in the system are modulated; that is, the information is transmitted by an ac carrier signal. Notice that the output controlled variable still behaves similar to that of the dc system if the two systems have the same control objective. In this case the modulated signals are demodulated by the low-pass characteristics of the control motor. Typical components of an ac control system are synchros, ac amplifiers, ac motors, gyroscopes, and accelerometers.

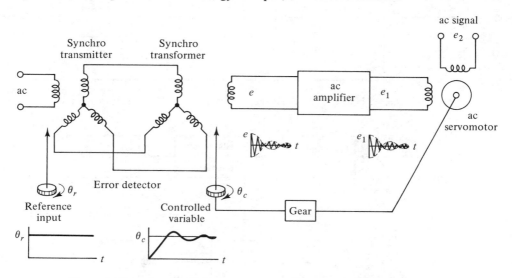

Figure 1-12 Schematic diagram of a typical ac closed-loop control system.

In practice, not all control systems are strictly the ac or the dc type. A system may incorporate a mixture of ac and dc components, using modulators and demodulators to match the signals at various points of the system.

Sampled-Data and Digital Control Systems

Sampled-data and digital control systems differ from the continuous-data systems in that the signals at one or more points of the system are in the form of either a pulse train or a digital code. Usually, sampled-data systems refer to a more general class of systems whose signals are in the form of pulse data, where a digital control system refers to the use of a digital computer or controller in the system. In this text the term "discrete-data control system" is used to describe both types of systems. For example,the printwheel control system shown in Fig. 1-6 is a typical discrete-data or digital control system, since the microprocessor receives and outputs digital data.

In general, a sampled-data system receives data or information only intermittently at specific instants of time. For instance, the error signal in a control system may be supplied only intermittently in the form of pulses, in which case the control system receives no information about the error signal during the periods between two consecutive pulses. Figure 1-13 illustrates how a typical sampled-data system operates. A continuous input signal $r(t)$ is applied to the system. The error signal $e(t)$ is sampled by a sampling device, the sampler, and the output of the sampler is a sequence of pulses. The sampling rate of the sampler may or may not be uniform. There are many advantages of incorporating sampling into a control system. One

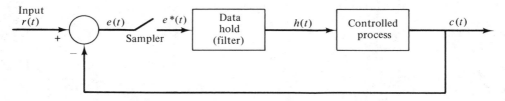

Figure 1-13   Block diagram of a sampled-data control system.

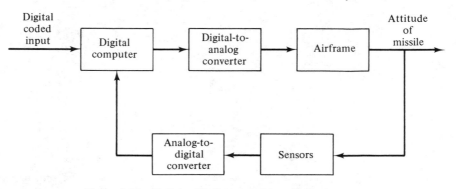

Figure 1-14   Digital autopilot system for a guided missile.

advantage easily understood provides time sharing of expensive equipment among several control channels.

Because digital computers provide many advantages in size and flexibility, computer control has become increasingly popular in recent years. Many airborne systems contain digital controllers that can pack several thousand discrete elements in a space no larger than the size of this book. Figure 1-14 shows the basic elements of a digital autopilot for a guided missile.

chapter two

Mathematical Foundation

2.1 INTRODUCTION

The studies of control systems rely to a great extent on the use of applied mathematics. Since the analysis and design of practical control systems have to deal with real problems, we cannot completely ignore the hardware and laboratory aspects of the problems. However, one of the major purposes of control systems studies is to develop a set of analytical tools so that the designer can arrive at reasonably predictable and reliable designs without depending completely on the drudgery of experimentation or strictly computer simulation.

For the study of classical control theory, which represents a good portion of this text, the required mathematical background includes such subjects as complex-variable theory, differential equations, Laplace transformation, z-transformation, and so on. Modern control theory, on the other hand, requires considerably more intensive mathematical background. In addition to the above-mentioned subjects, modern control theory is based on the foundation of matrix theory, set theory, linear algebra and transformation, variational calculus, mathematical programming, and other advanced mathematics.

In this chapter we present the background material that is needed for the analysis and design of control systems discussed in this text. Because of space limitations, the treatment of these mathematical subjects cannot be exhaustive. The reader who wishes to conduct an in-depth study of any of these subjects should refer to specialized books.

2.2 COMPLEX-VARIABLE CONCEPT

The classical control systems theory is based on the application of complex variables and functions of complex variables.

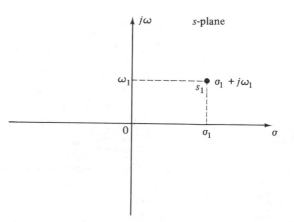

Figure 2-1 Complex *s*-plane.

Complex Variable

A complex variable s is considered to have two components: a real component σ, and an imaginary component ω. Graphically, the real component is represented by an axis in the horizontal direction, and the imaginary component is measured along the vertical axis, in the complex s-plane. In other words, a complex variable is always defined by a point in a complex plane that has a σ axis and a $j\omega$ axis. Figure 2-1 illustrates the complex s-plane, in which any arbitrary point, $s = s_1$, is defined by the coordinates $\sigma = \sigma_1$ and $\omega = \omega_1$, or simply $s_1 = \sigma_1 + j\omega_1$.

Functions of a Complex Variable

The function $G(s)$ is said to be a function of the complex variable s if for every value of s there is a corresponding value (or there are corresponding values) of $G(s)$. Since s is defined to have real and imaginary parts, the function $G(s)$ is also represented by its real and imaginary parts; that is,

$$G(s) = \operatorname{Re} G + j \operatorname{Im} G \qquad (2\text{-}1)$$

where $\operatorname{Re} G$ denotes the real part of $G(s)$ and $\operatorname{Im} G$ represents the imaginary part of G. Thus, the function $G(s)$ can also be represented by the complex G-plane whose horizontal axis represents $\operatorname{Re} G$ and whose vertical axis measures the imaginary component of $G(s)$. If for every value of s (every point in the s-plane) there is only one corresponding value for $G(s)$ [one corresponding point in the $G(s)$-plane], $G(s)$ is said to be a *single-valued function*, and the mapping (correspondence) from points in the s-plane onto points in the $G(s)$-plane is described as *single valued* (Fig. 2-2). However, there are many functions for which the mapping from the function plane to the complex-variable plane is not single-valued. For instance, given the function

$$G(s) = \frac{1}{s(s+1)} \qquad (2\text{-}2)$$

it is apparent that for each value of s there is only one unique corresponding value

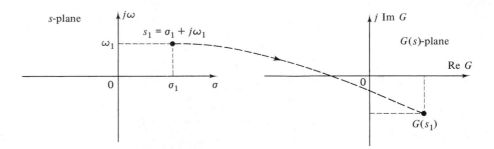

Figure 2-2 Single-valued mapping from the s-plane to the G(s)-plane.

for $G(s)$. However, the reverse is not true; for instance, the point $G(s) = \infty$ is mapped onto two points, $s = 0$ and $s = -1$, in the s-plane.

Analytic Function

A function $G(s)$ of the complex variable s is called an analytic function in a region of the s-plane if the function and all its derivatives exist in the region. For instance, the function given in Eq. (2-2) is analytic at every point in the s-plane except at the points $s = 0$ and $s = -1$. At these two points the value of the function is infinite. The function $G(s) = s + 2$ is analytic at every point in the finite s-plane.

Singularities and Poles of a Function

The *singularities* of a function are the points in the s-plane at which the function or its derivatives does not exist. A pole is the most common type of singularity and plays a very important role in the studies of the classical control theory.

The definition of a pole can be stated as: *If a function $G(s)$ is analytic and single valued in the neighborhood of s_i, except at s_i, it is said to have a pole of order r at $s = s_i$ if the limit*

$$\lim_{s \to s_i} \left[(s - s_i)^r G(s) \right]$$

has a finite, nonzero value. In other words, the denominator of $G(s)$ must include the factor $(s - s_i)^r$, so when $s = s_i$, the function becomes infinite. If $r = 1$, the pole at $s = s_i$ is called a *simple pole*. As an example, the function

$$G(s) = \frac{10(s + 2)}{s(s + 1)(s + 3)^2} \tag{2-3}$$

has a pole of order 2 at $s = -3$ and simple poles at $s = 0$ and $s = -1$. It can also be said that the function is analytic in the s-plane except at these poles.

Zeros of a Function

The definition of a zero of a function can be stated as: *If the function $G(s)$ is analytic at $s = s_i$, it is said to have a zero of order r at $s = s_i$ if the limit*

$$\lim_{s \to s_i} \left[(s - s_i)^{-r} G(s) \right] \tag{2-4}$$

has a finite, nonzero value. Or simply, $G(s)$ *has a zero of order r at $s = s_i$ if $1/G(s)$ has an rth-order pole at $s = s_i$.* For example, the function in Eq. (2-3) has a simple zero at $s = -2$.

If the function under consideration is a rational function of s, that is, a quotient of two polynomials of s, the total number of poles equals the total number of zeros, counting the multiple-order poles and zeros, if the poles and zeros at infinity and at zero are taken into account. The function in Eq. (2-3) has four finite poles at $s = 0, -1, -3, -3$; there is one finite zero at $s = -2$, but there are three zeros at infinity, since

$$\lim_{s \to \infty} G(s) = \lim_{s \to \infty} \frac{10}{s^3} = 0 \tag{2-5}$$

Therefore, the function has a total of four poles and four zeros in the entire s-plane.

2.3 DIFFERENTIAL EQUATIONS

Linear Ordinary Differential Equations

A large class of systems and phenomena in engineering are most conveniently formulated in terms of differential equations. These equations generally involve derivatives and integrals of the dependent variables with respect to the independent variable. For instance, a series electrical *RLC* (resistance–inductance–capacitance) network can be represented by the differential equation

$$Ri(t) + L\frac{di(t)}{dt} + \frac{1}{C} \int i(t)\, dt = v(t) \tag{2-6}$$

where R is the resistance, L the inductance, C the capacitance, $i(t)$ the current in the network, and $v(t)$ the applied voltage. In this case, $v(t)$ is the forcing function, t the independent variable, and $i(t)$ the dependent variable or the unknown which is to be determined by solving the differential equation.

Similarly, for a series mechanical mass–spring–damper system, the differential equation of the system can be written as

$$M\frac{d^2y(t)}{dt^2} + B\frac{dy(t)}{dt} + Ky(t) = f(t) \tag{2-7}$$

where $f(t)$ is the applied force, M the mass, B the damping coefficient, K the spring constant, and $y(t)$ the displacement.

Both equations illustrated above are defined as second-order differential equations, and we refer to the systems represented as second-order systems. Strictly, Eq. (2-6) should be referred to as an integrodifferential equation, since an integral is involved.

In general, the differential equation of an nth-order system is written

$$a_{n+1}\frac{dy^n(t)}{dt^n} + a_n\frac{dy^{n-1}(t)}{dt^{n-1}} + \cdots + a_2\frac{dy(t)}{dt} + a_1y(t) = f(t) \qquad (2\text{-}8)$$

The differential equations in Eqs. (2-6) through (2-8) are also known as linear ordinary differential equations if the coefficients $a_1, a_2, \ldots, a_{n+1}$ are not functions of $y(t)$, and since $y(t)$ and its derivatives are all of the first power.

In this book since we treat only systems that contain lumped parameters, the differential equations encountered are all of the ordinary type. However, for systems with distributed parameters, such as in the heat-transfer phenomenon, partial differential equations will result.

Nonlinear Differential Equations

Many physical systems are described by nonlinear differential equations. For instance, the differential equation that describes the motion of the pendulum shown in Fig. 2-3 is

$$ML\frac{d^2\theta(t)}{dt^2} + Mg\sin\theta(t) = 0 \qquad (2\text{-}9)$$

Since $\theta(t)$ appears as a sine function, Eq. (2-9) is nonlinear, and the system is customarily called a nonlinear system.

First-Order Differential Equations

In general, an nth-order differential equation can be decomposed into n first-order differential equations. Since, in principle, first-order differential equations are sim-

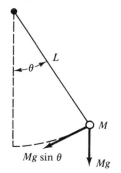

$Mg\sin\theta$ Mg **Figure 2-3** Simple pendulum.

pler to solve than higher-order ones, there are reasons why first-order differential equations are often used in the analytical studies of control systems.

For the second-order differential equation in Eq. (2-7) if we let

$$x_1(t) = y(t) \tag{2-10}$$

and

$$x_2(t) = \frac{dy(t)}{dt} = \frac{dx_1(t)}{dt} \tag{2-11}$$

the equation is written as

$$M\frac{dx_2(t)}{dt} = -Kx_1(t) - Bx_2(t) + f(t) \tag{2-12}$$

Now, rearranging Eqs. (2-11) and (2-12), we get two first-order differential equations:

$$\frac{dx_1(t)}{dt} = x_2(t) \tag{2-13}$$

$$\frac{dx_2(t)}{dt} = -\frac{K}{M}x_1(t) - \frac{B}{M}x_2(t) + \frac{1}{M}f(t) \tag{2-14}$$

For Eq. (2-6) we let

$$x_1(t) = \int i(t)\,dt \tag{2-15}$$

and

$$x_2(t) = \frac{dx_1(t)}{dt} = i(t) \tag{2-16}$$

Then Eq. (2-6) is decomposed into the following two first-order differential equations

$$\frac{dx_1(t)}{dt} = x_2(t) \tag{2-17}$$

$$\frac{dx_2(t)}{dt} = -\frac{1}{LC}x_1(t) - \frac{R}{L}x_2(t) + \frac{1}{L}v(t) \tag{2-18}$$

In a similar fashion, we can decompose Eq. (2-8) into n first-order differential equations. (Refer to Sec. 5.5.)

In control systems theory, the set of first-order differential equations are called the *state equations*, and $x_1, x_2, \ldots, x_n$ are called the *state variables*. Later in this chapter we show that the state equations can be written in a compact vector-matrix form.

2.4 LAPLACE TRANSFORM [3–5]

The Laplace transform is one of the mathematical tools used for the solution of
ordinary linear differential equations. In comparison with the classical method of
solving linear differential equations, the Laplace transform method has the following two attractive features:

1. The homogeneous equation and the particular integral are solved in one
   operation.
2. The Laplace transform converts the differential equation into an algebraic
   equation in s. It is then possible to manipulate the algebraic equation by
   simple algebraic rules to obtain the solution in the s-domain. The final
   solution is obtained by taking the inverse Laplace transform.

Definition of the Laplace Transform

Given the function $f(t)$ that satisfies the condition

$$\int_0^\infty \left| f(t)e^{-\sigma t} \right| dt < \infty \tag{2-19}$$

for some finite real σ, the Laplace transform of $f(t)$ is defined as

$$F(s) = \int_0^\infty f(t)e^{-st}\, dt \tag{2-20}$$

or

$$F(s) = \mathcal{L}[f(t)] \tag{2-21}$$

The variable s is referred to as the Laplace operator, which is a complex
variable; that is, $s = \sigma + j\omega$. The defining equation of Eq. (2-20) is also known as
the *one-sided Laplace transform*, as the integration is evaluated from 0 to ∞. This
simply means that all information contained in $f(t)$ prior to $t = 0$ is ignored or
considered to be zero. This assumption does not place any serious limitation on the
applications of the Laplace transform to linear system problems, since in the usual
time-domain studies, time reference is often chosen at the instant $t = 0$. Furthermore, for a physical system when an input is applied at $t = 0$, the response of the
system does not start sooner than $t = 0$; that is, response does not precede
excitation.

Strictly, the "one-sided" Laplace transform should be defined from $t = 0^-$ to
$t = \infty$. The symbol 0^- implies that the limit of $t \to 0$ is taken from the left side of
$t = 0$. This limit process will take care of situations under which the function $f(t)$
has a jump discontinuity or an impulse at $t = 0$. For the subjects treated in this text,
the defining equation of the Laplace transform in Eq. (2-20) is almost never used in
problem solving, since the transform expressions are either given or can be found
from the Laplace transform table. Thus, the fine point of using 0^- or 0^+ never needs

to be addressed. For simplicity, we shall, in general and without further justification, use $t = 0$ or $t = t_0$ as the initial time in all subsequent chapters.

The following examples serve as illustrations of how Eq. (2-20) may be used for the evaluation of the Laplace transform of a function $f(t)$.

■

Example 2-1

Let $f(t)$ be a unit step function that is defined to have a constant value of unity for $t > 0$ and a zero value for $t < 0$. Or,

$$f(t) = u_s(t) \tag{2-22}$$

Then the Laplace transform of $f(t)$ is

$$F(s) = \mathcal{L}[u_s(t)] = \int_0^\infty u_s(t) e^{-st} \, dt = -\frac{1}{s} e^{-st} \Big|_0^\infty = \frac{1}{s} \tag{2-23}$$

Of course, the Laplace transform given by Eq. (2-23) is valid if

$$\int_0^\infty |u_s(t) e^{-\sigma t}| \, dt = \int_0^\infty |e^{-\sigma t}| \, dt < \infty$$

which means that the real part of s, σ, must be greater than zero. However, in practice, we simply refer to the Laplace transform of the unit step function as $1/s$, and rarely do we have to be concerned about the region in which the transform integral converges absolutely.

Example 2-2

Consider the exponential function

$$f(t) = e^{-at} \qquad t \geq 0$$

where a is a constant.

The Laplace transform of $f(t)$ is written

$$F(s) = \int_0^\infty e^{-at} e^{-st} \, dt = -\frac{e^{-(s+a)t}}{s + a} \Big|_0^\infty = \frac{1}{s + a} \tag{2-24}$$

■

Inverse Laplace Transformation

The operation of obtaining $f(t)$ from the Laplace transform $F(s)$ is termed the *inverse Laplace transformation*. The inverse Laplace transform of $F(s)$ is denoted by

$$f(t) = \mathcal{L}^{-1}[F(s)] \tag{2-25}$$

and is given by the inverse Laplace transform integral

$$f(t) = \frac{1}{2\pi j} \int_{c-j\infty}^{c+j\infty} F(s) e^{st}\, ds \qquad (2\text{-}26)$$

where c is a real constant that is greater than the real parts of all the singularities of $F(s)$. Equation (2-26) represents a line integral that is to be evaluated in the s-plane. However, for most engineering purposes the inverse Laplace transform operation can be accomplished simply by referring to the Laplace transform table, such as the one given in Appendix B.

Important Theorems of the Laplace Transform

The applications of the Laplace transform in many instances are simplified by the utilization of the properties of the transform. These properties are presented in the following theorems, and no proofs are given.

1. **Multiplication by a Constant**
   The Laplace transform of the product of a constant k and a time function $f(t)$ is the constant k multiplied by the Laplace transform of $f(t)$; that is,

$$\mathcal{L}[kf(t)] = kF(s) \qquad (2\text{-}27)$$

   where $F(s)$ is the Laplace transform of $f(t)$.

2. **Sum and Difference**
   The Laplace transform of the sum (or difference) of two time functions is the sum (or difference) of the Laplace transforms of the time functions; that is,

$$\mathcal{L}[f_1(t) \pm f_2(t)] = F_1(s) \pm F_2(s) \qquad (2\text{-}28)$$

   where $F_1(s)$ and $F_2(s)$ are the Laplace transforms of $f_1(t)$ and $f_2(t)$, respectively.

3. **Differentiation**
   The Laplace transform of the first derivative of a time function $f(t)$ is s times the Laplace transform of $f(t)$ minus the limit of $f(t)$ as t approaches 0; that is,

$$\mathcal{L}\left[\frac{df(t)}{dt}\right] = sF(s) - \lim_{t \to 0} f(t) = sF(s) - f(0) \qquad (2\text{-}29)$$

In general, for higher-order derivatives,

$$\mathcal{L}\left[\frac{d^n f(t)}{dt^n}\right] = s^n F(s) - \lim_{t \to 0}\left[s^{n-1}f(t) + s^{n-2}\frac{df(t)}{dt} + \cdots + \frac{d^{n-1}f(t)}{dt^{n-1}}\right]$$

$$= s^n F(s) - s^{n-1}f(0) - s^{n-2}f^{(1)}(0) - \cdots - f^{(n-1)}(0) \qquad (2\text{-}30)$$

4. Integration

The Laplace transform of the first integral of a function $f(t)$ with respect to time is the Laplace transform of $f(t)$ divided by s; that is,

$$\mathcal{L}\left[\int_0^t f(\tau)\, d\tau\right] = \frac{F(s)}{s} \tag{2-31}$$

In general, for nth-order integration,

$$\mathcal{L}\left[\int_0^{t_1}\int_0^{t_2}\cdots\int_0^{t_n} f(\tau)\, d\tau\, dt_1\ldots dt_{n-1}\right] = \frac{F(s)}{s^n} \tag{2-32}$$

5. Shift in Time

The Laplace transform of $f(t)$ delayed by time T is equal to the Laplace transform of $f(t)$ multiplied by e^{-Ts}; that is,

$$\mathcal{L}\left[f(t - T)u_s(t - T)\right] = e^{-Ts}F(s) \tag{2-33}$$

where $u_s(t - T)$ denotes the unit step function, which is shifted in time to the right by T.

6. Initial-Value Theorem

If the Laplace transform of $f(t)$ is $F(s)$, then

$$\lim_{t \to 0} f(t) = \lim_{s \to \infty} sF(s) \tag{2-34}$$

if the time limit exists.

7. Final-Value Theorem

If the Laplace transform of $f(t)$ is $F(s)$ and if $sF(s)$ is analytic on the imaginary axis and in the right half of the s-plane, then

$$\lim_{t \to \infty} f(t) = \lim_{s \to 0} sF(s) \tag{2-35}$$

The final-value theorem is a very useful relation in the analysis and design of feedback control systems, since it gives the final value of a time function by determining the behavior of its Laplace transform as s tends to zero. However, the final-value theorem is not valid if $sF(s)$ contains any poles whose real part is zero or positive, which is equivalent to the analytic requirement of $sF(s)$ stated in the theorem. The following examples illustrate the care that one must take in applying the final-value theorem.

∎

Example 2-3

Consider the function

$$F(s) = \frac{5}{s(s^2 + s + 2)}$$

Since $sF(s)$ is analytic on the imaginary axis and in the right half of the s-plane, the final-value theorem may be applied. Therefore, using Eq. (2-35),

$$\lim_{t \to \infty} f(t) = \lim_{s \to 0} sF(s) = \lim_{s \to 0} \frac{5}{s^2 + s + 2} = \frac{5}{2} \tag{2-36}$$

Example 2-4

Consider the function

$$F(s) = \frac{\omega}{s^2 + \omega^2} \tag{2-37}$$

which is known to be the Laplace transform of $f(t) = \sin \omega t$. Since the function $sF(s)$ has two poles on the imaginary axis, the final-value theorem *cannot* be applied in this case. In other words, although the final-value theorem would yield a value of zero as the final value of $f(t)$, the result is erroneous.

■

8. Complex Shifting
The Laplace transform of $f(t)$ multiplied by $e^{\mp at}$, where a is a constant, is equal to the Laplace transform $F(s)$ with s replaced by $s \pm a$; that is,

$$\mathcal{L}\left[e^{\mp at}f(t)\right] = F(s \pm a) \tag{2-38}$$

9. Real Convolution (Complex Multiplication)
If the functions $f_1(t)$ and $f_2(t)$ have the Laplace transforms $F_1(s)$ and $F_2(s)$, respectively, and $f_1(t) = 0$, $f_2(t) = 0$, for $t < 0$, then

$$F_1(s)F_2(s) = \mathcal{L}\left[\int_0^t f_1(\tau)f_2(t - \tau)\, d\tau\right]$$
$$= \mathcal{L}\left[\int_0^t f_2(\tau)f_1(t - \tau)\, d\tau\right] \tag{2-39}$$
$$= \mathcal{L}\left[f_1(t) * f_2(t)\right]$$

*where the symbol " * " denotes convolution in the t-domain.* Thus, Eq. (2-39) shows that multiplication of two transformed functions in the complex s-domain is equivalent to the convolution of the two corresponding real functions of t in the t-domain. An important fact to remember is that the inverse Laplace transform of the product of two functions in the s-domain is *not* equal to the product of the two corresponding real functions in the t-domain; that is, in general,

$$\mathcal{L}^{-1}\left[F_1(s)F_2(s)\right] \neq f_1(t)f_2(t) \tag{2-40}$$

There is also a dual relation to the real convolution theorem, called the complex convolution, or real multiplication. Essentially, the theorem states that multiplication in the real t-domain is equivalent to convolution in the

Table 2-1 Theorems of Laplace Transforms

Multiplication by a constant	$\mathcal{L}[kf(t)] = kF(s)$
Sum and difference	$\mathcal{L}[f_1(t) \pm f_2(t)] = F_1(s) \pm F_2(s)$
Differentiation	$\mathcal{L}\left[\dfrac{df(t)}{dt}\right] = sF(s) - f(0)$
	$\mathcal{L}\left[\dfrac{d^n f(t)}{dt^n}\right] = s^n F(s) - s^{n-1}f(0) - s^{n-1}f^{(1)}(0)$ $- \cdots - sf^{(n-2)}(0) - f^{(n-1)}(0)$
	where
	$f^{(k)}(0) = \left.\dfrac{d^k f(t)}{dt^k}\right|_{t=0}$
Integration	$\mathcal{L}\left[\displaystyle\int_0^t f(\tau)\, d\tau\right] = \dfrac{F(s)}{s}$
	$\mathcal{L}\left[\displaystyle\int_0^{t_1}\int_0^{t_2}\cdots\int_0^{t_n} f(\tau)\, d\tau\, dt_1 \ldots dt_{n-1}\right] = \dfrac{F(s)}{s^n}$
Shift in time	$\mathcal{L}[f(t-T)u_s(t-T)] = e^{-Ts}F(s)$
Initial-value theorem	$\displaystyle\lim_{t\to 0} f(t) = \lim_{s\to\infty} sF(s)$
Final-value theorem	$\displaystyle\lim_{t\to\infty} f(t) = \lim_{s\to 0} sF(s)$ if $sF(s)$ does not have poles on or to the right of the imaginary axis
Complex shifting	$\mathcal{L}[e^{\mp at}f(t)] = F(s \pm a)$
Real convolution	$F_1(s)F_2(s) = \mathcal{L}\left[\displaystyle\int_0^t f_1(\tau)f_2(t-\tau)\, d\tau\right]$ $= \mathcal{L}\left[\displaystyle\int_0^t f_2(\tau)f_1(t-\tau)\, d\tau\right]$ $= \mathcal{L}[f_1(t) * f_2(t)]$

complex s-domain; that is,

$$\mathcal{L}\big[f_1(t)f_2(t)\big] = F_1(s) * F_2(s) \tag{2-41}$$

where $*$ denotes complex convolution.

Details of the complex convolution formula are not given here.

Table 2-1 summarizes the useful theorems of the Laplace transforms.

2.5 INVERSE LAPLACE TRANSFORM BY PARTIAL-FRACTION EXPANSION [7–11]

In a great majority of the problems in control systems, the evaluation of the inverse Laplace transform does not necessitate the use of the inversion integral of Eq. (2-26). The inverse Laplace transform operation involving rational functions can be carried out using a Laplace transform table and partial-fraction expansion.

When the Laplace transform solution of a differential equation is a rational function in s, it can be written

$$X(s) = \frac{P(s)}{Q(s)} \tag{2-42}$$

where $P(s)$ and $Q(s)$ are polynomials of s. It is assumed that the order of $Q(s)$ in s is greater than that of $P(s)$. The polynomial $Q(s)$ may be written

$$Q(s) = s^n + a_1 s^{n-1} + \cdots + a_{n-1}s + a_n \tag{2-43}$$

where $a_1, \ldots, a_n$ are real coefficients. The zeros of $Q(s)$ are either real or in complex-conjugate pairs, in simple or multiple order. The methods of partial-fraction expansion will now be given for the cases of simple poles, multiple-order poles, and complex poles, of $X(s)$.

Partial-Fraction Expansion When All the Poles of $X(s)$ Are Simple and Real

If all the poles of $X(s)$ are real and simple, Eq. (2-42) can be written

$$X(s) = \frac{P(s)}{Q(s)} = \frac{P(s)}{(s + s_1)(s + s_2) \cdots (s + s_n)} \tag{2-44}$$

where the poles $-s_1, -s_2, \ldots, -s_n$ are considered to be real numbers in the present case. Applying the partial-fraction expansion technique, Eq. (2-44) is written

$$X(s) = \frac{K_{s1}}{s + s_1} + \frac{K_{s2}}{s + s_2} + \cdots + \frac{K_{sn}}{s + s_n} \tag{2-45}$$

The coefficient, K_{si} $(i = 1, 2, \ldots, n)$, is determined by multiplying both sides of Eq. (2-44) or (2-45) by the factor $(s + s_i)$ and then setting s equal to $-s_i$. To find the coefficient K_{s1}, for instance, we multiply both sides of Eq. (2-44) by $(s + s_1)$ and let $s = -s_1$; that is,

$$K_{s1} = \left[(s + s_1) \frac{P(s)}{Q(s)} \right]_{s = -s_1} = \frac{P(-s_1)}{(s_2 - s_1)(s_3 - s_1) \cdots (s_n - s_1)} \tag{2-46}$$

As an illustrative example, consider the function

$$X(s) = \frac{5s + 3}{(s + 1)(s + 2)(s + 3)} \tag{2-47}$$

which is written in the partial-fractioned form

$$X(s) = \frac{K_{-1}}{s + 1} + \frac{K_{-2}}{s + 2} + \frac{K_{-3}}{s + 3} \tag{2-48}$$

The coefficients K_{-1}, K_{-2}, and K_{-3} are determined as follows:

$$K_{-1} = [(s + 1) X(s)]_{s=-1} = \frac{5(-1) + 3}{(2 - 1)(3 - 1)} = -1 \qquad (2\text{-}49)$$

$$K_{-2} = [(s + 2) X(s)]_{s=-2} = \frac{5(-2) + 3}{(1 - 2)(3 - 2)} = 7 \qquad (2\text{-}50)$$

$$K_{-3} = [(s + 3) X(s)]_{s=-3} = \frac{5(-3) + 3}{(1 - 3)(2 - 3)} = -6 \qquad (2\text{-}51)$$

Therefore, Eq. (2-48) becomes

$$X(s) = \frac{-1}{s + 1} + \frac{7}{s + 2} - \frac{6}{s + 3} \qquad (2\text{-}52)$$

Partial-Fraction Expansion When Some Poles of X(s) Are of Multiple Order

If r of the n poles of $X(s)$ are identical, or say, the pole at $s = -s_i$ is of multiplicity r, $X(s)$ is written

$$X(s) = \frac{P(s)}{Q(s)} = \frac{P(s)}{(s + s_1)(s + s_2) \cdots (s + s_i)^r (s + s_n)} \qquad (2\text{-}53)$$

Then $X(s)$ can be expanded as

$$X(s) = \frac{K_{s1}}{s + s_1} + \frac{K_{s2}}{s + s_2} + \cdots + \frac{K_{sn}}{s + s_n}$$

$$| \leftarrow (n - r) \text{ terms of simple poles} \rightarrow |$$

$$+ \frac{A_1}{s + s_i} + \frac{A_2}{(s + s_i)^2} + \cdots + \frac{A_r}{(s + s_i)^r} \qquad (2\text{-}54)$$

$$| \leftarrow r \text{ terms of repeated poles} \rightarrow |$$

The $n - r$ coefficients, which correspond to simple poles, $K_{s1}, K_{s2}, \ldots, K_{sn}$, may be evaluated by the method described by Eq. (2-46). The determination of the coefficients that correspond to the multiple-order poles is described below.

$$A_r = [(s + s_i)^r X(s)]_{s=-s_i} \qquad (2\text{-}55)$$

$$A_{r-1} = \frac{d}{ds} [(s + s_i)^r X(s)] \bigg|_{s=-s_i} \qquad (2\text{-}56)$$

$$A_{r-2} = \frac{1}{2!} \frac{d^2}{ds^2} [(s + s_i)^r X(s)] \bigg|_{s=-s_i} \qquad (2\text{-}57)$$

$$A_1 = \frac{1}{(r - 1)!} \frac{d^{r-1}}{ds^{r-1}} [(s + s_i)^r X(s)] \bigg|_{s=-s_i} \qquad (2\text{-}58)$$

■

Example 2-5

Consider the function

$$X(s) = \frac{1}{s(s+1)^3(s+2)} \tag{2-59}$$

Using the format of Eq. (2-54), $X(s)$ is written

$$X(s) = \frac{K_0}{s} + \frac{K_{-2}}{s+2} + \frac{A_1}{s+1} + \frac{A_2}{(s+1)^2} + \frac{A_3}{(s+1)^3} \tag{2-60}$$

Then the coefficients corresponding to the simple poles are

$$K_0 = [sX(s)]_{s=0} = \tfrac{1}{2} \tag{2-61}$$

$$K_{-2} = [(s+2)X(s)]_{s=-2} = \tfrac{1}{2} \tag{2-62}$$

and those of the third-order pole are

$$A_3 = \left[(s+1)^3 X(s)\right]\Big|_{s=-1} = -1 \tag{2-63}$$

$$A_2 = \frac{d}{ds}\left[(s+1)^3 X(s)\right]\Big|_{s=-1} = \frac{d}{ds}\left[\frac{1}{s(s+2)}\right]\Big|_{s=-1}$$

$$= \frac{-(2s+2)}{s^2(s+2)^2}\Big|_{s=-1} = 0 \tag{2-64}$$

and

$$A_1 = \frac{1}{2!}\frac{d^2}{ds^2}\left[(s+1)^3 X(s)\right]\Big|_{s=-1} = \frac{1}{2}\frac{d}{ds}\left[\frac{-2(s+1)}{s^2(s+2)^2}\right]\Big|_{s=-1}$$

$$= \left[\frac{-1}{s^2(s+2)^2} + \frac{2(s+1)}{s^2(s+2)^3} + \frac{2(s+1)}{s^3(s+2)^2}\right]\Big|_{s=-1} = -1 \tag{2-65}$$

The completed partial-fraction expansion is

$$X(s) = \frac{1}{2s} + \frac{1}{2(s+2)} - \frac{1}{s+1} - \frac{1}{(s+1)^3} \tag{2-66}$$

■

Partial-Fraction Expansion of Simple Complex-Conjugate Poles

The partial-fraction expansion of Eq. (2-45) is valid also for simple complex-conjugate poles. However, since complex-conjugate poles are more difficult to handle and are of special interest in control systems studies, they deserve separate treatment here.

Suppose that the rational function $X(s)$ of Eq. (2-42) contains a pair of complex poles:

$$s = -\alpha + j\omega \qquad \text{and} \qquad s = -\alpha - j\omega$$

Then the corresponding coefficients of these poles are

$$K_{-\alpha+j\omega} = (s + \alpha - j\omega) X(s)\big|_{s=-\alpha+j\omega} \tag{2-67}$$

$$K_{-\alpha-j\omega} = (s + \alpha + j\omega) X(s)\big|_{s=-\alpha-j\omega} \tag{2-68}$$

■

Example 2-6

Consider the function

$$X(s) = \frac{\omega_n^2}{s(s^2 + 2\zeta\omega_n s + \omega_n^2)} \tag{2-69}$$

Let us assume that the values of ζ and ω_n are such that the nonzero poles of $X(s)$ are complex numbers. Then $X(s)$ is expanded as follows:

$$X(s) = \frac{K_0}{s} + \frac{K_{-\alpha+j\omega}}{s + \alpha - j\omega} + \frac{K_{-\alpha-j\omega}}{s + \alpha + j\omega} \tag{2-70}$$

where

$$\alpha = \zeta\omega_n \tag{2-71}$$

and

$$\omega = \omega_n\sqrt{1 - \zeta^2} \tag{2-72}$$

The coefficients in Eq. (2-70) are determined as

$$K_0 = sX(s)\big|_{s=0} = 1 \tag{2-73}$$

$$K_{-\alpha+j\omega} = (s + \alpha - j\omega) X(s)\big|_{s=-\alpha+j\omega}$$

$$= \frac{\omega_n^2}{2j\omega(-\alpha + j\omega)} = \frac{\omega_n}{2\omega} e^{-j(\theta+\pi/2)} \tag{2-74}$$

where

$$\theta = \tan^{-1}\left[\frac{\omega}{-\alpha}\right] \tag{2-75}$$

Also,

$$K_{-\alpha-j\omega} = (s + \alpha + j\omega) X(s)\big|_{s=-\alpha-j\omega}$$

$$= \frac{\omega_n^2}{-2j\omega(-\alpha - j\omega)} = \frac{\omega_n}{2\omega} e^{j(\theta+\pi/2)} \tag{2-76}$$

The complete expansion is

$$X(s) = \frac{1}{s} + \frac{\omega_n}{2\omega} \left[\frac{e^{-j(\theta + \pi/2)}}{s + \alpha - j\omega} + \frac{e^{j(\theta + \pi/2)}}{s + \alpha + j\omega} \right] \qquad (2\text{-}77)$$

Taking the inverse Laplace transform on both sides of the last equation gives

$$x(t) = 1 + \frac{\omega_n}{2\omega} \left(e^{-j(\theta + \pi/2)} e^{(-\alpha + j\omega)t} + e^{j(\theta + \pi/2)} e^{(-\alpha - j\omega)t} \right)$$

$$= 1 + \frac{\omega_n}{\omega} e^{-\alpha t} \sin(\omega t - \theta) \qquad (2\text{-}78)$$

or

$$x(t) = 1 + \frac{1}{\sqrt{1 - \zeta^2}} e^{-\zeta \omega_n t} \sin\left(\omega_n \sqrt{1 - \zeta^2}\, t - \theta \right) \qquad (2\text{-}79)$$

where θ is given by Eq. (2-75).

■

2.6 APPLICATION OF LAPLACE TRANSFORM TO THE SOLUTION OF LINEAR ORDINARY DIFFERENTIAL EQUATIONS

With the aid of the theorems concerning Laplace transform given in Section 2.4 and a table of transforms, linear ordinary differential equations can be solved by the Laplace transform method. The advantages with the Laplace transform method are that, with the aid of a transform table the steps involved are all algebraic, and the homogeneous solution and the particular integral solution are obtained simultaneously.

Let us illustrate the method by several illustrative examples.

■

Example 2-7

Consider the differential equation

$$\frac{d^2 x(t)}{dt^2} + 3\frac{dx(t)}{dt} + 2x(t) = 5u_s(t) \qquad (2\text{-}80)$$

where $u_s(t)$ is the unit step function, which is defined as

$$u_s(t) = \begin{cases} 1 & t > 0 \\ 0 & t < 0 \end{cases} \qquad (2\text{-}81)$$

The initial conditions are $x(0) = -1$ and $x^{(1)}(0) = dx(t)/dt|_{t=0} = 2$. To solve the differential equation we first take the Laplace transform on both sides of Eq. (2-80); we have

$$s^2 X(s) - sx(0) - x^{(1)}(0) + 3sX(s) - 3x(0) + 2X(s) = \frac{5}{s} \qquad (2\text{-}82)$$

Substituting the values of $x(0)$ and $x^{(1)}(0)$ into Eq. (2-82) and solving for $X(s)$, we get

$$X(s) = \frac{-s^2 - s + 5}{s(s^2 + 3s + 2)} = \frac{-s^2 - s + 5}{s(s + 1)(s + 2)} \tag{2-83}$$

Equation (2-83) is expanded by partial-fraction expansion to give

$$X(s) = \frac{5}{2s} - \frac{5}{s + 1} + \frac{3}{2(s + 2)} \tag{2-84}$$

Now taking the inverse Laplace transform of Eq. (2-84), we get the complete solution as

$$x(t) = \tfrac{5}{2} - 5e^{-t} + \tfrac{3}{2}e^{-2t} \qquad t \geq 0 \tag{2-85}$$

The first term in Eq. (2-85) is the steady-state solution, and the last two terms are the transient solution. Unlike the classical method, which requires separate steps to give the transient and the steady-state solutions, the Laplace transform method gives the entire solution of the differential equation in one operation.

If only the magnitude of the steady-state solution is of interest, the final-value theorem may be applied. Thus,

$$\lim_{t \to \infty} x(t) = \lim_{s \to 0} sX(s) = \lim_{s \to 0} \frac{-s^2 - s + 5}{s^2 + 3s + 2} = \frac{5}{2} \tag{2-86}$$

where we have first checked and found that the function, $sX(s)$, has poles only in the left half of the s-plane.

Example 2-8

Consider the linear differential equation

$$\frac{d^2x(t)}{dt^2} + 34.5\frac{dx(t)}{dt} + 1000x(t) = 1000u_s(t) \tag{2-87}$$

where $u_s(t)$ is the unit step function. The initial values of $x(t)$ and $dx(t)/dt$ are assumed to be zero.

Taking the Laplace transform on both sides of Eq. (2-87) and applying zero initial conditions, we have

$$s^2X(s) + 34.5sX(s) + 1000X(s) = \frac{1000}{s} \tag{2-88}$$

Solving $X(s)$ from the last equation, we obtain

$$X(s) = \frac{1000}{s(s^2 + 34.5s + 1000)} \tag{2-89}$$

The poles of $X(s)$ are at $s = 0$, $s = -17.25 + j26.5$, and $s = -17.25 - j26.5$. Therefore,

Eq. (2-89) can be written as

$$X(s) = \frac{1000}{s(s + 17.25 - j26.5)(s + 17.25 + j26.5)} \tag{2-90}$$

One way of solving for $x(t)$ is to perform the partial-fraction expansion of Eq. (2-90), giving

$$X(s) = \frac{1}{s} + \frac{31.6}{2(26.5)}\left[\frac{e^{-j(\theta + \pi/2)}}{s + 17.25 - j26.5} + \frac{e^{j(\theta + \pi/2)}}{s + 17.25 + j26.5}\right] \tag{2-91}$$

where

$$\theta = \tan^{-1}\left(\frac{-26.5}{17.25}\right) = -56.9° \tag{2-92}$$

Then, using Eq. (2-78),

$$x(t) = 1 + 1.193\,e^{-17.25t}\sin(26.5t - \theta) \tag{2-93}$$

Another approach is to compare Eq. (2-89) with Eq. (2-69), so that

$$\omega_n = \pm\sqrt{1000} = \pm 31.6 \tag{2-94}$$

and

$$\zeta = 0.546 \tag{2-95}$$

and the solution to $x(t)$ is given directly by Eq. (2-79).

■

2.7 ELEMENTARY MATRIX THEORY [1, 2, 6]

In the study of modern control theory it is often desirable to use matrix notation to simplify complex mathematical expressions. The simplifying matrix notation may not reduce the amount of work required to solve the mathematical equations, but it usually makes the equations much easier to handle and manipulate.

As a motivation to the reason of using matrix notation, let us consider the following set of n simultaneous algebraic equations:

$$\begin{aligned}
a_{11}x_1 + a_{12}x_2 + \cdots + a_{1n}x_n &= y_1 \\
a_{21}x_1 + a_{22}x_2 + \cdots + a_{2n}x_n &= y_2 \\
&\cdots\cdots\cdots\cdots\cdots\cdots\cdots \\
a_{n1}x_1 + a_{n2}x_2 + \cdots + a_{nn}x_n &= y_n
\end{aligned} \tag{2-96}$$

We may use the matrix equation

$$\mathbf{Ax} = \mathbf{y} \tag{2-97}$$

as a simplified representation for Eq. (2-96).

The symbols **A**, **x**, and **y** are defined as matrices, which contain the coefficients and variables of the original equations as their elements. In terms of matrix algebra, which will be discussed later, Eq. (2-97) can be stated as: *The product of the matrices* **A** *and* **x** *is equal to the matrix* **y**. The three matrices involved here are defined to be

$$\mathbf{A} = \begin{bmatrix} a_{11} & a_{12} & \cdots & a_{1n} \\ a_{21} & a_{22} & \cdots & a_{2n} \\ \vdots & \vdots & & \vdots \\ a_{n1} & a_{n2} & \cdots & a_{nn} \end{bmatrix} \tag{2-98}$$

$$\mathbf{x} = \begin{bmatrix} x_1 \\ x_2 \\ \vdots \\ x_n \end{bmatrix} \tag{2-99}$$

$$\mathbf{y} = \begin{bmatrix} y_1 \\ y_2 \\ \vdots \\ y_n \end{bmatrix} \tag{2-100}$$

which are simply *bracketed arrays of coefficients and variables*. Thus, we can define a matrix as follows:

Definition of a Matrix

A matrix is a collection of elements arranged in a rectangular or square array. Several ways of representing a matrix are as follows:

$$\mathbf{A} = \begin{bmatrix} 0 & 3 & 10 \\ 1 & -2 & 0 \end{bmatrix}, \quad \mathbf{A} = \begin{pmatrix} 0 & 3 & 10 \\ 1 & -2 & 0 \end{pmatrix}$$

$$\mathbf{A} = \begin{Vmatrix} 0 & 3 & 10 \\ 1 & -2 & 0 \end{Vmatrix}, \quad \mathbf{A} = [a_{ij}]_{2,3}$$

In this text we use square brackets to represent the matrix. It is important to distinguish between a matrix and a determinant:

Matrix	*Determinant*
An array of numbers or elements with n rows and m columns.	*An array of numbers or elements with n rows and n columns (always square).*
Does not have a value, although a square matrix (n = m) has a determinant.	*Has a value.*

Some important definitions of matrices are given in the following.

Matrix Elements.   When a matrix is written

$$A = \begin{bmatrix} a_{11} & a_{12} & a_{13} \\ a_{21} & a_{22} & a_{23} \\ a_{31} & a_{32} & a_{33} \end{bmatrix} \tag{2-101}$$

a_{ij} is identified as the element in the ith *row* and the jth *column* of the matrix. As a rule, we always refer to the row first and the column last.

Order of a Matrix.   The order of a matrix refers to the total number of rows and columns of the matrix. For example, the matrix in Eq. (2-101) has three rows and three columns and, therefore, is called a 3×3 (three by three) matrix. In general, a matrix with n rows and m columns is termed "$n \times m$" or "n by m."

Square Matrix.   *A square matrix is one that has the same number of rows as columns.*

Column Matrix.   *A column matrix is one that has one column and more than one row, that is, an $m \times 1$ matrix, $m > 1$.*

Quite often, a column matrix is referred to as a *column vector* or simply an *m-vector* if there are m rows. The matrix in Eq. (2-99) is a typical column matrix that is $n \times 1$, or an *n-vector*.

Row Matrix.   *A row matrix is one that has one row and more than one column, that is, a $1 \times n$ matrix, where $n > 1$.* A row matrix can also be referred to as a *row vector*.

Diagonal Matrix.   *A diagonal matrix is a square matrix with $a_{ij} = 0$ for all $i \neq j$. Examples of a diagonal matrix are*

$$\begin{bmatrix} a_{11} & 0 & 0 \\ 0 & a_{22} & 0 \\ 0 & 0 & a_{33} \end{bmatrix}, \quad \begin{bmatrix} 5 & 0 \\ 0 & 3 \end{bmatrix} \tag{2-102}$$

Unity Matrix (Identity Matrix).   *A unity matrix is a diagonal matrix with all the elements on the main diagonal $(i = j)$ equal to 1.* A unity matrix is often designated by **I** or **U**. An example of a unity matrix is

$$I = \begin{bmatrix} 1 & 0 & 0 \\ 0 & 1 & 0 \\ 0 & 0 & 1 \end{bmatrix} \tag{2-103}$$

Null Matrix.   *A null matrix is one whose elements are all equal to zero;* for example,

$$\mathbf{0} = \begin{bmatrix} 0 & 0 & 0 \\ 0 & 0 & 0 \end{bmatrix} \tag{2-104}$$

Symmetric Matrix. *A symmetric matrix is a square matrix that satisfies the* condition

$$a_{ij} = a_{ji} \tag{2-105}$$

for all i and j. A symmetric matrix has the property that if its rows are interchanged with its columns, the same matrix is preserved. Two examples of symmetric matrices are

$$\begin{bmatrix} 6 & 5 & 1 \\ 5 & 0 & 10 \\ 1 & 10 & -1 \end{bmatrix} \quad \begin{bmatrix} 1 & -4 \\ -4 & 1 \end{bmatrix} \tag{2-106}$$

Determinant of a Matrix. With each square matrix a determinant having the same elements and order as the matrix may be defined. *The determinant of a square matrix* **A** *is designated by*

$$\det \mathbf{A} = \Delta_A = |\mathbf{A}| \tag{2-107}$$

As an illustrative example, consider the matrix

$$\mathbf{A} = \begin{bmatrix} 1 & 0 & -1 \\ 0 & 3 & 2 \\ -1 & 1 & 0 \end{bmatrix} \tag{2-108}$$

The determinant of **A** is

$$|\mathbf{A}| = \begin{vmatrix} 1 & 0 & -1 \\ 0 & 3 & 2 \\ -1 & 1 & 0 \end{vmatrix} = -5 \tag{2-109}$$

Singular Matrix. *A square matrix is said to be singular if the value of its determinant is zero.* On the other hand, if a square matrix has a nonzero determinant, it is called a *nonsingular matrix.*

When a matrix is singular, it usually means that not all the rows or not all the columns of the matrix are independent of each other. When the matrix is used to represent a set of algebraic equations, singularity of the matrix means that these equations are not independent of each other. As an illustrative example, let us consider the following set of equations:

$$2x_1 - 3x_2 + x_3 = 0$$
$$-x_1 + x_2 + x_3 = 0 \tag{2-110}$$
$$x_1 - 2x_2 + 2x_3 = 0$$

Note that in Eq. (2-110) the third equation is equal to the sum of the first two equations. Therefore, these three equations are not completely independent.

In matrix form, these equations may be represented by

$$\mathbf{Ax} = \mathbf{0}$$

where

$$\mathbf{A} = \begin{bmatrix} 2 & -3 & 1 \\ -1 & 1 & 1 \\ 1 & -2 & 2 \end{bmatrix} \tag{2-111}$$

$$\mathbf{x} = \begin{bmatrix} x_1 \\ x_2 \\ x_3 \end{bmatrix} \tag{2-112}$$

and $\mathbf{0}$ is a 3×1 null matrix. The determinant of $\mathbf{A}$ is

$$|\mathbf{A}| = \begin{vmatrix} 2 & -3 & 1 \\ -1 & 1 & 1 \\ 1 & -2 & 2 \end{vmatrix} = 4 - 3 + 2 - 1 - 6 + 4 = 0 \tag{2-113}$$

Therefore, the matrix $\mathbf{A}$ of Eq. (2-111) is singular. In this case the rows of $\mathbf{A}$ are dependent.

Transpose of a Matrix. *The transpose of a matrix $\mathbf{A}$ is defined as the matrix that is obtained by interchanging the corresponding rows and columns in $\mathbf{A}$.*

Let $\mathbf{A}$ be an $n \times m$ matrix which is represented by

$$\mathbf{A} = [a_{ij}]_{n,m} \tag{2-114}$$

Then the transpose of $\mathbf{A}$, denoted by $\mathbf{A}'$, is given by

$$\mathbf{A}' = \text{transpose of } \mathbf{A} = [a_{ji}]_{m,n} \tag{2-115}$$

Notice that the order of $\mathbf{A}$ is $n \times m$; the transpose of $\mathbf{A}$ has an order $m \times n$.

■

Example 2-9

As an example of the transpose of a matrix, consider the matrix

$$\mathbf{A} = \begin{bmatrix} 3 & 2 & 1 \\ 0 & -1 & 5 \end{bmatrix}$$

The transpose of $\mathbf{A}$ is given by

$$\mathbf{A}' = \begin{bmatrix} 3 & 0 \\ 2 & -1 \\ 1 & 5 \end{bmatrix}$$

■

Skew-Symmetric Matrix. *A skew-symmetric matrix is a square matrix that equals its negative transpose;* that is,

$$\mathbf{A} = -\mathbf{A}' \tag{2-116}$$

Some Operations of a Matrix Transpose

1. $(\mathbf{A}')' = \mathbf{A}$ (2-117)
2. $(k\mathbf{A})' = k\mathbf{A}'$, where k is a scalar   (2-118)
3. $(\mathbf{A} + \mathbf{B})' = \mathbf{A}' + \mathbf{B}'$ (2-119)
4. $(\mathbf{AB})' = \mathbf{B}'\mathbf{A}'$ (2-120)

Adjoint of a Matrix. Let $\mathbf{A}$ be a square matrix of order n. *The adjoint matrix of* $\mathbf{A}$, *denoted by* $\mathrm{adj}\,\mathbf{A}$, *is defined as*

$$\mathrm{adj}\,\mathbf{A} = [ij \text{ cofactor of } \det \mathbf{A}]'_{n,\,n} \qquad (2\text{-}121)$$

where the ij cofactor of the determinant of $\mathbf{A}$ *is the determinant obtained by omitting the* ith *row and the* jth *column of* $|\mathbf{A}|$ *and then multiplying it by* $(-1)^{i+j}$.

■

Example 2-10

As an example of determining the adjoint matrix, let us consider first a 2×2 matrix,

$$\mathbf{A} = \begin{bmatrix} a_{11} & a_{12} \\ a_{21} & a_{22} \end{bmatrix}$$

The determinant of $\mathbf{A}$ is

$$|\mathbf{A}| = \begin{vmatrix} a_{11} & a_{12} \\ a_{21} & a_{22} \end{vmatrix}$$

The $1,1$ cofactor, or the cofactor of the $(1,1)$ element of $|\mathbf{A}|$, is a_{22}; the $1,2$ cofactor is $-a_{21}$; the $2,1$ cofactor is $-a_{12}$; and the $2,2$ cofactor is a_{11}. Thus, from Eq. (2-121) the adjoint matrix of $\mathbf{A}$ is

$$\begin{aligned}
\mathrm{adj}\,\mathbf{A} &= \begin{bmatrix} 1,1 \text{ cofactor} & 1,2 \text{ cofactor} \\ 2,1 \text{ cofactor} & 2,2 \text{ cofactor} \end{bmatrix}' \\
&= \begin{bmatrix} 1,1 \text{ cofactor} & 2,1 \text{ cofactor} \\ 1,2 \text{ cofactor} & 2,2 \text{ cofactor} \end{bmatrix} \\
&= \begin{bmatrix} a_{22} & -a_{12} \\ -a_{21} & a_{11} \end{bmatrix}
\end{aligned} \qquad (2\text{-}122)$$

Example 2-11

As a second example of the adjoint matrix, consider

$$\mathbf{A} = \begin{bmatrix} a_{11} & a_{12} & a_{13} \\ a_{21} & a_{22} & a_{23} \\ a_{31} & a_{32} & a_{33} \end{bmatrix} \qquad (2\text{-}123)$$

Then

$$\begin{aligned}
\mathrm{adj}\,\mathbf{A} &= \begin{bmatrix} 1,1 \text{ cofactor} & 2,1 \text{ cofactor} & 3,1 \text{ cofactor} \\ 1,2 \text{ cofactor} & 2,2 \text{ cofactor} & 3,2 \text{ cofactor} \\ 1,3 \text{ cofactor} & 2,3 \text{ cofactor} & 3,3 \text{ cofactor} \end{bmatrix} \\
&= \begin{bmatrix} (a_{22}a_{33} - a_{23}a_{32}) & -(a_{12}a_{33} - a_{13}a_{32}) & (a_{12}a_{23} - a_{13}a_{22}) \\ -(a_{21}a_{33} - a_{23}a_{31}) & (a_{11}a_{33} - a_{13}a_{31}) & -(a_{11}a_{23} - a_{13}a_{21}) \\ (a_{21}a_{32} - a_{22}a_{31}) & -(a_{11}a_{32} - a_{12}a_{31}) & (a_{11}a_{22} - a_{12}a_{21}) \end{bmatrix}
\end{aligned} \qquad (2\text{-}124)$$

■

2.8 MATRIX ALGEBRA

When carrying out matrix operations it is necessary to define matrix algebra in the form of addition, subtraction, multiplication, division, and other necessary operations. It is important to point out at this stage that matrix operations are defined independently of the algebraic operations for scalar quantities.

Equality of Matrices

Two matrices **A** *and* **B** *are said to be equal to each other if they satisfy the following conditions*:

1. They are of the same order.
2. The corresponding elements are equal; that is,

$$a_{ij} = b_{ij} \qquad \text{for every } i \text{ and } j$$

For example,

$$\mathbf{A} = \begin{bmatrix} a_{11} & a_{12} \\ a_{21} & a_{22} \end{bmatrix} = \mathbf{B} = \begin{bmatrix} b_{11} & b_{12} \\ b_{21} & b_{22} \end{bmatrix} \tag{2-125}$$

implies that $a_{11} = b_{11}$, $a_{12} = b_{12}$, $a_{21} = b_{21}$, and $a_{22} = b_{22}$.

Addition of Matrices

Two matrices **A** and **B** can be added to form **A** + **B** if they are of the same order. Then

$$\mathbf{A} + \mathbf{B} = [a_{ij}]_{n,m} + [b_{ij}]_{n,m} = \mathbf{C} = [c_{ij}]_{n,m} \tag{2-126}$$

where

$$c_{ij} = a_{ij} + b_{ij} \tag{2-127}$$

for all i and j. The order of the matrices is preserved after addition.

■

Example 2-12

As an illustrative example, consider the two matrices

$$\mathbf{A} = \begin{bmatrix} 3 & 2 \\ -1 & 4 \\ 0 & -1 \end{bmatrix} \qquad \mathbf{B} = \begin{bmatrix} 0 & 3 \\ -1 & 2 \\ 1 & 0 \end{bmatrix}$$

which are of the same order. Then the sum of $\mathbf{A}$ and $\mathbf{B}$ is given by

$$\mathbf{C} = \mathbf{A} + \mathbf{B} = \begin{bmatrix} 3+0 & 2+3 \\ -1-1 & 4+2 \\ 0+1 & -1+0 \end{bmatrix} = \begin{bmatrix} 3 & 5 \\ -2 & 6 \\ 1 & -1 \end{bmatrix} \qquad (2\text{-}128)$$

∎

Matrix Subtraction

The rules governing the subtraction of matrices are similar to those of matrix addition. In other words, Eqs. (2-126) and (2-127) are true if all the plus signs are replaced by minus signs. Or,

$$\begin{aligned} \mathbf{C} = \mathbf{A} - \mathbf{B} &= [a_{ij}]_{n,m} - [b_{ij}]_{n,m} \\ &= [a_{ij}]_{n,m} + [-b_{ij}]_{n,m} \\ &= [c_{ij}]_{n,m} \end{aligned} \qquad (2\text{-}129)$$

where

$$c_{ij} = a_{ij} - b_{ij} \qquad (2\text{-}130)$$

for all i and j.

Associative Law of Matrix (Addition and Subtraction)

The associative law of scalar algebra still holds for matrix addition and subtraction. Therefore,

$$(\mathbf{A} + \mathbf{B}) + \mathbf{C} = \mathbf{A} + (\mathbf{B} + \mathbf{C}) \qquad (2\text{-}131)$$

Commutative Law of Matrix (Addition and Subtraction)

The commutative law for matrix addition and subtraction states that the following matrix relationship is true:

$$\begin{aligned} \mathbf{A} + \mathbf{B} + \mathbf{C} &= \mathbf{B} + \mathbf{C} + \mathbf{A} \\ &= \mathbf{A} + \mathbf{C} + \mathbf{B} \end{aligned} \qquad (2\text{-}132)$$

Matrix Multiplication

The matrices $\mathbf{A}$ and $\mathbf{B}$ may be multiplied together to form the product $\mathbf{AB}$ if they are *conformable*. This means that the number of columns of $\mathbf{A}$ must equal the number of rows of $\mathbf{B}$. In other words, let

$$\mathbf{A} = [a_{ij}]_{n,p}$$

$$\mathbf{B} = [b_{ij}]_{q,m}$$

Then **A** and **B** are conformable to form the product

$$C = AB = [a_{ij}]_{n,p}[b_{ij}]_{q,m} = [c_{ij}]_{n,m} \qquad (2\text{-}133)$$

if and only if $p = q$. The matrix **C** will have the same number of rows as **A** and the same number of columns as **B**.

It is important to note that **A** and **B** may be conformable for **AB**, but they may not be conformable for the product **BA**, unless in Eq. (2-133) n also equals m. This points out an important fact that the commutative law is not generally valid for matrix multiplication. It is also noteworthy that even though **A** and **B** are conformable for both **AB** and **BA**, usually $AB \neq BA$. In general, the following references are made with respect to matrix multiplication whenever they exist:

$$AB = A \text{ postmultiplied by } B$$

or

$$AB = B \text{ premultiplied by } A$$

Having established the condition for matrix multiplication, let us now turn to the rule of matrix multiplication. When the matrices **A** and **B** are conformable to form the matrix $C = AB$ as in Eq. (2-133), the ijth element of **C**, c_{ij}, is given by

$$c_{ij} = \sum_{k=1}^{p} a_{ik}b_{kj} \qquad (2\text{-}134)$$

for $i = 1, 2, \ldots, n$, and $j = 1, 2, \ldots, m$.   ∎

Example 2-13

Given the matrices

$$A = [a_{ij}]_{2,3} \qquad B = [b_{ij}]_{3,1}$$

we notice that these two matrices are conformable for the product **AB** but not for **BA**. Thus,

$$AB = \begin{bmatrix} a_{11} & a_{12} & a_{13} \\ a_{21} & a_{22} & a_{23} \end{bmatrix} \begin{bmatrix} b_{11} \\ b_{21} \\ b_{31} \end{bmatrix} \qquad (2\text{-}135)$$

$$= \begin{bmatrix} a_{11}b_{11} + a_{12}b_{21} + a_{13}b_{31} \\ a_{21}b_{11} + a_{22}b_{21} + a_{23}b_{31} \end{bmatrix}$$

Example 2-14

Given the matrices

$$A = \begin{bmatrix} 3 & -1 \\ 0 & 1 \\ 2 & 0 \end{bmatrix} \qquad B = \begin{bmatrix} 1 & 0 & -1 \\ 2 & 1 & 0 \end{bmatrix}$$

we notice that both **AB** and **BA** are conformable for multiplication.

$$\mathbf{AB} = \begin{bmatrix} 3 & -1 \\ 0 & 1 \\ 2 & 0 \end{bmatrix} \begin{bmatrix} 1 & 0 & -1 \\ 2 & 1 & 0 \end{bmatrix}$$

$$= \begin{bmatrix} (3)(1) + (-1)(2) & (3)(0) + (-1)(1) & (3)(-1) + (-1)(0) \\ (0)(1) + (1)(2) & (0)(0) + (1)(1) & (0)(-1) + (1)(0) \\ (2)(1) + (0)(2) & (2)(0) + (0)(1) & (2)(-1) + (0)(0) \end{bmatrix} \quad (2\text{-}136)$$

$$= \begin{bmatrix} 1 & -1 & -3 \\ 2 & 1 & 0 \\ 2 & 0 & -2 \end{bmatrix}$$

$$\mathbf{BA} = \begin{bmatrix} 1 & 0 & -1 \\ 2 & 1 & 0 \end{bmatrix} \begin{bmatrix} 3 & -1 \\ 0 & 1 \\ 2 & 0 \end{bmatrix}$$

$$= \begin{bmatrix} (1)(3) + (0)(0) + (-1)(2) & (1)(-1) + (0)(1) + (-1)(0) \\ (2)(3) + (1)(0) + (0)(2) & (2)(-1) + (1)(1) + (0)(0) \end{bmatrix} \quad (2\text{-}137)$$

$$= \begin{bmatrix} 1 & -1 \\ 6 & -1 \end{bmatrix}$$

Therefore, even though **AB** and **BA** may both exist, in general, they are not equal. In this case the products are not even of the same order.

■

Although the commutative law does not hold in general for matrix multiplication, the *associative* and the *distributive* laws are valid. For the distributive law, we state that

$$\mathbf{A(B + C)} = \mathbf{AB} + \mathbf{AC} \quad (2\text{-}138)$$

if the products are conformable.

For the associative law,

$$\mathbf{(AB)C} = \mathbf{A(BC)} \quad (2\text{-}139)$$

if the product is conformable.

Multiplication by a Scalar *k*

Multiplying a matrix **A** by any scalar k is equivalent to multiplying each element of **A** by k. Therefore, if $\mathbf{A} = [a_{ij}]_{n,\,m}$,

$$k\mathbf{A} = \left[ka_{ij} \right]_{n,\,m} \quad (2\text{-}140)$$

Inverse of a Matrix (Matrix Division)

In the algebra for scalar quantities, when we write

$$ax = y \quad (2\text{-}141)$$

it leads to

$$x = \frac{1}{a}y \tag{2-142}$$

or

$$x = a^{-1}y \tag{2-143}$$

Equations (2-142) and (2-143) are notationally equivalent.

   In matrix algebra, if

$$\mathbf{Ax} = \mathbf{y} \tag{2-144}$$

then it *may be possible* to write

$$\mathbf{x} = \mathbf{A}^{-1}\mathbf{y} \tag{2-145}$$

where $\mathbf{A}^{-1}$ denotes the "inverse of $\mathbf{A}$." The conditions that $\mathbf{A}^{-1}$ exists are:

1. $\mathbf{A}$ is a square matrix.
2. $\mathbf{A}$ must be nonsingular.

   If $\mathbf{A}^{-1}$ exists, it is given by

$$\mathbf{A}^{-1} = \frac{\text{adj } \mathbf{A}}{|\mathbf{A}|} \tag{2-146}$$

∎

Example 2-15

Given the matrix

$$\mathbf{A} = \begin{bmatrix} a_{11} & a_{12} \\ a_{21} & a_{22} \end{bmatrix} \tag{2-147}$$

the inverse of $\mathbf{A}$ is given by

$$\mathbf{A}^{-1} = \frac{\text{adj } \mathbf{A}}{|\mathbf{A}|} = \frac{\begin{bmatrix} a_{22} & -a_{12} \\ -a_{21} & a_{11} \end{bmatrix}}{a_{11}a_{22} - a_{12}a_{21}} \tag{2-148}$$

where for $\mathbf{A}$ to be nonsingular, $|\mathbf{A}| \neq 0$, or

$$a_{11}a_{22} - a_{12}a_{21} \neq 0 \tag{2-149}$$

   If we pay attention to the adjoint matrix of $\mathbf{A}$, which is the numerator of $\mathbf{A}^{-1}$, we see that for a 2×2 matrix, adj $\mathbf{A}$ is obtained by interchanging the two elements on the main diagonal and changing the signs of the elements on the off diagonal of $\mathbf{A}$.

Example 2-16

Given the matrix

$$\mathbf{A} = \begin{bmatrix} 1 & 1 & 0 \\ -1 & 0 & 2 \\ 1 & 1 & 1 \end{bmatrix} \qquad (2\text{-}150)$$

the determinant of **A** is

$$|\mathbf{A}| = \begin{vmatrix} 1 & 1 & 0 \\ -1 & 0 & 2 \\ 1 & 1 & 1 \end{vmatrix} = 1 \qquad (2\text{-}151)$$

Therefore, **A** has an inverse matrix, and is given by

$$\mathbf{A}^{-1} = \begin{bmatrix} -2 & -1 & 2 \\ 3 & 1 & -2 \\ -1 & 0 & 1 \end{bmatrix} \qquad (2\text{-}152)$$

∎

Some Properties of Matrix Inverse

1. $\mathbf{A}\mathbf{A}^{-1} = \mathbf{A}^{-1}\mathbf{A} = \mathbf{I}$ $\qquad (2\text{-}153)$
2. $(\mathbf{A}^{-1})^{-1} = \mathbf{A}$ $\qquad (2\text{-}154)$
3. In matrix algebra, in general,

$$\mathbf{A}\mathbf{B} = \mathbf{A}\mathbf{C} \qquad (2\text{-}155)$$

does not necessarily imply that $\mathbf{B} = \mathbf{C}$. The reader can easily construct an example to illustrate this property. However, if **A** is a square matrix, and is nonsingular, we can premultiply both sides of Eq. (2-155) by $\mathbf{A}^{-1}$. Then

$$\mathbf{A}^{-1}\mathbf{A}\mathbf{B} = \mathbf{A}^{-1}\mathbf{A}\mathbf{C} \qquad (2\text{-}156)$$

or

$$\mathbf{I}\mathbf{B} = \mathbf{I}\mathbf{C} \qquad (2\text{-}157)$$

which leads to

$$\mathbf{B} = \mathbf{C}$$

4. If **A** and **B** are square matrices and are nonsingular, then

$$(\mathbf{A}\mathbf{B})^{-1} = \mathbf{B}^{-1}\mathbf{A}^{-1} \qquad (2\text{-}158)$$

Rank of a Matrix

The rank of a matrix **A** is the maximum number of linearly independent columns of **A**; or, it is the order of the largest nonsingular matrix contained in **A**. Several examples on the rank of a matrix are as follows:

$$\begin{bmatrix} 0 & 1 \\ 0 & 0 \end{bmatrix} \text{ rank} = 1 \qquad \begin{bmatrix} 0 & 5 & 1 & 4 \\ 3 & 0 & 3 & 2 \end{bmatrix} \text{ rank} = 2$$

$$\begin{bmatrix} 3 & 9 & 2 \\ 1 & 3 & 0 \\ 2 & 6 & 1 \end{bmatrix} \text{ rank} = 2 \qquad \begin{bmatrix} 3 & 0 & 0 \\ 1 & 2 & 0 \\ 0 & 0 & 1 \end{bmatrix} \text{ rank} = 3$$

The following properties on rank are useful in the determination of the rank of a matrix. Given an $n \times m$ matrix $\mathbf{A}$,

1. Rank of $\mathbf{A}$ = Rank of $\mathbf{A}'$.
2. Rank of $\mathbf{A}$ = Rank of $\mathbf{A}'\mathbf{A}$.
3. Rank of $\mathbf{A}$ = Rank of $\mathbf{A}\mathbf{A}'$.

Properties 2 and 3 are useful in the determination of rank; since $\mathbf{A}'\mathbf{A}$ and $\mathbf{A}\mathbf{A}'$ are always square, the rank condition can be checked by evaluating the determinant of these matrices.

2.9 VECTOR-MATRIX FORM OF STATE EQUATIONS

The first-order differential equations called state equations in Section 2.3 can be conveniently expressed in vector-matrix form. For the two equations in Eqs. (2-13) and (2-14), we let the *state vector* be defined as the 2×1 matrix

$$\mathbf{x}(t) = \begin{bmatrix} x_1(t) \\ x_2(t) \end{bmatrix} \qquad (2\text{-}159)$$

Then the state equations in vector-matrix form are

$$\dot{\mathbf{x}}(t) = \frac{d\mathbf{x}(t)}{dt} = \mathbf{A}\mathbf{x}(t) + \mathbf{B}u(t) \qquad (2\text{-}160)$$

where

$$\mathbf{A} = \begin{bmatrix} 0 & 1 \\ -\dfrac{K}{M} & -\dfrac{B}{M} \end{bmatrix} \qquad (2\text{-}161)$$

$$\mathbf{B} = \begin{bmatrix} 0 \\ \dfrac{1}{M} \end{bmatrix} \qquad (2\text{-}162)$$

and $u(t) = f(t)$.

Similarly, Eq. (2-160) is valid for Eqs. (2-17) and (2-18) with

$$\mathbf{A} = \begin{bmatrix} 0 & 1 \\ -\dfrac{1}{LC} & -\dfrac{R}{L} \end{bmatrix} \tag{2-163}$$

$$\mathbf{B} = \begin{bmatrix} 0 \\ \dfrac{1}{L} \end{bmatrix} \tag{2-164}$$

and $u(t) = v(t)$.

In general, for a linear nth-order system for which there are n state variables with n state equations, p inputs, Eq. (2-160) is still valid. In this case, $\mathbf{x}(t)$ and $\dot{\mathbf{x}}(t)$ are $n \times 1$ column matrices, $\mathbf{A}$ is $n \times n$, $\mathbf{B}$ is $n \times p$, and $\mathbf{u}(t)$ is the $p \times 1$ input vector (boldface for vector and matrix quantities).

The vector-matrix form can also be applied to nonlinear state equations. The general expression is

$$\frac{d\mathbf{x}(t)}{dt} = \mathbf{f}[\mathbf{x}(t), \mathbf{u}(t), t] \tag{2-165}$$

2.10 DIFFERENCE EQUATIONS

Because digital computers are frequently used as controllers in modern control systems, it is important to establish equations that relate digital and discrete-time signals. Just as differential equations are used to represent systems with analog signals, difference equations are used for systems with discrete or digital data. Difference equations are also used to approximate differential equations, since the former are more easily programmed on a digital computer, and are generally easier to solve.

As an example of digital approximation, we can use a forward difference to approximate the derivative of a function at a given point; that is,

$$\left. \frac{dy(t)}{dt} \right|_{t=kT} \cong \frac{y[(k+1)T] - y(kT)}{T} \tag{2-166}$$

where T is chosen to be some small value that will render Eq. (2-166) a good approximation. Using Eq. (2-166), we can approximate the first-order differential equation

$$\frac{dy(t)}{dt} + ay(t) = f(t) \tag{2-167}$$

at $t = kT$ by

$$\frac{y[(k+1)T] - y(kT)}{T} + ay(kT) = f(kT) \tag{2-168}$$

After rearranging, Eq. (2-168) becomes

$$\frac{1}{T}y[(k+1)T] + \left(a - \frac{1}{T}\right)y(kT) = f(kT) \qquad (2\text{-}169)$$

which is recognized as a first-order difference equation.

In general, a linear nth-order difference equation with constant coefficients can be written as

$$a_{n+1}y(k+n) + a_n y(k+n-1) + \cdots + a_2 y(k+1) + a_1 y(k) = f(k) \qquad (2\text{-}170)$$

where $y(i)$, $i = k, k+1, \ldots, k+n$, denotes the discrete dependent variable y at the ith instant if the independent variable is time. In general, the independent variable can represent any real quantity.

Similar to the case of the analog systems, it is convenient to use a set of first-order difference equations (state equations) to represent a high-order difference equation. For the difference equation in Eq. (2-170), if we let

$$x_1(k) = y(k)$$
$$x_2(k) = x_1(k+1) = y(k+1)$$
$$\vdots$$
$$x_{n-1}(k) = x_{n-2}(k+1) = y(k+n-2) \qquad (2\text{-}171)$$

then the equation is written as

$$x_n(k+1) = -\frac{a_1}{a_{n+1}}x_1(k) - \frac{a_2}{a_{n+1}}x_2(k) - \cdots - \frac{a_n}{a_{n+1}}x_n(k) + \frac{1}{a_{n+1}}f(k)$$
$$(2\text{-}172)$$

The first $n-1$ state equations are taken directly from Eq. (2-171), and the final one is given by Eq. (2-172). Writing these n first-order difference state equations in vector-matrix form, we have

$$\mathbf{x}[(k+1)T] = \mathbf{A}x(kT) + \mathbf{B}u(kT) \qquad (2\text{-}173)$$

where

$$\mathbf{x}(kT) = \begin{bmatrix} x_1(kT) \\ x_2(kT) \\ \vdots \\ x_n(kT) \end{bmatrix} \qquad (2\text{-}174)$$

is the $n \times 1$ state vector, and

$$
\mathbf{A} = \begin{bmatrix}
0 & 1 & 0 & \cdots & 0 \\
0 & 0 & 1 & \cdots & 0 \\
0 & 0 & 0 & \cdots & 0 \\
\cdot\cdot & \cdot\cdot\cdot & \cdot\cdot\cdot & \cdots & \cdot\cdot \\
0 & 0 & 0 & \cdots & 1 \\
-\dfrac{a_1}{a_{n+1}} & -\dfrac{a_2}{a_{n+1}} & -\dfrac{a_3}{a_{n+1}} & \cdots & -\dfrac{a_n}{a_{n+1}}
\end{bmatrix}
\tag{2-175}
$$

$$
\mathbf{B} = \begin{bmatrix}
0 \\
0 \\
0 \\
\vdots \\
0 \\
\dfrac{1}{a_{n+1}}
\end{bmatrix}
\tag{2-176}
$$

2.11 THE z-TRANSFORM [12]

In Section 2.6 we have shown how the Laplace transform can be used to solve linear ordinary differential equations. For linear difference equations and linear systems with discrete or digital data, the z-transform is more appropriate to use.

Let us first consider the analysis of a discrete-data system that is represented by the block diagram of Fig. 2-4. One way of describing the discrete nature of the signals is to consider that the input and the output of the system are sequences of numbers. These numbers appear at uniform intervals of time T. Thus, the input sequence and the output sequence may be represented by $u(kT)$ and $c(kT)$, respectively, $k = 0, 1, 2, \ldots$.

In order to represent the input sequence and output sequence by time-domain expressions, we introduce an impulse train such that the numbers are represented by the strengths of the impulses at the corresponding time instants. Thus, the input sequence is expressed as the impulse train:

$$
u^*(t) = \sum_{k=0}^{\infty} u(kT)\delta(t - kT)
\tag{2-177}
$$

A similar expression can be written for the output sequence.

Another important type of discrete-data system is the sampled-data system. A sampled-data system is one in which some of the signals are continuous at certain parts of the system but are converted to discrete data by devices such as an

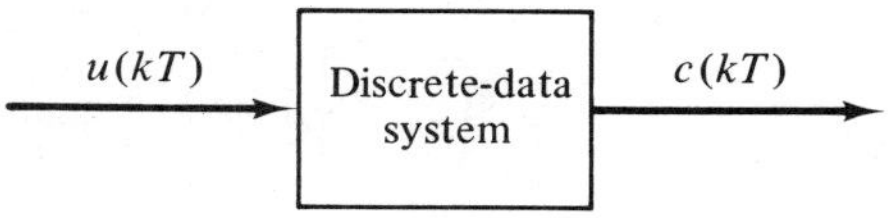

Figure 2-4 Block diagram of a discrete-data system.

$u(kT)$ → | Discrete-data system | → $c(kT)$

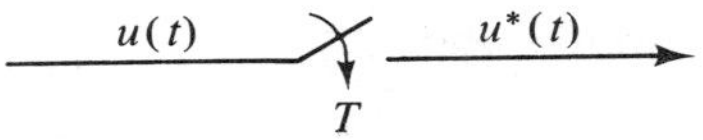

Figure 2-5 Ideal sampler.

analog-to-digital converter (A/D). The formation of discrete data from continuous data can be represented by samplers. Figure 2-5 shows the schematic diagram of a uniform-rate sampler that converts the continuous-data signal $u(t)$ into a discrete-data signal $u*(t)$; the latter is described by Eq. (2-177). The time duration between the closings of the sampler is T and is called the *sampling period* (sec). Since the impulse $\delta(t - kT)$ has a zero pulse width, the sampler so defined closes only for an infinitesimally small time duration. Therefore, such a sampler is not real, and is called an *ideal sampler*.

Now we are ready to investigate the application of transform methods to discrete-data systems. Taking the Laplace transform on both sides of Eq. (2-177), we have

$$U*(s) = \sum_{k=0}^{\infty} u(kT)e^{-kTs} \tag{2-178}$$

The fact that Eq. (2-178) contains the exponential term e^{-kTs} reveals the difficulty of using Laplace transform for the general treatment of discrete-data systems, since the transfer function relations will no longer be algebraic as in the continuous-data case. Although it is conceptually simple to perform inverse Laplace transform on algebraic transfer relations, it is not a simple matter to perform inverse Laplace transform on transcendental functions. One simple fact is that the commonly used Laplace transform tables do not have entries with transcendental functions in s. This necessitates the use of the z-transform. Our motivation here for the generation of the z-transform is simply to convert transcendental functions in s into algebraic ones in z. The definition of z-transform is given with this objective in mind.

Definition of the *z*-Transform

The z-transform is defined as

$$z = e^{Ts} \tag{2-179}$$

where s is the Laplace transform variable and T is the sampling period. Equation (2-179) also leads to

$$s = \frac{1}{T}\ln z \tag{2-180}$$

Using Eq. (2-179), the expression in Eq. (2-178) is written

$$U*\left(s = \frac{1}{T}\ln z\right) = U(z) = \sum_{k=0}^{\infty} u(kT)z^{-k} \tag{2-181}$$

or

$$U(z) = z\text{-transform of } u*(t)$$
$$= \mathfrak{z}[u*(t)]$$
$$= [\text{Laplace transform of } u*(t)]_{s=1/T \ln z} \qquad (2\text{-}182)$$

Therefore, we have treated the z-transform as simply a change in variable, $z = e^{Ts}$. The following examples illustrate some of the simple z-transform operations.

∎

Example 2-17

Consider the sequence

$$r(kT) = e^{-akT}, \qquad k = 0, 1, 2, \dots \qquad (2\text{-}183)$$

where a is a constant.

From Eq. (2-177),

$$r*(t) = \sum_{k=0}^{\infty} e^{-akT}\delta(t - kT) \qquad (2\text{-}184)$$

Then

$$R*(s) = \sum_{k=0}^{\infty} e^{-akT}e^{-kTs} \qquad (2\text{-}185)$$

Multiply both sides of Eq. (2-185) by $e^{-(s+a)T}$ and subtract the resulting equation from Eq. (2-185); we can show easily that $R*(s)$ can be written in a closed form,

$$R*(s) = \frac{1}{1 - e^{-(s+a)T}} \qquad (2\text{-}186)$$

for

$$|e^{-(\sigma+a)T}| < 1 \qquad (2\text{-}187)$$

where σ is the real part of s. Then the z-transform of $r*(t)$ is

$$R(z) = \frac{1}{1 - e^{-aT}z^{-1}} = \frac{z}{z - e^{-aT}} \qquad (2\text{-}188)$$

for $|e^{-aT}z^{-1}| < 1$.

Example 2-18

In Example 2-17, if $a = 0$, we have

$$r(kT) = 1, \qquad k = 0, 1, 2, \dots \qquad (2\text{-}189)$$

which represents a sequence of numbers all equal to unity. Then

$$R^*(s) = \sum_{k=0}^{\infty} e^{-kTs} \qquad\qquad (2\text{-}190)$$

$$R(z) = \sum_{k=0}^{\infty} z^{-k} = 1 + z^{-1} + z^{-2} + z^{-3} + \cdots \qquad\qquad (2\text{-}191)$$

This expression is written in closed form as

$$R(z) = \frac{1}{1 - z^{-1}} \qquad |z^{-1}| < 1 \qquad\qquad (2\text{-}192)$$

or

$$R(z) = \frac{z}{z - 1} \qquad |z^{-1}| < 1 \qquad\qquad (2\text{-}193)$$

■

In general, the z-transforms of more complex functions are obtained by use of the same procedure as described in the preceding two examples. If a time function $r(t)$ is given as the starting point, the procedure of finding its z-transform is to first form the sequence $r(kT)$ and then use Eq. (2-181) to get $R(z)$. An equivalent interpretation of this step is to send the signal $r(t)$ through an ideal sampler whose output is $r^*(t)$. We then take the Laplace transform of $r^*(t)$ to give $R^*(s)$ as in Eq. (2-178), and $R(z)$ is obtained by substituting z for e^{Ts}.

Table 2-2 gives the z-transforms of some of the functions commonly used in systems analysis. A more extensive table may be found in the literature [12].

Inverse z-Transformation

Just as in the Laplace transformation, one of the major objectives of the z-transformation is that algebraic manipulations can be made first in the z-domain, and then the final time response is determined by the inverse z-transformation. In general, the inverse z-transformation of $R(z)$ can yield information only on $r(kT)$, not on $r(t)$. In other words, the z-transform carries information only in a discrete fashion. When the time signal $r(t)$ is sampled by the ideal sampler, only information on the signal at the sampling instants, $t = kT$, is retained. With this in mind, the inverse z-transformation can be affected by one of the following three methods:

1. The partial-fraction expansion method.
2. The power-series method.
3. The inversion formula.

Table 2-2   Table of z-Transforms

Laplace Transform	Time Function	z-Transform
1	Unit impulse $\delta(t)$	1
$\dfrac{1}{s}$	Unit step $u_s(t)$	$\dfrac{z}{z-1}$
$\dfrac{1}{1-e^{-Ts}}$	$\delta_T(t) = \displaystyle\sum_{n=0}^{\infty} \delta(t-nT)$	$\dfrac{z}{z-1}$
$\dfrac{1}{s^2}$	t	$\dfrac{Tz}{(z-1)^2}$
$\dfrac{1}{s^3}$	$\dfrac{t^2}{2}$	$\dfrac{T^2 z(z+1)}{2(z-1)^3}$
$\dfrac{1}{s^{n+1}}$	$\dfrac{t^n}{n!}$	$\displaystyle\lim_{a\to 0} \dfrac{(-1)^n}{n!} \dfrac{\partial^n}{\partial a^n}\left(\dfrac{z}{z-e^{-aT}}\right)$
$\dfrac{1}{s+a}$	e^{-at}	$\dfrac{z}{z-e^{-aT}}$
$\dfrac{1}{(s+a)^2}$	te^{-at}	$\dfrac{Tze^{-aT}}{(z-e^{-aT})^2}$
$\dfrac{a}{s(s+a)}$	$1-e^{-at}$	$\dfrac{(1-e^{-aT})z}{(z-1)(z-e^{-aT})}$
$\dfrac{\omega}{s^2+\omega^2}$	$\sin \omega t$	$\dfrac{z\sin \omega T}{z^2 - 2z\cos \omega T + 1}$
$\dfrac{\omega}{(s+a)^2+\omega^2}$	$e^{-at}\sin \omega t$	$\dfrac{ze^{-aT}\sin \omega T}{z^2 - 2ze^{-aT}\cos \omega T + e^{-2aT}}$
$\dfrac{s}{s^2+\omega^2}$	$\cos \omega t$	$\dfrac{z(z-\cos \omega T)}{z^2 - 2z\cos \omega T + 1}$
$\dfrac{s+a}{(s+a)^2+\omega^2}$	$e^{-at}\cos \omega t$	$\dfrac{z^2 - ze^{-aT}\cos \omega T}{z^2 - 2ze^{-aT}\cos \omega T + e^{-2aT}}$

Partial-Fraction Expansion Method. The z-transform function $R(z)$ is expanded by partial-fraction expansion into a sum of simple recognizable terms, and the z-transform table is used to determine the corresponding $r(kT)$. In carrying out the partial-fraction expansion, there is a slight difference between the z-transform and the Laplace transform procedures. With reference to the z-transform table, we note that practically all the transform functions have the term z in the numerator. Therefore, we should expand $R(z)$ into the form

$$R(z) = \frac{K_1 z}{z - e^{-aT}} + \frac{K_2 z}{z - e^{-bT}} + \cdots \tag{2-194}$$

For this, we should first expand $R(z)/z$ into fractions and then multiply z across to obtain the final desired expression. The following example will illustrate this recommended procedure.

■

Example 2-19

Given the z-transform function

$$R(z) = \frac{(1 - e^{-aT})z}{(z - 1)(z - e^{-aT})} \tag{2-195}$$

it is desired to find the inverse z-transform.

Expanding $R(z)/z$ by partial-fraction expansion, we have

$$\frac{R(z)}{z} = \frac{1}{z - 1} - \frac{1}{z - e^{-aT}} \tag{2-196}$$

Thus,

$$R(z) = \frac{z}{z - 1} - \frac{z}{z - e^{-aT}} \tag{2-197}$$

From the z-transform table of Table 2-2 the corresponding inverse z-transform of $R(z)$ is found to be

$$r(kT) = 1 - e^{-akT} \tag{2-198}$$

■

Power-Series Method. The z-transform $R(z)$ is expanded into a power series in powers of z^{-1}. In view of Eq. (2-181), the coefficient of z^{-k} is the value of $r(t)$ at $t = kT$, or simply $r(kT)$. For example, for the $R(z)$ in Eq. (2-195), we expand it into a power series in powers of z^{-1} by long division; then we have

$$R(z) = (1 - e^{-aT})z^{-1} + (1 - e^{-2aT})z^{-2} + (1 - e^{-3aT})z^{-3}$$
$$+ \cdots + (1 - e^{-akT})z^{-k} + \cdots \tag{2-199}$$

or

$$R(z) = \sum_{k=0}^{\infty} (1 - e^{-akT})z^{-k} \tag{2-200}$$

Thus,

$$r(kT) = 1 - e^{-akT} \tag{2-201}$$

which is the same result as in Eq. (2-198).

Inversion Formula. The time sequence $r(kT)$ may be determined from $R(z)$ by use of the inversion formula,

$$r(kT) = \frac{1}{2\pi j} \oint_{\Gamma} R(z)z^{k-1} \, dz \tag{2-202}$$

which is a contour integration [5] along the path Γ, where Γ is a circle of radius $|z| = e^{cT}$ centered at the origin in the z-plane, and c is of such a value that all the poles of $R(z)$ are inside the circle.

One way of evaluating the contour integration of Eq. (2-202) is by use of the residue theorem of complex variable theory. Equation (2-202) may be written

$$r(kT) = \frac{1}{2\pi j} \oint_{\Gamma} R(z) z^{k-1} \, dz$$

$$= \sum \text{residues of } R(z) z^{k-1} \text{ at the poles of } R(z) z^{k-1}$$

(2-203)

For simple poles, the residues of $R(z)z^{k-1}$ at the pole $z = z_j$ is obtained as

$$\text{residue of } R(z)z^{k-1} \text{ at the pole } z_j = (z - z_j) R(z) z^{k-1} \big|_{z=z_j} \quad (2\text{-}204)$$

Now let us consider the same function used in Example 2-19. The function $R(z)$ of Eq. (2-195) has two poles: $z = 1$ and $z = e^{-aT}$. Using Eq. (2-203), we have

$$r(kT) = \left[\text{residue of } R(z)z^{k-1} \text{ at } z = 1 \right] + \left[\text{residue of } R(z)z^{k-1} \text{ at } z = e^{-aT} \right]$$

$$= \frac{(1 - e^{-aT})z^k}{(z - e^{-aT})} \bigg|_{z=1} + \frac{(1 - e^{-aT})z^k}{(z - 1)} \bigg|_{z=e^{-aT}}$$

(2-205)

$$= 1 - e^{-akT}$$

which again agrees with the result obtained earlier.

Some Important Theorems of the z-Transformation

Some of the commonly used theorems of the z-transform are stated in the following without proof. Just as in the case of the Laplace transform, these theorems are useful in many aspects of the z-transform analysis.

1. Addition and Subtraction
If $r_1(kT)$ and $r_2(kT)$ have z-transforms $R_1(z)$ and $R_2(z)$, respectively, then

$$\mathfrak{z}[r_1(kT) \pm r_2(kT)] = R_1(z) \pm R_2(z) \quad (2\text{-}206)$$

2. Multiplication by a Constant

$$\mathfrak{z}[ar(kT)] = a\mathfrak{z}[r(kT)] = aR(z) \quad (2\text{-}207)$$

where a is a constant.

3. Real Translation

$$\mathfrak{z}[r(kT - nT)] = z^{-n}R(z) \quad (2\text{-}208)$$

and

$$\mathfrak{z}[r(kT + nT)] = z^n\left[R(z) - \sum_{k=0}^{n-1} r(kT)z^{-k}\right] \tag{2-209}$$

where n is a positive integer. Equation (2-208) represents the z-transform of a time sequence that is shifted to the right by nT, and Eq. (2-209) denotes that of a time sequence shifted to the left by nT. The reason the right-hand side of Eq. (2-209) is not $z^n R(z)$ is because the z-transform, similar to the Laplace transform, is defined only for $k \geq 0$. Thus, the second term on the right-hand side of Eq. (2-209) simply represents the sequence that is lost after it is shifted to the left by nT.

4. Complex Translation

$$\mathfrak{z}\left[e^{\mp akT}r(kT)\right] = R(ze^{\pm aT}) \tag{2-210}$$

5. Initial-Value Theorem

$$\lim_{k \to 0} r(kT) = \lim_{z \to \infty} R(z) \tag{2-211}$$

if the limit exists.

6. Final-Value Theorem

$$\lim_{k \to \infty} r(kT) = \lim_{z \to 1} (1 - z^{-1})R(z) \tag{2-212}$$

if the function, $(1 - z^{-1})R(z)$, has no poles on or outside the unit circle centered at the origin in the z-plane, $|z| = 1$.

7. Real Convolution

$$\begin{aligned}
F_1(z)F_2(z) &= \mathfrak{z}\left[\sum_{k=0}^{N} f_1(k)f_2(N - k)\right] \\
&= \mathfrak{z}\left[\sum_{k=0}^{N} f_2(k)f_1(N - k)\right] = \mathfrak{z}[f_1(k) * f_2(k)]
\end{aligned} \tag{2-213}$$

where "$*$" denotes real convolution in the discrete time domain. Thus, we see that just as in the Laplace transformation, the z-transform of the product of two real functions $f_1(k)$ and $f_2(k)$ is *not* equal to the product of the z-transforms $F_1(z)$ and $F_2(z)$. One exception to this is if one of the two functions is the integral delay, e^{-NTs}; then

$$\mathfrak{z}[e^{-NTs}F(s)] = \mathfrak{z}[e^{-NTs}]\mathfrak{z}[F(s)] = z^{-N}F(z) \tag{2-214}$$

Table 2-3 summarizes the theorems on z-transforms that are given above. The following examples illustrate the usefulness of some of these theorems.

Table 2-3   Theorems of z-Transforms

Addition and subtraction	$\mathfrak{z}[r_1(kT) \pm r_2(kT)] = R_1(z) \pm R_2(z)$
Multiplication by a constant	$\mathfrak{z}[ar(kT)] = a\mathfrak{z}[r(kT)] = aR(z)$
Real translation	$\mathfrak{z}[r(kT - nT)] = z^{-n}R(z)$
	$\mathfrak{z}[r(kT + nT)] = z^n R(z) - \sum_{k=0}^{n-1} r(kT)z^{-k}$
	where n = positive integer
Complex translation	$\mathfrak{z}[e^{\mp akT}r(kT)] = R(ze^{\pm aT})$
Initial-value theorem	$\lim_{k \to 0} r(kT) = \lim_{z \to \infty} R(z)$
Final-value theorem	$\lim_{k \to \infty} r(kT) = \lim_{z \to 1} (1 - z^{-1})R(z)$ if $(1 - z^{-1})R(z)$ has no poles on or outside the unit circle $|z| = 1$.
Real convolution	$F_1(z)F_2(z) = \mathfrak{z}\left[\sum_{k=0}^{N} f_1(k)f_2(N - k)\right]$ $= \mathfrak{z}\left[\sum_{k=0}^{N} f_2(k)f_1(N - k)\right]$ $= \mathfrak{z}[f_1(k) * f_2(k)]$

■

Example 2-20

Apply the complex translation theorem to find the z-transform of $f(t) = te^{-at}$, $t \geq 0$.
Let $r(t) = t$, $t \geq 0$; then

$$R(z) = \mathfrak{z}[tu_s(t)] = \mathfrak{z}(kT) = \frac{Tz}{(z - 1)^2} \tag{2-215}$$

Using the complex translation theorem, we obtain

$$F(z) = \mathfrak{z}[te^{-at}u_s(t)] = R(ze^{aT}) = \frac{Tze^{-aT}}{(z - e^{-aT})^2} \tag{2-216}$$

Example 2-21

Given the function

$$R(z) = \frac{0.792z^2}{(z - 1)(z^2 - 0.416z + 0.208)} \tag{2-217}$$

determine the value of $r(kT)$ as k approaches infinity.
Since

$$(1 - z^{-1})R(z) = \frac{0.792z}{z^2 - 0.416z + 0.208} \tag{2-218}$$

does not have any pole on or outside the unit circle $|z| = 1$ in the z-plane, the final-value theorem of the z-transform can be applied. Hence,

$$\lim_{k \to \infty} r(kT) = \lim_{z \to 1} \frac{0.792z}{z^2 - 0.416z + 0.208} = 1 \qquad (2\text{-}219)$$

This result is easily checked by expanding $R(z)$ in powers of z^{-1}:

$$R(z) = 0.792z^{-1} + 1.121z^{-2} + 1.091z^{-3} + 1.013z^{-4} + 0.986z^{-5}$$
$$+ 0.981z^{-6} + 0.998z^{-7} + \cdots \qquad (2\text{-}220)$$

It is apparent that the coefficients of this power series converge rapidly to the final value of unity.

∎

2.12 APPLICATION OF THE z-TRANSFORM TO THE SOLUTION OF LINEAR DIFFERENCE EQUATIONS

The z-transform can be used for the purpose of solving linear difference equations. Let us consider the first-order unforced difference equation

$$x(k + 1) + x(k) = 0 \qquad (2\text{-}221)$$

To solve this equation, we take the z-transform on both sides of the equation. By this we mean that we multiply both sides of the equation by z^{-k} and taking the sum from $k = 0$ to $k = \infty$. We have

$$\sum_{k=0}^{\infty} x(k + 1)z^{-k} + \sum_{k=0}^{\infty} x(k)z^{-k} = 0 \qquad (2\text{-}222)$$

The second term on the left side of the last equation is recognized to be the z-transform of $x(k)$, $X(z)$. The first term is conditioned by letting $n = k + 1$. Then

$$\sum_{k=0}^{\infty} x(k + 1)z^{-k} = \sum_{n=1}^{\infty} x(n)z^{-n+1}$$
$$= z \sum_{n=1}^{\infty} x(n)z^{-n} = z[X(z) - x(0)] \qquad (2\text{-}223)$$

Note that we could have obtained this same result by using the shifting theorem of Eq. (2-209).

Substituting Eq. (2-223) in Eq. (2-222) and solving for $X(z)$, we get

$$X(z) = \frac{z}{z + 1} x(0) \qquad (2\text{-}224)$$

Taking the inverse z-transform on both sides of Eq. (2-224) by expanding the right-hand side of the equation into a power series in z^{-1}, we have

$$X(z) = (1 - z^{-1} + z^{-2} - z^{-3} + \cdots)x(0) \tag{2-225}$$

Thus, $x(k)$ is given as the coefficient of z^{-k} times $x(0)$, or

$$x(k) = (-1)^k x(0) \tag{2-226}$$

$n = 0, 1, 2, \ldots$.

In a similar fashion we can obtain the general solution of the nth-order difference state equation

$$\mathbf{x}(k + 1) = \mathbf{Ax}(k) + \mathbf{Bu}(k) \tag{2-227}$$

where $\mathbf{x}(k) = n \times 1$ state vector
$\mathbf{u}(k) = p \times 1$ input vector ($p \leq n$)
$\mathbf{A} = n \times n$ coefficient matrix with constant elements
$\mathbf{B} = n \times p$ coefficient matrix with constant elements

Taking the z-transform on both sides of Eq. (2-227), and using the shifting property of Eq. (2-209), we have

$$z\mathbf{X}(z) - z\mathbf{x}(0) = \mathbf{AX}(z) + \mathbf{BU}(z) \tag{2-228}$$

where

$$\mathbf{X}(z) = \begin{bmatrix} X_1(z) \\ X_2(z) \\ \vdots \\ X_n(z) \end{bmatrix} \tag{2-229}$$

$$X_i(z) = \sum_{k=0}^{\infty} x_i(k)z^{-k} \qquad i = 1, 2, \ldots, n \tag{2-230}$$

$$\mathbf{U}(z) = \begin{bmatrix} U_1(z) \\ U_2(z) \\ \vdots \\ U_p(z) \end{bmatrix} \tag{2-231}$$

$$U_j(z) = \sum_{k=0}^{\infty} u_j(k)z^{-k} \qquad j = 1, 2, \ldots, p \tag{2-232}$$

Solving for $\mathbf{X}(z)$ from Eq. (2-228), we get

$$\mathbf{X}(z) = (z\mathbf{I} - \mathbf{A})^{-1} z\mathbf{x}(0) + (z\mathbf{I} - \mathbf{A})^{-1}\mathbf{BU}(z) \tag{2-233}$$

The inverse z-transform of Eq. (2-233) is written

$$\mathbf{x}(k) = {}_{3}^{-1}\left[(z\mathbf{I} - \mathbf{A})^{-1}z\right]\mathbf{x}(0) + {}_{3}^{-1}\left[(z\mathbf{I} - \mathbf{A})^{-1}\mathbf{B}U(z)\right] \qquad (2\text{-}234)$$

Since the inverse z-transform of $(z\mathbf{I} - \mathbf{A})^{-1}$ is known to be a function of k, we can designate it as $\phi(k)$. Thus, let

$$\phi(k) = {}_{3}^{-1}\left[(z\mathbf{I} - \mathbf{A})^{-1}z\right] \qquad (2\text{-}235)$$

Using the real convolution of z-transform of Eq. (2-213) and the shifting theorem of Eq. (2-208), the second term on the right side of Eq. (2-234) is written

$$_{3}^{-1}\left[(z\mathbf{I} - \mathbf{A})^{-1}\mathbf{B}U(z)\right] = \sum_{k=0}^{N-1} \phi(N - k - 1)\mathbf{B}\mathbf{u}(k) \qquad (2\text{-}236)$$

Thus, the solution of the nth-order difference state equation in Eq. (2-227) is

$$\mathbf{x}(N) = \phi(N)\mathbf{x}(0) + \sum_{k=0}^{N-1} \phi(N - k - 1)\mathbf{B}\mathbf{u}(k) \qquad (2\text{-}237)$$

for $N = 0, 1, 2, \ldots, N - 1$. Equation (2-237) is also known as the *state transition equation*.

REFERENCES

Complex Variables, Laplace Transforms, and Matrix Algebra

1. F. B. HILDEBRAND, *Methods of Applied Mathematics*, 2d ed., Prentice-Hall, Inc., Englewood Cliffs, N.J., 1965.
2. R. BELLMAN, *Introduction to Matrix Analysis*, McGraw-Hill Book Company, New York, 1960.
3. B. C. KUO, *Linear Networks and Systems*, McGraw-Hill Book Company, New York, 1967.
4. R. LEGROS and A. V. J. MARTIN, *Transform Calculus for Electrical Engineers*, Prentice-Hall, Inc., Englewood Cliffs, N.J., 1961.
5. C. R. WYLIE, JR., *Advanced Engineering Mathematics*, 2nd ed., McGraw-Hill Book Company, New York, 1960.
6. S. BARNETT, "Matrices, Polynomials, and Linear Time-Invariant Systems," *IEEE Trans. Automatic Control*, Vol. AC-18, pp. 1–10, Feb. 1973.

Partial-Fraction Expansion

7. D. HAZONY and J. RILEY, "Evaluating Residues and Coefficients of High Order Poles," *IRE Trans. Automatic Control*, Vol. AC-4, pp. 132–136, Nov. 1959.
8. C. POTTLE, "On the Partial Fraction Expansion of a Rational Function with Multiple Poles by Digital Computer," *IEEE Trans. Circuit Theory*, Vol. CT-11, pp. 161–162, Mar. 1964.

9. M. I. Younis, "A Quick Check on Partial Fraction Expansion Coefficients," *IEEE Trans. Automatic Control*, Vol. AC-11, pp. 318–319, Apr. 1966.

10. N. Ahmed and K. R. Rao, "Partial Fraction Expansion of Rational Functions with One High-Order Pole," *IEEE Trans. Automatic Control*, Vol. AC-13, p. 133, Feb. 1968.

11. B. O. Watkins, "A Partial Fraction Algorithm," *IEEE Trans. Automatic Control*, Vol. AC-16, pp. 489–491, Oct. 1971.

Sampled-Data and Discrete-Data Control Systems

12. B. C. Kuo, *Digital Control Systems*, Holt, Rinehart and Winston, New York, 1980.

PROBLEMS

2.1. Find the poles and zeros of the following functions (include the ones at infinity):

   (a) $G(s) = \dfrac{5(s+1)}{s^2(s+2)(s+5)}$

   (b) $G(s) = \dfrac{s^2(s+1)}{(s+2)(s^2+3s+2)}$

   (c) $G(s) = \dfrac{K(s+2)}{s(s^2+s+1)}$

   (d) $G(s) = \dfrac{Ke^{-2s}}{(s+1)(s+2)}$

2.2. Find the Laplace transforms of the following functions:

   (a) $g(t) = te^{-2t}$

   (b) $g(t) = t \cos 5t$

   (c) $g(t) = e^{-t}\sin \omega t$

   (d) $g(t) = \displaystyle\sum_{k=0}^{\infty} g(kT)\delta(t-kT); \quad \delta(t) =$ unit impulse function

2.3. Find the Laplace transforms of the functions in Fig. 2P-3.

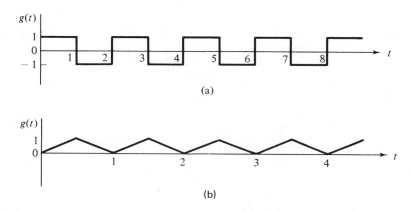

(a)

(b)

Figure 2P-3

2.4. Find the Laplace transform of the following function:

$$f(t) = \begin{cases} 0 & t < 1 \\ t + 1 & 1 \le t < 3 \\ 4 & 3 \le t \le 4 \\ 0 & 4 < t \end{cases}$$

2.5. Solve the following differential equation by means of the Laplace transformation:

$$\frac{d^2 f(t)}{dt^2} + 5\frac{df(t)}{dt} + 4f(t) = e^{-t}u_s(t)$$

Assume that all the initial conditions are zero.

2.6. Find the inverse Laplace transforms of the following functions:

(a) $G(s) = \dfrac{1}{(s + 2)(s + 3)}$

(b) $G(s) = \dfrac{1}{(s + 1)^2(s + 4)}$

(c) $G(s) = \dfrac{10}{s(s^2 + 4)(s + 1)}$

(d) $G(s) = \dfrac{2(s + 1)}{s(s^2 + s + 2)}$

2.7. Carry out the following matrix sums and differences:

(a) $\begin{bmatrix} 3 & 6 \\ 0 & -5 \end{bmatrix} + \begin{bmatrix} 7 & 0 \\ -3 & 10 \end{bmatrix}$

(b) $\begin{bmatrix} 15 \\ -1 \\ 3 \end{bmatrix} - \begin{bmatrix} 20 \\ -4 \\ 5 \end{bmatrix}$

(c) $\begin{bmatrix} \dfrac{1}{s} & 0 & s + 1 \\ 5 & \dfrac{1}{s - 3} & s^2 \end{bmatrix} + \begin{bmatrix} 0 & 10 & -s \\ s & \dfrac{1}{s} & 0 \end{bmatrix}$

2.8. Determine if the following matrices are conformable for the products **AB** and **BA**. Find the valid products.

(a) $\mathbf{A} = \begin{bmatrix} 1 \\ 0 \\ 3 \end{bmatrix} \qquad \mathbf{B} = \begin{bmatrix} 6 & 0 & 1 \end{bmatrix}$

(b) $\mathbf{A} = \begin{bmatrix} 2 & -1 \\ 3 & 0 \end{bmatrix} \qquad \mathbf{B} = \begin{bmatrix} 10 & 0 & 9 \\ -1 & -1 & 0 \end{bmatrix}$

2.9. Express the following set of algebraic equations in matrix form:

$$5x_1 + x_2 - x_3 = 1$$
$$-x_1 + 3x_2 - x_3 = 1$$
$$3x_1 - 7x_2 - 2x_3 = 0$$

2.10. Express the following set of differential equations in the form $\dot{\mathbf{x}}(t) = \mathbf{A}\mathbf{x}(t) + \mathbf{B}\mathbf{u}(t)$:

$$\dot{x}_1(t) = -x_1(t) + x_2(t)$$
$$\dot{x}_2(t) = -2x_2(t) - 3x_3(t) + u_1(t)$$
$$\dot{x}_3(t) = -x_1(t) - 5x_2(t) - 3x_3(t) + u_2(t)$$

2.11. Find the inverse of the following matrices:

(a) $\mathbf{A} = \begin{bmatrix} 2 & 5 \\ 10 & -1 \end{bmatrix}$

(b) $\mathbf{A} = \begin{bmatrix} 3 & 0 & -1 \\ -2 & 1 & 2 \\ 0 & 1 & -1 \end{bmatrix}$

(c) $\mathbf{A} = \begin{bmatrix} 1 & 3 & 4 \\ -1 & 1 & 0 \\ -1 & 0 & -1 \end{bmatrix}$

2.12. Determine the ranks of the following matrices:

(a) $\begin{bmatrix} 3 & 2 \\ 7 & 1 \\ 0 & 3 \end{bmatrix}$

(b) $\begin{bmatrix} 2 & 4 & 0 & 8 \\ 1 & 2 & 6 & 3 \end{bmatrix}$

2.13. The following signals are sampled by an ideal sampler with a sampling period of T seconds. Determine the output of the sampler, $f^*(t)$, and find the Laplace transform of $f^*(t)$, $F^*(s)$. Express $F^*(s)$ in closed form.
(a) $f(t) = te^{-at}$
(b) $f(t) = e^{-at}\sin \omega t$

2.14. Determine the z-transform of the following functions:

(a) $G(s) = \dfrac{1}{(s + a)^n}$

(b) $G(s) = \dfrac{1}{s(s + 5)^2}$

(c) $G(s) = \dfrac{1}{s^3(s + 2)}$

(d) $g(t) = t^2e^{-2t}$

(e) $g(t) = t \sin \omega t$

2.15. Find the inverse z-transform of

$$G(z) = \frac{10z(z + 1)}{(z - 1)(z^2 + z + 1)}$$

by means of the following methods:
(a) the inversion formula
(b) partial-fraction expansion

2.16. Consider that a new car is purchased with a loan of P_0 dollars over a period of N months, at a monthly interest rate of r percent. The principal and interest are to be paid back in N equal payments of u dollars each.

(a) Show that the difference equation that describes the loan process can be written as

$$P(k + 1) = (1 + r)P(k) - u \qquad (1)$$

where $P(k)$ = amount owed after the kth period
$P(k + 1)$ = amount owed after the $(k + 1)$st period, $k = 0, 1, 2, \ldots, N$
$P(0) = P_0$ = initial amount borrowed
$P(N) = 0$ (after N periods, owe nothing)
The last two conditions are also known as the boundary conditions.

(b) The difference equation in Eq. (1) can be solved by a recursive method; that is, start with $k = 0$, and then use $k = 1, 2, \ldots$ in the equation and substitute successively. Show that the solution of the equation is

$$u = \frac{(1 + r)^N P_0 r}{(1 + r)^N - 1}$$

(c) Consider that $P_0 = \$5000$, $r = 0.01$ (1 percent per month), $N = 36$ months. Find u, the monthly payment.

2.17. The time signal $f(t) = (5e^{-t} + e^{-2t})u_s(t)$, where $u_s(t)$ is the unit step function, is sampled by an ideal sampler.

(a) Find the output of the sampler, $f^*(t)$.

(b) Find the pulse transform $F^*(s)$, and express it in closed form.

(c) Find the z-transform $F(z)$ and express it in closed form.

chapter three

Transfer Function, Block Diagram, and Signal Flow Graph

3.1 INTRODUCTION

An important first step in the analysis and design of a control system is the mathematical description and modeling of the process that is to be controlled. A mathematical model of the controlled process is essential if the controller is to be designed by analytical means.

In general, given a controlled process, the set of variables that identify the dynamic characteristics of the process should first be defined. For instance, consider a motor used for control purposes.[1] We may identify the applied voltage, current in the armature windings, developed torque on the rotor shaft, angular displacement and velocity of the rotor, and others if necessary, as the system variables. These variables are interrelated through established physical laws which lead to mathematical equations of various forms. Depending upon the nature of the process, as well as the operating condition of the system, some or all of the system equations may be linear or nonlinear, time-varying or time-invariant. In addition, these equations can be algebraic equations, differential equations, or difference equations, or a combination of these.

The physical laws that govern the principles of operation of systems in real life are often quite complex, and realistic characterization of the systems may require nonlinear and/or time-varying equations that are very difficult to solve. For practical reasons, in order to establish a class of applicable analysis and design tools for control systems, assumptions and approximations are made to the physical systems, whenever possible, so that these systems can be studied using linear systems theory. There are generally two ways of justifying the linear systems approach. One is that the system is basically linear, or the system is operated in the linear region so that the conditions of linearity are satisfied. The second is that the

[1] Refer to Chapter 4 on the various types of motors used for control systems.

system is basically nonlinear, but in order to apply the linear theory of analysis and design, we linearize the system about a nominal operating point, such as that described in Sec. 4.8. However, it should be kept in mind that the analysis is applicable only for the range of the variables in which the linearization is valid.

3.2 IMPULSE RESPONSE AND TRANSFER FUNCTIONS OF LINEAR SYSTEMS

Impulse Response

Consider that a linear time-invariant system has the input $r(t)$ and the output $c(t)$. The system can be characterized by its *impulse response* $g(t)$ which is defined as the output when the input is a unit impulse function $\delta(t)$. Once the impulse response of a linear system is known, the output of the system, $c(t)$, with any input, $r(t)$, can be found by using the *transfer function*.

Transfer Function (Single-Input Single-Output Systems)

The transfer function of a linear time-invariant system is defined as the Laplace transform of the impulse response, with all the initial conditions set to zero.

Let $G(s)$ denote the transfer function of a single-input single-output system with input $r(t)$ and output $c(t)$. Then, the transfer function $G(s)$ is defined as

$$G(s) = \mathcal{L}[g(t)] \tag{3-1}$$

The transfer function $G(s)$ is related to the Laplace transforms of the input and output through the following relation:

$$G(s) = \frac{C(s)}{R(s)} \tag{3-2}$$

with all the initial conditions set to zero, where

$$R(s) = \mathcal{L}[r(t)] \tag{3-3}$$

and

$$C(s) = \mathcal{L}[c(t)] \tag{3-4}$$

Although the transfer function of a linear system is defined in terms of the impulse response, in practice, the input–output relation of a linear time-invariant system with continuous-data input is often described by a differential equation, so that it is more appropriate to derive a general-form transfer function directly from the differential equation. Let us consider that the input–output relation of a linear time-invariant system is described by the following nth-order differential equation with constant real coefficients:

$$\frac{d^n c(t)}{dt^n} + a_n \frac{d^{n-1} c(t)}{dt^{n-1}} + \cdots + a_2 \frac{dc(t)}{dt} + a_1 c(t)$$

$$= b_{m+1} \frac{d^m r(t)}{dt^m} + b_m \frac{d^{m-1} r(t)}{dt^{m-1}} + \cdots + b_2 \frac{dr(t)}{dt} + b_1 r(t) \quad (3\text{-}5)$$

The coefficients $a_1, a_2, \ldots, a_n$ and $b_1, b_2, \ldots, b_{m+1}$ are real constants, and $n \geq m$. Once the input $r(t)$ for $t \geq t_0$ and the initial conditions of $c(t)$ and the derivatives of $c(t)$ are specified at the initial time $t = t_0$, the output response $c(t)$ for $t \geq t_0$ is determined by solving Eq. (3-5). However, from the standpoint of analyzing and designing a linear system, the method of using differential equations exclusively to describe the system is quite cumbersome. Thus, the differential equation of Eq. (3-5) is seldom used in its original form for the analysis and design of control systems. It is important to point out that although efficient subroutines are available on digital computers for the solution of high-order differential equations, the basic philosophy of linear control systems theory is that of developing analysis and design techniques that would avoid the exact solutions of the system differential equations, except when computer-simulation solutions are desired.

To obtain the transfer function of the linear system that is represented by Eq. (3-5), we simply take the Laplace transform on both sides of the equation, and assume *zero initial conditions*. The result is

$$\left(s^n + a_n s^{n-1} + \cdots + a_2 s + a_1 \right) C(s)$$

$$= \left(b_{m+1} s^m + b_m s^{m-1} + \cdots + b_2 s + b_1 \right) R(s) \quad (3\text{-}6)$$

The transfer function between $r(t)$ and $c(t)$ is given by

$$G(s) = \frac{C(s)}{R(s)} = \frac{b_{m+1} s^m + b_m s^{m-1} + \cdots + b_2 s + b_1}{s^n + a_n s^{n-1} + \cdots + a_2 s + a_1} \quad (3\text{-}7)$$

We can summarize the properties of the transfer function as follows:

1. Transfer function is defined only for a linear time-invariant system. It is meaningless for nonlinear systems.
2. The transfer function between an input variable and an output variable of a system is defined as the Laplace transform of the impulse response. Alternately, the transfer function between a pair of input and output variables is the ratio of the Laplace transform of the output to the Laplace transform of the input.
3. When defining the transfer function, all initial conditions of the system are set to zero.
4. The transfer function is independent of the input of the system.

5. Transfer function is expressed only as a function of the complex variable s. It is not a function of the real variable, time, or any other variable that is used as the independent variable. When a system is subject to discrete-time or digital input, it may be more convenient to model the system by difference equations; then, the transfer function becomes a function of the complex variable z, when z-transform is used.

Transfer Function (Multivariable Systems)

The definition of transfer function is easily extended to a system with a multiple number of inputs and outputs. A system of this type is often referred to as the *multivariable system*. In a multivariable system, a differential equation of the form of Eq. (3-5) may be used to describe the relationship between a pair of input and output variables. When dealing with the relationship between one input and one output, it is assumed that all other inputs are set to zero. Since the principle of superposition is valid for linear systems, the total effect on any output variable due to all the inputs acting simultaneously is obtained by adding up the outputs due to each input acting alone.

Examples on multivariable systems are plentiful in practice. For example, in the control of the speed of a motor, $\omega(t)$, by controlling the input voltage, $v(t)$, subjecting to an external disturbance torque $T_d(t)$, the system is considered to have two inputs in $v(t)$ and $T_d(t)$, and one output in $\omega(t)$. Another example on a multivariable system is the idle-speed control system of an automobile engine. In this case, the two inputs are the amounts of fuel and air intake to the engine, and the output is the idle speed of the engine. In the control of a turbo-propeller engine, the input variables are the fuel rate and the propeller blade angle. The output variables are the speed of rotation of the engine and the turbine-inlet temperature. In general, either one of the outputs is affected by the changes in both inputs. For instance, when the blade angle of the propeller is increased, the speed of the rotation of the engine will decrease and the temperature usually increases. The following transfer function relations may be determined from tests performed on the system, and after linearization:

$$C_1(s) = G_{11}(s)R_1(s) + G_{12}(s)R_2(s) \tag{3-8}$$

$$C_2(s) = G_{21}(s)R_1(s) + G_{22}(s)R_2(s) \tag{3-9}$$

where $C_1(s)$ = transformed variable of speed of rotation
        $C_2(s)$ = transformed variable of turbine-inlet temperature
        $R_1(s)$ = transformed variable of fuel rate
        $R_2(s)$ = transformed variable of propeller blade angle
All these variables are assumed to be measured from some reference levels.

Since Eqs. (3-8) and (3-9) are written with the assumption that the system is linear, the principle of superposition holds. Therefore, $G_{11}(s)$ represents the transfer function between the fuel rate and the speed of rotation of the engine with the propeller blade angle held at the reference value; that is, $R_2(s) = 0$. Similar statements can be made for the other transfer functions.

In general, if a linear system has p inputs and q outputs, the transfer function between the ith output and the jth input is defined as

$$G_{ij}(s) = \frac{C_i(s)}{R_j(s)} \tag{3-10}$$

with $R_k(s) = 0$, $k = 1, 2, \ldots, p$, $k \neq j$. Note that Eq. (3-10) is defined with only the jth input in effect, while the other inputs are set to zero. The ith output transform of the system is related to all the input transforms by

$$C_i(s) = G_{i1}(s)R_1(s) + G_{i2}(s)R_2(s) + \cdots + G_{ip}(s)R_p(s)$$

$$= \sum_{j=1}^{p} G_{ij}(s)R_j(s) \qquad (i = 1, 2, \ldots, q) \tag{3-11}$$

where $G_{ij}(s)$ is defined in Eq. (3-10).

It is convenient to represent Eq. (3-11) by matrix equation

$$\mathbf{C}(s) = \mathbf{G}(s)\mathbf{R}(s) \tag{3-12}$$

where

$$\mathbf{C}(s) = \begin{bmatrix} C_1(s) \\ C_2(s) \\ \vdots \\ C_q(s) \end{bmatrix} \tag{3-13}$$

is a $q \times 1$ matrix, called the *transformed output vector*;

$$\mathbf{R}(s) = \begin{bmatrix} R_1(s) \\ R_2(s) \\ \vdots \\ R_p(s) \end{bmatrix} \tag{3-14}$$

is a $p \times 1$ matrix, called the *transformed input vector*;

$$\mathbf{G}(s) = \begin{bmatrix} G_{11}(s) & G_{12}(s) & \cdots & G_{1p}(s) \\ G_{21}(s) & G_{22}(s) & \cdots & G_{2p}(s) \\ \vdots & & & \\ G_{q1}(s) & G_{q2}(s) & \cdots & G_{qp}(s) \end{bmatrix} \tag{3-15}$$

is a $q \times p$ matrix, called the *transfer function matrix*.

As an example on the derivation of the transfer functions of a simple multivariable system, consider that the differential equations of a dc motor are given as[2]

$$v(t) = Ri(t) + L\frac{di(t)}{dt} \tag{3-16}$$

$$T(t) = J\frac{d\omega(t)}{dt} + B\omega(t) + T_L(t) \tag{3-17}$$

where $v(t)$ = applied voltage to the armature
        $i(t)$ = armature current
        R = armature resistance
        L = armature inductance
        J = motor inertia of the rotor
        B = viscous frictional coefficient
        $T(t)$ = torque developed by the motor
     $T_L(t)$ = load or disturbance torque
      $\omega(t)$ = velocity of motor shaft
The torque developed by the motor is related to the armature current through

$$T(t) = K_i i(t) \tag{3-18}$$

where K_i is the torque constant.

To find the transfer functions between the inputs, $v(t)$ and $T_L(t)$, and the output, $\omega(t)$, we take the Laplace transform on both sides of Eqs. (3-16) through (3-18). Assuming zero initial conditions, we get

$$V(s) = (R + Ls)I(s) \tag{3-19}$$
$$T(s) = (B + Js)\Omega(s) + T_L(s) \tag{3-20}$$
$$T(s) = K_i I(s) \tag{3-21}$$

where $V(s)$, $I(s)$, $T(s)$, $\Omega(s)$, and $T_L(s)$ are the Laplace transforms of $v(t)$, $i(t)$, $T(t)$, $\omega(t)$, and $T_L(t)$, respectively.

Eliminating $T(s)$ and $I(s)$ from the last three equations, we get, by solving for $\Omega(s)$,

$$\Omega(s) = \frac{K_i}{(B + Js)(R + Ls)}V(s) - \frac{1}{B + Js}T_L(s) \tag{3-22}$$

This equation can be written as

$$C(s) = G_{11}(s)R_1(s) + G_{12}(s)R_2(s) \tag{3-23}$$

---

[2] Refer to Chapter 4 for more details on the mathematical modeling of dc motors.

where

$$C(s) = \Omega(s)$$

$$R_1(s) = V(s)$$

$$R_2(s) = T_L(s)$$

$$G_{11}(s) = \frac{K_i}{(B + Js)(R + Ls)}$$

$$G_{12}(s) = \frac{-1}{B + Js}$$

$G_{11}(s)$ is regarded as the transfer function between the input voltage and the motor velocity when the load torque is zero; $G_{12}(s)$ is regarded as the transfer function between the load torque and the motor velocity when the input voltage is zero.

3.3 BLOCK DIAGRAMS [1]

Because of its simplicity and versatility, *block diagram* is often used by control engineers to portray systems of all types. A block diagram can be used simply to represent the composition and interconnection of a system. Or, it can be used, together with transfer functions, to represent the cause-and-effect relationships throughout the system. For instance, the block diagram of Fig. 3-1 represents a turbine-driven hydraulic power system for an aircraft. The main components of the system include a pressure-compensated hydraulic pump, an air-driven pump, an electronic speed controller, and a control valve. The block diagram in the figure depicts how these components are interconnected.

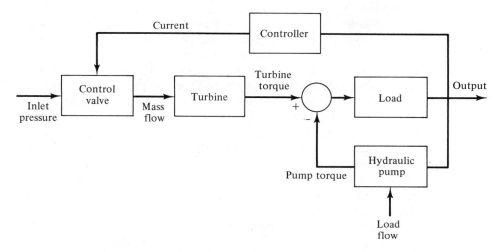

Figure 3-1 Block diagram of a turbine-driven hydraulic power system.

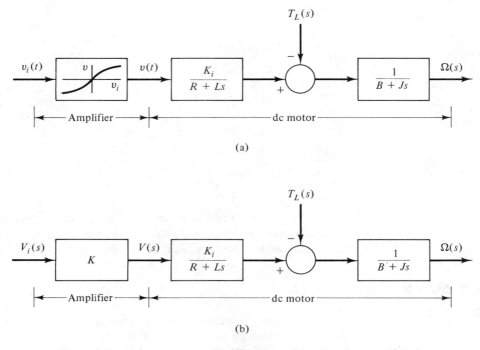

(a)

(b)

Figure 3-2 Block diagrams of a dc motor system. (a) Amplifier with a nonlinear gain characteristic. (b) Amplifier with a linear characteristic.

If the mathematical and functional relationships of all the system elements are known, the block diagram can be used as a reference for the analytical or the computer solution of the system. Furthermore, if all the system elements are assumed to be linear, the transfer function for the overall system can be obtained by means of block diagram algebra.

The essential point is that block diagram can be used to portray nonlinear as well as linear systems. For example, the input–output relations of the dc motor with load torque described above may be represented by the block diagram shown in Fig. 3-2(a). In the figure, the input voltage to the motor is shown to be obtained from the output of a power amplifier which realistically has a nonlinear input–output relation. The motor is assumed to be linear—or, more appropriately, we should say that it is operated in the linear regions of its characteristics—and its dynamics are represented by the transfer function relation of Eq. (3-22). Figure 3-2(b) illustrates the same system but with the amplifier characteristic approximated by a constant gain. In this case, the overall system is linear.

It should be noted that an overall transfer function relation cannot be defined for the nonlinear system in Fig. 3-2(a). The nonlinear amplifier gain can only describe the relationship between the time variables $v_i(t)$ and $v(t)$, and no transfer function exists between the Laplace transform variables $V_i(s)$ and $V(s)$. In the case shown in Fig. 3-2(b), the transfer relation between $V_i(s)$ and $V(s)$ is

$$V(s) = KV_i(s) \tag{3-24}$$

Block Diagrams of Control Systems

We shall now define some block diagram elements used frequently in control systems and the block diagram algebra. One of the important components of a feedback control system is the sensing device that acts as a function point for signal comparisons. The physical components involved are the potentiometer, synchro, resolver, differential amplifier, multiplier, and other signal-processing transducers. In general, the sensing devices perform simple mathematical operations such as addition, subtraction, multiplication, and sometimes combinations of these. The block diagram elements of these operations are illustrated as shown in Fig. 3-3. The addition and subtraction operations in Figs. 3-3(a), (b), and (c) are linear, so that the input and output variables of these block diagram elements can be time–domain variables or Laplace transform variables. Thus, in Fig. 3-3(a), the block diagram implies

$$e(t) = r(t) - c(t) \tag{3-25}$$

or

$$E(s) = R(s) - C(s) \tag{3-26}$$

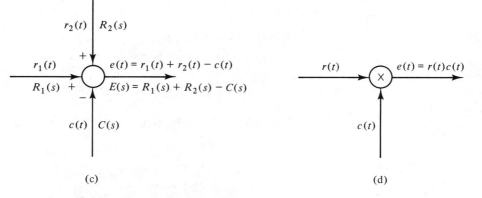

Figure 3-3 Block diagram elements of typical sensing devices of control systems. (a) Subtraction. (b) Addition. (c) Addition and subtraction. (d) Multiplication.

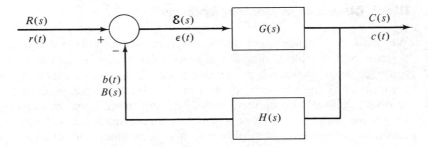

Figure 3-4 Basic block diagram of a feedback control system.

The multiplication operation depicted by Fig. 3-3(d) is nonlinear, so that the input–output relation portrayed has meaning only in the real (time) domain; therefore,

$$e(t) = r(t)c(t) \tag{3-27}$$

The block diagram element in this case *does not* imply

$$E(s) = R(s)C(s)$$

From the complex-convolution theorem of Laplace transform we can identify that the Laplace transform of Eq. (3-27) yields $E(s) = R(s) * C(s)$, where the symbol $*$ in this case represents complex convolution of $R(s)$ and $C(s)$.

Figure 3-4 shows the block diagram of a linear feedback control system. The following terminology often used in control systems is defined with reference to the block diagram:

$$r(t), R(s) = \text{reference input}$$
$$c(t), C(s) = \text{output signal (controlled variable)}$$
$$b(t), B(s) = \text{feedback signal}$$
$$\epsilon(t), \mathcal{E}(s) = \text{actuating signal}$$
$$e(t), E(s) = R(s) - C(s) = \text{error signal}$$
$$G(s) = \frac{C(s)}{\mathcal{E}(s)} = \begin{array}{l}\text{open-loop transfer function or} \\ \text{forward-path transfer function}\end{array}$$
$$M(s) = \frac{C(s)}{R(s)} = \text{closed-loop transfer function}$$
$$H(s) = \text{feedback-path transfer function}$$
$$G(s)H(s) = \text{loop transfer function}$$

The closed-loop transfer function, $M(s) = C(s)/R(s)$, can be expressed as a function of $G(s)$ and $H(s)$. From Fig. 3-4 we write

$$C(s) = G(s)\mathcal{E}(s) \tag{3-28}$$

and

$$B(s) = H(s)C(s) \tag{3-29}$$

The actuating signal is written

$$\mathcal{E}(s) = R(s) - B(s) \tag{3-30}$$

Substituting Eq. (3-30) into Eq. (3-28) yields

$$C(s) = G(s)R(s) - G(s)B(s) \tag{3-31}$$

Substituting Eq. (3-29) into Eq. (3-31) gives

$$C(s) = G(s)R(s) - G(s)H(s)C(s) \tag{3-32}$$

Solving $C(s)$ from the last equation, the closed-loop transfer function of the system is given by

$$M(s) = \frac{C(s)}{R(s)} = \frac{G(s)}{1 + G(s)H(s)} \tag{3-33}$$

In general, a practical control system may contain more than one feedback loop, and the evaluation of the transfer function from the block diagram by means of the algebraic method described above may be tedious. Although in principle the block diagram of a system with one input and one output can always be reduced to the basic single-loop form of Fig. 3-4, the algebraic steps involved in the reduction process may again be quite tedious. We shall show later that the transfer function of any linear system can be obtained directly from its block diagram by use of the signal flow graph gain formula.

Block Diagrams and Transfer Functions of Multivariable Systems

In this section we shall illustrate the block diagram and matrix representations of multivariable systems.

Two block diagram representations of a multiple-variable system with p inputs and q outputs are shown in Fig. 3-5(a) and (b). In Fig. 3-5(a) the individual input and output signals are designated, whereas in the block diagram of Fig. 3-5(b), the multiplicity of the inputs and outputs is denoted by vectors. The case of Fig. 3-5(b) is preferable in practice because of its simplicity.

Figure 3-6 shows the block diagram of a multivariable feedback control system. The transfer function relationship between the input and the output of the system is obtained by using matrix algebra:

$$\mathbf{C}(s) = \mathbf{G}(s)\mathcal{E}(s) \tag{3-34}$$

$$\mathcal{E}(s) = \mathbf{R}(s) - \mathbf{B}(s) \tag{3-35}$$

$$\mathbf{B}(s) = \mathbf{H}(s)\mathbf{C}(s) \tag{3-36}$$

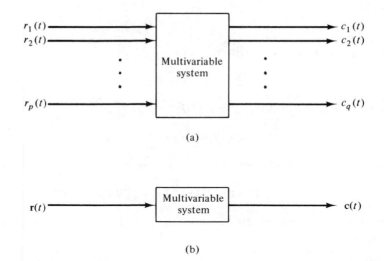

(a)

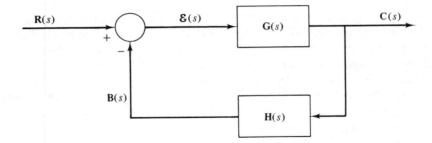

(b)

Figure 3-5   Block diagram representations of a multivariable system.

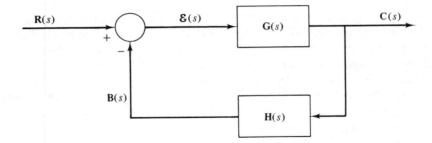

Figure 3-6   Block diagram of a multivariable feedback control system.

where $\mathbf{C}(s)$ is the $q \times 1$ output vector, $\mathcal{E}(s)$, $\mathbf{R}(s)$, and $\mathbf{B}(s)$ are all $p \times 1$ vectors, and $\mathbf{G}(s)$ and $\mathbf{H}(s)$ are $q \times p$ and $p \times q$ transfer function matrices, respectively. Substituting Eq. (3-36) into Eq. (3-35) and then from Eq. (3-35) into Eq. (3-34) yields

$$\mathbf{C}(s) = \mathbf{G}(s)\mathbf{R}(s) - \mathbf{G}(s)\mathbf{H}(s)\mathbf{C}(s) \qquad (3\text{-}37)$$

Solving for $\mathbf{C}(s)$ from Eq. (3-37) gives

$$\mathbf{C}(s) = [\mathbf{I} + \mathbf{G}(s)\mathbf{H}(s)]^{-1}\mathbf{G}(s)\mathbf{R}(s) \qquad (3\text{-}38)$$

provided that $\mathbf{I} + \mathbf{G}(s)\mathbf{H}(s)$ is nonsingular.

It should be mentioned that although the development of the input–output relationship here is similar to that of the single input–output case, in the present situation it is improper to speak of the ratio $\mathbf{C}(s)/\mathbf{R}(s)$ since $\mathbf{C}(s)$ and $\mathbf{R}(s)$ are matrices. However, it is still possible to define the closed-loop transfer matrix as

$$\mathbf{M}(s) = \left[\mathbf{I} + \mathbf{G}(s)\mathbf{H}(s)\right]^{-1}\mathbf{G}(s) \tag{3-39}$$

Then Eq. (3-38) is written

$$\mathbf{C}(s) = \mathbf{M}(s)\mathbf{R}(s) \tag{3-40}$$

■

Example 3-1

Consider that the forward-path transfer function matrix and the feedback-path transfer function matrix of the system shown in Fig. 3-6 are

$$\mathbf{G}(s) = \begin{bmatrix} \dfrac{1}{s+1} & -\dfrac{1}{s} \\ 2 & \dfrac{1}{s+2} \end{bmatrix} \tag{3-41}$$

and

$$\mathbf{H}(s) = \begin{bmatrix} 1 & 0 \\ 0 & 1 \end{bmatrix} \tag{}$$

respectively.

The closed-loop transfer matrix of the system is given by Eq. (3-39) and is evaluated as follows:

$$\mathbf{I} + \mathbf{G}(s)\mathbf{H}(s) = \begin{bmatrix} 1 + \dfrac{1}{s+1} & -\dfrac{1}{s} \\ 2 & 1 + \dfrac{1}{s+2} \end{bmatrix} = \begin{bmatrix} \dfrac{s+2}{s+1} & -\dfrac{1}{s} \\ 2 & \dfrac{s+3}{s+2} \end{bmatrix} \tag{3-42}$$

The closed-loop transfer matrix is

$$\mathbf{M}(s) = \left[\mathbf{I} + \mathbf{G}(s)\mathbf{H}(s)\right]^{-1}\mathbf{G}(s) = \dfrac{1}{\Delta} \begin{bmatrix} \dfrac{s+3}{s+2} & \dfrac{1}{s} \\ -2 & \dfrac{s+2}{s+1} \end{bmatrix} \begin{bmatrix} \dfrac{1}{s+1} & -\dfrac{1}{s} \\ 2 & \dfrac{1}{s+2} \end{bmatrix} \tag{3-43}$$

where

$$\Delta = \dfrac{s+2}{s+1}\dfrac{s+3}{s+2} + \dfrac{2}{s} = \dfrac{s^2 + 5s + 2}{s(s+1)} \tag{3-44}$$

Thus

$$\mathbf{M}(s) = \dfrac{s(s+1)}{s^2 + 5s + 2} \begin{bmatrix} \dfrac{3s^2 + 9s + 4}{s(s+1)(s+2)} & -\dfrac{1}{s} \\ 2 & \dfrac{3s+2}{s(s+1)} \end{bmatrix} \tag{3-45}$$

■

3.4 SIGNAL FLOW GRAPHS [2]

A signal flow graph may be regarded as a simplified notation for a block diagram, although it was originally introduced by S. J. Mason [2] as a cause-and-effect representation of linear systems. In general, besides the difference in the physical appearances of the signal flow graph and the block diagram, we may regard the signal flow graph to be constrained by more rigid mathematical relationships, whereas the rules of using the block diagram notation are far more flexible and less stringent.

A signal flow graph may be defined as a graphical means of portraying the input–output relationships between the variables of a set of linear algebraic equations.

Consider that a linear system is described by the set of N algebraic equations

$$y_j = \sum_{k=1}^{N} a_{kj} y_k \qquad j = 1, 2, \ldots, N \tag{3-46}$$

It should be pointed out that these N equations are written in the form of cause-and-effect relations:

$$j\text{th effect} = \sum_{k=1}^{N} (\text{gain from } k \text{ to } j)(k\text{th cause}) \tag{3-47}$$

or simply

$$\text{output} = \sum (\text{gain})(\text{input}) \tag{3-48}$$

This is the single most important axiom in the construction of the set of algebraic equations from which a signal flow graph is drawn.

In the case when a system is represented by a set of integrodifferential equations, we must first transform them into Laplace transform equations and then rearrange the latter into the form of Eq. (3-46), or

$$Y_j(s) = \sum_{k=1}^{N} G_{kj}(s) Y_k(s) \qquad j = 1, 2, \ldots, N \tag{3-49}$$

When constructing a signal flow graph, junction points or *nodes* are used to represent the variables y_j and y_k. The nodes are connected together by line segments called *branches*, according to the cause-and-effect equations. The branches have associated branch gains and directions. A signal can transmit through a branch only in the direction of the arrow. In general, given a set of equations such as those of Eq. (3-46) or Eq. (3-49), the construction of the signal flow graph is basically a matter of following through the cause-and-effect relations relating each variable in

Figure 3-7 Signal flow graph of $y_2 = a_{12}y_1$.

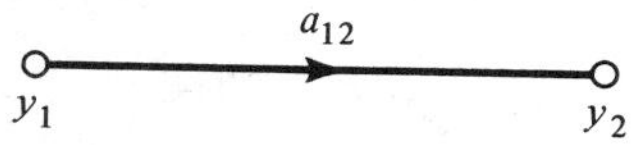

terms of itself and the other variables. For instance, consider that a linear system is represented by the simple equation

$$y_2 = a_{12}y_1 \tag{3-50}$$

where y_1 is the input variable, y_2 the output variable, and a_{12} the gain or transmittance between the two variables. The signal-flow-graph representation of Eq. (3-50) is shown in Fig. 3-7. Notice that the branch directing from node y_1 to node y_2 expresses the dependence of y_2 upon y_1. It should be reiterated that Eq. (3-50) and Fig. 3-7 represent only the dependence of the output variable upon the input variable, not the reverse. An important consideration in the application of signal flow graphs is that the branch between the two nodes y_1 and y_2 should be integrated as a unilateral amplifier with gain a_{12}, so that when a signal of one unit is applied at the input y_1, the signal is multiplied by a_{12} and a signal of strength a_{12} is delivered at node y_2. Although algebraically Eq. (3-50) can be rewritten

$$y_1 = \frac{1}{a_{12}}y_2 \tag{3-51}$$

the signal flow graph of Fig. 3.7 does not imply this relationship. If Eq. (3-51) is valid as a cause-and-effect equation in the physical sense, a new signal flow graph must be drawn.

As another illustrative example, consider the following set of algebraic equations:

$$
\begin{aligned}
y_2 &= a_{12}y_1 + a_{32}y_3 \\
y_3 &= a_{23}y_2 + a_{43}y_4 \\
y_4 &= a_{24}y_2 + a_{34}y_3 + a_{44}y_4 \\
y_5 &= a_{25}y_2 + a_{45}y_4
\end{aligned} \tag{3-52}
$$

The signal flow graph for these equations is constructed step by step as shown in Fig. 3-8, although the indicated sequence of steps is not unique. The nodes representing the variables y_1, y_2, y_3, y_4, and y_5 are located in order from left to right. The first equation states that y_2 depends upon two signals, $a_{12}y_1$ and $a_{32}y_3$; the signal flow graph representing this equation is drawn as shown in Fig. 3-8(a). The second equation states that y_3 depends upon $a_{23}y_2$ and $a_{43}y_4$; therefore, on the signal flow graph of Fig. 3-8(a), a branch of gain a_{23} is drawn from node y_2 to y_3, and a branch of gain a_{43} is drawn from y_4 to y_3 with the directions of the branches indicated by the arrows, as shown in Fig. 3-8(b). Similarly, with the consideration of the third equation, Fig. 3-8(c) is obtained. Finally, when the last equation of Eq. (3-52) is portrayed, the complete signal flow graph is shown in Fig. 3-8(d). The branch that begins from the node y_4 and ends at y_4 is called a loop, and with a gain a_{44}, represents the dependence of y_4 upon itself.

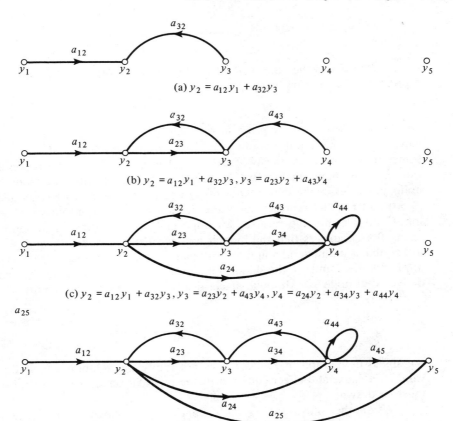

(a) $y_2 = a_{12}y_1 + a_{32}y_3$

(b) $y_2 = a_{12}y_1 + a_{32}y_3$, $y_3 = a_{23}y_2 + a_{43}y_4$

(c) $y_2 = a_{12}y_1 + a_{32}y_3$, $y_3 = a_{23}y_2 + a_{43}y_4$, $y_4 = a_{24}y_2 + a_{34}y_3 + a_{44}y_4$

(d) Complete signal flow graph

Figure 3-8   Step-by-step construction of the signal flow graph for Eq. (3-52). (a) $y_2 = a_{12}y_1 + a_{32}y_3$. (b) $y_2 = a_{12}y_1 + a_{32}y_3$, $y_3 = a_{23}y_2 + a_{43}y_4$. (c) $y_2 = a_{12}y_1 + a_{32}y_3$, $y_3 = a_{23}y_2 + a_{43}y_4$, $y_4 = a_{24}y_2 + a_{34}y_3 + a_{44}y_4$. (d) Complete signal flow graph.

3.5 SUMMARY OF BASIC PROPERTIES OF SIGNAL FLOW GRAPHS

At this point it is best to summarize some of the important properties of the signal flow graph.

1. A signal flow graph applies only to linear systems.
2. The equations based on which a signal flow graph is drawn must be algebraic equations in the form of effects as functions of causes.

3. Nodes are used to represent variables. Normally, the nodes are arranged from left to right, following a succession of causes and effects through the system.

4. Signals travel along branches only in the direction described by the arrows of the branches.

5. The branch directing from node y_k to y_j represents the dependence of the variable y_j upon y_k, but not the reverse.

6. A signal y_k traveling along a branch between nodes y_k and y_j is multiplied by the gain of the branch, a_{kj}, so that a signal $a_{kj}y_k$ is delivered at node y_j.

3.6 DEFINITIONS FOR SIGNAL FLOW GRAPHS

In addition to the branches and nodes defined earlier for the signal flow graph, the following terms are useful for the purposes of identification and reference.

Input Node (Source). An input node is a node that has only outgoing branches. (Example: node y_1 in Fig. 3-8.)

Output Node (Sink). An output node is a node which has only incoming branches. (Example: node y_5 in Fig. 3-8.) However, this condition is not always readily met by an output node. For instance, the signal flow graph shown in Fig. 3-9(a) does not have any node that satisfies the condition of an output node. However, it may be necessary to regard nodes y_2 and/or y_3 as output nodes. In order to meet the definition requirement, we may simply introduce branches with

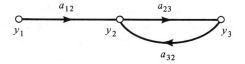

(a) Original signal flow graph

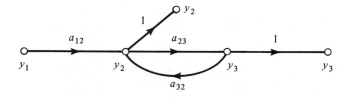

(b) Modified signal flow graph

Figure 3-9 Modification of a signal flow graph so that y_2 and y_3 satisfy the requirement as output nodes. (a) Original signal flow graph. (b) Modified signal flow graph.

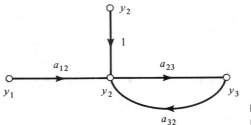

Figure 3-10 Erroneous way to make the node y_2 an input node.

unity gains and additional variables y_2 and y_3, as shown in Fig. 3-9(b). Notice that in the modified signal flow graph it is equivalent that the equations $y_2 = y_2$ and $y_3 = y_3$ are added. In general, we can state that any noninput node of a signal flow graph can always be made an output node by the aforementioned operation. However, we cannot convert a noninput node into an input node by using a similar operation. For instance, node y_2 of the signal flow graph of Fig. 3-9(a) does not satisfy the definition of an input node. If we attempt to convert it into an input node by adding an incoming branch of unity gain from another identical node y_2, the signal flow graph of Fig. 3-10 would result. However, the equation that portrays the relationship at node y_2 now reads

$$y_2 = y_2 + a_{12}y_1 + a_{32}y_3 \tag{3-53}$$

which is different from the original equation, as written from Fig. 3-9(a),

$$y_2 = a_{12}y_1 + a_{32}y_3 \tag{3-54}$$

Since the only proper way that a signal flow graph can be drawn is from a set of cause-and-effect equations, that is, with the causes on the right side of the equation and the effects on the left side of the equation, we must transfer y_2 to the right side of Eq. (3-54) if it were to be an input. Rearranging Eq. (3-54), the two equations originally for the signal flow graph of Fig. 3-9 now become

$$y_1 = \frac{1}{a_{12}}y_2 - \frac{a_{32}}{a_{12}}y_3 \tag{3-55}$$

$$y_3 = a_{23}y_2 \tag{3-56}$$

The signal flow graph for these two equations is shown in Fig. 3-11, with y_2 as an input node.

Path. A path is any collection of a continuous succession of branches traversed in the same direction. The definition of a path is entirely general since it does not prevent any node from being traversed more than once. Therefore, as simple as the signal flow graph of Fig. 3-9(a) is, it may have numerous paths.

Forward Path. A forward path is a path that starts at an input node and ends at an output node and along which no node is traversed more than once. For example, in the signal flow graph of Fig. 3-8(d), y_1 is the input node, and there are

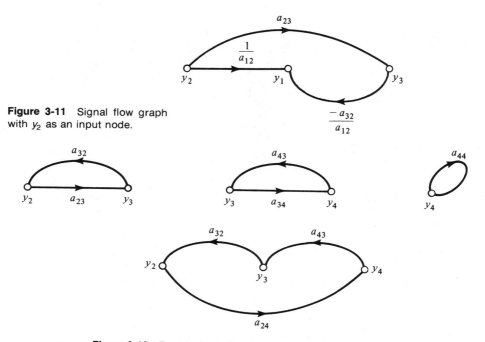

Figure 3-11 Signal flow graph with y_2 as an input node.

Figure 3-12 Four loops in the signal flow graph of Fig. 3-8(d).

four possible output nodes in y_2, y_3, y_4, and y_5. The forward path between y_1 and y_2 is simply the connecting branch between y_1 and y_2. There are two forward paths between y_1 and y_3; one contains the branches from y_1 to y_2 to y_3, and the other one contains the branches from y_1 to y_2 to y_4 (through the branch with gain a_{24}) and then back to y_3 (through the branch with gain a_{43}). The reader may determine the two forward paths between y_1 and y_4. Similarly, there are three forward paths between y_1 and y_5.

Loop. A loop is a path that originates and terminates on the same node and along which no other node is encountered more than once. For example, there are four loops in the signal flow graph of Fig. 3-8(d). These are shown in Fig. 3-12.

Path Gain. The product of the branch gains encountered in traversing a path is called the path gain. For example, the path gain for the path $y_1 - y_2 - y_3 - y_4$ in Fig. 3-8(d) is $a_{12}a_{23}a_{34}$.

Forward-Path Gain. Forward-path gain is defined as the path gain of a forward path.

Loop Gain. Loop gain is defined as the path gain of a loop. For example, the loop gain of the loop $y_2 - y_4 - y_3 - y_2$ in Fig. 3.12 is $a_{24}a_{43}a_{32}$.

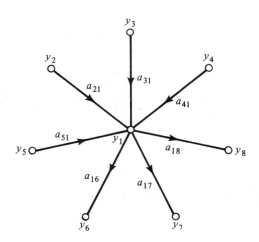

Figure 3-13 Node as a summing point and as a transmitting point.

3.7 SIGNAL-FLOW-GRAPH ALGEBRA

Based on the properties of the signal flow graph, we can outline the following manipulation and algebra of the signal flow graph.

1. The value of the variable represented by a node is equal to the sum of all the signals entering the node. Therefore, for the signal flow graph of Fig. 3-13 the value of y_1 is equal to the sum of the signals transmitted through all the incoming branches; that is,

$$y_1 = a_{21}y_2 + a_{31}y_3 + a_{41}y_4 + a_{51}y_5 \qquad (3\text{-}57)$$

2. The value of the variable represented by a node is transmitted through all branches leaving the node. In the signal flow graph of Fig. 3-13, we have

$$y_6 = a_{16}y_1$$
$$y_7 = a_{17}y_1$$
$$y_8 = a_{18}y_1 \qquad (3\text{-}58)$$

3. Parallel branches in the same direction connecting two nodes can be replaced by a single branch with gain equal to the sum of the gains of the parallel branches. An example of this case is illustrated in Fig. 3-14.

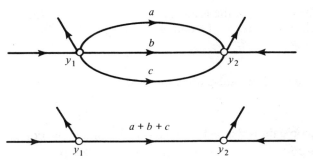

Figure 3-14 Signal flow graph with parallel paths replaced by one with a single branch.

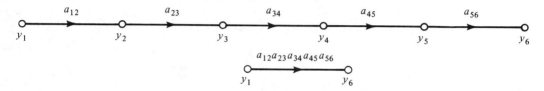

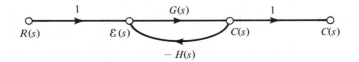

Figure 3-15 Signal flow graph with cascaded unidirectional branches replaced by a single branch.

Figure 3-16 Signal flow graph of a feedback control system.

4. A series connection of unidirectional branches, as shown in Fig. 3-15, can be replaced by a single branch with gain equal to the product of the branch gains.

5. *Signal flow graph of a feedback control system.* Figure 3-16 shows the signal flow graph of a feedback control system whose block diagram is given in Fig. 3-4. Therefore, the signal flow graph may be regarded as a simplified notation for the block diagram. Writing the equations for the signals at the nodes $\mathcal{E}(s)$ and $C(s)$ we have

$$\mathcal{E}(s) = R(s) - H(s)C(s) \tag{3-59}$$

and

$$C(s) = G(s)\mathcal{E}(s) \tag{3-60}$$

The closed-loop transfer function is obtained from these two equations,

$$\frac{C(s)}{R(s)} = \frac{G(s)}{1 + G(s)H(s)} \tag{3-61}$$

For complex signal flow graphs we do not need to rely on algebraic manipulation to determine the input–output relation. In Section 3.9 a general gain formula is introduced which allows the determination of the gain between an input node and an output by mere inspection.

3.8 EXAMPLES OF THE CONSTRUCTION OF SIGNAL FLOW GRAPHS

It was emphasized earlier that the construction of a signal flow graph of a physical system depends upon first writing the equations of the system in the cause-and-effect form. In this section we give two simple illustrative examples. Owing to the lack of background on systems at this early stage, we are using two electric networks as

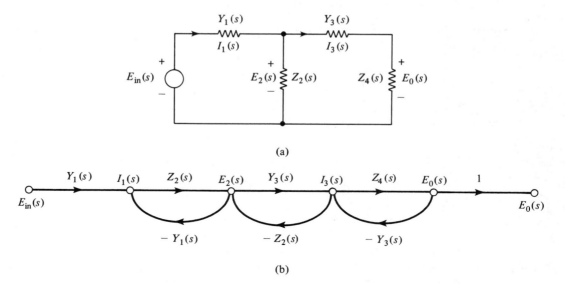

(a)

(b)

Figure 3-17   (a) Passive ladder network. (b) A signal flow graph for the network.

examples. More elaborate cases will be discussed in Chapter 4, where the modeling of systems is formally covered.

■

Example 3-2

The passive network shown in Fig. 3-17(a) is considered to consist of R, L, and C elements so that the network elements can be represented by impedance functions, $Z(s)$ and admittance functions, $Y(s)$. The Laplace transform of the input voltage is denoted by $E_{in}(s)$ and that of the output voltage is $E_0(s)$. In this case it is more convenient to use the branch currents and node voltages designated as shown in Fig. 3-17(a). Then one set of independent equations representing cause-and-effect relation is

$$I_1(s) = [E_{in}(s) - E_2(s)]Y_1(s) \tag{3-62}$$

$$E_2(s) = [I_1(s) - I_3(s)]Z_2(s) \tag{3-63}$$

$$I_3(s) = [E_2(s) - E_0(s)]Y_3(s) \tag{3-64}$$

$$E_0(s) = Z_4(s)I_3(s) \tag{3-65}$$

With the variables $E_{in}(s)$, $I_1(s)$, $E_2(s)$, $I_3(s)$, and $E_0(s)$ arranged from left to right in order, the signal flow graph of the network is constructed as shown in Fig. 3-17(b).

It is noteworthy that in the case of network analysis, the cause-and-effect equations that are most convenient for the construction of a signal flow graph are neither the loop equations nor the node equations. Of course, this does not mean that we cannot construct a signal flow graph using the loop or the node equations. For instance, in Fig. 3-17(a), if we let $I_1(s)$ and $I_3(s)$ be the loop currents of the two loops, the loop equations are

$$E_{in}(s) = [Z_1(s) + Z_2(s)]I_1(s) - Z_2(s)I_3(s) \tag{3-66}$$

$$0 = -Z_2(s)I_1(s) + [Z_2(s) + Z_3(s) + Z_4(s)]I_3(s) \tag{3-67}$$

$$E_0(s) = Z_4(s)I_3(s) \tag{3-68}$$

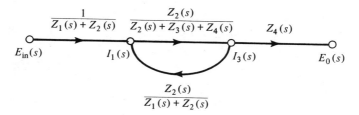

Figure 3-18 Signal flow graph of the network in Fig. 3-17(a) using the loop equations as a starting point.

However, Eqs. (3-66) and (3-67) should be rearranged, since only effect variables can appear on the left-hand sides of the equations. Therefore, solving for $I_1(s)$ from Eq. (3-66) and $I_3(s)$ from Eq. (3-67), we get

$$I_1(s) = \frac{1}{Z_1(s) + Z_2(s)} E_{in}(s) + \frac{Z_2(s)}{Z_1(s) + Z_2(s)} I_3(s) \tag{3-69}$$

$$I_3(s) = \frac{Z_2(s)}{Z_2(s) + Z_3(s) + Z_4(s)} I_1(s) \tag{3-70}$$

Now, Eqs. (3-68), (3-69), and (3-70) are in the form of cause-and-effect equations. The signal flow graph portraying these equations is drawn as shown in Fig. 3-18. This exercise also illustrates that the signal flow graph of a system is not unique.

Example 3-3

Let us consider the *RLC* network shown in Fig. 3-19(a). We shall define the current $i(t)$ and the voltage $e_c(t)$ as the dependent variables of the network. Writing the voltage across the inductance and the current in the capacitor, we have the following differential equations:

$$L\frac{di(t)}{dt} = e_1(t) - Ri(t) - e_c(t) \tag{3-71}$$

$$C\frac{de_c(t)}{dt} = i(t) \tag{3-72}$$

However, we cannot construct a signal flow graph using these two equations since they are differential equations. In order to arrive at algebraic equations, we divide Eqs. (3-71) and (3-72) by L and C, respectively. When we take the Laplace transform, we have

$$sI(s) = i(0) + \frac{1}{L}E_1(s) - \frac{R}{L}I(s) - \frac{1}{L}E_c(s) \tag{3-73}$$

$$sE_c(s) = e_c(0) + \frac{1}{C}I(s) \tag{3-74}$$

where $i(0)$ is the initial current and $e_c(0)$ is the initial voltage at $t = 0$. In these last two equations, $e_c(0)$, $i(0)$, and $E_1(s)$ are the input variables. There are several possible ways of constructing the signal flow graph for these equations. One way is to solve for $I(s)$ from Eq.

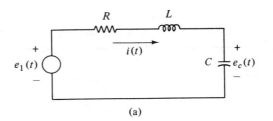

(a)

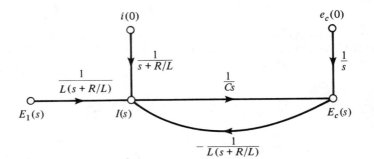

(b)

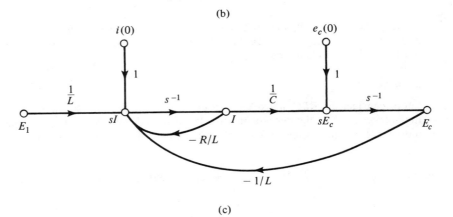

(c)

Figure 3-19 (a) *RLC* network. (b) Signal flow graph. (c) Alternative signal flow graph.

(3-73) and $E_c(s)$ from Eq. (3-74); we get

$$I(s) = \frac{1}{s + (R/L)} i(0) + \frac{1}{L[s + (R/L)]} E_1(s) - \frac{1}{L[s + (R/L)]} E_c(s) \quad (3\text{-}75)$$

$$E_c(s) = \frac{1}{s} e_c(0) + \frac{1}{Cs} I(s) \quad (3\text{-}76)$$

The signal flow graph using the last equations is drawn as shown in Fig. 3-19(b).

The signal flow graph in Fig. 3-19(b) is of analytical value only. In other words, we can solve for $I(s)$ and $E_c(s)$ from the signal flow graph in terms of the inputs, $e_c(0)$, $i(0)$, and $E_1(s)$, but the value of the signal flow graph would probably end here. As an alternative, we can use Eqs. (3-73) and (3-74) directly, and define $I(s)$, $E_c(s)$, $sI(s)$, and $sE_c(s)$ as the noninput variables. These four variables are related by the equations

$$I(s) = s^{-1}[sI(s)] \tag{3-77}$$

$$E_c(s) = s^{-1}[sE_c(s)] \tag{3-78}$$

The significance of using s^{-1} is that it represents pure integration in the time domain. Now, a signal flow graph using Eqs. (3-73), (3-74), (3-77), and (3-78) is constructed as shown in Fig. 3-19(c). Notice that in this signal flow graph the Laplace transform variable appears only in the form of s^{-1}. Therefore, this signal flow graph may be used as a basis for analog or digital computer solution of the problem. Signal flow graphs in this form are defined as the *state diagrams* [5].

∎

3.9 GENERAL GAIN FORMULA FOR SIGNAL FLOW GRAPHS [3]

Given a signal flow graph or a block diagram, it is usually a tedious task to solve for its input–output relationships by analytical means. Fortunately, there is a general gain formula available which allows the determination of the input–output relationship of a signal flow graph by mere inspection. The general gain formula is

$$M = \frac{y_{\text{out}}}{y_{\text{in}}} = \sum_{k=1}^{N} \frac{M_k \Delta_k}{\Delta} \tag{3-79}$$

where

M = gain between y_{in} and y_{out}

y_{out} = output node variable

y_{in} = input node variable

N = total number of forward paths

M_k = gain of the k th forward path

$\Delta = 1 - \sum_m P_{m1} + \sum_m P_{m2} - \sum_m P_{m3} + \cdots$

P_{mr} = gain product of the m th possible combination of r nontouching[3] loops

$\tag{3-80}$

---

[3] Two parts of a signal flow graph are said to be nontouching if they do not share a common node.

or

$$\Delta = 1 - \text{(sum of all individual loop gains)} + \text{(sum of gain products of all possible combinations of two nontouching loops)} - \text{(sum of the gain products of all possible combinations of three nontouching loops)} + \cdots \qquad (3\text{-}81)$$

$\Delta_k = $ the Δ for the part of the signal flow graph which is nontouching with the k th forward path

This general formula may seem formidable to use at first glance. However, the only complicated term in the gain formula is Δ; but in practice, systems having a large number of nontouching loops are rare. An error that is frequently made with regard to the gain formula is the condition under which it is valid. It must be emphasized that *the gain formula can be applied only between an input node and an output node.*

■

Example 3-4

Consider the signal flow graph of Fig. 3-16. We wish to find the transfer function $C(s)/R(s)$ by use of the gain formula, Eq. (3-79). The following conclusions are obtained by inspection from the signal flow graph:

1. There is only one forward path between $R(s)$ and $C(s)$, and the forward-path gain is

$$M_1 = G(s) \qquad (3\text{-}82)$$

2. There is only one loop; the loop gain is

$$P_{11} = -G(s)H(s) \qquad (3\text{-}83)$$

3. There are no nontouching loops since there is only one loop. Furthermore, the forward path is in touch with the only loop. Thus $\Delta_1 = 1$, and $\Delta = 1 - P_{11} = 1 + G(s)H(s)$.

By use of Eq. (3-79), the transfer function of the system is obtained as

$$\frac{C(s)}{R(s)} = \frac{M_1 \Delta_1}{\Delta} = \frac{G(s)}{1 + G(s)H(s)} \qquad (3\text{-}84)$$

which agrees with the result obtained in Eq. (3-61).

Example 3-5

Consider, in Fig. 3-71(b) that the functional relation between E_{in} and E_0 is to be determined by use of the general gain formula. The signal flow graph is redrawn in Fig. 3-20(a). The following conclusions are obtained by inspection from the signal flow graph:

1. There is only one forward path between E_{in} and E_0, as shown in Fig. 3-20(b). The forward-path gain is

$$M_1 = Y_1 Z_2 Y_3 Z_4 \qquad (3\text{-}85)$$

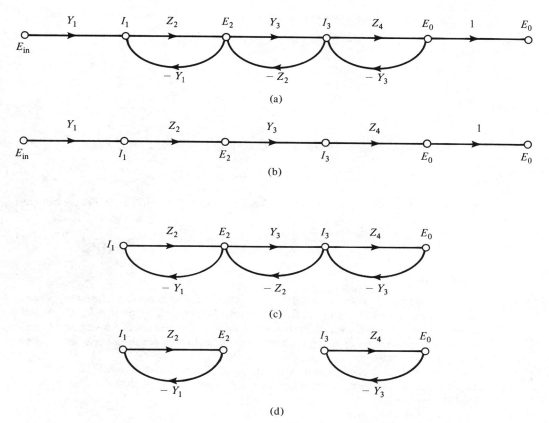

Figure 3-20 (a) Signal flow graph of the passive network in Fig. 3-17(a). (b) Forward path between E_{in} and E_0. (c) Three individual loops. (d) Two non-touching loops.

2. There are three individual loops, as shown in Fig. 3-20(c); the loop gains are

$$P_{11} = -Z_2 Y_1 \tag{3-86}$$

$$P_{21} = -Z_2 Y_3 \tag{3-87}$$

$$P_{31} = -Z_4 Y_3 \tag{3-88}$$

3. There is one pair of nontouching loops, as shown in Fig. 3-20(d); the loop gains of these two loops are

$$-Z_2 Y_1 \quad \text{and} \quad -Z_4 Y_3$$

Thus,

$$P_{12} = \text{product of gains of the first (and only) possible}$$
$$\text{combination of two nontouching loops} = Z_2 Z_4 Y_1 Y_3 \tag{3-89}$$

4. There are no three nontouching loops, four nontouching loops, and so on; thus,

$$P_{m3} = 0, \qquad P_{m4} = 0, \ldots$$

From Eq. (3-80),

$$\Delta = 1 - (P_{11} + P_{21} + P_{31}) + P_{12}$$
$$= 1 + Z_2 Y_1 + Z_2 Y_3 + Z_4 Y_3 + Z_2 Z_4 Y_1 Y_3 \tag{3-90}$$

5. All the three feedback loops are in touch with the forward path; thus,

$$\Delta_1 = 1 \tag{3-91}$$

Substituting the quantities in Eqs. (3-85) through (3-91) into Eq. (3-79), we obtain

$$\frac{E_0}{E_{\text{in}}} = \frac{M_1 \Delta_1}{\Delta} = \frac{Y_1 Y_3 Z_2 Z_4}{1 + Z_2 Y_1 + Z_2 Y_3 + Z_4 Y_3 + Z_2 Z_4 Y_1 Y_3} \tag{3-92}$$

Example 3-6

Consider the signal flow graph of Fig. 3-19(c). It is desired to find the relationships between I and the three inputs, E_1, $i(0)$, and $e_c(0)$. Similar relationship is desired for E_c. Since the system is linear, the principle of superposition applies. The gain between one input and one output is determined by applying the gain formula to the two variables while setting the rest of the inputs to zero.

The signal flow graph is redrawn as shown in Fig. 3-21(a). Let us first consider I as the output variable. The forward paths between each input and I are shown in Fig. 3-21(b), (c), and (d), respectively.

The signal flow graph has two loops; the Δ is given by

$$\Delta = 1 + \frac{R}{L} s^{-1} + \frac{1}{LC} s^{-2} \tag{3-93}$$

All the forward paths are in touch with the two loops; thus $\Delta_1 = 1$ for all cases.

Considering each input separately, we have

$$\frac{I}{E_1} = \frac{(1/L)s^{-1}}{\Delta} \qquad i(0) = 0 \qquad e_c(0) = 0 \tag{3-94}$$

$$\frac{I}{i(0)} = \frac{s^{-1}}{\Delta} \qquad E_1 = 0 \qquad e_c(0) = 0 \tag{3-95}$$

$$\frac{I}{e_c(0)} = \frac{-(1/L)s^{-2}}{\Delta} \qquad i(0) = 0 \qquad E_1 = 0 \tag{3-96}$$

When all three inputs are applied simultaneously, we write

$$I = \frac{1}{\Delta} \left[\frac{1}{L} s^{-1} E_1 + s^{-1} i(0) - \frac{1}{L} s^{-2} e_c(0) \right] \tag{3-97}$$

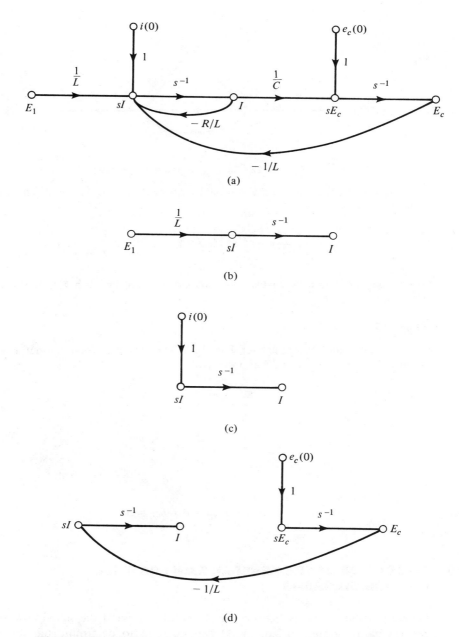

Figure 3-21 (a) Signal flow graph of the *RLC* network in Fig. 3-19(a). (b) Forward path between E_1 and *I*. (c) Forward path between $i(0)$ and *I*. (d) Forward path between $e_c(0)$ and *I*.

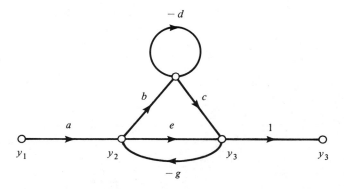

Figure 3-22   Signal flow graph for Example 3-7.

In a similar fashion, the reader should verify that when E_c is considered as the output variable, we have

$$E_c = \frac{1}{\Delta}\left[\frac{1}{LC}s^{-2}E_1 + \frac{1}{C}s^{-2}i(0) + s^{-1}\left(1 + \frac{R}{L}s^{-1}\right)e_c(0)\right] \qquad (3\text{-}98)$$

Notice that the loop between the nodes sI and I is not in touch with the forward path between $e_c(0)$ and E_c.

Example 3-7

Consider the signal flow graph of Fig. 3-22. The following input–output relations are obtained by use of the general gain formula:

$$\frac{y_2}{y_1} = \frac{a(1 + d)}{\Delta} \qquad (3\text{-}99)$$

$$\frac{y_3}{y_1} = \frac{ae(1 + d) = abc}{\Delta} \qquad (3\text{-}100)$$

where

$$\Delta = 1 + eg + d + bcg + deg \qquad (3\text{-}101)$$

■

3.10 APPLICATION OF THE GENERAL GAIN FORMULA TO BLOCK DIAGRAMS

Because of the similarity between the block diagram and the signal flow graph, the general gain formula in Eq. (3-79) can be used to determine the input–output relationships of either. In general, given a block diagram of a linear system we can apply the gain formula directly to it. However, in order to be able to identify all the loops and nontouching parts clearly, sometimes it may be helpful if an equivalent signal flow graph is drawn for a block diagram before applying the gain formula.

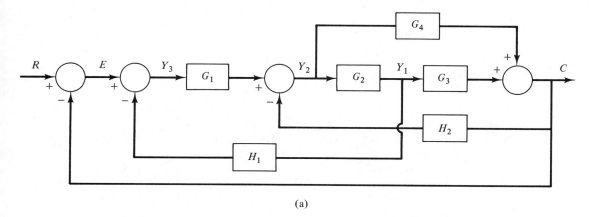

(a)

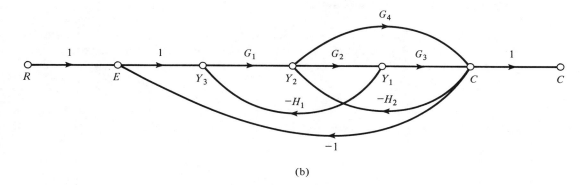

(b)

Figure 3-23 (a) Block diagram of a control system. (b) Equivalent signal flow graph.

To illustrate how the signal flow graph and the block diagram are related, the equivalent models of a control system are shown in Fig. 3-23. Note that since a node on the signal flow graph is interpreted as a summing point of all incoming signals to the node, the negative feedback paths in this case are represented by assigning negative gains to the feedback paths.

The closed-loop transfer function of the system is obtained by applying Eq. (3-79) to either the block diagram or the signal flow graph:

$$\frac{C(s)}{R(s)} = \frac{G_1G_2G_3 + G_1G_4}{1 + G_1G_2H_1 + G_2G_3H_2 + G_1G_2G_3 + G_4H_2 + G_1G_4} \quad (3\text{-}102)$$

Similarly,

$$\frac{E(s)}{R(s)} = \frac{1 + G_1G_2H_1 + G_2G_3H_2 + G_4H_2}{\Delta} \quad (3\text{-}103)$$

$$\frac{Y_3(s)}{R(s)} = \frac{1 + G_2G_3H_2 + G_4H_2}{\Delta} \quad (3\text{-}104)$$

where

$$\Delta = 1 + G_1G_2H_1 + G_2G_3H_2 + G_1G_2G_3 + G_4H_2 + G_1G_4 \qquad (3\text{-}105)$$

3.11 STATE DIAGRAM

The signal flow graph discussed in Section 3.4 applies only to algebraic equations. In this section we introduce the methods of the *state diagram*, which represents an extension of the signal flow graph to portray state equations and differential equations. The important significance of the state diagram is that it forms a close relationship among the state equations, computer simulation, and transfer functions. A state diagram is constructed following all the rules of the signal flow graph. Therefore, the state diagram may be used for solving linear systems either analytically or by computers.

Basic Analog Computer Elements

Before taking up the subject of state diagrams, it is useful to discuss the basic elements of an analog computer. The fundamental linear operations that can be performed on an analog computer are *multiplication by a constant*, *addition* and *integration*. These are discussed separately in the following.

    Multiplication by a Constant. Multiplication of a machine variable by a constant is done by potentiometers and amplifiers. Let us consider the operation

$$x_2(t) = ax_1(t) \qquad (3\text{-}106)$$

where a is a constant. If a lies between zero and unity, a potentiometer is used to realize the operation of Eq. (3-106). An operational amplifier is used to simulate Eq. (3-106) if a is a negative integer less than -1. The negative value of a considered is due to the fact that there is usually an $180°$ phase shift between the output and the input of an operational amplifier. The computer block diagram symbols of the potentiometer and the operational amplifier are shown in Figs. 3-24 and 3-25, respectively.

    Algebraic Sum of Two or More Variables. The algebraic sum of two or more machine variables may be obtained by means of the operational amplifier. Amplifi-

Figure 3-24 Analog computer block diagram symbol of a potentiometer.

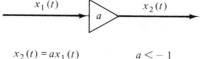

Figure 3-25 Analog computer block diagram of an operational amplifier.

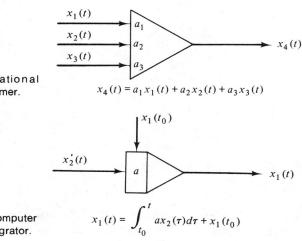

Figure 3-26 Operational amplifier used as a summer.

$$x_4(t) = a_1 x_1(t) + a_2 x_2(t) + a_3 x_3(t)$$

Figure 3-27 Analog computer block diagram of an integrator.

$$x_1(t) = \int_{t_0}^{t} a x_2(\tau) d\tau + x_1(t_0)$$

cation may be accompanied by algebraic sum. For example, Fig. 3-26 illustrates the analog computer block diagram of a summing operational amplifier which portrays the following equation:

$$x_4(t) = a_1 x_1(t) + a_2 x_2(t) + a_3 x_3(t) \tag{3-107}$$

Integration. The integration of a machine variable of an analog computer is achieved by means of a computer element called the *integrator*. If $x_1(t)$ is the output of the integrator with initial condition $x_1(t_0)$ given at $t = t_0$ and $x_2(t)$ is the input, the integrator performs the following operations:

$$x_1(t) = \int_{t_0}^{t} a x_2(\tau) \, d\tau + x_1(t_0) \qquad a \le 1 \tag{3-108}$$

The block diagram symbol of the integrator is shown in Fig. 3-27. The integrator can also serve simultaneously as a summing and amplification device.

We shall now show that these analog computer operations can be portrayed by signal flow graphs which are called state diagrams because the state variables are involved.

First consider the multiplication of a variable by a constant; we take the Laplace transform on both sides of Eq. (3-106). We have

$$X_2(s) = aX_1(s) \tag{3-109}$$

The signal flow graph of Eq. (3-109) is shown in Fig. 3-28.

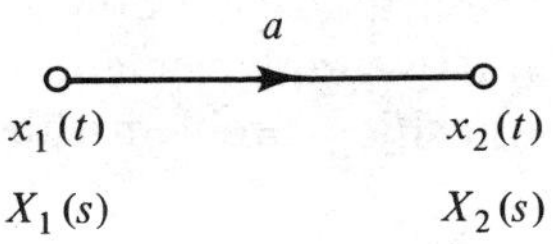

Figure 3-28 Signal-flow-graph representation of $x_2(t) = ax_1(t)$ or $X_2(s) = aX_1(s)$.

$$x_1(t) \qquad\qquad x_2(t)$$
$$X_1(s) \qquad\qquad X_2(s)$$

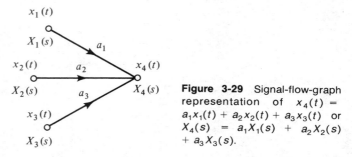

Figure 3-29 Signal-flow-graph representation of $x_4(t) = a_1 x_1(t) + a_2 x_2(t) + a_3 x_3(t)$ or $X_4(s) = a_1 X_1(s) + a_2 X_2(s) + a_3 X_3(s)$.

For the summing operation of Eq. (3-107), the Laplace transform equation is

$$X_4(s) = a_1 X_1(s) + a_2 X_2(s) + a_3 X_3(s) \qquad (3\text{-}110)$$

The signal-flow-graph representation of the last equation is shown in Fig. 3-29. It is important to note that the variables in the signal flow graphs of Figs. 3-28 and 3-29 may be in the time domain or the Laplace transform domain. Since the branch gains are constants in these cases, the equations are algebraic in both domains.

For the integration operation, we take the Laplace transform on both sides of Eq. (3-108). In this case the transform operation is necessary, since the signal-flow-graph algebra does not handle integration in the time domain. We have

$$
\begin{aligned}
X_1(s) &= a\mathcal{L}\left(\int_{t_0}^{t} x_2(\tau)\, d\tau \right) + \frac{x_1(t_0)}{s} \\
&= a\mathcal{L}\left(\int_{0}^{t} x_2(\tau)\, d\tau - \int_{0}^{t_0} x_2(\tau)\, d\tau \right) + \frac{x_1(t_0)}{s} \qquad (3\text{-}111) \\
&= \frac{aX_2(s)}{s} + \frac{x_1(t_0)}{s} - a\mathcal{L}\left(\int_{0}^{t_0} x_2(\tau)\, d\tau \right)
\end{aligned}
$$

However, since the past history of the integrator is represented by $x_2(t_0)$, and the state transition starts from $\tau = t_0$, $x_2(\tau) = 0$ for $0 < \tau < t_0$. Thus, Eq. (3-111) becomes

$$X_1(s) = \frac{aX_2(s)}{s} + \frac{x_1(t_0)}{s} \qquad \tau \geq t_0 \qquad (3\text{-}112)$$

It should be emphasized that Eq. (3-112) is defined only for the period $\tau \geq t_0$. Therefore, the inverse Laplace transform of $X_1(s)$ in Eq. (3-112) will lead to $x_1(t)$ of Eq. (3-108).

Equation (3-112) is now algebraic and can be represented by a signal flow graph as shown in Fig. 3-30. An alternative signal flow graph for Eq. (3-112) is shown in Fig. 3-31.

Thus, we have established a correspondence between the simple analog computer operations and the signal-flow-graph representations. Since, as shown in Fig. 3-31, these signal-flow-graph elements may include initial conditions and can be used to solve state transition problems, they form the basic elements of the state diagram.

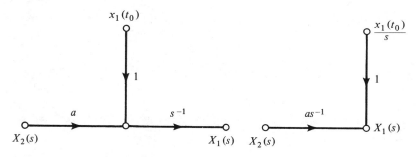

Figure 3-30 Signal-flow-graph representation of $X_1(s) = [aX_2(s)/s] + [x_1(t_0)/s]$.

Figure 3-31 Signal-flow-graph representation of $X_1(s) = [aX_2(s)/s] + [x_1(t_0)/s]$.

Before embarking on several illustrative examples on state diagrams, let us point out the important usages of the state diagrams.

1. A state diagram can be constructed directly from the system's differential equation. This allows the determination of the state variables and the state equations once the differential equation of the system is given.
2. A state diagram can be constructed from the system's transfer function. This step is defined as the decomposition of transfer functions (Section 5.11).
3. The state diagram can be used for the programming of the system on an analog computer.
4. The state diagram can be used for the simulation of the system on a digital computer.
5. The state transition equation in the Laplace transform domain may be obtained from the state diagram by means of the signal-flow-graph gain formula.
6. The transfer functions of a system can be obtained from the state diagram.
7. The state equations and the output equations can be determined from the state diagram.

The details of these techniques are given below.

From Differential Equation to State Diagram

When a linear system is described by a high-order differential equation, a state diagram can be constructed from these equations, although a direct approach is not always the most convenient. Consider the following differential equation:

$$\frac{d^n c}{dt^n} + a_n \frac{d^{n-1}c}{dt^{n-1}} + \cdots + a_2 \frac{dc}{dt} + a_1 c = r \tag{3-113}$$

To construct a state diagram using this equation, we rearrange the equation to read

$$\frac{d^n c}{dt^n} = -a_n \frac{d^{n-1} c}{dt^{n-1}} - \cdots - a_2 \frac{dc}{dt} - a_1 c + r \qquad (3\text{-}114)$$

Let us use the following symbols to simplify the representation of the derivatives of c:

$$c^{(i)} = \frac{d^i c}{dt^i} \qquad i = 1, 2, \ldots, n \qquad (3\text{-}115)$$

Frequently, $\dot{c}$ is used in the literature to represent dc/dt.

Now the variables $r, c, c^{(1)}, c^{(2)}, \ldots, c^{(n)}$ are represented by nodes arranged as shown in Fig. 3-32(a). In terms of Laplace transform, these variables are denoted by $R(s), C(s), sC(s), s^2 C(s), \ldots, s^n C(s)$, respectively.

As the next step, the nodes in Fig. 3-32(a) are connected by branches to portray Eq. (3-114), resulting Fig. 3-32(b). Since the variables $c^{(i)}$ and $c^{(i-1)}$ are related through integration with respect to time, they can be interconnected by a branch with gain s^{-1} and the elements of Fig. 3-31 can be used. Therefore, the complete state diagram is drawn as shown in Fig. 3-32(c).

Notice that in Fig. 3-32 the output variables of the integrators are defined as the state variables $x_1, x_2, \ldots, x_n$. This turns out to be a natural choice of state variables once the state diagram is constructed.

When the differential equation is that of Eq. (3-5), with derivatives of the input on the right side, the problem of drawing the state diagram directly is not so straightforward. We shall show later that, in general, it is more convenient to obtain the transfer function from the differential equation first and then obtain the state diagram through decomposition (Section 5.11).

∎

Example 3-8

Consider the following differential equation:

$$\frac{d^2 c}{dt^2} + 3 \frac{dc}{dt} + 2c = r \qquad (3\text{-}116)$$

Equating the highest-ordered term of Eq. (3-116) to the rest of the terms, we have

$$\frac{d^2 c}{dt^2} = -2c - 3 \frac{dc}{dt} + r \qquad (3\text{-}117)$$

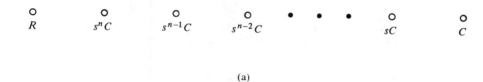

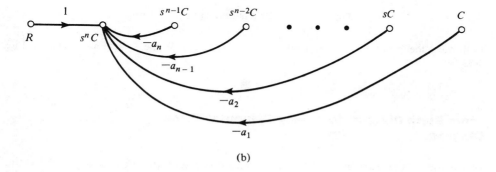

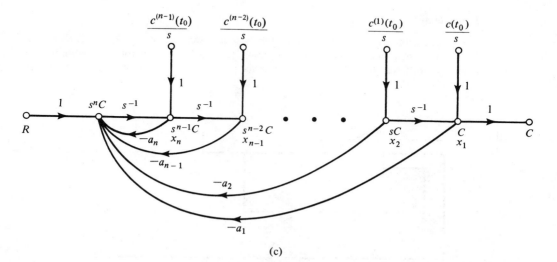

Figure 3-32 State diagram representation of the differential equation of Eq. (3-113).

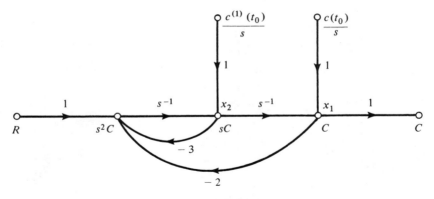

Figure 3-33   State diagram for Eq. (3-116).

Following the procedure as outlined, the state diagram of the system is shown in Fig. 3-33.

■

From State Diagram to Analog Computer Block Diagram

It was mentioned earlier that the state diagram is essentially a block diagram for the programming of an analog computer, except for the phase reversal through amplification and integration.

■

Example 3-9

An analog computer block diagram of the system described by Eq. (3-116) is shown in Fig. 3-34 The final practical version of the computer block diagram for programming may be somewhat different from the one shown, since amplitude and time scaling may be necessary.

■

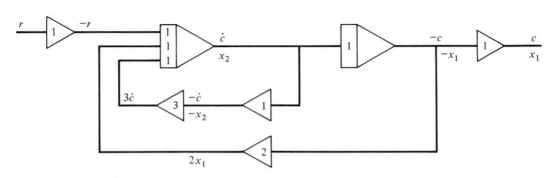

Figure 3-34   Analog-computer block diagram for the system described by Eq. (3-116).

From State Diagram to Digital Computer Simulation

The solution of differential equations by FORTRAN on the digital computer has been well established. However, from the standpoint of programming, a convenient way of modeling a system on the digital computer is by CSMP (Continuous System Modeling Program) [8] or ACSL (Advanced Continuous Simulation Language) [9]. In many respects CSMP or AC serves the same purpose as an analog computer program, except that the scaling problem is practically eliminated. The state diagram or the state equations form a natural basis for the solution by CSMP. The following examples illustrate typical CSMP statements for the mathematical equations listed:

Mathematical Equations	*CSMP Statements*
$c = a_1 x_1 + a_2 x_2$	$C = A1 * X1 + A2 * X2$
$y = \dfrac{x_1}{2}$	$Y = X1/2.$
$x_1 = \displaystyle\int_0^t x_2(\tau)\, d\tau + x_1(0)$	$X1 = \text{INTGRL}(X2, X10)$

■

Example 3-10

From the state diagram of Fig. 3-33, the following equations are written:

$$c = \int \dot{c}\, dt \qquad (3\text{-}118)$$

$$\dot{c} = \int \ddot{c}\, dt \qquad (3\text{-}119)$$

$$\ddot{c} = r - 3\dot{c} - 2c \qquad (3\text{-}120)$$

Let the variables of these equations be denoted by

$$c = C \qquad c(0) = C0$$
$$\dot{c} = C1 \qquad \dot{c}(0) = C10$$
$$\ddot{c} = C2$$
$$r = R$$

on the CSMP. Then the main program of the CSMP representation of the system is given as follows:

$$C = \text{INTGRL}(C1, C0) \qquad (3\text{-}121)$$
$$C1 = \text{INTGRL}(C2, C10) \qquad (3\text{-}122)$$
$$C2 = R - 3.*C1 - 2.*C \qquad (3\text{-}123)$$

■

From State Diagram to Transfer Function

The transfer function between an input and an output is obtained from the state diagram by setting all other inputs and all initial states to zero.

■

Example 3-11

Consider the state diagram of Fig. 3-33. The transfer function $C(s)/R(s)$ is obtained by applying the gain formula between these two nodes and setting $x_1(t_0) = 0$ and $x_2(t_0) = 0$. Therefore,

$$\frac{C(s)}{R(s)} = \frac{1}{s^2 + 3s + 2} \qquad (3\text{-}124)$$

The characteristic equation of the system is

$$s^2 + 3s + 2 = 0 \qquad (3\text{-}125)$$

■

From State Diagram to State Equations

When the state diagram of a system is already given, the state equations and the output equations can be obtained directly from the state diagram by use of the gain formula.

The left side of the state equation contains the first-order time derivative of the state variable $\dot{x}_i(t)$. The right side of the equation contains the state variables and the input variables. There are no Laplace operator s or initial state variables in a state equation. Therefore, to obtain state equations from the state diagram, we should disregard all the initial states and all the integrator branches with gains s^{-1}. To avoid confusion, the initial states and the branches with the gain s^{-1} can actually be eliminated from the state diagram. The state diagram of Fig. 3-33 is simplified as described above, and the result is shown in Fig. 3-35. Then, using $\dot{x}_1$ and $\dot{x}_2$ as output nodes and x_1, x_2, and r as input nodes, and applying the gain

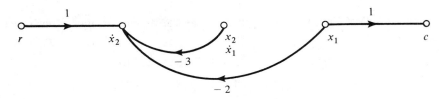

Figure 3-35   State diagram of Fig. 3-33 with the initial states and the integrator branches eliminated.

formula between these nodes, the state equations are written directly:

$$\frac{dx_1}{dt} = x_2$$

$$\frac{dx_2}{dt} = -2x_1 - 3x_2 + r$$

(3-126)

∎

Example 3-12

As another example of illustrating the determination of the state equations from the state diagram, consider the state diagram shown in Fig. 3-36(a). This illustration will emphasize the

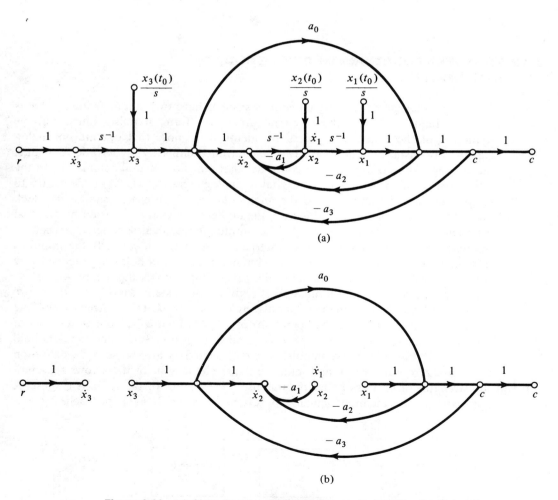

(a)

(b)

Figure 3-36 (a) State diagram. (b) State diagram in (a) with all initial states and integrators eliminated.

importance of using the gain formula. Figure 3-36(b) shows the state diagram with the initial states and the integrators being eliminated. Notice that in this case the state diagram in Fig. 3-36(b) still contains a loop. Applying the gain formula to the diagram of Fig. 3-36(b) with $\dot{x}_1$, $\dot{x}_2$, and $\dot{x}_3$ as the output node variables and r, x_1, x_2, and x_3 as the input nodes, the state equations are determined as follows:

$$
\begin{bmatrix} \dfrac{dx_1}{dt} \\[2ex] \dfrac{dx_2}{dt} \\[2ex] \dfrac{dx_3}{dt} \end{bmatrix} = \begin{bmatrix} 0 & 1 & 0 \\[2ex] \dfrac{-(a_2 + a_3)}{1 + a_0 a_3} & -a_1 & \dfrac{1 - a_0 a_2}{1 + a_0 a_3} \\[2ex] 0 & 0 & 0 \end{bmatrix} \begin{bmatrix} x_1 \\[2ex] x_2 \\[2ex] x_3 \end{bmatrix} + \begin{bmatrix} 0 \\[2ex] 0 \\[2ex] 1 \end{bmatrix} r \qquad (3\text{-}127)
$$

■

3.12 TRANSFER FUNCTIONS OF DISCRETE-DATA SYSTEMS [7]

In general, discrete-data or digital control systems have two unique features. One is that the input signals to these systems are in the form of pulse trains, and the dynamics of the system or process are identical to those that operate on analog signals. For instance, a dc motor can be controlled either by a controller that puts out analog signals or by a digital controller that sends out digital commands. In the latter case an interface such as a digital-to-analog converter (D/A) is necessary to couple the digital component to the analog devices. Figure 3-37 shows the block diagram of such a system. In this case, the digital operation is modeled by an ideal sampler with a sampling period T. The output of the ideal sampler is a train of impulses. The data hold acts as an interface or filter which converts the impulses into an analog signal. One of the most commonly used data holds in practice is the *zero-order hold* (z.o.h.). Functionally, the z.o.h. simply holds the magnitude of the signal carried by the impulse function at a given time instant, say kT, for the entire sampling period T, until the next impulse arrives at $t = (k + 1)T$. Analytically, the z.o.h. is also used to model the operations of a D/A. Figure 3-38 illustrates a set of typical signals represented by $r(t)$, $r^*(t)$, and $h(t)$ of Fig. 3-37 when the data hold is a z.o.h. In Fig. 3-38(b) the impulses are represented by arrows, since by definition they have zero pulse width and infinite height. The lengths of the arrows represent the strengths of or areas under the impulses. The output of the z.o.h. is a staircase approximation of the input signal $r(t)$. It is easy to see that $h(t)$ approaches $r(t)$ as

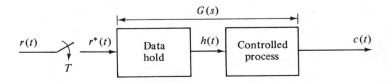

Figure 3-37 Block diagram of a discrete-data system.

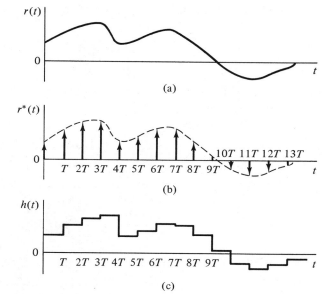

Figure 3-38 (a) Input signal to ideal sampler. (b) Output signal of ideal sampler. (c) Output signal of z.o.h.

the sampling period T approaches zero; that is,

$$\lim_{T \to 0} h(t) = r(t) \tag{3-128}$$

However, the limit of $r^*(t)$, the impulse train, as T approaches zero *does not* have any physical meaning, and

$$\lim_{T \to 0} r^*(t) \neq r(t) \tag{3-129}$$

Another situation often encountered in a discrete-data or digital control system is that the process receives discrete or digital data and sends out signals in the same form, such as in the case of a microcomputer. Figure 3-39 shows the block diagram representation of this type of system.

There are several ways of deriving the transfer function representation of the system of Fig. 3-37. The following derivation is based on the Fourier series representation of the input $r^*(t)$. We begin by writing

$$r^*(t) = r(t)\delta_T(t) \tag{3-130}$$

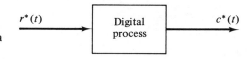

Figure 3-39 Block diagram of a digital process.

where $\delta_T(t)$ is the unit impulse train,

$$\delta_T(t) = \sum_{k=-\infty}^{\infty} \delta(t - kT) \tag{3-131}$$

Since $\delta_T(t)$ is a periodic function with period T, it can be expressed as a Fourier series,

$$\delta_T(t) = \sum_{n=-\infty}^{\infty} C_n e^{j2\pi nt/T} \tag{3-132}$$

where C_n is the Fourier coefficient, and is given by

$$C_n = \frac{1}{T} \int_0^T \delta_T(t) e^{-jn\omega_s t} \, dt \tag{3-133}$$

where

$$\omega_s = 2\pi/T \tag{3-134}$$

is defined as the *sampling frequency* in rad/sec.

Since the impulse function is defined as a pulse with a width of δ and a height of $1/\delta$, and $\delta \to 0$, C_n is written

$$
\begin{aligned}
C_n &= \lim_{\delta \to 0} \frac{1}{T\delta} \int_0^\delta e^{-jn\omega_s t} \, dt \\
&= \lim_{\delta \to 0} \frac{1 - e^{-jn\omega_s \delta}}{jn\omega_s T\delta} = \frac{1}{T}
\end{aligned} \tag{3-135}
$$

Substituting Eq. (3-135) in Eq. (3-132), we have

$$\delta_T(t) = \frac{1}{T} \sum_{n=-\infty}^{\infty} e^{jn\omega_s t} \tag{3-136}$$

Thus, Eq. (3-130) becomes

$$r^*(t) = \frac{1}{T} \sum_{n=-\infty}^{\infty} r(t) e^{jn\omega_s t} \tag{3-137}$$

Now taking the Laplace transform on both sides of Eq. (3-137), and using the complex shifting property of Eq. (2-38), we get

$$R^*(s) = \frac{1}{T} \sum_{n=-\infty}^{\infty} R(s - jn\omega_s) \tag{3-138}$$

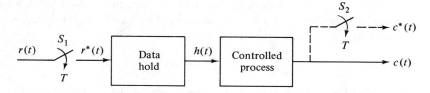

Figure 3-40 Discrete-data system with fictitious sampler.

or

$$R^*(s) = \frac{1}{T} \sum_{n=-\infty}^{\infty} R(s + jn\omega_s) \qquad (3\text{-}139)$$

since the summation in Eq. (3-138) is symmetric. Equation (3-139) represents an alternative expression of $R^*(s)$ to that in Eq. (2-178).

One of the interesting properties of $R^*(s)$ is

$$R^*(s + jm\omega_s) = R^*(s) \qquad (3\text{-}140)$$

for m = an integer.

Now we are ready to derive the transfer function for the discrete-data system in Fig. 3-37. The Laplace transform of the system output $c(t)$ is written

$$C(s) = G(s)R^*(s) \qquad (3\text{-}141)$$

Although in principle the output $c(t)$ is obtained by taking the inverse Laplace transform on both sides of Eq. (3-141), in reality this step is difficult to carry out because $G(s)$ and $R^*(s)$ represent different types of functions. To alleviate this problem, we apply a fictitious sampler at the output of the system, as shown in Fig. 3-40. The fictitious sampler S_2 has the same sampling period T and is synchronized to the original sampler S_1. The sampled form of $c(t)$ is $c^*(t)$. Applying Eq. (3-139) to $c^*(t)$, and using Eq. (3-141), we get

$$\begin{aligned} C^*(s) &= \frac{1}{T} \sum_{n=-\infty}^{\infty} C(s + jn\omega_s) \\ &= \frac{1}{T} \sum_{n=-\infty}^{\infty} G(s + jn\omega_s)R^*(s + jn\omega_s) \end{aligned} \qquad (3\text{-}142)$$

In view of the relationship in Eq. (3-139), Eq. (3-142) becomes

$$\begin{aligned} C^*(s) &= R^*(s)\frac{1}{T} \sum_{n=-\infty}^{\infty} G(s + jn\omega_s) \\ &= R^*(s)G^*(s) \end{aligned} \qquad (3\text{-}143)$$

where $G^*(s)$ is defined the same way as $R^*(s)$.

Now that all the functions in Eq. (3-143) are in sampled form, we can take the z-transform on both sides of the equation. This involves simply the substitution of $z = e^{Ts}$ in Eq. (3-143), and we have

$$C(z) = R(z)G(z) \qquad (3\text{-}144)$$

where $G(z)$ is defined as the "z-transfer function" of the linear process, and is given by

$$G(z) = \sum_{k=0}^{\infty} g(kT)z^{-k} \qquad (3\text{-}145)$$

Thus, for the discrete-data systems of Figs. 3-37 and 3-40, the z-transform of the output is equal to the product of the z-transform of the transfer function of the linear process and the z-transform of the input.

The transfer function relations in Eqs. (3-143) and (3-144) are directly applicable to the all-digital system shown in Fig. 3-39.

Transfer Functions of Discrete-Data Systems with Cascaded Elements

The transfer function representation of discrete-data systems with cascaded elements is slightly more involved than that for continuous-data systems, because of the variation of having or not having any samplers in between the elements. Figure 3-41 illustrates two different situations of a discrete-data system that contains two

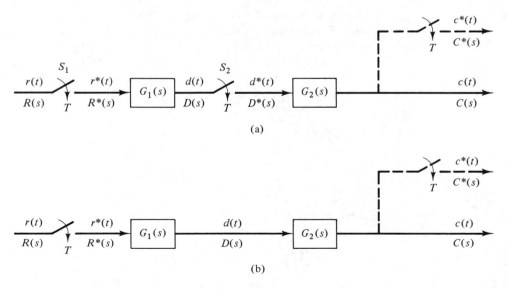

(a)

(b)

Figure 3-41 (a) Discrete-data system with cascaded elements and sampler separates the two elements. (b) Discrete-data system with cascaded elements and no sampler in between.

cascaded elements. In the system of Fig. 3-41(a), the two elements are separated by a sampler S_2 which is synchronized to and has the same period as the sampler S_1. The two elements with transfer functions $G_1(s)$ and $G_2(s)$ of the system in Fig. 3-41(b) are connected directly together. In discrete-data systems, it is important to distinguish these two cases when deriving the pulse transfer functions.

Let us consider first the system of Fig. 3-41(a). The output of $G_1(s)$ is written

$$D(s) = G_1(s)R*(s) \tag{3-146}$$

and the system output is

$$C(s) = G_2(s)D*(s) \tag{3-147}$$

Taking the pulse transform on both sides of Eq. (3-146) and substituting the result into Eq. (3-147) yields

$$C(s) = G_2(s)G_1^*(s)R*(s) \tag{3-148}$$

Then, taking the pulse transform on both sides of the last equation gives

$$C*(s) = G_2^*(s)G_1^*(s)R*(s) \tag{3-149}$$

where we have made use of the relation in Eq. (3-139). The corresponding z-transform expression of Eq. (3-148) is

$$C(z) = G_2(z)G_1(z)R(z) \tag{3-150}$$

We conclude that the z-transform of two linear elements separated by a sampler is equal to the product of the z-transforms of the two individual transfer functions.

The Laplace transform of the output of the system in Fig. 3-41(b) is

$$C(s) = G_1(s)G_2(s)R*(s) \tag{3-151}$$

The pulse transform of Eq. (3-151) is

$$C*(s) = [G_1(s)G_2(s)]*R*(s) \tag{3-152}$$

where

$$[G_1(s)G_2(s)]* = \frac{1}{T}\sum_{n=-\infty}^{\infty} G_1(s+jn\omega_s)G_2(s+jn\omega_s) \tag{3-153}$$

Notice that since $G_1(s)$ and $G_2(s)$ are not separated by a sampler, they have to be treated as one element when taking the pulse transform. For simplicity, we define the following notation:

$$\begin{aligned}[G_1(s)G_2(s)]* &= G_1G_2^*(s) \\ &= G_2G_1^*(s)\end{aligned} \tag{3-154}$$

Then Eq. (3-152) becomes

$$C^*(s) = G_1 G_2^*(s) R^*(s) \tag{3-155}$$

Taking the z-transform on both sides of Eq. (3-155) gives

$$C(z) = G_1 G_2(z) R(z) \tag{3-156}$$

where $G_1 G_2(z)$ is defined as the z-transform of the product of $G_1(s)$ and $G_2(s)$, and it should be treated as a single function.

It is important to note that, in general,

$$G_1 G_2^*(s) \neq G_1^*(s) G_2^*(s) \tag{3-157}$$

and

$$G_1 G_2(z) \neq G_1(z) G_2(z) \tag{3-158}$$

Therefore, we conclude that the z-transform of two cascaded elements with no sampler in between is equal to the z-transform of the product of the transfer functions of the two elements.

Transfer Function of the Zero-Order Hold

Since most discrete-data system modeling includes zero-order hold devices, the situation that occurs frequently is that the z.o.h. is connected directly in cascade with a linear process with transfer function $G_p(s)$. Based on the definition of the z.o.h. given earlier, its impulse response is shown in Fig. 3-42, and its transfer function is

$$G_h(s) = \mathcal{L}[g_h(t)] = \frac{1 - e^{-Ts}}{s} \tag{3-159}$$

Thus, if the z.o.h. is connected in cascade with a linear process with transfer function $G_p(s)$, as shown in Fig. 3-43, the z-transform of the combination is written as

$$\begin{aligned} G(z) &= \mathfrak{z}\left[G_h(s) G_p(s)\right] \\ &= \mathfrak{z}\left[\frac{1 - e^{-Ts}}{s} G_p(s)\right] \end{aligned} \tag{3-160}$$

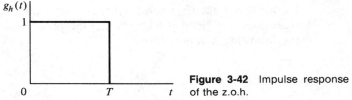

Figure 3-42 Impulse response of the z.o.h.

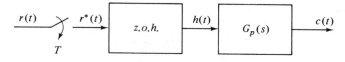

Figure 3-43 Linear process cascaded with a z.o.h.

Using the subtraction property and the time-delay property of z-transforms, Eq. (3-160) is simplified to

$$G(z) = (1 - z^{-1}) \mathcal{Z} \left[\frac{G_p(s)}{s} \right]$$ (3-161)

Transfer Functions of Closed-Loop Discrete-Data Systems

In this section the transfer functions of simple closed-loop discrete-data systems are derived by algebraic means. Consider the closed-loop system shown in Fig. 3-44. The output transform is

$$C(s) = G(s)E^*(s)$$ (3-162)

The Laplace transform of the continuous error function is

$$E(s) = R(s) - H(s)C(s)$$ (3-163)

Substituting Eq. (3-162) into Eq. (3-163) yields

$$E(s) = R(s) - G(s)H(s)E^*(s)$$ (3-164)

Taking the pulse transform on both sides of the last equation and solving for $E^*(s)$

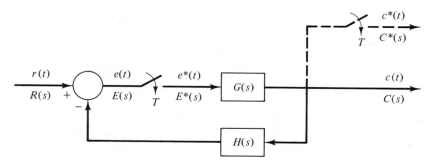

Figure 3-44 Closed-loop discrete-data system.

gives

$$E^*(s) = \frac{R^*(s)}{1 + GH^*(s)} \qquad (3\text{-}165)$$

The output transform $C(s)$ is obtained by substituting $E^*(s)$ from Eq. (3-165) into Eq. (3-162); we have

$$C(s) = \frac{G(s)}{1 + GH^*(s)} R^*(s) \qquad (3\text{-}166)$$

Now taking the pulse transform on both sides of Eq. (3-166) gives

$$C^*(s) = \frac{G^*(s)}{1 + GH^*(s)} R^*(s) \qquad (3\text{-}167)$$

In this case it is possible to define the pulse transfer function between the input and the output of the closed-loop system as

$$\frac{C^*(s)}{R^*(s)} = \frac{G^*(s)}{1 + GH^*(s)} \qquad (3\text{-}168)$$

The z-transfer function of the system is

$$\frac{C(z)}{R(z)} = \frac{G(z)}{1 + GH(z)} \qquad (3\text{-}169)$$

We shall show in the following that although it is possible to define a transfer function for the closed-loop system of Fig. 3-44, in general, this may not be possible for all discrete-data systems. Let us consider the system shown in Fig. 3-45. The

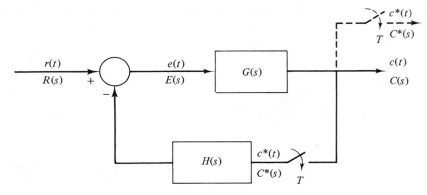

Figure 3-45 Closed-loop discrete-data system.

output transforms, $C(s)$ and $C(z)$, are derived as follows:

$$C(s) = G(s)E(s) \tag{3-170}$$

$$E(s) = R(s) - H(s)C^*(s) \tag{3-171}$$

Substituting Eq. (3-171) into Eq. (3-170) yields

$$C(s) = G(s)R(s) - G(s)H(s)C^*(s) \tag{3-172}$$

Taking the pulse transform on both sides of Eq. (3-172) and solving for $C^*(s)$, we have

$$C^*(s) = \frac{GR^*(s)}{1 + GH^*(s)} \tag{3-173}$$

Note that the input and the transfer function $G(s)$ are now combined as one function, $GR^*(s)$, and the two cannot be separated. In this case we cannot define a transfer function in the form of $C^*(s)/R^*(s)$.

The z-transform of the output is determined directly from Eq. (3-173) to be

$$C(z) = \frac{GR(z)}{1 + GH(z)} \tag{3-174}$$

where it is important to note that

$$GR(z) = \mathfrak{z}[G(s)R(s)] \tag{3-175}$$

and

$$GH(z) = \mathfrak{z}[G(s)H(s)] \tag{3-176}$$

To determine the transform of the continuous output, $C(s)$, we substitute $C^*(s)$ from Eq. (3-173) into Eq. (3-172). We have

$$C(s) = G(s)R(s) - \frac{G(s)H(s)}{1 + GH^*(s)}GR^*(s) \tag{3-177}$$

Although we have been able to arrive at the input–output transfer function and transfer relation of the systems in Figs. 3-44 and 3-45 by algebraic means without difficulty, for more complex system configurations, the algebraic method may become tedious. The signal-flow-graph method is extended to the analysis of discrete-data systems; the reader may refer to the literature [7].

REFERENCES

Block Diagram and Signal Flow Graphs

1. T. D. GRAYBEAL, "Block Diagram Network Transformation," *Elec. Eng.*, Vol. 70, pp. 985–990, 1951.
2. S. J. MASON, "Feedback Theory—Some Properties of Signal Flow Graphs," *Proc. IRE*, Vol. 41, No. 9, pp. 1144–1156, Sept. 1953.
3. S. J. MASON, "Feedback Theory—Further Properties of Signal Flow Graphs," *Proc. IRE*, Vol. 44, No. 7, pp. 920–926, July 1956.
4. L. P. A. ROBICHAUD, M. BOISVERT, and J. ROBERT, *Signal Flow Graphs and Applications*, Prentice-Hall, Inc., Englewood Cliffs, N.J., 1962.
5. B. C. KUO, *Linear Networks and Systems*, McGraw-Hill Book Company, New York, 1967.
6. N. AHMED, "On Obtaining Transfer Functions from Gain-Function Derivatives," *IEEE Trans. Automatic Control*, Vol. AC-12, p. 229, Apr. 1967.

Signal Flow Graphs of Sampled-Data Systems

7. B. C. KUO, *Digital Control Systems*, Holt, Rinehart, and Winston, New York, 1980.

CSMP (Continuous System Modeling Program)

8. *System/360 Continuous System Modeling Program (360A-CX-16X) User's Manual*, Technical Publications Dept., International Business Machines Corporation, White Plains, N.Y.
9. *Advanced Continuous Simulation Language (ACSL) User Guide/Reference Manual*, 3rd ed., Mitchell and Gauthier, Associates, Concord, Mass., 1981.

PROBLEMS

3.1. The following differential equations represent linear time-invariant systems, where $r(t)$ denotes the input and $c(t)$ the output. Find the transfer function $C(s)/R(s)$ for each of the systems.

(a) $\dfrac{d^3c(t)}{dt^3} + \dfrac{d^2c(t)}{dt^2} + 5\dfrac{dc(t)}{dt} + 2c(t) = 3\dfrac{dr(t)}{dt} + 2r(t)$

(b) $\dfrac{d^4c(t)}{dt^4} + 5\dfrac{d^2c(t)}{dt^2} + 3\dfrac{dc(t)}{dt} + 2c(t) = 6r(t)$

(c) $\dfrac{d^3c(t)}{dt^3} + 10\dfrac{dc(t)}{dt} + c(t) + 2\displaystyle\int_0^t c(\tau)\,d\tau = \dfrac{dr(t)}{dt} + 2r(t)$

(d) $2\dfrac{d^2c(t)}{dt^2} + 2\dfrac{dc(t)}{dt} + c(t) = r(t) + 2r(t-1)$

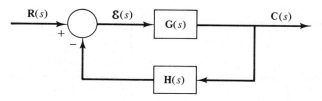

Figure 3P-3

3.2. A linear time-variant multivariable system is described by the following set of differential equations.

$$\frac{d^2c_1(t)}{dt^2} + 5\frac{dc_1(t)}{dt} + 3c_2(t) = r_1(t)$$

$$\frac{d^2c_2(t)}{dt^2} + 3c_1(t) - c_2(t) = r_2(t) + \frac{dr_1(t)}{dt}$$

Find the following transfer functions:

$$\left.\frac{C_1(s)}{R_1(s)}\right|_{R_2=0} \qquad \left.\frac{C_2(s)}{R_1(s)}\right|_{R_2=0} \qquad \left.\frac{C_1(s)}{R_2(s)}\right|_{R_1=0} \qquad \left.\frac{C_2(s)}{R_2(s)}\right|_{R_1=0}$$

3.3. The block diagram of a multivariable feedback control system is shown in Fig. 3P-3. The transfer function matrices of the system are

$$\mathbf{G}(s) = \begin{bmatrix} \dfrac{1}{s+1} & \dfrac{2}{s(s+2)} \\ \dfrac{5}{s} & 10 \end{bmatrix} \qquad \mathbf{H}(s) = \begin{bmatrix} 1 & 0 \\ 0 & 1 \end{bmatrix}$$

Find the closed-loop transfer function matrix of the system.

3.4. A multivariable system with two inputs and two outputs is shown in Fig. 3P-4. Determine the following transfer functions:

$$\left.\frac{C_1(s)}{R_1(s)}\right|_{R_2=0} \qquad \left.\frac{C_2(s)}{R_1(s)}\right|_{R_2=0} \qquad \left.\frac{C_1(s)}{R_2(s)}\right|_{R_1=0} \qquad \left.\frac{C_2(s)}{R_2(s)}\right|_{R_1=0}$$

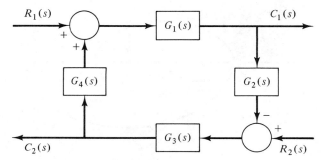

Figure 3P-4

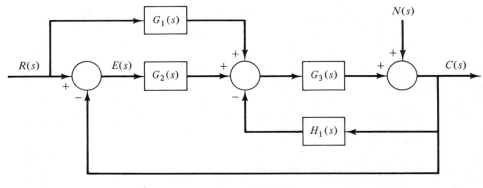

Figure 3P-6

Write the transfer function relations of the system in the form of

$$C(s) = G(s)R(s)$$

3.5. Draw a signal flow graph for the following set of algebraic equations.
  (a) $x_1 = -x_2 - 3x_3 + 2$
  $x_2 = 5x_1 - x_2 - x_3$
  $x_3 = 4x_1 + x_2 - 5x_3 + 1$
  (b) $2x_1 + 5x_2 + x_3 = -1$
  $x_1 - 2x_2 + x_3 = 1$
  $x_2 + 2x_3 = 0$
  These equations should first be converted into the cause-and-effect form. Show that
  there are many possible signal flow graphs for these equations.

3.6. Draw an equivalent signal flow graph for the block diagram shown in Fig. 3P-6. Find
  the following transfer functions using the Mason's gain formula. You should be able to
  get the results by applying the gain formula directly to the block diagram also.
  Compare the answers to those obtained from the equivalent signal flow graph.

$$\left.\frac{C(s)}{R(s)}\right|_{N=0} \qquad \left.\frac{C(s)}{N(s)}\right|_{R=0} \qquad \left.\frac{E(s)}{R(s)}\right|_{N=0} \qquad \left.\frac{E(s)}{N(s)}\right|_{R=0}$$

3.7. Find the gains y_6/y_1, y_3/y_1 and y_5/y_2 for the signal flow graph shown in Fig. 3P-7.

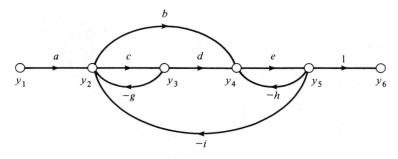

Figure 3P-7

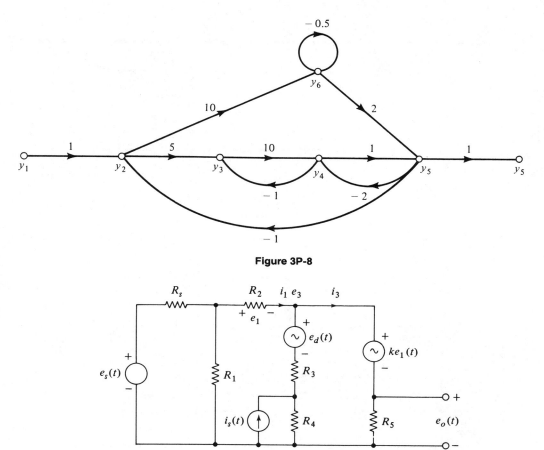

Figure 3P-8

Figure 3P-9

3.8. Find the gains y_5/y_1 and y_2/y_1 from the signal flow graph shown in Fig. 3P-8.

3.9. Signal flow graphs are used to solve a variety of electrical network problems. Shown in Fig. 3P-9 is the equivalent circuit of an electronic circuit. The voltage source $e_d(t)$ represents a disturbance voltage. The objective is to find the value of the constant k so that the output voltage $e_0(t)$ is not affected by $e_d(t)$. To solve the problem, it is best to first write a set of cause-and-effect equations for the network. This involves a combination of node and loop equations. Then construct a signal flow graph using these equations; find the gain $e_0(t)/e_d(t)$ with all other inputs set to zero. For $e_d(t)$ to not affect $e_0(t)$, set $e_0(t)/e_d(t)$ to zero.

3.10. Show that the two systems in Fig. 3P-10(a) and (b) are *not* equivalent.

3.11. The block diagram of a feedback control system is shown in Fig. 3P-11.

   (a) Apply Mason's gain formula directly to the block diagram and find the transfer function

$$\left. \frac{C(s)}{R(s)} \right|_{N=0}$$

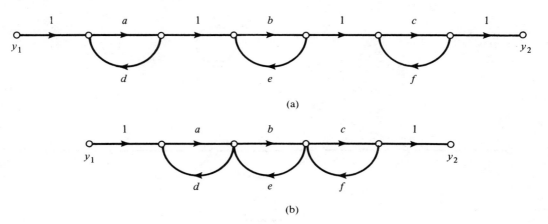

(a)

(b)

Figure 3P-10

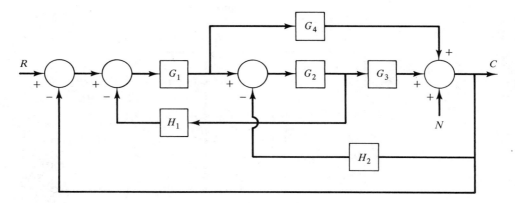

Figure 3P-11

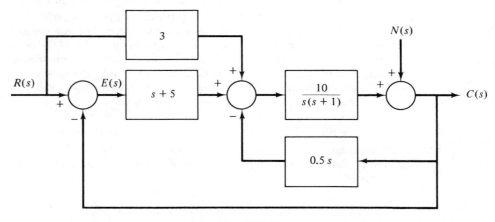

Figure 3P-12

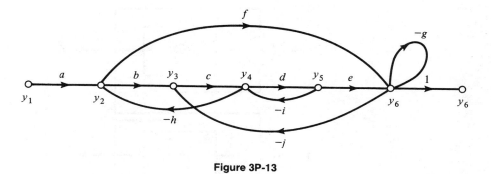

Figure 3P-13

(b) Find the relation among the transfer functions $G_1(s)$, $G_2(s)$, $G_3(s)$, $G_4(s)$, $H_1(s)$, and $H_2(s)$ so that the output $C(s)$ is not affected by the disturbance signal $N(s)$.

3.12. The block diagram of a feedback control system is shown in Fig. 3P-12.
  (a) Find the transfer function $C(s)/E(s)$, $N(s) = 0$.
  (b) Find the transfer function $C(s)/R(s)$, $N(s) = 0$.
  (c) Find the transfer function $C(s)/N(s)$, $R(s) = 0$.
  (d) Find the transfer expression of the output $C(s)$ in terms of the two inputs $R(s)$ and $N(s)$ when both inputs are applied.
  The best way to solve this problem is to draw a signal flow graph for the system and then apply Mason's gain formula to the graph. You may apply the gain formula directly to the block diagram, but care should be taken in tracing through the paths.

3.13. Find the input–output transfer functions relations for the signal flow graph shown in Fig. 3P-13.

  (a) $\dfrac{y_6}{y_1}$ **(b)** $\dfrac{y_2}{y_1}$ **(c)** $\dfrac{y_6}{y_2}$

3.14. Find the following gain relations for the signal flow graph shown in Fig. 3P-14.

  (a) $\left.\dfrac{y_7}{y_1}\right|_{y_8=0}$ **(b)** $\left.\dfrac{y_7}{y_8}\right|_{y_1=0}$

  (c) $\left.\dfrac{y_7}{y_4}\right|_{y_8=0}$ **(d)** $\left.\dfrac{y_7}{y_4}\right|_{y_1=0}$

  Comment on the results in parts (c) and (d).

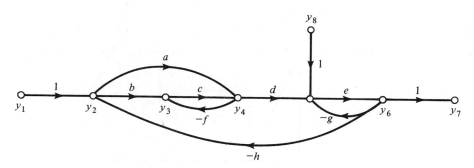

Figure 3P-14

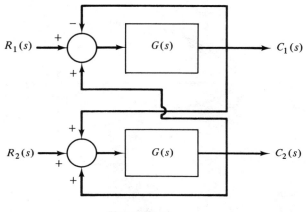

Figure 3P-15

3.15. (a) For the block diagram shown in Fig. 3P-15, draw an equivalent signal flow graph.
   (b) Find the Δ of the system by means of Mason's gain formula.
   (c) Find the following transfer function relations:

$$\left.\frac{C_1(s)}{R_1(s)}\right|_{R_2=0} \qquad \left.\frac{C_1(s)}{R_2(s)}\right|_{R_1=0} \qquad \left.\frac{C_2(s)}{R_1(s)}\right|_{R_2=0} \qquad \left.\frac{C_2(s)}{R_2(s)}\right|_{R_1=0}$$

   (d) Express the transfer function relation in matrix form, $\mathbf{C}(s) = \mathbf{G}(s)\mathbf{R}(s)$ where

$$\mathbf{C}(s) = \begin{bmatrix} C_1(s) \\ C_2(s) \end{bmatrix} \qquad \mathbf{R}(s) = \begin{bmatrix} R_1(s) \\ R_2(s) \end{bmatrix}$$

   Express $\mathbf{G}(s)$ as a 2×2 matrix in terms of $G(s)$.

3.16. A linear feedback control system has the block diagram shown in Fig. 3P-16. Let
   r = input, e = error, N = noise, and c = output.

$$G(s) = \frac{K(s + 2)}{s(s + 1)}$$

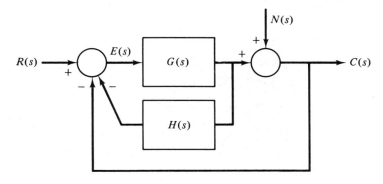

Figure 3P-16

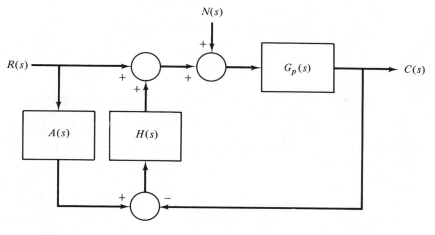

Figure 3P-17

(a) Find the transfer function $H(s)$ so that the output $c(t)$ is not affected by the noise; that is,

$$\left.\frac{C(s)}{N(s)}\right|_{r=0} = 0$$

(b) With $H(s)$ as determined in part (a), find the value of K so that the steady-state value of $e(t)$ is equal to 0.1 when the input is a unit ramp function, $r(t) = tu_s(t)$, $R(s) = 1/s^2$, $N = 0$.

3.17. The block diagram of a feedback control system is shown in Fig. 3P-17. $G_p(s)$ is the transfer function of the process, $R(s)$ is the reference input, and $A(s)$ and $H(s)$ represent controllers.

(a) Derive

$$\left.\frac{C(s)}{R(s)}\right|_{N=0} \quad \text{and} \quad \left.\frac{C(s)}{N(s)}\right|_{R=0}$$

Find $C(s)/R(s)|_{N=0}$ if $A(s) = G_p(s)$.

(b) Let

$$G_p(s) = A(s) = \frac{100}{(s+1)(s+2)}$$

Find the output response $c(t)$ when $N(t) = 0$, $r(t) =$ unit step function.

(c) For $G_p(s)$ and $A(s)$ as given in part (b), select a $H(s)$ such that when $N(t)$ is a unit step function, with $r(t) = 0$, $\lim_{t \to \infty} c(t) = 0$.

3.18. Determine the z-transform of the following sequences.

(a) $f(k) = \begin{cases} 1 & \text{for } k = 0 \text{ and even integers} \\ -1 & \text{for } k = \text{odd integers} \end{cases}$

(b) $f(kT) = kTe^{-5kT}$

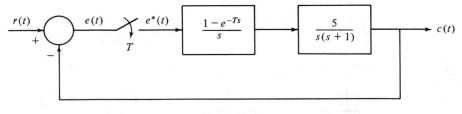

Figure 3P-21

3.19. A linear time-invariant digital control system has an output that is described by the time sequence

$$c(kT) = 1 - e^{-5kT} \qquad k = 0,1,2,\ldots$$

when the system is subject to an input sequence described by $r(kT) = 1$ for all $k \geq 0$. Find the transfer function $C(z)/R(z)$ of the system.

3.20. Find the z-transform of the function

$$F(s) = \frac{5}{s^2(s+1)}$$

Apply partial-fraction expansion to $F(s)$ first and then use the transform table.

3.21. Find the transfer function $C(z)/R(z)$ for the sampled-data system shown in Fig. 3P-21.

3.22. Find the transfer function $C(z)/R(z)$ of the digital control system shown in Fig. 3P-22. The sampling period is 1 sec.

3.23. Find the transfer function $C(z)/R(z)$ of the digital control system shown in Fig. 3P-23. The sampling period is 1 sec.

3.24. Find the transfer function $C(z)/R(z)$ of the digital control system shown in Fig. 3P-24. The sampling period is 0.5 sec.

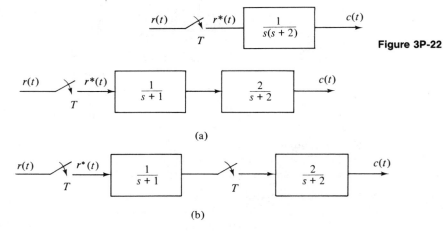

Figure 3P-23

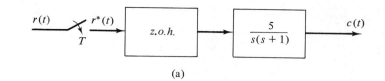

(a)

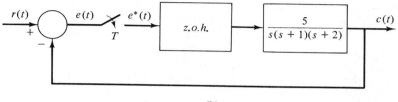

(b)

Figure 3P-24

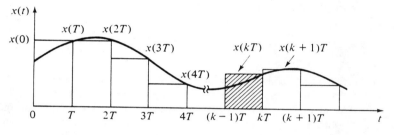

(a)

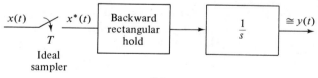

(b)

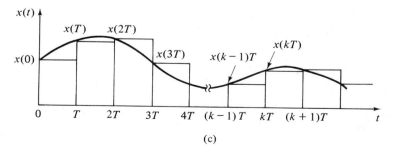

(c)

Figure 3P-25

3.25. It is well known that the transfer function of an analog integrator is represented by

$$G(s) = \frac{Y(s)}{X(s)} = \frac{1}{s}$$

where $X(s)$ is the Laplace transform of the input signal $x(t)$, and $Y(s)$ is the Laplace transform of the output of the integrator, $y(t)$. There are many ways of implementing integration digitally. In the basic computer course the rectangular integration is described. Figure 3P-25(a) illustrates the rectangular integration principle. As shown, the continuous signal $x(t)$ is approximated by a staircase signal; T is the sampling period. The integral of $x(t)$ which is the area under $x(t)$ is approximated by the area under the rectangular approximation signal.

(a) Let $y(kT)$ denote the digital approximation of the integral of $x(t)$ from $t = 0$ to $t = kT$. Then $y(kT)$ can be written as

$$y(kT) = y[(k - 1)T] + Tx(kT) \tag{1}$$

where $y[(k - 1)T]$ denotes the area under $x(t)$ from $t = 0$ to $t = (k - 1)T$. Take the z-transform on both sides of Eq. (1) and show that the transfer function of the digital integrator (by rectangular approximation) is

$$G(z) = \frac{Y(z)}{X(z)} = \frac{Tz}{z - 1} \tag{2}$$

(b) The rectangular integration operation described in Fig. 3P-25(a) can also be interpreted as a sample-and-hold operation. As shown in Fig. 3P-25(a), the signal $x(t)$ is first sent through an ideal sampler. The output of the sampler is the sequence $x(0), x(T), \ldots, x(kT), \ldots$, or an impulse train that carries the sequence of numbers. These numbers are then sent through a "backward" hold device to give the staircase signal; that is, the value of $x(kT)$ at $t = kT$ is held backward in time to give the rectangle of height $x(kT)$ during the time interval from $(k - 1)T$ to kT. The block diagram representation of the process is shown in Fig. 3P-25(b). As an alternative, we can use a "forward" rectangular hold, as shown in Fig. 3P-25(c). Find the transfer function $G(z)$ of such a rectangular integrator.

chapter four

Mathematical Modeling of Physical Systems

4.1 INTRODUCTION

One of the most important tasks in the analysis and design of control systems is mathematical modeling of the systems. In the preceding chapters we have introduced a number of well-known methods of modeling linear systems. The two most common methods are the transfer function approach and the state equation approach. The transfer function method is valid only for linear time-invariant systems, whereas the state equations are first-order differential equations that can be applied to portray linear as well as nonlinear systems. Since in reality all physical systems are nonlinear to some extent, in order to use transfer functions and linear state equations the system must first be linearized, or its range of operation must be confined to a linear range.

Although the analysis and design of linear control systems have been well developed, their counterparts for nonlinear systems are usually quite complex. Therefore, the control systems engineer often has the task of determining not only how to accurately describe a system mathematically, but, more important, how to make proper assumptions and approximations, whenever necessary, so that the system may be adequately characterized by a linear mathematical model.

It is important to point out that the modern control engineer should place special emphasis on the mathematical modeling of the system so that the analysis and design problems can be adaptable for computer solutions. Therefore, the main objectives of this chapter are:

1. To demonstrate the mathematical modeling of control systems and components.
2. To demonstrate how the modeling will lead to computer solutions.

The modeling of many system components and control systems will be illustrated in this chapter. However, the emphasis is placed on the approach to the problem, and no attempt is being made to cover all possible types of systems encountered in practice.

Since a significant portion of the material presented in this chapter deals with the writing of the state equations of a dynamic system, we shall first introduce the basic concept of state.

To begin with the state-variable approach, we should first define the state of a system. As the word implies, the *state* of a system refers to the *past*, *present*, and *future* conditions of the system. It is interesting to note that a simple example is the "State of the Union" speech given by the president of the United States every year. In this case, the entire system encompasses all elements of the government, society, economy, and so on. In general, the state can be described by a set of numbers, a curve, an equation, or something that is more abstract in nature. From a mathematical sense it is convenient to define a set of *state variables* and *state equations* to portray systems. There are some basic ground rules regarding the definition of a state variable and what constitutes a state equation. Consider that the set of variables, $x_1(t), x_2(t), \ldots, x_n(t)$ is chosen to describe the dynamic characteristics of a system. Let us define these variables as the state variables of the system. Then, these state variables must satisfy the following conditions:

1. At any time $t = t_0$, the state variables, $x_1(t_0), x_2(t_0), \ldots, x_n(t_0)$ define the *initial states* of the system at the selected initial time.
2. Once the inputs of the system for $t \geq t_0$ and the initial states defined above are specified, the state variables should completely define the future behavior of the system.

Therefore, we may define the state variables as follows:

Definition of State Variables. *The state variables of a system are defined as a minimal set of variables, $x_1(t), x_2(t), \ldots, x_n(t)$ such that knowledge of these variables at any time t_0, plus information on the input excitation subsequently applied, is sufficient to determine the state of the system at any time $t > t_0$.*

One should not confuse the state variables with the outputs of a system. An output of a system is a variable that can be *measured*, but a state variable does not always, and often does not, satisfy this requirement. However, an output variable is usually defined as a function of the state variables.

The concept and definition of state given above are applied to the formulation of state equations of electrical and mechanical systems in the following sections.

4.2 EQUATIONS OF ELECTRICAL NETWORKS

The classical way of writing equations of electrical network is based on the loop method and the node method, which are formulated from the two laws of Kirchhoff. However, although Kirchhoff's voltage law and current law are simple to follow, the

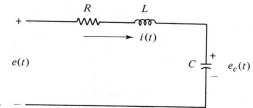

Figure 4-1 *RLC* network.

resulting loop and node equations are not natural for computer solution. A modern method of writing network equations is the state-variable method, which is briefly illustrated in Section 2.3. Since the networks encountered in most control systems are of rather simple complexity, we shall present the subject here only at the introductory level. More detailed discussions on the state equations of electrical networks may be found in texts on network theory.

Let us again consider the *RLC* network shown in Fig. 4-1. A practical way is to assign the current in the inductor L, $i(t)$, and the voltage across the capacitor C, $e_c(t)$, as the state variables. The reason for this choice is because the state variables are directly related to the energy-storage elements of a system. In this case, the inductor is a storage for kinetic energy, and the capacitor is a storage of electric potential energy. By assigning $i(t)$ and $e_c(t)$ as state variables, we have a complete description of the past history (via the initial values of the state variables), and present and future states of the network.

The state equations for the network in Fig. 4-1 are written by first equating the current in C and the voltage across L in terms of the state variables and the applied voltage $e(t)$. Therefore, we have

$$\text{Current in } C: \quad C\frac{de_c(t)}{dt} = i(t) \tag{4-1}$$

$$\text{Voltage across } L: \quad L\frac{di(t)}{dt} = -e_c(t) - Ri(t) + e(t) \tag{4-2}$$

The state equations in vector-matrix form are then written as

$$\begin{bmatrix} \dfrac{de_c(t)}{dt} \\[2ex] \dfrac{di(t)}{dt} \end{bmatrix} = \begin{bmatrix} 0 & \dfrac{1}{C} \\[2ex] -\dfrac{1}{L} & -\dfrac{R}{L} \end{bmatrix} \begin{bmatrix} e_c(t) \\[2ex] i(t) \end{bmatrix} + \begin{bmatrix} 0 \\[2ex] \dfrac{1}{L} \end{bmatrix} e(t) \tag{4-3}$$

$\blacksquare$

Example 4-1

As another example of writing the state equations of an electric network, consider the network shown in Fig. 4-2. According to the foregoing discussion, the voltage across the capacitor e_c and the currents in the inductors i_1 and i_2 are assigned as state variables, as shown in Fig. 4-2.

The state equations of the network are obtained by writing the voltages across the inductors and the currents in the capacitor in terms of the three state variables. The state

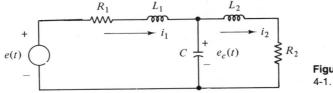

Figure 4-2 Network in Example 4-1.

equations are

$$L_1 \frac{di_1(t)}{dt} = -R_1 i_1(t) - e_c(t) + e(t) \tag{4-4}$$

$$L_2 \frac{di_2(t)}{dt} = -R_2 i_2(t) + e_c(t) \tag{4-5}$$

$$C \frac{de_c(t)}{dt} = i_1(t) - i_2(t) \tag{4-6}$$

Rearranging the constant coefficients, the state equations are written in the following canonical form:

$$\begin{bmatrix} \dfrac{di_1(t)}{dt} \\[2mm] \dfrac{di_2(t)}{dt} \\[2mm] \dfrac{de_c(t)}{dt} \end{bmatrix} = \begin{bmatrix} -\dfrac{R_1}{L_1} & 0 & -\dfrac{1}{L_1} \\[2mm] 0 & -\dfrac{R_2}{L_2} & \dfrac{1}{L_2} \\[2mm] \dfrac{1}{C} & -\dfrac{1}{C} & 0 \end{bmatrix} \begin{bmatrix} i_1(t) \\[2mm] i_2(t) \\[2mm] e_c(t) \end{bmatrix} + \dfrac{1}{L_1} \begin{bmatrix} 1 \\[2mm] 0 \\[2mm] 0 \end{bmatrix} e(t) \tag{4-7}$$

∎

4.3 MODELING OF MECHANICAL SYSTEM ELEMENTS [3]

Most feedback control systems contain mechanical as well as electrical components. From a mathematical viewpoint, the descriptions of electrical and mechanical elements are analogous. In fact, we can show that given an electrical device, there is usually an analogous mechanical counterpart, and vice versa. The analogy, of course, is a mathematical one; that is, two systems are analogous to each other if they are described mathematically by similar equations.

The motion of mechanical elements can be described in various dimensions as translational, rotational, or a combination of both. The equations governing the motions of mechanical systems are often directly or indirectly formulated from Newton's law of motion.

Translational Motion

The motion of translation is defined as a motion that takes place along a straight line. The variables that are used to describe translational motion are acceleration, velocity, and displacement.

Newton's law of motion states that *the algebraic sum of forces acting on a rigid body in a given direction is equal to the product of the mass of the body and its acceleration in the same direction.* The law can be expressed as

$$\Sigma \text{forces} = Ma \tag{4-8}$$

where M denotes the mass and a is the acceleration in the direction considered.

For translational motion, the following elements are usually involved:

1. *Mass*. Mass is considered as an indication of the property of an element which stores the kinetic energy of translational motion. It is analogous to inductance of electrical networks. If W denotes the weight of a body, then M is given by

$$M = \frac{W}{g} \tag{4-9}$$

where g is the acceleration of the body due to gravity of the acceleration of free fall. ($g = 32.174$ ft/sec² in the British unit system, and $g = 9.8066$ m/sec² in the SI unit system.)

The consistent sets of basic units in the British and SI systems are as follows:

Units	Mass M	Acceleration	Force
SI	kilogram (kg)	m/sec²	newton (N)
British	slug	ft/sec²	pound (lb force)

Conversion factors between these and other secondary units are as follows:

Force: 1 N = 0.2248 lb (force) = 3.5969 oz (force)

Mass: 1 kg = 1000 g = 2.2046 lb (mass)
                        = 35.274 oz (mass)
                        = 0.06852 slug

Distance: 1 m = 3.2808 ft = 39.37 in.
            1 in. = 25.4 mm
            1 ft = 0.3048 m

Figure 4-3 illustrates the situation where a force is acting on a body with mass

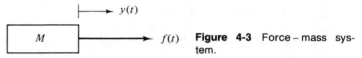

Figure 4-3 Force – mass system.

M. The force equation is written

$$f(t) = Ma(t) = M\frac{d^2y(t)}{dt^2} = M\frac{dv(t)}{dt} \qquad (4\text{-}10)$$

where $y(t)$ represents displacement, $v(t)$ the velocity, and $a(t)$ is the acceleration, all referenced in the direction of the applied force.

2. *Linear spring.* A linear spring in practice may be an actual spring or the compliance of a cable or a belt. In general, a spring is considered to be an element that stores potential energy. It is analogous to a capacitor in electric networks. In practice, all springs are nonlinear to some extent. However, if the deformation of a spring is small, its behavior may be approximated by a linear relationship,

$$f(t) = Ky(t) \qquad (4\text{-}11)$$

where K is the spring constant, or simply stiffness.

The two basic unit systems for the spring constant are:

Units	Spring Constant K
SI	N/m
British	lb/ft

Equation (4-11) implies that the force acting on the spring is directly proportional to the displacement (deformation) of the spring. The model representing a linear spring element is shown in Fig. 4-4.

If the spring is preloaded with a preload tension of T, then Eq. (4-11) should be modified to

$$f(t) - T = Ky(t) \qquad (4\text{-}12)$$

Friction for Translation Motion.   Whenever there is motion or tendency of motion between two elements, frictional forces exist. The frictional forces encountered in physical systems are usually of a nonlinear nature. The characteristics of the frictional forces between two contacting surfaces often depend on such factors as the composition of the surfaces, the pressure between the surfaces, their relative velocity, and others, so that an exact mathematical description of the frictional force

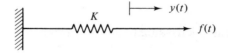

Figure 4-4 Force – spring system.

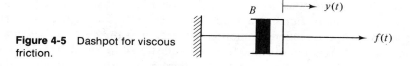

Figure 4-5 Dashpot for viscous friction.

is difficult. However, for practical purposes, frictional forces can be divided into three basic categories: viscous friction, static friction, and Coulomb friction. These are discussed separately in detail next.

1. *Viscous friction.* Viscous friction represents a retarding force that is a linear relationship between the applied force and velocity. The schematic diagram element for friction is often represented by a dashpot such as that shown in Fig. 4-5. The mathematical expression of viscous friction is

$$f(t) = B\frac{dy(t)}{dt} \tag{4-13}$$

where B is the viscous frictional coefficient.

The dimensions of B in the two unit systems are as follows:

Units	Frictional Coefficient B
SI	N/m/sec
British	lb/ft/sec

Figure 4-6(a) shows the functional relation between the viscous frictional force and velocity.

2. *Static friction.* Static friction represents a retarding force that tends to prevent motion from beginning. The static frictional force can be represented by the expression

$$f(t) = \pm(F_s)_{\dot{y}=0} \tag{4-14}$$

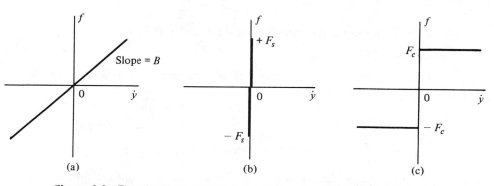

Figure 4-6 Functional relationships of linear and nonlinear frictional forces. (a) Viscous friction. (b) Static friction. (c) Coulomb friction.

where $(F_s)_{\dot{y}=0}$ is defined as the static frictional force that exists only when the body is stationary but has a tendency of moving. The sign of the friction depends on the direction of motion or the initial direction of velocity. The force–velocity relation of static friction is illustrated in Fig. 4-6(b). Notice that once motion begins, the static frictional force vanishes, and other frictions take over.

3. *Coulomb friction*. Coulomb friction is a retarding force that has a constant amplitude with respect to the change in velocity, but the sign of the frictional force changes with the reversal of the direction of velocity. The mathematical relation for the Coulomb friction is given by

$$f(t) = F_c \left(\frac{dy}{dt} \middle/ \left| \frac{dy}{dt} \right| \right) \tag{4-15}$$

where F_c is the Coulomb friction coefficient. The functional description of the friction to velocity relation is shown in Fig. 4-6(c).

Rotational Motion

The rotational motion of a body may be defined as motion about a fixed axis. The variables generally used to describe the motion of rotation are torque; angular acceleration, α; angular velocity, ω; and angular displacement, θ. The following elements are usually involved with the rotational motion.

Inertia. Inertia, J, is considered as an indication of the property of an element which stores the kinetic energy of rotational motion. The inertia of a given element depends on the geometric composition about the axis of rotation and its density.

For instance, the inertia of a circular disk or a circular shaft about its geometric axis is given by

$$J = \tfrac{1}{2} M r^2 \tag{4-16}$$

where M is the mass of the disk or shaft and r is its radius.

Example 4-2

Given a disk that is 1 in. in radius, 0.25 in. thick, and weighing 5 oz, its inertia is

$$J = \frac{1}{2} \frac{W r^2}{g} = \frac{1}{2} \frac{(5 \text{ oz})(1 \text{ in.})^2}{386 \text{ in./sec}^2} \tag{4-17}$$

$$= 0.00647 \text{ oz-in.-sec}^2$$

Usually, the density of the material is given in weight per unit volume. Then for a circular disk or shaft it can be shown that the inertia is proportional to the fourth power of the radius and the first power of the thickness or length. Therefore, if the weight W is

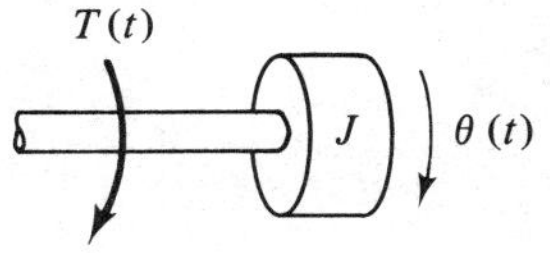

Figure 4-7 Torque – inertia system.

expressed as

$$W = \rho(\pi r^2 h) \qquad (4\text{-}18)$$

where ρ is the density in weight per unit volume, r the radius, and h the thickness or length, Eq. (4-17) is written

$$J = \frac{1}{2} \frac{\rho \pi h r^4}{g} = 0.00406 \rho h r^4 \qquad (4\text{-}19)$$

where h and r are in inches.

For steel, ρ is 4.53 oz/in.^{3}; Eq. (4-19) becomes

$$J = 0.0184 h r^4 \qquad (4\text{-}20)$$

For aluminum, ρ is 1.56 oz/in.^{3}; Eq. (4-19) becomes

$$J = 0.00636 h r^4 \qquad (4\text{-}21)$$

■

When a torque is applied to a body with inertia J, as shown in Fig. 4-7, the torque equation is written

$$T(t) = J\alpha(t) = J\frac{d\omega(t)}{dt} = J\frac{d^2\theta(t)}{dt^2} \qquad (4\text{-}22)$$

The SI and British units for the quantities in Eq. (4-22) are tabulated as follows:

Units	Inertia	Torque	Angular Displacement
SI	kg-m^2	N-m	radian
		dyne-cm	radian
British	slug-ft^2 or lb-ft-sec^2	lb-ft	
	oz-in.-sec^2	oz-in.	radian

The following conversion factors are often found useful:

Angular Displacement

$$1 \text{ rad} = \frac{180}{\pi} = 57.3°$$

Angular Velocity

$$1 \text{ rpm} = \frac{2\pi}{60} = 0.1047 \text{ rad/sec}$$

$$1 \text{ rpm} = 6 \text{ deg/sec}$$

Torque

$$1 \text{ g-cm} = 0.0139 \text{ oz-in.}$$
$$1 \text{ lb-ft} = 192 \text{ oz-in.}$$
$$1 \text{ oz-in.} = 0.00521 \text{ lb-ft}$$

Inertia

$$1 \text{ g-cm}^2 = 1.417 \times 10^{-5} \text{ oz-in.-sec}^2$$
$$1 \text{ lb-ft-sec}^2 = 192 \text{ oz-in.-sec}^2$$
$$1 \text{ oz-in.-sec}^2 = 386 \text{ oz-in.}^2$$
$$1 \text{ g-cm-sec}^2 = 980 \text{ g-cm}^2$$
$$1 \text{ lb-ft-sec}^2 = 32.2 \text{ lb-ft}^2$$

Torsional Spring. As with the linear spring for translational motion, a torsional spring constant K, in torque per unit angular displacement, can be devised to represent the compliance of a rod or a shaft when it is subject to an applied torque. Figure 4-8 illustrates a simple torque–spring system that can be represented by the equation

$$T(t) = K\theta(t) \tag{4-23}$$

The units of K in the SI and British systems are as follows:

Units	Spring Constant K
SI	N-m/rad
British	ft-lb/rad

If the torsional spring is preloaded by a preload torque of TP, Eq. (4-23) is modified to

$$T(t) - TP = K\theta(t) \tag{4-24}$$

Friction for Rotational Motion. The three types of friction described for translational motion can be carried over to the motion of rotation. Therefore, Eqs.

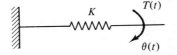

Figure 4-8 Torque – torsional spring system.

(4-13), (4-14), and (4-15) can be replaced, respectively, by their counterparts:

$$T(t) = B \frac{d\theta(t)}{dt} \tag{4-25}$$

$$T(t) = \pm(F_s)_{\theta=0} \tag{4-26}$$

$$T(t) = F_c \left(\frac{d\theta}{dt} \middle/ \left| \frac{d\theta}{dt} \right| \right) \tag{4-27}$$

where B is the viscous frictional coefficient in torque per unit angular velocity, $(F_s)_{\theta=0}$ is the static friction, and F_c is the Coulomb friction torque.

Relation Between Translational and Rotational Motions

In motion control problems it is often necessary to convert rotational motion into a translational one. For instance, a load may be controlled to move along a straight line through a rotary motor and screw assembly, such as that shown in Fig. 4-9. Figure 4-10 shows a similar situation in which a rack and pinion is used as the mechanical linkage. Another common system in motion control is the control of a mass through a pulley by a rotary prime mover, such as that shown in Fig. 4-11. The systems shown in Figs. 4-9, 4-10, and 4-11 can all be represented by a simple system with an equivalent inertia connected directly to the drive motor. For instance, the

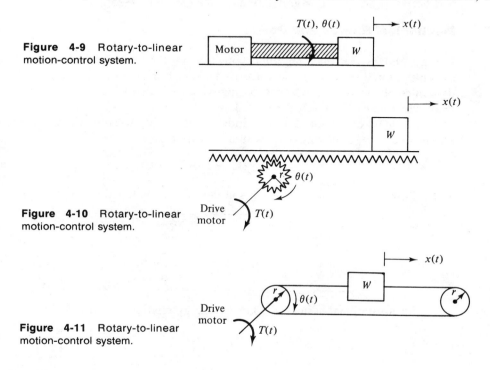

Figure 4-9 Rotary-to-linear motion-control system.

Figure 4-10 Rotary-to-linear motion-control system.

Figure 4-11 Rotary-to-linear motion-control system.

mass in Fig. 4-11 can be regarded as a point mass which moves about the pulley, which has a radius r. Disregarding the inertia of the pulley, the equivalent inertia that the motor sees is

$$J = Mr^2 = \frac{W}{g} r^2 \tag{4-28}$$

If the radius of the pinion in Fig. 4-10 is r, the equivalent inertia that the motor sees is also given by Eq. (4-28).

Now consider the system of Fig. 4-9. The lead of the screw, L, is defined as the linear distance which the mass travels per revolution of the screw. In principle, the two systems in Fig. 4-10 and 4-11 are equivalent. In Fig. 4-10, the distance traveled by the mass per revolution of the pinion is $2\pi r$. Therefore, using Eq. (4-28) as the equivalent inertia for the system of Fig. 4-9,

$$J = \frac{W}{g} \left(\frac{L}{2\pi} \right)^2 \tag{4-29}$$

where in the British system

$$J = \text{inertia } (\text{oz-in.-sec}^2)$$
$$W = \text{weight } (\text{oz})$$
$$L = \text{screw lead } (\text{in.})$$
$$g = \text{gravitational force } (386.4 \text{ in./sec}^2)$$

Mechanical Energy and Power

Energy and power play an important role in the design of electromechanical systems. Stored energy in the form of kinetic and potential energy controls the dynamics of the system, whereas dissipative energy usually is spent in the form of heat, which must be closely controlled.

The mass or inertia of a body indicates its ability to store kinetic energy. The kinetic energy of a moving mass with a velocity v is

$$W_k = \tfrac{1}{2} M v^2 \tag{4-30}$$

The following consistent sets of units are given for the kinetic energy relation:

Units	Energy	Mass	Velocity
SI	joule or N-m	$N/m/sec^2$	m/sec
British	ft-lb	$lb/ft/sec^2$ (slug)	ft/sec

For a rotational system, the kinetic energy relation is written

$$W_k = \tfrac{1}{2} J \omega^2 \tag{4-31}$$

where J is the moment of inertia and ω the angular velocity. The following units are given for a rotational kinetic energy:

Units	Energy	Inertia	Angular Velocity
SI	joule or N-m	kg-m^2	rad/sec
British	oz-in.	oz-in.-sec^2	rad/sec

Potential energy stored in a mechanical element represents the amount of work required to change the configuration. For a linear spring that is deformed by y in length, the potential energy stored in the spring is

$$W_p = \tfrac{1}{2}Ky^2 \tag{4-32}$$

where K is the spring constant. For a torsional spring, the potential energy stored is given by

$$W_p = \tfrac{1}{2}K\theta^2 \tag{4-33}$$

When dealing with a frictional element, the form of energy differs from the previous two cases in that the energy represents a loss or dissipation by the system in overcoming the frictional force. Power is the time rate of doing work. Therefore, the power dissipated in a frictional element is the product of force and velocity; that is,

$$P = fv \tag{4-34}$$

Since $f = Bv$, where B is the frictional coefficient, Eq. (4-34) becomes

$$P = Bv^2 \tag{4-35}$$

The SI unit for power is in newton-m/sec or watts (W).

In the British unit system, power is represented in ft-lb/sec or horsepower (hp). Furthermore,

$$\begin{aligned} 1 \text{ hp} &= 746 \text{ W} \\ &= 550 \text{ ft-lb/sec} \end{aligned} \tag{4-36}$$

Since power is the rate at which energy is being dissipated, the energy dissipated in a frictional element is

$$W_d = B \int v^2 \, dt \tag{4-37}$$

Gear Trains, Levers, and Timing Belts

A gear train, a lever, or a timing belt over pulleys is a mechanical device that transmits energy from one part of a system to another in such a way that force, torque, speed, and displacement are altered. These devices may also be regarded as

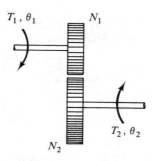

T_1, θ_1 N_1

T_2, θ_2

N_2 **Figure 4-12**   Gear train.

matching devices used to attain maximum power transfer. Two gears are shown coupled together in Fig. 4-12. The inertia and friction of the gears are neglected in the ideal case considered.

The relationships between the torques T_1 and T_2, angular displacements θ_1 and θ_2, and the teeth numbers N_1 and N_2, of the gear train are derived from the following facts:

1. The number of teeth on the surface of the gears is proportional to the radii r_1 and r_2 of the gears; that is,

$$r_1 N_2 = r_2 N_1 \tag{4-38}$$

2. The distance traveled along the surface of each gear is the same. Therefore,

$$\theta_1 r_1 = \theta_2 r_2 \tag{4-39}$$

3. The work done by one gear is equal to that of the other since there is assumed to be no loss. Thus,

$$T_1 \theta_1 = T_2 \theta_2 \tag{4-40}$$

If the angular velocities of the two gears, ω_1 and ω_2, are brought into the picture, Eqs. (4-38) through (4-40) lead to

$$\frac{T_1}{T_2} = \frac{\theta_2}{\theta_1} = \frac{N_1}{N_2} = \frac{\omega_2}{\omega_1} = \frac{r_1}{r_2} \tag{4-41}$$

In practice, real gears do have inertia and friction between the coupled gear teeth which often cannot be neglected. An equivalent representation of a gear train

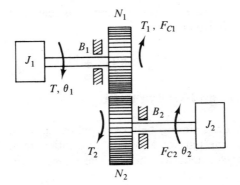

N_1

T_1, F_{C1}

B_1

J_1

T, θ_1

B_2

J_2

T_2

$F_{C2}\ \theta_2$

N_2

Figure 4-13 Gear train with friction and inertia.

with viscous friction, Coulomb friction, and inertia considered as lumped elements is shown in Fig. 4-13. The following variables and parameters are defined for the gear train:

$$T = \text{applied torque}$$
$$\theta_1, \theta_2 = \text{angular displacements}$$
$$T_1, T_2 = \text{torque transmitted to gears}$$
$$J_1, J_2 = \text{inertia of gears}$$
$$N_1, N_2 = \text{number of teeth}$$
$$F_{c1}, F_{c2} = \text{Coulomb friction coefficients}$$
$$B_1, B_2 = \text{viscous frictional coefficients}$$

The torque equation for gear 2 is written

$$T_2(t) = J_2 \frac{d^2\theta_2(t)}{dt^2} + B_2 \frac{d\theta_2(t)}{dt} + F_{c2}\frac{\dot{\theta}_2}{|\dot{\theta}_2|} \tag{4-42}$$

The torque equation on the side of gear 1 is

$$T(t) = J_1 \frac{d^2\theta_1(t)}{dt^2} + B_1 \frac{d\theta_1(t)}{dt} + F_{c1}\frac{\dot{\theta}_1}{|\dot{\theta}_1|} + T_1(t) \tag{4-43}$$

By the use of Eq. (4-41), Eq. (4-42) is converted to

$$T_1(t) = \frac{N_1}{N_2}T_2(t) = \left(\frac{N_1}{N_2}\right)^2 J_2 \frac{d^2\theta_1(t)}{dt^2} + \left(\frac{N_1}{N_2}\right)^2 B_2 \frac{d\theta_1(t)}{dt} + \frac{N_1}{N_2}F_{c2}\frac{\dot{\theta}_2}{|\dot{\theta}_2|} \tag{4-44}$$

Equation (4-44) indicates that it is possible to reflect inertia, friction (and compliance), torque, speed, and displacement from one side of a gear train to the other.

Therefore, the following quantities are obtained when reflecting from gear 2 to gear 1:

$$\text{Inertia: } \left(\frac{N_1}{N_2}\right)^2 J_2$$

$$\text{Viscous frictional coefficient: } \left(\frac{N_1}{N_2}\right)^2 B_2$$

$$\text{Torque: } \frac{N_1}{N_2}T_2$$

$$\text{Angular displacement: } \frac{N_2}{N_1}\theta_2$$

$$\text{Angular velocity: } \frac{N_2}{N_1}\omega_2$$

$$\text{Coulomb frictional torque: } \frac{N_1}{N_2}F_{c2}\frac{\omega_2}{|\omega_2|}$$

If torsional spring effect were present, the spring constant is also multiplied by $(N_1/N_2)^2$ in reflecting from gear 2 to gear 1. Now, substituting Eq. (4-44) into Eq. (4-43), we get

$$T(t) = J_{1e}\frac{d^2\theta_1(t)}{dt^2} + B_{1e}\frac{d\theta_1(t)}{dt} + T_F \qquad (4\text{-}45)$$

where

$$J_{1e} = J_1 + \left(\frac{N_1}{N_2}\right)^2 J_2 \qquad (4\text{-}46)$$

$$B_{1e} = B_1 + \left(\frac{N_1}{N_2}\right)^2 B_2 \qquad (4\text{-}47)$$

$$T_F = F_{c1}\frac{\dot{\theta}_1}{|\dot{\theta}_1|} + \frac{N_1}{N_2}F_{c2}\frac{\dot{\theta}_2}{|\dot{\theta}_2|} \qquad (4\text{-}48)$$

■

Example 4-3

Given a load that has inertia of 0.05 oz-in.-sec² and a Coulomb friction torque of 2 oz-in., find the inertia and frictional torque reflected through a 1 : 5 gear train ($N_1/N_2 = \frac{1}{5}$ with N_2 on the load side). The reflected inertia on the side of N_1 is $(\frac{1}{5})^2 \times 0.05 = 0.002$ oz-in.-sec². The reflected Coulomb friction is $(\frac{1}{5})2 = 0.4$ oz-in.

■

Timing belts and chain drives serve the same purposes as the gear train except that they allow the transfer of energy over a longer distance without using an excessive number of gears. Figure 4-14 shows the diagram of a belt or chain drive between two pulleys. Assuming that there is no slippage between the belt and the pulleys, it is easy to see that Eq. (4-41) still applies to this case. In fact, the reflection or transmittance of torque, inertia, friction, and so on, is similar to that of a gear train.

The lever system shown in Fig. 4-15 transmits translational motion and force in the same way that gear trains transmit rotational motion. The relation between the forces and distances is

$$\frac{f_1}{f_2} = \frac{l_2}{l_1} = \frac{x_2}{x_1} \qquad (4\text{-}49)$$

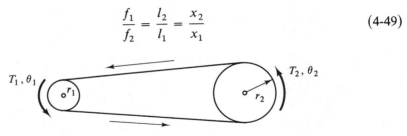

Figure 4-14 Belt or chain drive.

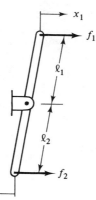

Figure 4-15 Lever system.

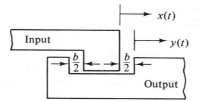

Figure 4-16 Physical model of backlash between two mechanical elements.

Backlash and Dead Zone

Backlash and dead zone usually play an important role in gear trains and similar mechanical linkages. In a great majority of situations, backlash may give rise to undesirable oscillations and instability in control systems. In addition, it has a tendency to wear down the mechanical elements. Regardless of the actual mechanical elements, a physical model of backlash or dead zone between an input and an output member is shown in Fig. 4-16. The model can be used for a rotational system as well as for a translational system. The amount of backlash is $b/2$ on either side of the reference position.

In general, the dynamics of the mechanical linkage with backlash depend upon the relative inertia-to-friction ratio of the output member. If the inertia of the output member is very small compared with that of the input member, the motion is

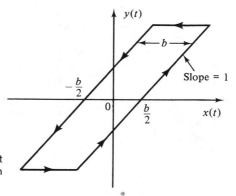

Figure 4-17 Input – output characteristic of backlash with negligible output inertia.

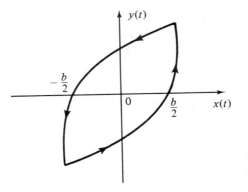

Figure 4-18 Input – output displacement characteristic of a backlash element without friction.

controlled predominantly by friction. This means that the output member will not coast whenever there is no contact between the two members. When the output is driven by the input, the two members will travel together until the input member reverses its direction; then the output member will stand still until the backlash is taken up on the other side, at which time it is assumed that the output member instantaneously takes on the velocity of the input member. The transfer characteristic between the input and the output displacements of a backlash element with negligible output inertia is shown in Fig. 4-17.

The transfer characteristic between the input and the output displacement of a backlash element with negligible output friction is shown in Fig. 4-18.

4.4 EQUATIONS OF MECHANICAL SYSTEMS [3]

The equations of a linear mechanical system are written by first constructing a model of the system containing interconnected linear elements, and then the system equations are written by applying Newton's law of motion to the free-body diagram.

■

Example 4-4

Let us consider the mechanical system shown in Fig. 4-19(a). The free-body diagram of the system is shown in Fig. 4-19(b). The force equation of the system is written

$$f(t) = M\frac{d^2y(t)}{dt^2} + B\frac{dy(t)}{dt} + Ky(t) \tag{4-50}$$

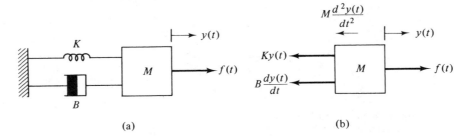

(a)   (b)

Figure 4-19   (a) Mass – spring – friction system. (b) Free-body diagram.

This second-order differential equation can be decomposed into two first-order state equations. Let us assign $x_1 = y$ and $x_2 = dy/dt$ as the state variables. Then Eq. (4-50) is written

$$\frac{dx_1(t)}{dt} = x_2(t) \tag{4-51}$$

$$\frac{dx_2(t)}{dt} = -\frac{K}{M}x_1(t) - \frac{B}{M}x_2(t) + \frac{1}{M}f(t) \tag{4-52}$$

It is not difficult to see that this mechanical system is analogous to a series RLC electric network. With this analogy it is simple to formulate the state equations directly from the mechanical system using a different set of state variables. If we consider that mass is analogous to inductance, and the spring constant K is analogous to the inverse of capacitance, $1/C$, it is logical to assign $v(t)$, the velocity, and $f_k(t)$, the force acting on the spring, as state variables, since the former is analogous to the current in an inductor and the latter is analogous to the voltage across a capacitor.

Then the state equations of the system can be written from the following equations:

$$\text{Force on mass: } M\frac{dv(t)}{dt} = -Bv(t) - f_k(t) + f(t) \tag{4-53}$$

$$\text{Velocity of spring: } \frac{1}{K}\frac{df_k(t)}{dt} = v(t) \tag{4-54}$$

Notice that the first state equation is similar to writing the equation on the voltage across an inductor; the second is like that of the current through a capacitor.

This simple example illustrates the fact that the state equations and state variables of a dynamic system are not unique.

Example 4-5

As a second example of writing equations for mechanical systems, consider the system shown in Fig. 4-20(a). Since the spring is deformed when it is subjected to the force $f(t)$, two displacements, y_1 and y_2, must be assigned to the end points of the spring. The free-body diagrams of the system are given in Fig. 4-20(b). From these free-body diagrams the force equations of the system are written

$$f(t) = K[y_1(t) - y_2(t)] \tag{4-55}$$

$$K[y_1(t) - y_2(t)] = M\frac{d^2y_2(t)}{dt^2} + B\frac{dy_2(t)}{dt} \tag{4-56}$$

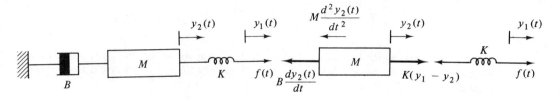

(a)   (b)

Figure 4-20 Mechanical system for Example 4-5. (a) Mass – spring – friction system. (b) Free-body diagrams.

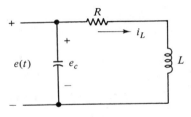

Figure 4-21 Electric network analogous to the mechanical system of Fig. 4-20.

Now let us write the state equations of the system. Since the differential equation of the system is already available in Eq. (4-56), the most direct way is to decompose this equation into two first-order differential equations.

Therefore, letting $x_1(t) = y_2(t)$ and $x_2(t) = dy_2(t)/dt$, Eqs. (4-55) and (4-56) give

$$\frac{dx_1(t)}{dt} = x_2(t) \tag{4-57}$$

$$\frac{dx_2(t)}{dt} = -\frac{B}{M}x_2(t) + \frac{1}{M}f(t) \tag{4-58}$$

As an alternative, we can assign the velocity $v(t)$ of the body with mass M as one state variable, and the force $f_k(t)$ on the spring as the other state variable, so we have

$$\frac{dv(t)}{dt} = -\frac{B}{M}v(t) + \frac{1}{M}f_k(t) \tag{4-59}$$

and

$$f_k(t) = f(t) \tag{4-60}$$

One may wonder at this point if the two equations in Eqs. (4-59) and (4-60) are correct as state equations, since it seems that only Eq. (4-59) is a state equation, but we do have two state variables in $v(t)$ and $f_k(t)$. Why do we need only one state equation here, whereas Eqs. (4-57) and (4-58) clearly are two independent state equations? The situation is better explained (at least for electrical engineers) by referring to the analogous electric network of the system, shown in Fig. 4-21. It is clear that although the network has two reactive elements in L and C and thus there should be two state variables, the capacitance in this case is a "redundant" element, since $e_c(t)$ is equal to the applied voltage $e(t)$. However, the equations in Eqs. (4-59) and (4-60) can provide only the solution to the velocity of M once $f(t)$ is specified. If we need to find the displacement $y_1(t)$ at the point where $f(t)$ is applied, we have to use the relation

$$y_1(t) = \frac{f_k(t)}{K} + y_2(t) = \frac{f(t)}{K} + \int_0^t v(\tau)\, d\tau + y_2(0) \tag{4-61}$$

where $y_2(0)$ is the initial displacement of the body with mass M. On the other hand, we can solve for $y_2(t)$ from the two state equations of Eqs. (4-57) and (4-58), and then $y_1(t)$ is determined from Eq. (4-55).

Example 4-6

In this example the equations for the mechanical system in Fig. 4-22(a) are to be written. Then we are to draw state diagrams and derive transfer functions for the system.

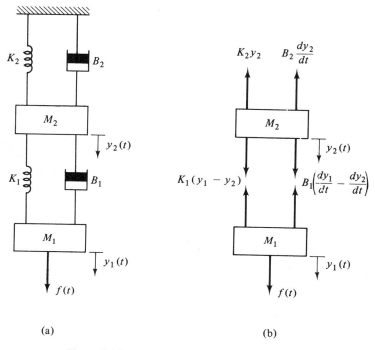

(a)   (b)

Figure 4-22 Mechanical system for Example 4-6.

The free-body diagrams for the two masses are shown in Fig. 4-22(b), with the reference directions of the displacements y_1 and y_2 as indicated. The Newton's force equations for the system are written directly from the free-body diagram:

$$f(t) = M_1 \frac{d^2 y_1(t)}{dt^2} + B_1 \left[\frac{dy_1(t)}{dt} - \frac{dy_2(t)}{dt} \right] + K_1 [y_1(t) - y_2(t)] \qquad (4\text{-}62)$$

$$0 = -B_1 \left[\frac{dy_1(t)}{dt} - \frac{dy_2(t)}{dt} \right] - K_1 [y_1(t) - y_2(t)]$$

$$+ M_2 \frac{d^2 y_2(t)}{dt^2} + B_2 \frac{dy_2(t)}{dt} + K_2 y_2(t) \qquad (4\text{-}63)$$

We may now decompose these two second-order simultaneous differential equations into four state equations by defining the following state variables:

$$x_1 = y_1 \qquad (4\text{-}64)$$

$$x_2 = \frac{dy_1}{dt} = \frac{dx_1}{dt} \qquad (4\text{-}65)$$

$$x_3 = y_2 \qquad (4\text{-}66)$$

$$x_4 = \frac{dy_2}{dt} = \frac{dx_3}{dt} \qquad (4\text{-}67)$$

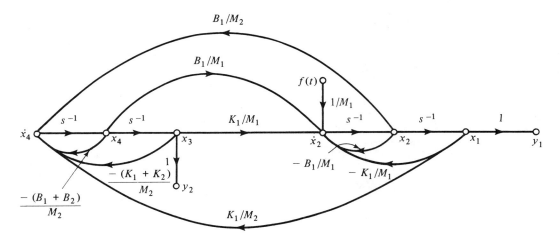

Figure 4-23   State diagram for the mechanical system of Fig. 4-22.

Equations (4-65) and (4-67) form two state equations naturally; the other two are obtained by substituting Eqs. (4-64) through (4-67) into Eqs. (4-62) and (4-63), and rearranging, we have

$$\frac{dx_1}{dt} = x_2 \tag{4-68}$$

$$\frac{dx_2}{dt} = -\frac{K_1}{M_1}(x_1 - x_3) - \frac{B_1}{M_1}(x_2 - x_4) + \frac{1}{M_1}f(t) \tag{4-69}$$

$$\frac{dx_2}{dt} = x_4 \tag{4-70}$$

$$\frac{dx_4}{dt} = \frac{K_1}{M_2}x_1 + \frac{B_1}{M_2}x_2 - \frac{K_1 + K_2}{M_2}x_3 - \frac{1}{M_2}(B_1 + B_2)x_4 \tag{4-71}$$

If we are interested in the displacements y_1 and y_2, the output equations are written

$$y_1(t) = x_1(t) \tag{4-72}$$

$$y_2(t) = x_3(t) \tag{4-73}$$

The state diagram of the system, according to the equations written above, is drawn as shown in Fig. 4-23. The transfer functions $Y_1(s)/F(s)$ and $Y_2(s)/F(s)$ are obtained from the state diagram by applying Mason's gain formula. The reader should verify the following results (make sure that all the loops and nontouching loops are taken into account):

$$\frac{Y_1(s)}{F(s)} = \frac{M_2 s^2 + (B_1 + B_2)s + (K_1 + K_2)}{\Delta} \tag{4-74}$$

$$\frac{Y_2(s)}{F(s)} = \frac{B_1 s + K_1}{\Delta} \tag{4-75}$$

where

$$\Delta = M_1 M_2 s^4 + \left[M_1 (B_1 + B_2) + B_1 M_2 \right] s^3 + \left[M_1 (K_1 + K_2) + K_1 M_2 + B_1 B_2 \right] s^2$$
$$+ (K_1 B_2 + B_1 K_2) s + K_1 K_2 \tag{4-76}$$

The state equations can also be written directly from the diagram of the mechanical system. The state variables are assigned as $v_1 = dy_1/dt$, $v_2 = dy_2/dt$, and the forces on the two springs, f_{K1} and f_{K2}. Then, if we write the forces acting on the masses and the velocities of the springs as functions of the four state variables and the external force, the state equations are:

$$\text{Force on } M_1: \ M_1 \frac{dv_1}{dt} = -B_1 v_1 + B_1 v_2 - f_{K1} + f \tag{4-77}$$

$$\text{Force on } M_2: \ M_2 \frac{dv_2}{dt} = B_1 v_1 - (B_1 + B_2) v_2 + f_{K1} - f_{K2} \tag{4-78}$$

$$\text{Velocity on } K_1: \ \frac{df_{K1}}{dt} = K_1 (v_1 - v_2) \tag{4-79}$$

$$\text{Velocity on } K_2: \ \frac{df_{K2}}{dt} = K_2 v_2 \tag{4-80}$$

Example 4-7

The rotational system shown in Fig. 4-24(a) consists of a disk mounted on a shaft that is fixed at one end. The moment of inertia of the disk about its axis is J. The edge of the disk is riding on a surface, and the viscous friction coefficient between the two surfaces is B. The inertia of the shaft is negligible, but the stiffness is K.

Assume that a torque is applied to the disk as shown; then the torque or moment equation about the axis of the shaft is written from the free-body diagram of Fig. 4-24(b):

$$T(t) = J \frac{d^2\theta(t)}{dt^2} + B \frac{d\theta(t)}{dt} + K\theta(t) \tag{4-81}$$

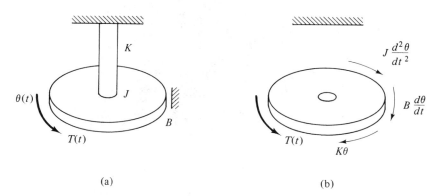

(a)   (b)

Figure 4-24 Rotational system for Example 4-7.

Notice that this system is analogous to the translational system of Fig. 4-19. The state equations may be written by defining the state variables as $x_1(t) = \theta(t)$ and $dx_1(t)/dt = x_2(t)$. The reader may carry out the next step of writing the state equations as an exercise.

■

4.5 SENSORS AND ENCODERS IN CONTROL SYSTEMS

Sensors and encoders are important components in feedback control systems. In open-loop control systems encoders are often used to monitor the performance of the system. In closed-loop control systems, sensors and encoders are used to feedback signals for control purposes. These components are also used for the identification of an unknown or changing process.

In this section some of the sensors and encoders that are commonly used in control systems are described; their principle of operation and applications are discussed. These devices include potentiometers, optical encoders, and electromagnetic transducers. In systems that require extremely high-precision measurements and detections, laser and optical devices are used. These subjects are not explored in this text.

Potentiometer

A potentiometer is an electromechanical transducer that converts mechanical energy into electrical energy. The input to the device is in the form of a mechanical displacement, either linear or rotational. When a voltage is applied across the fixed

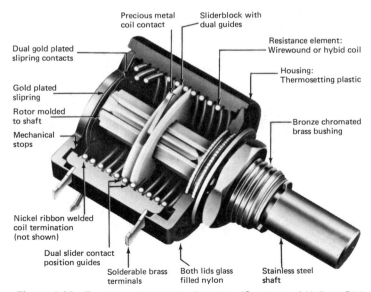

Figure 4-25 Ten-turn rotary potentiometer. (Courtesy of Helipot Division of Beckman Instruments, Inc.)

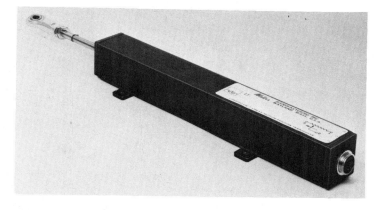

Figure 4-26 Linear motion potentiometer with built-in operational amplifier. (Courtesy of Waters Manufacturing, Inc.)

terminals of the potentiometer, the output voltage, which is measured across the variable terminals, is proportional to the input displacement either linearly or according to some nonlinear relation.

Rotary potentiometers are available commercially in single-revolution or multi-revolution form. Some of the potentiometers have limited motion, such as one or more revolutions, and some have unlimited rotational motion. The potentiometers commonly are made with wire-wound or conductive plastic resistance elements. Figure 4-25 shows the photo of the cutaway view of a rotary potentiometer, and Fig. 4-26 shows the photo of a linear potentiometer which also contains a built-in operational amplifier. For precision control, the conductive plastic potentiometer is more preferable, since it has infinite resolution, long rotational life, good output smoothness, and low static noise. Typical resistance tolerance of a standard potentiometer is between ± 5 and ± 10 percent, and special high-quality units can have a tolerance of ± 1 percent. The linearity of the resistance is in the range of ± 0.25 to ± 0.5 percent.

Figure 4-27 shows the equivalent circuit representation of a potentiometer, linear or rotary. Since the voltage across the variable terminal and reference is proportional to the shaft displacement of the potentiometer, when a voltage is applied across the fixed terminals, the device can be used to indicate the absolute position of a system or the relative position of two mechanical outputs.

Figure 4-28(a) shows the arrangement when the housing of the potentiometer is fixed at reference; the output voltage $e(t)$ will be proportional to the shaft position $\theta_c(t)$, in the case of a rotary motion. Then

$$e(t) = K_s \theta_c(t) \tag{4-82}$$

Figure 4-27 Circuit representation of potentiometer.

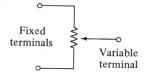

Fixed terminals

Variable terminal

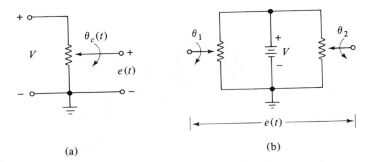

(a)   (b)

Figure 4-28 (a) Potentiometer used as position indicator. (b) Two
potentiometers used as position sensing of two shafts.

where K_s is the proportional constant. For an N-turn potentiometer the total
displacement of the variable arm is $2\pi N$ radians. The proportional constant K_s is
given by

$$K_s = \frac{V}{2\pi N} \text{ V/rad} \qquad (4\text{-}83)$$

where V is the magnitude of the reference voltage applied to the fixed terminals. A
more flexible arrangement is obtained by using two potentiometers connected in
parallel, as shown in Fig. 4-28(b). This arrangement allows the comparison of two
remotely located shaft positions. The output voltage is taken across the variable
terminals of the two potentiometers and is given by

$$e(t) = K_s[\theta_1(t) - \theta_2(t)] \qquad (4\text{-}84)$$

Figure 4-29 illustrates the block diagram representations of the setups in Fig. 4-28.
    In dc motor control systems potentiometers are sometimes used for position
feedback. Figure 4-30(a) shows the schematic diagram of a typical dc motor
position-control system. The potentiometers are used in the feedback path to
compare the actual load position with the desired reference position. If there is a
discrepancy between the load position and the reference input, an error signal is
generated by the potentiometers which will drive the motor in such a way that this
error is minimized quickly. As shown in Fig. 4-30(a), the error signal is amplified by
a dc amplifier whose output drives the armature of a permanent-magnet dc motor.
Typical waveforms of the signals in the system when the input $\theta_r(t)$ is a step

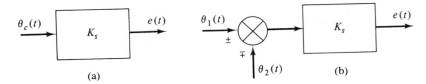

(a)   (b)

Figure 4-29 Block diagram representations of potentiometer arrangements
in Fig. 4-28.

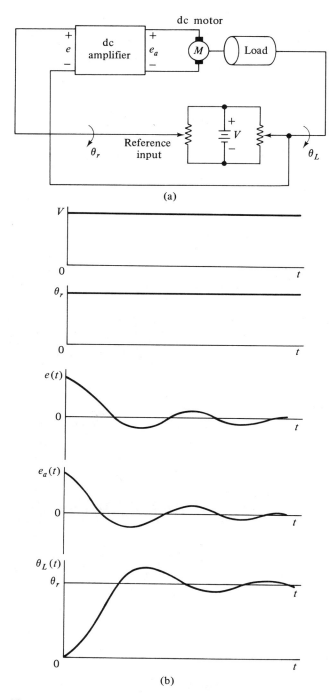

(a)

(b)

Figure 4-30 (a) DC motor position-control system with potentiometers as error sensor. (b) Typical waveforms of signals in
the control system of (a).

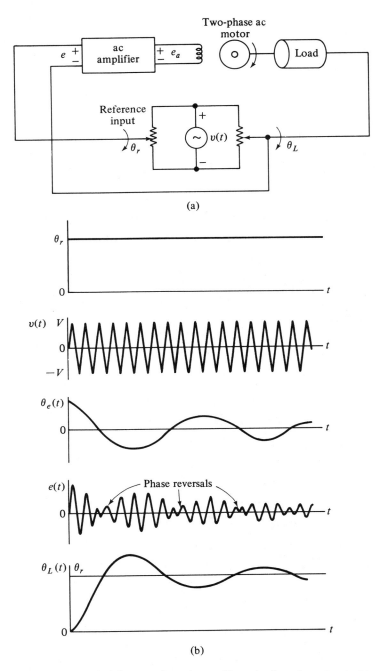

Figure 4-31 (a) AC control system with potentiometers as error detector. (b) Typical waveforms of signals in the control system of (a).

function are shown in Fig. 4-30(b). Note that the electric signals are all unmodulated. In control systems terminology, a dc signal usually refers to an unmodulated signal. On the other hand, an ac signal in control systems refers to signals that are modulated by a modulation process. These definitions are different from those commonly used in electrical engineering, where dc simply refers to unidirectional and ac indicates alternating.

Figure 4-31(a) illustrates a control system which serves essentially the same purpose as that of the system in Fig. 4-30(a), except that ac signals prevail. In this case the voltage applied to the error detector is sinusoidal. The frequency of this signal is usually much higher than that of the signal that is being transmitted through the system. Typical signals of the ac control system are shown in Fig. 4-31(b). The signal $v(t)$ is referred to as the carrier signals whose frequency is ω_c, or

$$v(t) = V \sin \omega_c t \tag{4-85}$$

Analytically, the output of the error sensor is given by

$$e(t) = K_s \theta_e(t) v(t) \tag{4-86}$$

where $\theta_e(t)$ is the difference between the input displacement and the load displacement, $\theta_e(t) = \theta_r(t) - \theta_L(t)$. For the $\theta_e(t)$ shown in Fig. 4-31(b), $e(t)$ becomes a *suppressed-carrier modulated* signal. A reversal in phase of $e(t)$ occurs whenever the signal crosses the zero-magnitude axis. This reversal in phase causes the ac motor to reverse in direction according to the desired sense of correction of the error $\theta_e(t)$. The name "suppressed-carrier modulation" stems from the fact that when a signal $\theta_e(t)$ is modulated by a carrier signal $v(t)$ according to Eq. (4-86), the resultant signal $e(t)$ no longer contains the original carrier frequency ω_c. To illustrate this, let us assume that $\theta_e(t)$ is also a sinusoid given by

$$\theta_e(t) = \sin \omega_s(t) \tag{4-87}$$

where normally, $\omega_s \ll \omega_c$. By use of familiar trigonometric relations, substituting Eqs. (4-84) and (4-85) into Eq. (4-86) we get

$$e(t) = \tfrac{1}{2} K_s V \left[\cos(\omega_c - \omega_s)t - \cos(\omega_c + \omega_s)t \right] \tag{4-88}$$

Therefore, $e(t)$ no longer contains the carrier frequency ω_c or the signal frequency ω_s, but it does have the two sidebands $\omega_c + \omega_s$ and $\omega_c - \omega_s$.

Interestingly enough, when the modulated signal is transmitted through the system, the motor acts as a demodulator, so that the displacement of the load will be of the same form as the dc signal before modulation. This is clearly seen from the waveforms of Fig. 4-31(b).

It should be pointed out that a control system need not contain all-dc or all-ac components. It is quite common to couple a dc component to an ac component through a modulator or an ac device to a dc device through a demodulator. For instance, the dc amplifier of the system in Fig. 4-30(a) may be replaced by an ac amplifier that is preceded by a modulator and followed by a demodulator.

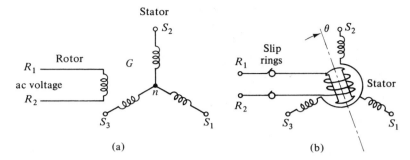

Figure 4-32   Schematic diagrams of a synchro transmitter.

Synchros

Synchros are used widely in control systems as detectors and encoders because of their ruggedness in construction and high reliability. Basically, a synchro is a rotary device that operates on the same principle as a transformer and produces a correlation between an angular position and a voltage or set of voltages. Therefore, synchros are ac devices. There are many types of synchros and the applications vary. In this section only the *synchro transmitter* and the *synchro control transformer* are discussed.

Synchro Transmitter.   A synchro transmitter has a Y-connected stator winding which resembles the stator of a three-phase induction motor. The rotor is a salient-pole, dumbbell-shaped magnet with a single winding. The schematic diagram of a synchro transmitter is shown in Fig. 4-32. A single-phase ac voltage is applied to the rotor through two slip rings. The symbol G is often used to designate a synchro transmitter, which is sometimes also known as a synchro generator.

Let the ac voltage applied to the rotor of a synchro transmitter be

$$e_r(t) = E_r \sin \omega_c t \tag{4-89}$$

When the rotor is at the position of $\theta = 0°$ with reference to Fig. 4-32, which is defined as the *electric zero*, the voltage induced across the stator winding between S_2 and the neutral n is maximum and is written

$$e_{S_2 n}(t) = K E_r \sin \omega t \tag{4-90}$$

where K is a proportional constant. The voltages across the terminals $S_1 n$ and $S_3 n$ are

$$e_{S_1 n}(t) = K E_r \cos 240° \sin \omega t = -0.5 K E_r \sin \omega t \tag{4-91}$$

$$e_{S_3 n}(t) = K E_r \cos 120° \sin \omega t = -0.5 K E_r \sin \omega t \tag{4-92}$$

Then the terminal voltages of the stator are

$$e_{S_1 S_2} = e_{S_1 n} - e_{S_2 n} = -1.5 K E_r \sin \omega t$$

$$e_{S_2 S_3} = e_{S_2 n} - e_{S_3 n} = 1.5 K E_r \sin \omega t \tag{4-93}$$

$$e_{S_3 S_1} = e_{S_3 n} - e_{S_1 n} = 0 \tag{4-94}$$

The foregoing equations show that, despite the similarity between the construction of the stator of a synchro and that of a three-phase machine, there are only single-phase voltages induced in the stator.

Consider now that the rotor of the synchro transmitter is at an angle of θ with reference to the electric zero, as shown in Fig. 4-32. The voltages in each stator winding will vary as a function of the cosine of the rotor displacement θ; that is, the voltage magnitudes are

$$E_{S_1 n} = K E_r \cos(\theta - 240°) \tag{4-95}$$

$$E_{S_2 n} = K E_r \cos \theta \tag{4-96}$$

$$E_{S_3 n} = K E_r \cos(\theta - 120°) \tag{4-97}$$

The magnitudes of the stator terminal voltages become

$$E_{S_1 S_2} = E_{S_1 n} - E_{S_2 n} = \sqrt{3} K E_r \sin(\theta + 240°) \tag{4-98}$$

$$E_{S_2 S_3} = E_{S_2 n} - E_{S_3 n} = \sqrt{3} K E_r \sin(\theta + 120°) \tag{4-99}$$

$$E_{S_3 S_1} = E_{S_3 n} - E_{S_1 n} = \sqrt{3} K E_r \sin \theta \tag{4-100}$$

A plot of these terminal voltages as a function of the rotor shaft position is shown in Fig. 4-33. Notice that each rotor position corresponds to one unique set of

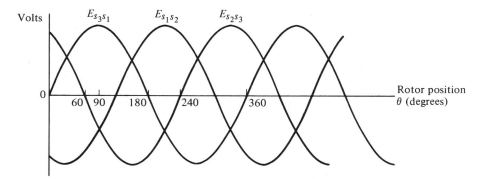

Figure 4-33 Variation of the terminal voltages of a synchro transmitter as a function of the rotor position. θ is measured counterclockwise from the electric zero.

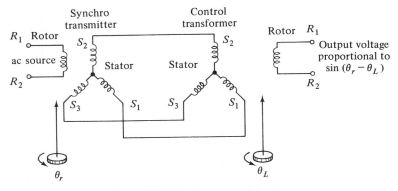

Figure 4-34   Synchro error detector.

stator voltages. This leads to the use of the synchro transmitter to identify angular positions by measuring and identifying the set of voltages at the three stator terminals.

Synchro Control Transformer.   Since the function of an error detector is to convert the difference of two shaft positions into an electrical signal, a single synchro transmitter is apparently inadequate. A typical arrangement of a synchro error detector involves the use of two synchros: a transmitter and a control transformer, as shown in Fig. 4-34.

Basically, the principle of operation of a synchro control transformer is identical to that of the synchro transmitter, except that the rotor is cylindrically shaped so that the air-gap flux is uniformly distributed around the rotor. This feature is essential for a control transformer, since its rotor terminals are usually connected to an amplifier or similar electrical device, in order that the latter sees a constant impedance. The change in the rotor impedance with rotations of the shaft position should be minimized.

The symbol CT is often used to designate a synchro control transformer.

Referring to the arrangement shown in Fig. 4-34, the voltage given by Eqs. (4-98), (4-99), and (4-100) are now impressed across the corresponding stator terminals of the control transformer. When the rotor positions of the two synchros are in perfect alignment, the voltage generated across the terminals of the rotor windings is zero. When the two rotor shafts are not in alignment, the rotor voltage of the CT is approximately a sine function of the difference between the two shaft angles, as shown in Fig. 4-35.

From Fig. 4-35 it is apparent that the synchro error detector is a nonlinear device. However, for small angular deviations of up to 15 degrees in the vicinity of the two null positions, the rotor voltage of the control transformer is approximately proportional to the difference between the positions of the rotors of the transmitter and the control transformer. Therefore, for small deviations, the transfer function of the synchro error detector can be approximated by a constant K_S:

$$K_S = \frac{E}{\theta_r - \theta_L} = \frac{E}{\theta_e} \tag{4-101}$$

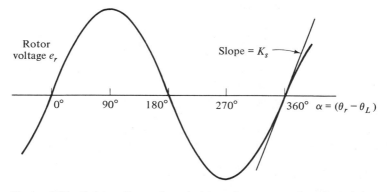

Figure 4-35 Rotor voltage of control transformer as a function of the difference of rotor positions.

where E = error voltage

θ_r = shaft position of synchro transmitter, degrees

θ_L = shaft position of synchro control transformer, degrees

θ_e = error in shaft positions

K_s = sensitivity of the error detector, volts per degree

The schematic diagram of a positional control system employing a synchro error detector is shown in Fig. 4-36. The purpose of the control system is to make the controlled shaft follow the angular displacement of the reference input shaft as closely as possible. The rotor of the control transformer is mechanically connected to the controlled shaft, and the rotor of the synchro transmitter is connected to the reference input shaft. When the controlled shaft is aligned with the reference shaft, the error voltage is zero and the motor does not turn. When an angular misalignment exists, an error voltage of relative polarity appears at the amplifier input, and the output of the amplifier will drive the motor in such a direction as to reduce the

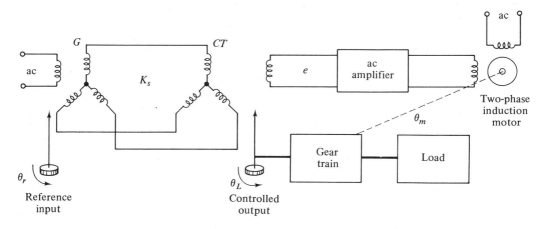

Figure 4-36 Alternating-current control system employing synchro error detector.

error. For small deviations between the controlled and the reference shafts, the synchro error detector can be represented by the constant K_s given by Eq. (4-101).

From the characteristic of the error detector shown in Fig. 4-35, it is clear that K_s has opposite signs at the two null positions. However, in closed-loop systems, only one of the two null positions is a true null; the other one corresponds to an unstable operating point.

Suppose that, in the system shown in Fig. 4-36, the synchro positions are close to the true null and the controlled shaft lags behind the reference shaft; a positive error voltage will cause the motor to turn in the proper direction to correct this lag. But if the synchros are operating close to the false null, for the same lag between θ_r and θ_L, the error voltage is negative and the motor is driven in the direction that will increase the lag. A larger lag in the controlled shaft position will increase the magnitude of the error voltage still further and cause the motor to rotate in the same direction, until the true null position is reached.

In reality, the error signal at the rotor terminals of the synchro control transformer may be represented as a function of time. If the ac signal applied to the rotor terminals of the transmitter is denoted by $\sin \omega_c t$, where ω_c is known as the carrier frequency, the error signal is given by

$$e(t) = K_s \theta_e(t) \sin \omega_c t \qquad (4\text{-}102)$$

Therefore, as explained earlier, $e(t)$ is again a suppressed-carrier modulated signal.

In the preceding two sections, although we have illustrated the potentiometers as being used in a dc (unmodulated) control system, and the synchros in an ac (modulated) control system with an ac motor, in general, the applications of these sensing devices can be quite broad and versatile. For example, the output signal of a synchro control transformer can be demodulated and the dc signal is used to drive a dc motor for control purposes. In modern control systems the outputs of these sensing devices can all be conditioned by interfacing devices and fed into microcomputers for digital control.

Tachometers

Tachometers are electromechanical devices that convert mechanical energy into electrical energy. The device works essentially as a generator with the output voltage proportional to the magnitude of the angular velocity.

In control systems most of the tachometers used are of the dc variety (i.e., the output voltage is a dc signal). Dc tachometers are used in control systems in many ways; they can be used as velocity indicators to provide shaft-speed readout or to provide velocity feedback for speed control or stabilization. Figure 4-37 shows the block diagram of a typical velocity-control system in which the tachometer output is compared with the reference voltage, which represents the desired speed. The difference between the two signals, or the error, is amplified and used to drive the motor, so that the speed will eventually reach the desired value. In this type of application the accuracy of the tachometer is highly critical, as the accuracy of the speed control depends on it.

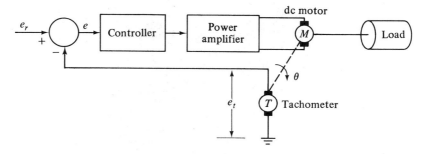

Figure 4-37 Velocity-control system with tachometer feedback.

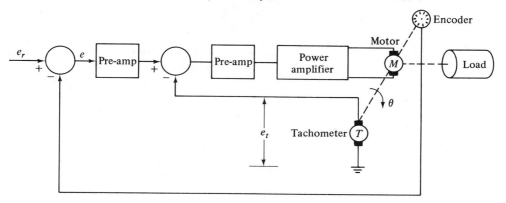

Figure 4-38 Position-control system with tachometer feedback.

In a position-control system, velocity feedback is often used to improve the stability or the damping of the overall system. Figure 4-38 shows the block diagram of such an application. In this case the tachometer feedback forms an inner loop to improve the damping characteristics of the system, and the accuracy of the tachometer is not so critical for this type of application.

The third and most traditional use of dc tachometers is in providing visual speed readout of a rotating shaft. Tachometers used in this capacity are generally connected directly to a voltmeter calibrated in rpm. The most common type of dc tachometer contains an iron-core rotor. The magnetic field is provided by permanent magnet, and no external supply voltage is necessary. The windings on the rotor (armature) are connected to the commutator segments, and the output voltage is taken across a pair of brushes that ride on the commutator segments.

To reduce the inertia of the rotor assembly, moving-coil tachometers contain a rotor that is ironless. In this case the rotor with the armature windings is shaped in the form of a cup and is cantilevered between the permanent magnet poles and the inner iron structure; the latter is needed for the completion of the flux paths. Figure 4-39 shows the moving-coil rotor with the commutator, and the housing-and-magnet assembly.

Since the brushes and the commutator require occasional maintainance, and thus have limited life, the dc tachometer is also available in the brushless configuration. In this case commutation is accomplished by optical or magnetic encoders.

Figure 4-39 Low-inertia moving-coil tachometer armature, housing-and-magnet assembly.

For an ac tachometer a sinusoidal voltage of rated value is applied to the primary winding. A secondary winding is placed at a 90-degree angle mechanically with respect to the primary winding. When the rotor shaft is rotated, the magnitude of the sinusoidal output voltage is proportional to the rotor speed. The phase of the voltage is dependent on the direction of rotation. An ac tachometer can be used in a dc control system by using a phase-sensitive demodulator to convert the ac output to dc. On the other hand, the output of a dc tachometer can be modulated for the control of ac system components if necessary.

Mathematical Modeling of Tachometers. Regardless of the type of a tachometer, its basic characteristic is that the output voltage is proportional to the rotor speed. Thus, the dynamics of the tachometer can be represented by the equation

$$e_t(t) = K_t \frac{d\theta(t)}{dt} = K_t \omega(t) \qquad (4\text{-}103)$$

where $e_t(t)$ is the output voltage, $\theta(t)$ the rotor displacement in radians, $\omega(t)$ the rotor velocity in rad/sec, and K_t the tachometer constant in V/rad/sec. The value of K_t is usually given as a catalog parameter in volts per 1000 rpm (V/krpm). A typical value for K_t is 6 V/krpm.

The transfer function of a tachometer is obtained by taking the Laplace transform on both sides of Eq. (4-103). We have

$$\frac{E_t(s)}{\Theta(s)} = K_t s \qquad (4\text{-}104)$$

where the rotor displacement $\Theta(s)$ is considered as the input and the voltage $E_t(s)$ is the output.

It should be noted that Eq. (4-103) neglects the ripples in the tachometer output voltage. These ripples are caused by the transition of the brushes between commutation segments, armature eccentricity, or other high-frequency sources, such as electromagnetic induction.

Figure 4-40 Rotary incremental encoder. (Courtesy DISC Instruments, Inc.)

Incremental Encoder

One type of encoder that is frequently found in modern control systems converts linear or rotary displacement into digitally coded or pulse signals. The encoders that output a digital signal are known as the *absolute encoders*. In the simplest terms the absolute encoder provides as output a *distinct* digital code indicative of each particular least significant increment of resolution. The *incremental encoder*, on the other hand, provides a pulse for each increment of resolution but does not make distinction between the increments. In practice, the choice of which type of encoder to use depends on economics and control objectives. For the most part, the need for absolute encoders has much to do with the concern for data loss during power failure or the application involving periods of mechanical motion without the readout under power. However, the incremental encoder's simplicity in construction, low cost, ease of application, and versatility have made it by far one of the most popular encoders in control systems.

Incremental encoders are available in rotary and linear forms. Figures 4-40 and 4-41 show the photographs of typical rotary and linear incremental encoders, respectively.

A typical incremental encoder has four basic parts: a light source, a rotary disk, a stationary mask, and a sensor, as shown in Fig. 4-42. The disk has alternate

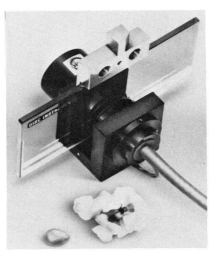

Figure 4-41 Linear incremental encoder. (Courtesy DISC Instruments, Inc.)

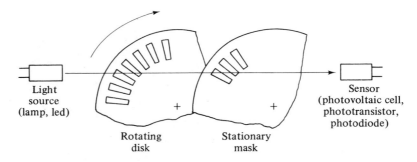

Figure 4-42 Typical incremental optomechanics.

opaque and transparent sectors. Any pair of these sectors represents an incremental period. The mask is used to pass or block the light beam between the light source and the photosensor located behind the mask. For encoders with a relatively low resolution, the mask is not necessary. For fine-resolution encoders (up to thousands of increments per revolution), a multiple-slit mask is often used to maximize reception of the shutter light.

For most control applications the low-level signals obtained directly from the optoelectronic sensors of the encoder are not adequate for control or signal processing. Therefore, amplifiers and waveform-shaping circuits are often incorporated in the encoder package.

The waveforms of the sensor outputs are generally triangular or sinusoidal, depending on the resolution required. Square-wave signals compatible with digital logic are derived by using a linear amplifier followed by a comparator.

Figure 4-43(a) shows a typical rectangular output waveform of a single-channel incremental encoder. In this case pulses are produced for both directions of shaft rotation. A dual-channel encoder with two sets of output pulses is necessary for direction sensing and other control functions. When the phase of the two output

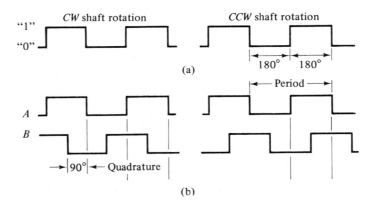

Figure 4-43 (a) Typical rectangular output wave form of a single-channel encoder device (bidirectional). (b) Typical dual-channel encoder signals in quadrature (bidirectional).

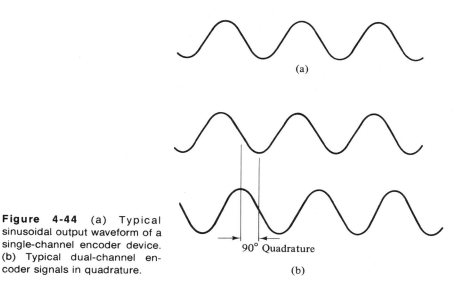

Figure 4-44 (a) Typical sinusoidal output waveform of a single-channel encoder device. (b) Typical dual-channel encoder signals in quadrature.

pulse trains is 90 degrees apart electrically, the two signals are said to be in quadrature, as shown in Fig. 4-43(b); the signals uniquely define 0-to-1 and 1-to-0 logic transitions with respect to the direction of rotation of the encoder disk, so that a direction-sensing logic circuit can be constructed to decode the signals. Figure 4-44 shows the single-channel output and the quadrature outputs with sinusoidal waveforms. Just as in the case of the synchros, the sinusoidal signals from the incremental encoder can be used for fine-position control in feedback control systems.

The following example illustrates some applications of the incremental encoder in control systems.

■

Example 4-8

Consider an incremental encoder that generates two sinusoidal signals in quadrature as the encoder disk rotates. The output signals of the two channels are shown in Fig. 4-45 over one cycle. Note that the two encoder signals generate four zero crossings per cycle. These zero crossings may be used for position indication, position control, or speed measurements in control systems. Let us assume that the encoder shaft is coupled directly to the rotor shaft of a motor which directly drives the printwheel of an electronic typewriter or word processor. The printwheel has 96 character positions on its periphery, and the encoder has 480 cycles. Thus, there are $480 \times 4 = 1920$ zero crossings per revolution. For the 96-character printwheel, this corresponds to $1920/96 = 20$ zero crossings per character; that is, there are 20 zero crossings between two adjacent characters.

One way of measuring the velocity of the printwheel is to count the number of pulses generated by an electronic clock which occur between consecutive zero crossings of the encoder outputs. Let us assume that a 500 kHz clock is used (i.e., the clock generates 500,000 pulses/sec). If the counter counts, say, 500 clock pulses while the encoder rotates from one

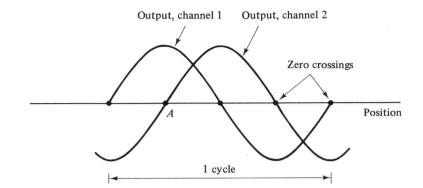

Figure 4-45 One cycle of the output signals of a dual-channel incremental encoder.

zero crossing to the next, the shaft speed is

$$\frac{500,000 \ (\text{pulses/sec})}{500 \ (\text{pulses/zero crossing})} = 1000 \ \text{zero crossings/sec}$$

$$= \frac{1000 \ (\text{zero crossings/sec})}{1920 \ (\text{zero crossings/rev})} = 0.52083 \ \text{rev/sec}$$

$$= 31.25 \ \text{rpm} \qquad\qquad (4\text{-}105)$$

The encoder arrangement described can be used for fine-position control of the printwheel. Let the zero crossing A of the waveforms in Fig. 4-45 correspond to a character position on the printwheel (the next character position is 20 zero crossings away), and the point corresponds to a stable equilibrium point. The coarse-position control of the system must first drive the printwheel position to within one zero crossing on either side of position A, then using the slope of the sine wave at position A, the position control system should null the error quickly.

■

4.6 DC MOTORS IN CONTROL SYSTEMS

Direct-current motors are one of the most widely used prime movers in industry today. Years ago a majority of the small servomotors used for control purposes were of the ac variety. As we shall see, ac motors are more difficult to control, especially for position control, and their characteristics are quite nonlinear, which makes the analytical task more difficult. Dc motors, on the other hand, are more expensive, because of the brushes and commutators, and variable-flux dc motors are suitable only for certain types of control applications. Before permanent-magnet technology was fully developed, the torque per unit volume or weight of a dc motor with a permanent-magnet (PM) field was far from desirable. Today, with the development of the rare-earth magnet, it is possible to achieve very high torque-to-volume PM dc motors. Furthermore, the advances made in brush-and-commutator technology have

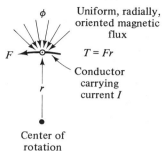

Figure 4-46 Torque production in a dc motor.

made these wearable parts practically maintenance-free. Advanced manufacturing techniques have also produced dc motors with ironless rotors and rotors with very low inertia, thus achieving very high torque-to-inertia ratios, and the low-time-constant properties have opened new applications for dc motors in computer peripheral equipment such as tape drives, printers, disk-pack drives, and word processors, as well as in the machine-tool industry.

Basic Operational Principles of DC Motors

The dc motor is basically a torque transducer that converts electric energy into mechanical energy. The torque developed on the motor shaft is directly proportional to the field flux and the armature current. As shown in Fig. 4-46, a current-carrying conductor is established in a magnetic field with flux ϕ, and the conductor is located at a distance r from the center of rotation. The relationship among the developed torque, flux ϕ, and current i_a is

$$T_m = K_m \phi i_a \tag{4-106}$$

where T_m is the motor torque (N-m), ϕ the magnetic flux (webers), i_a the armature current (amperes), and K_m is a proportional constant.

In addition to the torque developed by the arrangement shown in Fig. 4-46, when the conductor moves in the magnetic field, a voltage is generated across its terminals. This voltage, which is proportional to the shaft velocity, tends to oppose the current flow. The relationship between this back emf and the shaft velocity is

$$e_b = K_m \phi \omega_m \tag{4-107}$$

where e_b denotes the back emf (volts), and ω_m is the shaft velocity (rad/sec) of the motor. Equations (4-106) and (4-107) form the basis of dc motor operation.

Basic Classifications of DC Motors

DC motors can be classified into several broad categories based on the way the magnetic field is produced, and on the basic design and construction of the armature. In terms of the magnetic field, dc motors can be classified as *variable-magnetic-flux motors* and the *constant-magnetic-flux motors*. In variable-magnetic-flux

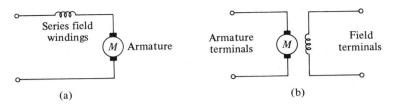

Figure 4-47 (a) Series dc motor armature and field connection. (b) DC motor with separately variable field.

motors, the magnetic field is produced by field windings that are connected to external sources. These motors are also divided into two subclasses: the *series-field motor*, in which the field winding is connected in series with the armature, as shown in Fig. 4-47(a), and the *separately excited-field motor*, in which the field winding is separate from the armature, as shown in Fig. 4-47(b).

For the series-field type, since the magnetic flux in the motor is proportional to the field current, which varies, we have a relationship between the torque and speed which is generally nonlinear. Thus, series-field-type dc motors are useful only for specific applications where high torque at low speeds are called for. The motor torque usually drops off rapidly as the motor speed increases.

For the separately excited dc motor, since the magnetic flux is independent of the armature current, it can be controlled externally over a wide range.

The constant-magnetic-flux dc motor is also known as the permanent-magnet (PM) dc motor. In this case, the magnetic field is produced by a permanent magnet and is constant. This allows the torque-speed characteristics of the motor to be relatively linear.

PM dc motors can be further classified according to commutation scheme and armature design. Conventional dc motors have mechanical brushes and commutators, while there are dc motors in which the commutation is done electronically; this type of motor is called the *brushless dc motor*.

According to the armature construction, the PM dc motor can be broken down into three types of armature design: *iron core*, *surface wound*, and *moving-coil* motors.

Iron-Core PM DC Motors.   The rotor and stator configuration of an iron-core PM dc motor is shown in Fig. 4-48. The permanent-magnet material can be barium ferrite, Alnico, or a rare-earth compound. The magnetic flux produced by the permanent magnet passes through a laminated rotor structure which contains slots. The armature conductors are placed in the rotor slots. This type of dc motor is characterized by relatively high rotor inertia, high inductance, low cost, and high reliability.

Surface-Wound DC Motors.   Figure 4-49 shows the rotor construction of a surface-wound PM dc motor. The armature conductors are bonded to the surface of a cylindrical rotor structure, which is made of laminated disks fastened to the motor shaft. Since no slots are used on the rotor in this design, the armature has no "cogging" effect. Since the conductors are laid out in the air gap between the rotor

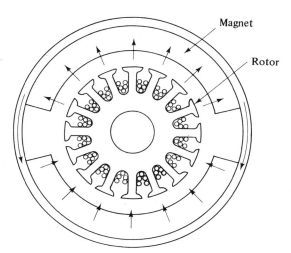

Figure 4-48 Cross-sectional view of a permanent-magnet iron-core dc motor.

disks and the permanent magnet field, this type of motor has lower inductance than that of the iron-core structure. However, because the air gap between the magnetic and the low-inductance rotor is larger than in the iron-core motor, a larger magnet is required in order to provide a magnetic flux equivalent to that of the iron-core motor. Therefore, surface-wound dc motors are more expensive to produce and have larger outside diameters than equivalent iron-core motor.

Moving-Coil DC Motors. Moving-coil motors are designed to have very small moments of inertia and very low armature inductance. This is achieved by placing the armature conductors in the air gap between a stationary flux return path

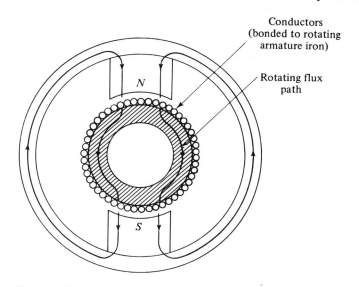

Figure 4-49 Cross-sectional view of a surface-wound permanent-magnet dc motor.

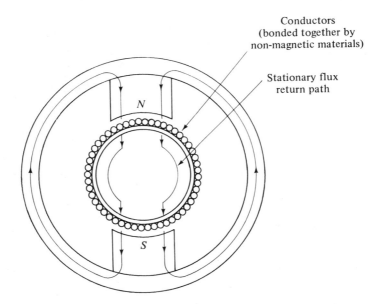

Figure 4-50 Cross-sectional view of a moving-coil permanent-magnet dc motor.

and the permanent magnet structure, as shown in Fig. 4-50. In the case of Fig. 4-50, the conductor structure is supported by nonmagnetic material—usually epoxy resins and fiberglass—to form a hollow cylinder. One end of the cylinder forms a hub, which is attached to the motor shaft. A cross-sectional view of such a motor is shown in Fig. 4-51. Since all unnecessary elements have been removed from the armature of the moving-coil motor, its moment of inertia is very low. However, it has a larger air gap than the two motors discussed earlier, and therefore requires an even larger magnetic structure than do the other two types of motors to produce an equivalent air-gap flux. Since the conductors in the moving-coil armature are not in

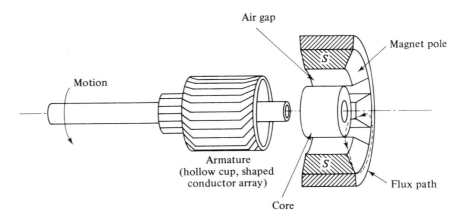

Figure 4-51 Cross-sectional side view of a moving-coil dc motor.

direct contact with iron, the motor inductance is very low; values of less than 100 μH are common in this type of motor. The low-inertia and low-inductance properties make the moving-coil motor the best actuator choice for high-performance control systems.

Mathematical Modeling of DC Motors

Since dc motors of various types are being used extensively in control systems, for analytical purposes it is essential that we establish a mathematical model for the motor. In the following paragraphs we develop the mathematical models of the separately excited and the PM dc motors.

Separately Excited DC Motor. The circuit diagram of a separately excited dc motor is shown in Fig. 4-52. The armature is modeled as a circuit with a resistance R_a in series with an inductance L_a, and a voltage source e_b representing the generated voltage (back emf) in the armature when the rotor rotates. The wound field is represented by a resistance R_f in series with an inductance L_f. The air-gap flux is designated by ϕ. The following summary on the variables and parameters is given:

$$e_a(t) = \text{armature voltage}$$

$$e_f(t) = \text{field voltage}$$

$$R_a = \text{armature resistance}$$

$$e_b(t) = \text{back emf}$$

$$R_f = \text{field resistance}$$

$$L_a = \text{armature inductance}$$

$$L_f = \text{field inductance}$$

$$i_a(t) = \text{armature current}$$

$$i_f(t) = \text{field current}$$

$$K_i = \text{torque constant}$$

$$K_b = \text{back emf constant}$$

$$\phi(t) = \text{magnetic flux}$$

$$T_m(t) = \text{torque developed by motor}$$

$$J_m = \text{rotor inertia of motor}$$

$$B_m = \text{viscous frictional coefficient}$$

$$\theta_m(t) = \text{rotor angular displacement}$$

$$\omega_m(t) = \text{rotor angular velocity}$$

$$T_L(t) = \text{load torque}$$

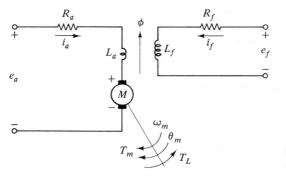

Figure 4-52 Model of a separately excited dc motor.

With reference to the circuit diagram of Fig. 4-52, the control is applied at the armature terminals in the form of the applied voltage $e_a(t)$, and we assume that $e_f(t)$ is applied sufficiently long so that the field current $i_f(t)$ is constant. For linear analysis we assume further that:

1. The air-gap flux is proportional to the field current; that is,

$$\phi(t) = K_f i_f(t) = K_f I_f = \text{constant} \qquad (4\text{-}108)$$

2. The torque developed by the motor is proportional to the air-gap flux and the armature current. Thus,

$$
\begin{aligned}
T_m(t) &= K_m \phi(t) i_a(t) \\
&= K_m K_f I_f i_a(t)
\end{aligned}
\qquad (4\text{-}109)
$$

Since $K_m K_f I_f$ is constant, Eq. (4-109) is written

$$T_m(t) = K_i i_a(t) \qquad (4\text{-}110)$$

where K_i is the torque constant in N-m/amp, lb-ft/amp, or oz-in./amp. Starting with the control input voltage $e_a(t)$, the cause-and-effect equations for the system of Fig. 4-52 are written

$$\frac{di_a(t)}{dt} = \frac{1}{L_a} e_a(t) - \frac{R_a}{L_a} i_a(t) - \frac{1}{L_a} e_b(t) \qquad (4\text{-}111)$$

$$T_m(t) = K_i i_a(t) \qquad (4\text{-}112)$$

$$e_b(t) = K_b \frac{d\theta_m(t)}{dt} = K_b \omega_m(t) \qquad (4\text{-}113)$$

$$\frac{d^2\theta_m(t)}{dt^2} = \frac{1}{J_m} T_m(t) - \frac{1}{J_m} T_L(t) - \frac{B_m}{J_m} \frac{d\theta_m(t)}{dt} \qquad (4\text{-}114)$$

where $T_L(t)$ denotes the load torque. In general, $T_L(t)$ represents a torque that the motor has to overcome in order to have motion; $T_L(t)$ can be a constant frictional torque such as Coulomb friction.

Equations (4-111) through (4-114) are written by considering that $e_a(t)$ is the cause of all causes; Eq. (4-111) considers that $di_a(t)/dt$ is the immediate effect due to $e_a(t)$, then in Eq. (4-112) $i_a(t)$ causes the torque $T_m(t)$ to be generated, Eq. (4-112) defines the back emf, and finally, in Eq. (4-114) the torque causes the angular displacement θ_m.

The state variables of the system can be defined as θ_m, ω_m, and i_a, not necessarily in this order. By direct substitution and eliminating all the nonstate variables from Eqs. (4-111) through (4-114), the state equations of the dc motor system are written in vector-matrix form:

$$\begin{bmatrix} \dfrac{di_a(t)}{dt} \\[2ex] \dfrac{d\omega_m(t)}{dt} \\[2ex] \dfrac{d\theta_m(t)}{dt} \end{bmatrix} = \begin{bmatrix} -\dfrac{R_a}{L_a} & -\dfrac{K_b}{L_a} & 0 \\[2ex] \dfrac{K_i}{J_m} & -\dfrac{B_m}{J_m} & 0 \\[2ex] 0 & 1 & 0 \end{bmatrix} \begin{bmatrix} i_a(t) \\[2ex] \omega_m(t) \\[2ex] \theta_m(t) \end{bmatrix} + \begin{bmatrix} \dfrac{1}{L_a} \\[2ex] 0 \\[2ex] 0 \end{bmatrix} e_a(t) - \begin{bmatrix} 0 \\[2ex] \dfrac{1}{J_m} \\[2ex] 0 \end{bmatrix} T_L(t)$$

(4-115)

Notice that in this case $T_L(t)$ is treated as a second input in the state equations.

The state diagram of the system is drawn as shown in Fig. 4-53, using Eq. (4-115). The transfer function between the motor displacement and the input voltage is obtained from the state diagram as

$$\frac{\Theta_m(s)}{E_a(s)} = \frac{K_i}{L_a J_m s^3 + (R_a J_m + B_m L_a)s^2 + (K_b K_i + R_a B_m)s}$$

(4-116)

where T_L has been set to zero.

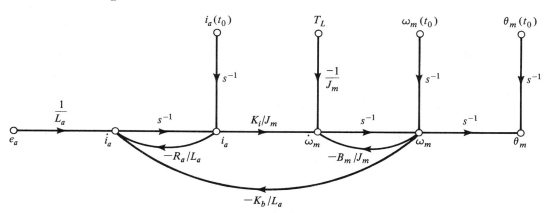

Figure 4-53 State diagram of a dc motor.

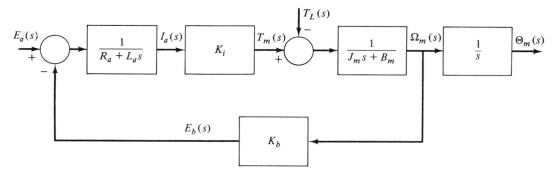

Figure 4-54 Block diagram of a dc motor system.

Figure 4-54 shows a block diagram representation of the dc motor system. The advantage of using the block diagram is that it gives a clear picture of the transfer function relation between each block of the system. Since an s can be factored out from the denominator of Eq. (4-116), the significance of the transfer function $\Theta_m(s)/E_a(s)$ is that the dc motor is essentially an integrating device between these two variables. This is expected, since if e_a is a constant input, the output motor displacement will behave as the output of an integrator; that is, it will increase linearly with time.

Although a dc motor by itself is basically an open-loop system, the state diagram of Fig. 4-53 and the block diagram of Fig. 4-54 show that the motor has a "built-in" feedback loop caused by the back emf. Physically, the back emf represents the feedback of a signal that is proportional to the negative of the speed of the motor. As seen from Eq. (4-116), the back-emf constant K_b represents an added term to the resistance R_a and viscous-friction coefficient B_m. Therefore, the back-emf effect is equivalent to an "electrical friction" which tends to improve the stability of the motor and, in general, the stability of the system in which the motor is used as an actuator.

Relationship between K_i and K_b. Although functionally the torque constant K_i and the back-emf constant K_b are two separate parameters, for a given motor, their values are closely related. To show the relationship, we write the mechanical power developed in the motor armature as

$$P = e_b(t)i_a(t) \qquad \text{watts} \tag{4-117}$$

The mechanical power is also expressed as

$$P = T_m(t)\omega_m(t) \qquad \text{watts} \tag{4-118}$$

where in SI units, $T_m(t)$ is in N-m and $\omega_m(t)$ is in rad/sec. Now substituting Eqs. (4-110) and (4-113) in Eq. (4-117), we get

$$P = T_m(t)\omega_m(t) = K_b\omega_m(t)\frac{T_m(t)}{K_i} \tag{4-119}$$

from which

$$K_b \left(\frac{V}{\text{rad/sec}} \right) = K_i (\text{N-m/A}) \qquad (4\text{-}120)$$

Thus, we see that in SI unit system, the values of K_b and K_i are identical if K_b is represented in V/rad/sec.

For the British unit system, we convert Eq. (4-117) into horsepower (hp); that is,

$$P = \frac{e_b(t)i_a(t)}{746} \qquad \text{hp} \qquad (4\text{-}121)$$

In terms of torque and angular velocity, P is written

$$P = \frac{T_m(t)\omega_m(t)}{550} \qquad \text{hp} \qquad (4\text{-}122)$$

where $T_m(t)$ is in ft-lb and $\omega_m(t)$ is in rad/sec. Using Eqs. (4-110) and (4-113) and equating the last two equations, we get

$$\frac{K_b\omega_m(t)T_m(t)}{746K_i} = \frac{T_m(t)\omega_m(t)}{550} \qquad (4\text{-}123)$$

Thus,

$$K_b = \frac{746}{550}K_i = 1.3564K_i \qquad (4\text{-}124)$$

where K_b is in V/rad/sec and K_i is in ft-lb/A.

We can determine the torque constant of the dc motor using torque-speed curves. A typical set of torque-speed curves for various applied voltages is shown in Fig. 4-55. The rated voltage is denoted by E_r. At no load, $T_m = 0$, the speed is given by the intersect on the abscissa for a given applied voltage. Then the back-emf constant K_b is given by

$$K_b = \frac{E_r}{\Omega_r} \qquad (4\text{-}125)$$

where in this case the rated values are used for voltage and angular velocity.

When the motor is stalled, the blocked-rotor torque at the rated voltage is designated by $T_0(t)$. Let

$$k = \frac{\text{blocked-rotor torque at rated voltage}}{\text{rated voltage}} = \frac{T_0}{E_r} \qquad (4\text{-}126)$$

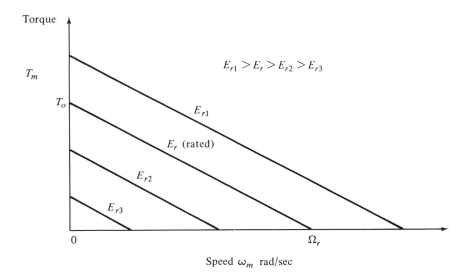

Figure 4-55 Typical torque-speed curves of a dc motor.

Also,

$$T_0(t) = K_i i_a(t) = \frac{K_i}{R_a} E_r \qquad (4\text{-}127)$$

Therefore, from the last two equations, the torque constant is determined:

$$K_i = kR_a \qquad (4\text{-}128)$$

Permanent-Magnet DC Motor. For a PM dc motor, the air-gap flux is generated by the permanent-magnet field. Since in the development above it is assumed that $\phi(t)$ is constant, the equations starting from Eq. (4-110) are equally valid for the PM dc motor.

4.7 TWO-PHASE INDUCTION MOTORS

For low-power applications in control systems, ac motors are sometimes used because of their rugged construction. Most ac motors used in control systems are of the two-phase induction type, which generally are rated from a fraction of a watt up to a few hundred watts (fractional horsepower). The frequency of the motor is normally rated at 60, 400, or 1000 Hz. The use of high-frequency ac actuators is often preferred in airborne systems owing to the immunization from noise.

A schematic diagram of a two-phase induction motor is shown in Fig. 4-56. The motor consists of a stator with two distributed windings displaced 90 electrical degrees apart. Under normal operating conditions in control applications, a fixed voltage from a constant-voltage source is applied to one phase, the *fixed* or *reference phase*. The other phase, the *control phase*, is energized by a voltage that is 90 degrees out of phase with respect to the voltage of the fixed phase. The

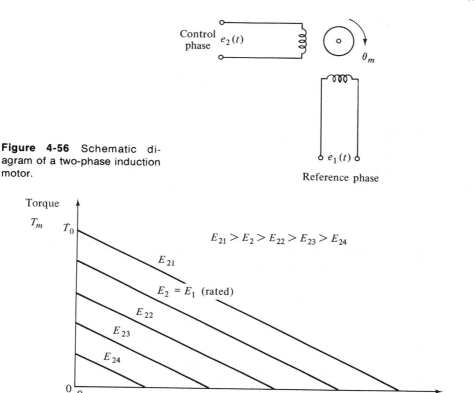

Figure 4-56 Schematic diagram of a two-phase induction motor.

Figure 4-57 Typical linearized torque-speed curves of a two-phase induction motor.

control-phase voltage is usually supplied from a servo amplifier, and the voltage has a variable amplitude and polarity. The direction of rotation of the motor reverses when the control phase signal changes its sign.

Unlike that of a dc motor, the torque-speed curve of a two-phase induction motor is quite nonlinear. However, for linear analysis, it is generally considered an acceptable practice to approximate the torque-speed curves of a two-phase induction motor by straight lines, such as those shown in Fig. 4-57. These curves are assumed to be straight lines parallel to the torque-speed curve at a rated control voltage ($E_2 = E_1$ = rated value), and they are equally spaced for equal increments of the control voltage.

The state equations of the motor are determined as follows. Let k be the *blocked-rotor torque at rated voltage* per unit control voltage; that is,

$$k = \frac{\text{blocked-rotor torque at } E_2 = E_1}{\text{rated control voltage } E_1} = \frac{T_0}{E_1} \qquad (4\text{-}129)$$

Let m be a negative number which represents the slope of the linearized torque-speed curve shown in Fig. 4-57. Then

$$m = -\frac{\text{blocked-rotor torque}}{\text{no-load speed}} = -\frac{T_0}{\Omega_0} \tag{4-130}$$

For any torque T_m, the family of straight lines in Fig. 4-56 is represented by the equation

$$T_m(t) = m\omega_m(t) + ke_2(t) \tag{4-131}$$

where $\omega_m(t)$ is the speed of the motor and $e_2(t)$ the control voltage. Now, if we designate $\omega_m(t)$ as a state variable, one of the state equations may be obtained from

$$J_m\frac{d\omega_m(t)}{dt} = -B_m\omega_m(t) + T_m(t) \tag{4-132}$$

Substituting Eq. (4-131) into Eq. (4-132) and recognizing that $\theta_m(t)$ is the other state variable, we have the two state equations

$$\frac{d\theta_m(t)}{dt} = \omega_m(t) \tag{4-133}$$

$$\frac{d\omega_m(t)}{dt} = \frac{1}{J_m}(m - B_m)\omega_m + \frac{k}{J_m}e_2(t) \tag{4-134}$$

The state diagram of the two-phase induction motor is shown in Fig. 4-58. The transfer function of the motor between the control voltage and the motor displacement is obtained as

$$\frac{\theta_m(s)}{E_2(s)} = \frac{k}{(B_m - m)s[1 + J_m/(B_m - m)s]} \tag{4-135}$$

or

$$\frac{\theta_m(s)}{E_2(s)} = \frac{K_m}{s(1 + \tau_m s)} \tag{4-136}$$

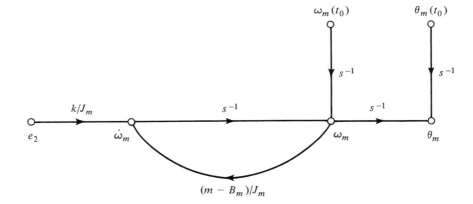

Figure 4-58   State diagram of the two-phase induction motor.

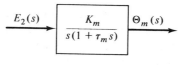

(a)

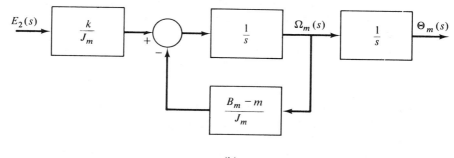

(b)

Figure 4-59 Block diagram representations of the two-phase induction motor.

where

$$K_m = \frac{k}{B_m - m} = \text{motor gain constant} \qquad (4\text{-}137)$$

$$\tau_m = \frac{J_m}{B_m - m} = \text{motor time constant} \qquad (4\text{-}138)$$

Since m is a negative number, the equations above show that the effect of the slope of the torque-speed curve is to add more friction to the motor, thus improving the damping or stability of the motor. Therefore, the slope of the torque-speed curve of a two-phase induction motor is analogous to the back emf effect of a dc motor. However, if m is a positive number, negative damping occurs for $m > B_m$, it can be shown that the motor becomes unstable.

Figure 4-59 shows the block diagram representations of the linearized two-phase induction motor. The block diagram in Fig. 4-59(a) represents the motor as a second-order open-loop system with the transfer function given by Eq. (4-136). Figure 4-59(b) gives the block diagram equivalent of the state diagram in Fig. 4-58. Again, as pointed out above, if $m > B_m$, the feedback becomes positive.

4.8 LINEARIZATION OF NONLINEAR SYSTEMS

From the discussions given in the preceding sections we should realize that most components and actuators found in physical systems have nonlinear characteristics. In practice, we may find that some devices have moderate nonlinear characteristics, or the nonlinear properties would occur if they are driven into certain operating regions. For these devices, the modeling by linear system models may give quite accurate analytical results over a relatively wide range of operating conditions. However, there are numerous physical devices, such as the incremental encoders and the two-phase induction motor discussed earlier, which possess strong nonlinear characteristics. For these devices, strictly, a linearized model is valid only for a limited range of operation, and often only at the operating point at which the linearization is carried out. More important, when a nonlinear system is linearized at an operating point, the linear model may contain time-varying elements.

Let us represent a nonlinear system by the following vector-matrix state equations:

$$\frac{d\mathbf{x}(t)}{dt} = \mathbf{f}[\mathbf{x}(t), \mathbf{r}(t)] \tag{4-139}$$

where $\mathbf{x}(t)$ represents the $n \times 1$ state vector, $\mathbf{r}(t)$ the $p \times 1$ input vector, and $\mathbf{f}[\mathbf{x}(t), \mathbf{r}(t)]$ denotes an $n \times 1$ function vector. In general, $\mathbf{f}$ is a function of the state vector and the input vector.

Being able to represent a nonlinear and/or time-varying system by state equations is a distinct advantage of the state-variable approach over the transfer function method, since the latter is strictly defined for only linear time-invariant systems.

As a simple illustrative example, the following state equations are nonlinear:

$$\frac{dx_1(t)}{dt} = x_1(t) + x_2^2(t)$$

$$\frac{dx_2(t)}{dt} = x_1(t) + r(t) \tag{4-140}$$

Since nonlinear systems are usually difficult to analyze and design, it would be desirable to perform a linearization whenever the situation justifies.

A linearization process that depends on expanding the nonlinear state equation into a Taylor series about a nominal operating point or trajectory is now described. All the terms of the Taylor series of order higher than the first are discarded, and linear approximation of the nonlinear state equation at the nominal point results.

Let the nominal operating trajectory be denoted by $\mathbf{x}_0(t)$, which corresponds to the nominal input $\mathbf{r}_0(t)$ and some fixed initial states. Expanding the nonlinear state equation of Eq. (4-139) into a Taylor series about $\mathbf{x}(t) = \mathbf{x}_0(t)$ and neglecting all the higher-order terms yields

$$\dot{x}_i(t) = f_i(\mathbf{x}_0, \mathbf{r}_0) + \sum_{j=1}^{n} \frac{\partial f_i(\mathbf{x}, \mathbf{r})}{\partial x_j}\bigg|_{\mathbf{x}_0, \mathbf{r}_0} (x_j - x_{0j})$$

$$+ \sum_{j=1}^{p} \frac{\partial f_i(\mathbf{x}, \mathbf{r})}{\partial r_j}\bigg|_{\mathbf{x}_0, \mathbf{r}_0} (r_j - r_{0j}) \tag{4-141}$$

$i = 1, 2, \ldots, n$. Let

$$\Delta x_i = x_i - x_{0i} \tag{4-142}$$

and

$$\Delta r_j = r_j - r_{0j} \tag{4-143}$$

Then

$$\Delta \dot{x}_i = \dot{x}_i - \dot{x}_{0i} \tag{4-144}$$

Since

$$\dot{x}_{0i} = f_i(\mathbf{x}_0, \mathbf{r}_0) \tag{4-145}$$

Equation (4-141) is written

$$\Delta \dot{x}_i = \sum_{j=1}^{n} \frac{\partial f_i(\mathbf{x}, \mathbf{r})}{\partial x_j} \bigg|_{\mathbf{x}_0, \mathbf{r}_0} \Delta x_j + \sum_{j=1}^{p} \frac{\partial f_i(\mathbf{x}, \mathbf{r})}{\partial r_j} \bigg|_{\mathbf{x}_0, \mathbf{r}_0} \Delta r_j \tag{4-146}$$

Equation (4-146) may be written in vector-matrix form,

$$\Delta \dot{\mathbf{x}} = \mathbf{A}^* \Delta \mathbf{x} + \mathbf{B}^* \Delta \mathbf{r} \tag{4-147}$$

where

$$\mathbf{A}^* = \begin{bmatrix} \dfrac{\partial f_1}{\partial x_1} & \dfrac{\partial f_1}{\partial x_2} & \cdots & \dfrac{\partial f_1}{\partial x_n} \\ \dfrac{\partial f_2}{\partial x_1} & \dfrac{\partial f_2}{\partial x_2} & \cdots & \dfrac{\partial f_2}{\partial x_n} \\ \cdots & \cdots & \cdots & \cdots \\ \dfrac{\partial f_n}{\partial x_1} & \dfrac{\partial f_n}{\partial x_2} & \cdots & \dfrac{\partial f_n}{\partial x_n} \end{bmatrix} \tag{4-148}$$

$$\mathbf{B}^* = \begin{bmatrix} \dfrac{\partial f_1}{\partial r_1} & \dfrac{\partial f_1}{\partial r_2} & \cdots & \dfrac{\partial f_1}{\partial r_p} \\ \dfrac{\partial f_2}{\partial r_1} & \dfrac{\partial f_2}{\partial r_2} & \cdots & \dfrac{\partial f_2}{\partial r_p} \\ \cdots & \cdots & \cdots & \cdots \\ \dfrac{\partial f_n}{\partial r_1} & \dfrac{\partial f_n}{\partial r_2} & \cdots & \dfrac{\partial f_n}{\partial r_p} \end{bmatrix} \tag{4-149}$$

where it should be reiterated that $\mathbf{A}^*$ and $\mathbf{B}^*$ are evaluated at the nominal point. Thus, we have linearized the nonlinear system of Eq. (4-139) at a nominal operating point. However, in general, although Eq. (4-147) is linear, the elements of $\mathbf{A}^*$ and $\mathbf{B}^*$ may be time varying.

The following examples serve to illustrate the linearization procedure just described.

■

Example 4-9

Figure 4-60 shows the block diagram of a control system with a saturation nonlinearity. The state equations of the system are

$$\dot{x}_1 = f_1 = x_2 \tag{4-150}$$

$$\dot{x}_2 = f_2 = u \tag{4-151}$$

where the input-output relation of the saturation nonlinearity is represented by

$$u = (1 - e^{-K|x_1|})\,\mathrm{SGN}\,x_1 \tag{4-152}$$

where

$$\mathrm{SGN}\,x_1 = \begin{cases} +1 & x_1 > 0 \\ -1 & x_1 < 0 \end{cases} \tag{4-153}$$

Substituting Eq. (4-152) into Eq. (4-151) and using Eq. (4-146), we have the linearized state equation

$$\Delta \dot{x}_1 = \frac{\partial f_1}{\partial x_2}\Delta x_2 = \Delta x_2 \tag{4-154}$$

$$\Delta \dot{x}_2 = \frac{\partial f_2}{\partial x_1}\Delta x_1 = Ke^{-K|x_{01}|}\Delta x_1 \tag{4-155}$$

where x_{01} denotes a nominal value of x_1. Notice that the last two equations are linear and are valid only for small signals. In vector-matrix form, these linearized state equations are written as

$$\begin{bmatrix} \Delta \dot{x}_1 \\ \Delta \dot{x}_2 \end{bmatrix} = \begin{bmatrix} 0 & 1 \\ a & 0 \end{bmatrix}\begin{bmatrix} \Delta x_1 \\ \Delta x_2 \end{bmatrix} \tag{4-156}$$

where

$$a = Ke^{-K|x_{01}|} = \text{constant} \tag{4-157}$$

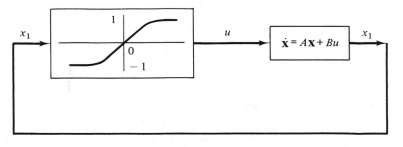

Figure 4-60 Nonlinear control system.

It is of interest to check the significance of the linearization. If x_{01} is chosen to be at the origin of the nonlinearity, $x_{01} = 0$, then $a = K$; Eq. (4-155) becomes

$$\Delta \dot{x}_2 = K \Delta x_1 \tag{4-158}$$

Thus the linearized model is equivalent to having a linear amplifier with a constant gain K. On the other hand, if x_{01} is a large number, the nominal operating point will lie on the saturated portion of the nonlinearity, and $a = 0$. This means than any small variation in x_1 (small Δx_1) will give rise to practically no change in $\Delta \dot{x}_2$.

Example 4-10

In Example 4-9 the linearized system turns out to be time invariant. In general, linearization of a nonlinear system often leads to a linear time-varying system. Consider the following nonlinear system:

$$\dot{x}_1 = \frac{-1}{x_2^2} \tag{4-159}$$

$$\dot{x}_2 = u x_1 \tag{4-160}$$

We would like to linearize these equations about the nominal trajectory $[x_{01}(t), x_{02}(t)]$, which is the solution to the equations with the initial conditions $x_1(0) = x_2(0) = 1$ and the input $u(t) = 0$.

Integrating both sides of Eq. (4-160), we have

$$x_2 = x_2(0) = 1 \tag{4-161}$$

Then Eq. (4-159) gives

$$x_1 = -t + 1 \tag{4-162}$$

Therefore, the nominal trajectory about which Eqs. (4-159) and (4-160) are to be linearized is described by

$$x_{01}(t) = -t + 1 \tag{4-163}$$
$$x_{02}(t) = 1 \tag{4-164}$$

Now evaluating the coefficients of Eq. (4-146), we get

$$\frac{\partial f_1}{\partial x_1} = 0 \qquad \frac{\partial f_1}{\partial x_2} = \frac{2}{x_2^3} \qquad \frac{\partial f_2}{\partial x_1} = u \qquad \frac{\partial f_2}{\partial u} = x_1$$

Equation (4-146) gives

$$\Delta \dot{x}_1 = \frac{2}{x_{02}^3} \Delta x_2 \tag{4-165}$$

$$\Delta \dot{x}_2 = u_0 \Delta x_1 + x_{01} \Delta u \tag{4-166}$$

Substituting Eqs. (4-163) and (4-164) into Eqs. (4-165) and (4-166), the linearized equations are written as

$$\begin{bmatrix} \Delta \dot{x}_1 \\ \Delta \dot{x}_2 \end{bmatrix} = \begin{bmatrix} 0 & 2 \\ 0 & 0 \end{bmatrix} \begin{bmatrix} \Delta x_1 \\ \Delta x_2 \end{bmatrix} + \begin{bmatrix} 0 \\ 1 - t \end{bmatrix} \Delta u \tag{4-167}$$

which is a set of linear state equations with time-varying coefficients.

Example 4-11

The state equations of a two-phase permanent-magnet step motor [9, 10] are given as

$$\frac{di_a(t)}{dt} = \frac{1}{L} \left[v_a(t) - Ri_a(t) + K_b \omega_m \sin N_r \theta_m \right] \tag{4-168}$$

$$\frac{di_b(t)}{dt} = \frac{1}{L} \left[v_b(t) - Ri_b(t) - K_b \omega_m \cos N_r \theta_m \right] \tag{4-169}$$

$$\frac{d\omega_m(t)}{dt} = \frac{1}{J} \left[-K_m i_a(t) \sin N_r \theta_m + K_m i_b \cos N_r \theta_m - B \omega_m \right] \tag{4-170}$$

$$\frac{d\theta_m(t)}{dt} = \omega_m(t) \tag{4-171}$$

where $v_a(t)$ = voltage applied to phase a
$v_b(t)$ = voltage applied to phase b
$i_a(t)$ = current in phase a
$i_b(t)$ = current in phase b
$\omega_m(t)$ = angular velocity of motor
$\theta_m(t)$ = angular displacement of motor
R = stator-winding resistance per phase
L = stator-winding inductance per phase
B = viscous friction coefficient
J = rotor inertia
K_b = back-emf constant
K_m = torque constant
N_r = number of teeth on rotor

Clearly, the first three state equations are nonlinear.

Let us linearize these nonlinear state equations about the following operating point:

$$\theta_m^*(t) = \theta_m^* = \text{constant} \tag{4-172}$$

$$\omega_m^*(t) = 0 \tag{4-173}$$

Since $d\omega_m^*(t)/dt$ is zero, Eq. (4-170) becomes

$$-i_a^* \sin N_r \theta_m^* + i_b^* \cos N_r \theta_m^* = \frac{B}{K_m} \omega_m^* \tag{4-174}$$

where i_a^* and i_b^* denote the nominal solution of i_a and i_b, respectively. Equation (4-174) may have more than one solution, but one obvious set of solutions is

$$i_a^* = -\frac{B\omega_m^*}{K_m}\sin N_r\theta_m^* \tag{4-175}$$

$$i_b^* = \frac{B\omega_m^*}{K_m}\cos N_r\theta_m^* \tag{4-176}$$

Now evaluating the elements of A^* and B^*, we have

$$\frac{\partial f_1}{\partial i_a} = -\frac{R}{L} \qquad \frac{\partial f_1}{\partial i_b} = 0 \qquad \frac{\partial f_1}{\partial \omega_m} = \frac{K_b}{L}\sin N_r\theta_m \qquad \frac{\partial f_1}{\partial \theta_m} = \frac{K_b}{L}\omega_m N_r\cos N_r\theta_m$$

$$\frac{\partial f_2}{\partial i_a} = 0 \qquad \frac{\partial f_2}{\partial i_b} = -\frac{R}{L} \qquad \frac{\partial f_2}{\partial \omega_m} = -\frac{K_b}{L}\cos N_r\theta_m \qquad \frac{\partial f_2}{\partial \theta_m} = \frac{K_b}{L}\omega_m N_r\sin N_r\theta_m$$

$$\frac{\partial f_3}{\partial i_a} = -\frac{K_m}{J}\sin N_r\theta_m \qquad \frac{\partial f_3}{\partial i_b} = \frac{K_m}{J}\cos N_r\theta_m \qquad \frac{\partial f_3}{\partial \omega_m} = -\frac{B}{J}$$

$$\frac{\partial f_3}{\partial \theta_m} = -\frac{K_m}{J}N_r(i_a\cos N_r\theta_m + i_b\sin N_r\theta_m) \qquad \frac{\partial f_4}{\partial i_a} = 0 \qquad \frac{\partial f_4}{\partial i_b} = 0$$

$$\frac{\partial f_4}{\partial \omega_m} = 1 \qquad \frac{\partial f_4}{\partial \theta_m} = 0$$

Substituting these coefficients into Eqs. (4-148) and (4-149), evaluating at the nominal trajectories of θ_m^* and ω_m^*, and using the identities of Eqs. (4-175) and (4-176), we get the linearized system of equations as

$$\begin{bmatrix} \Delta\dot{i}_a \\ \Delta\dot{i}_b \\ \Delta\dot{\omega}_m \\ \Delta\dot{\theta}_m \end{bmatrix} = \begin{bmatrix} -\dfrac{R}{L} & 0 & \dfrac{K_b}{L}\sin N_r\theta_m^* & 0 \\ 0 & -\dfrac{R}{L} & -\dfrac{K_b}{L}\cos N_r\theta_m^* & 0 \\ -\dfrac{K_m}{J}\sin N_r\theta_m^* & \dfrac{K_m}{J}\cos N_r\theta_m^* & -\dfrac{B}{J} & 0 \\ 0 & 0 & 1 & 0 \end{bmatrix} \begin{bmatrix} \Delta i_a \\ \Delta i_b \\ \Delta\omega_m \\ \Delta\theta_m \end{bmatrix}$$

$$+ \begin{bmatrix} \dfrac{1}{L} & 0 \\ 0 & \dfrac{1}{L} \\ 0 & 0 \\ 0 & 0 \end{bmatrix} \begin{bmatrix} \Delta v_a \\ \Delta v_b \end{bmatrix} \tag{4-177}$$

∎

4.9 SYSTEMS WITH TRANSPORTATION LAGS

Thus far the systems considered have all had transfer functions that are quotients of polynomials. In practice, pure time delays may be encountered in various types of systems, especially systems with hydraulic, pneumatic, or mechanical transmissions.

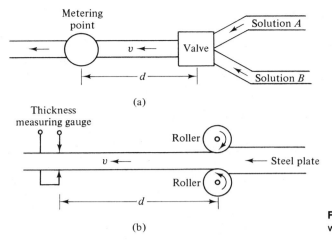

Figure 4-61 Physical systems with transportation lags.

Systems with computer control also have time delays, since it takes time for the computer to execute numerical operations. In these systems the output will not begin to respond to an input until after a given time interval. Figure 4-61 illustrates examples in which transportation lags or pure time delays are observed. Figure 4-61(a) outlines an arrangement in which two different fluids are to be mixed in appropriate proportions. To assure that a homogeneous solution is measured, the monitoring point is located some distance from the mixing point. A transportation lag therefore exists between the mixing point and the place where the change in concentration is detected. If the rate of flow of the mixed solution is v inches per second and d is the distance between the mixing and the metering points, the time lag is given by

$$T = \frac{d}{v} \quad \text{sec} \tag{4-178}$$

If it is assumed that the concentration at the mixing point is $c(t)$ and that it is reproduced without change T seconds later at the monitoring point, the measured quantity is

$$b(t) = c(t - T) \tag{4-179}$$

The Laplace transform of Eq. (4-179) is

$$B(s) = e^{-Ts}C(s) \tag{4-180}$$

Thus, the transfer function between $b(t)$ and $c(t)$ is

$$\frac{B(s)}{C(s)} = e^{-Ts} \tag{4-181}$$

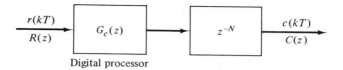

Figure 4-62 Digital control system with pure time delay.

The arrangement shown in Fig. 4-61(b) may be thought of as a thickness control of the rolling of steel plates. As in the case above, the transfer function between the thickness at the rollers and the measuring point is given by Eq. (4-181).

Other examples of transportation lags are found in human beings as control systems where action and reaction are always accompanied by pure time delays. The operation of the sample-and-hold device of a sampled-data system closely resembles a pure time delay; it is sometimes approximated by a simple time-lag term, e^{-Ts}.

In terms of state variables, a system with pure time delay can no longer be described by the matrix-state equation

$$\frac{d\mathbf{x}(t)}{dt} = \mathbf{A}\mathbf{x}(t) + \mathbf{B}\mathbf{u}(t) \tag{4-182}$$

A general state description of a system containing time lags is given by the following matrix differential-difference equation:

$$\frac{d\mathbf{x}(t)}{dt} = \sum_{i=1}^{p} \mathbf{A}_i\mathbf{x}(t - T_i) + \sum_{j=1}^{q} \mathbf{B}_j\mathbf{u}(t - T_j) \tag{4-183}$$

where T_i and T_j are fixed time delays. In this case Eq. (4-183) represents a general situation where time delays may exist on both the inputs as well as the states.

Figure 4-62 shows the block diagram of a digital control system. The block with the transfer function z^{-N} represents pure time delay resulting from the time required to execute the digital processing. In this case, N is an integer, so that the time delay is approximated as an integral multiple of the sampling period T.

REFERENCES

State-Variable Analysis of Electric Networks

1. B. C. Kuo, *Linear Circuits and Systems*, McGraw-Hill Book Company, New York, 1967.
2. R. Rohrer, *Circuit Analysis: An Introduction to the State Variable Approach*, McGraw-Hill Book Company, New York, 1970.

Mechanical Systems

3. R. Cannon, *Dynamics of Physical Systems*, McGraw-Hill Book Company, New York, 1967.

Control System Components

4. W. R. AHRENDT, *Servomechanism Practice*, McGraw-Hill Book Company, New York, 1954.

5. J. E. GIBSON and F. B. TUTEUR, *Control System Components*, McGraw-Hill Book Company, New York, 1958.

Two-Phase Induction Motor

6. W. A. STEIN and G. J. THALER, "Effect of Nonlinearity in a 2-Phase Servomotor," *AIEE Trans.*, Vol. 73, Part II, pp. 518–521, 1954.

7. B. C. KUO, "Studying the Two-Phase Servomotor," *Instrument Soc. Amer. J.*, Vol. 7, No. 4, pp. 64–65, Apr. 1960.

DC Motors

8. B. C. KUO and J. TAL, eds., *Incremental Motion Control*, Vol. 1, *DC Motors and Control Systems*, SRL Publishing Co., Champaign, Ill., 1979.

Step Motors

9. B. C. KUO, ed., *Incremental Motion Control*, Vol. 2, *Step Motors and Control Systems*, SRL Publishing Co., Champaign, Ill., 1980.

10. B. C. KUO, *Theory and Applications of Step Motors*, West Publishing Co., St. Paul, Minn., 1974.

PROBLEMS

4.1. Write the force or torque equations for the mechanical systems shown in Fig. 4P-1. Write the state equations from the force or torque equations.

4.2. Write a set of state equations for the mechanical system shown in Fig. 4P-2. On the first try, one will probably end up with four state equations with the state variables defined as θ_2, ω_2, θ_1, and ω_1. However, it is apparent that there are only three energy-storage elements in J_1, K, and J_2, so the system has a minimum order of three.

    (a) Write the state equations in vector-matrix form with the state variables defined as above.

    (b) Redefine the state variables so that there are only three state equations.

    (c) Draw state diagrams for both cases.

    (d) Derive the transfer function $\Omega_2(s)/T(s)$ for each case, and compare the results.

4.3. For the system shown in Fig. 4P-3, determine the transfer function $E_0(s)/T_m(s)$. The potentiometer rotates through 10 turns, and the voltage applied across the potentiometer terminals is E volts.

4.4. Write the torque equations of the gear-train system shown in Fig. 4P-4. The moments of inertia of the gears and shafts are J_1, J_2, and J_3. $T(t)$ is the applied torque. N denotes the number of gear teeth. Assume rigid shafts.

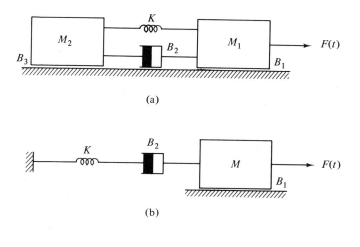

(a)

(b)

(c)

Figure 4P-1

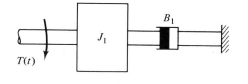

Figure 4P-2

θ_1, ω_1 θ_2, ω_2

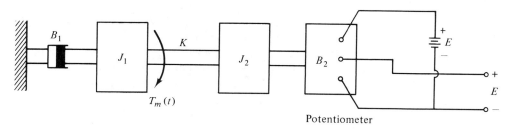

Potentiometer

B_2 = viscous friction coefficient of
potentiometer contact

Figure 4P-3

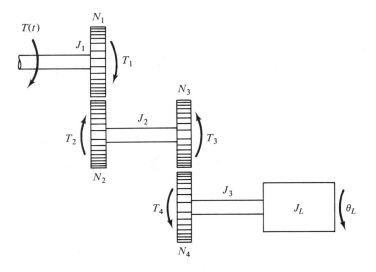

Figure 4P-4

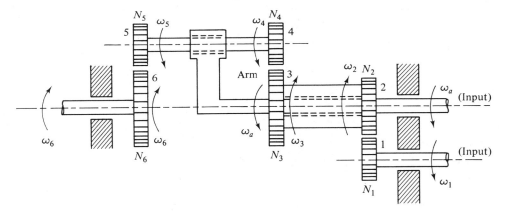

Figure 4P-5

4.5. Figure 4P-5 shows the diagram of an epicyclic gear train.
  (a) Using the reference directions of the angular velocity variables as indicated, write algebraic equations that relate these variables.
  (b) Draw a signal flow graph to relate among the inputs ω_a and ω_1 and the output ω_6.
  (c) Find the transfer function relation between ω_6 and ω_1 and ω_a.

4.6. The block diagram of the automatic braking control of a high-speed train is shown in Fig. 4P-6(a), where

$$V_r = \text{voltage representing desired speed}$$
$$v = \text{velocity of train}$$
$$K = \text{amplifier gain} = 100$$
$$M = \text{mass of train} = 5 \times 10^4 \text{ lb/ft/sec}^2$$
$$K_i = \text{tachometer constant} = 0.15 \text{ V/ft/sec}$$
$$e_t = K_t v$$

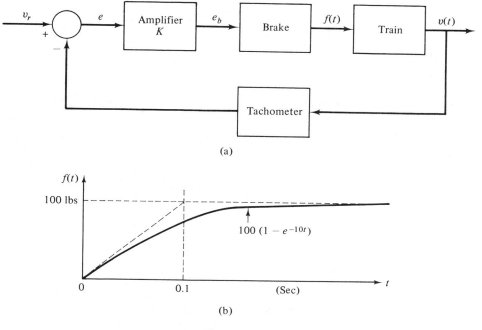

(a)

(b)

Figure 4P-6

The force characteristics of the brake are shown in Fig. 4P-6(b) when $e_b = 1$ V. (Neglect all frictional forces.)

(a) Draw a block diagram of the system and include the transfer function of each block.

(b) Determine the closed-loop transfer function between V_t and velocity v of the train.

(c) If the steady-state velocity of the train is to be maintained at 20 ft/sec, what should be the value of V_t?

4.7. The voltage equation of a dc motor is written as

$$e(t) = Ri(t) + L\frac{di}{dt} + K_b\omega(t) \tag{1}$$

where $e(t)$ = terminal voltage of armature
$i(t)$ = armature current
R = armature resistance
L = armature inductance
K_b = back-emf constant
$\omega(t)$ = motor velocity

Taking the Laplace transform on both sides of Eq. (1) and rearranging, we get

$$\Omega(s) = \frac{E(s) - (R + Ls)I(s)}{K_b} \tag{2}$$

which shows that the velocity information can be generated by feeding back the armature voltage and current. The block diagram in Fig. 4P-7 shows a dc motor system with voltage and current feedbacks for speed control.

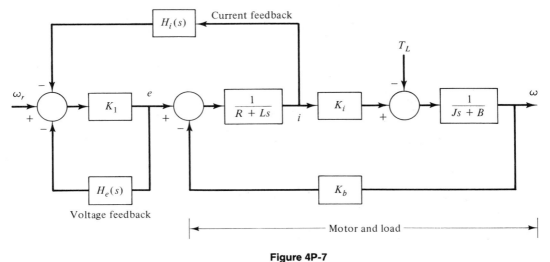

Figure 4P-7

(a) Let K_1 be a very high-gain amplifier. Show that when $H_i(s)/H_e(s) = -(R + Ls)$ the motor velocity $\omega(t)$ is totally independent of the load disturbance torque T_L.

(b) Find the transfer function between $\Omega(s)$ and $\Omega_r(s)$ $(T_L = 0)$ when $H_i(s)$ and $H_e(s)$ are selected as in part (a).

4.8. Figure 4P-8 illustrates a winding process of newsprint. The system parameters and variables are defined as follows:

$$e_a = \text{applied voltage}$$
$$R = \text{armature resistance of dc motor}$$
$$L = \text{armature inductance of dc motor}$$
$$i_a = \text{armature current}$$
$$K_b = \text{back emf of dc motor}$$
$$T_m = \text{motor torque} = K_m i_a$$
$$J_m = \text{motor inertia}$$
$$B_m = \text{motor friction coefficient}$$
$$J_L = \text{inertia of windup reel}$$
$$\omega_m = \text{angular velocity of dc motor}$$
$$\omega = \text{angular velocity of windup reel}$$
$$T_L = \text{torque at the windup reel}$$
$$r = \text{effective radius of windup reel}$$
$$V_w = \text{linear velocity of web at windup reel}$$
$$T = \text{tension}$$
$$V_s = \text{linear velocity of web at input pinch rolls}$$

Assume that the linear velocity at the input pinch rolls, V_s, is constant. The elasticity of the web material is assumed to satisfy Hooke's law; that is, the distortion of the

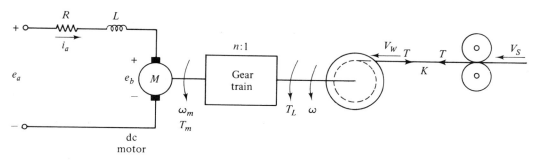

Figure 4P-8

material is directly proportional to the force applied, and the proportional constant is K (force/displacement).

(a) Write the nonlinear state equations for the system using i_a, ω_m, and T as state variables.

(b) Assuming that r is constant, draw a state diagram of the system with e_a and V_s as inputs.

4.9. The schematic diagram of a steel rolling process is shown in Fig. 4P-9.

(a) Describe the system by a set of differential-difference equation of the form of Eq. (4-183).

(b) Derive the transfer function between $c(t)$ and $r(t)$.

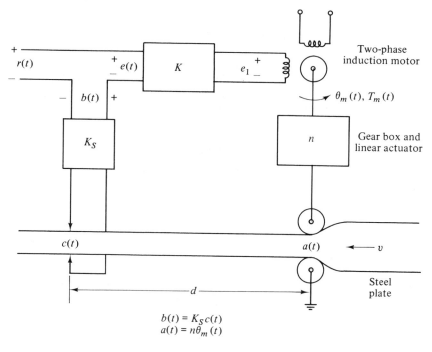

$$b(t) = K_S c(t)$$
$$a(t) = n\theta_m(t)$$

Figure 4P-9

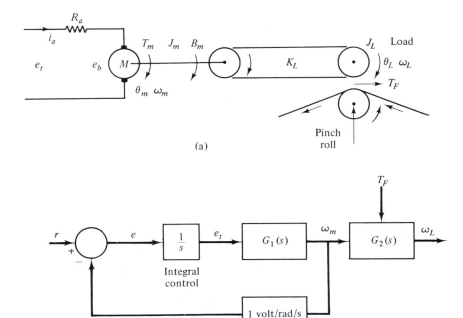

(a)

(b)   Feedback transducer

Figure 4P-10

4.10. Figure 4P-10(a) shows an industrial process in which a dc motor drives a capstan and tape assembly. The objective is to drive the tape at a certain constant speed. Another tape driven by a separate source is made to be in contact with the primary tape by the action of a pinch roll over certain periods of time. When the two tapes are in contact, we may consider that a constant frictional torque of T_F is seen at the load. The following system parameters are defined:

e_t = applied motor voltage, volts

i_a = armature current, amperes

e_b = back emf voltage = $K_b \omega_m$, volts

K_b = back emf constant = 0.052 V/rad/sec

K_i = torque constant = 10 oz-in./A

T_m = torque, oz-in.

θ_m = motor displacement, rad

ω_m = motor speed, rad/sec

R_a = motor resistance = 1 Ω

J_m = motor inertia = 0.1 oz-in./rad/sec^2 (includes capstan inertia)

B_m = motor viscous friction = 0.1 oz-in./rad/sec

K_L = spring constant of tape = 100 oz-in./rad (converted to rotational)

J_L = load inertia = 0.1 oz-in./rad/sec^2

(a) Write the equations of the system in vector-matrix state equation form.
(b) Draw a state diagram for the system.
(c) Derive the transfer function for $\Omega_L(s)$ with $E_t(s)$ and $T_F(s)$ as inputs.
(d) If a constant voltage $e_t(t) = 10$ V is applied to the motor, find the steady-state speed of the motor in rpm when the pinch roll is not activated. What is the steady-state speed of the load?
(e) When the pinch roll is activated, making the two tapes in contact, the constant friction torque T_F is 1 oz-in. Find the change in the steady-state speed ω_L when $e_t = 10$ V.
(f) To overcome the effect of the frictional torque T_F it is suggested that a closed-loop system should be formed as shown by the block diagram in Fig. 4P-10(b). In this case the motor speed is fed back and compared with the reference input. The closed-loop system should give accurate speed control, and the integral control should give better regulation to the frictional torque. Draw a state diagram for the closed-loop system.
(g) Determine the steady-state speed of the load when the input is 1 V. First consider that the pinch roll is not activated, and then is activated.

4.11. This problem deals with the attitude control of a guided missile. When traveling through the atmosphere, a missile encounters aerodynamic forces that usually tend to cause instability in the attitude of the missile. The basic concern from the flight control standpoint is the lateral force of the air, which tends to rotate the missile about its center of gravity. If the missile centerline is not aligned with the direction in which the center of gravity C is traveling, as shown in Fig. 4P-11 with the angle θ (θ is also called the angle of attack), a side force is produced by the resistance of the air through which the missile is traveling. The total force F_α may be considered to be centered at the center of pressure P. As shown in Fig. 4P-11, this side force has a tendency to cause the missile to tumble, especially if the point P is in front of the center of gravity C. Let the angular acceleration of the missile about the point C, due to the side force, be denoted by α_F. Normally, α_F is directly proportional to the angle of attack θ and is given by

$$\alpha_F = a\theta$$

where a is a constant described by

$$a = \frac{K_F d_1}{J}$$

K_F is a constant that depends on such parameters as dynamic pressure, velocity of the

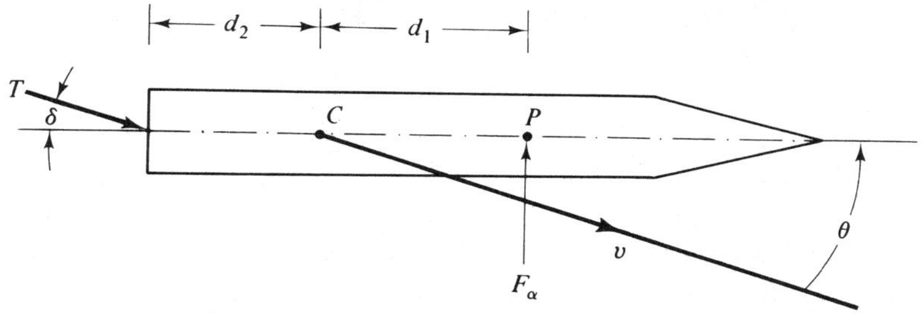

Figure 4P-11

missile, air density, and so on, and

$$J = \text{missile moment of inertia about } C$$
$$d_1 = \text{distance between } C \text{ and } P$$

The main object of the flight control system is to provide the stabilizing action to counter the effect of the side force. One of the standard control means is to use gas injection at the tail of the missile to deflect the direction of the rocket engine thrust T_s, as shown in Fig. 4P-11.

(a) Write a torque differential equation to relate among T, δ, θ, and the system parameters. Assume that δ is very small.

(b) Assume that T_s is constant and find the transfer function $\theta(s)/\delta(s)$ for small δ.

(c) Repeat (a) and (b) with the points C and P interchanged.

4.12. The following equations describe the motion of an electric train in a traction system:

$$\dot{x}(t) = v(t)$$
$$\dot{v}(t) = -k(v) - g(x) + T(t)$$

where $x(t)$ = linear displacement of train
$v(t)$ = linear velocity of train
$k(v)$ = train resistance force [odd function of v, with the properties $k(0) = 0$ and $dk(v)/dv > 0$]
$g(x)$ = force due to gravity for a nonlevel track or due to curvature of track
$T(t)$ = tractive force

The electric motor that provides the traction force is described by the following relations:

$$e(t) = K_b\phi(t)v(t) + Ri_a(t)$$
$$T(t) = K_m\phi(t)i_a(t)$$

where R = armature resistance
$\phi(t)$ = magnetic flux = $K_f i_f(t)$
$e(t)$ = applied voltage
K_m, K_b = proportional constants
$i_a(t)$ = armature current
$i_f(t)$ = field current

(a) Consider that the motor is a dc series motor so that $i_a(t) = i(t)$; $g(x) = 0$, $k(v) = Bv(t)$, and $R = 0$. The voltage $e(t)$ is the input. Show that the system is described by the following set of nonlinear state equations:

$$\dot{x}(t) = v(t)$$
$$\dot{v}(t) = -Bv(t) + \frac{K_m}{K_b^2 K_f v^2(t)} e^2(t)$$

(b) Consider that $i_a(t) = i_f(t)$ is the input and derive the state equations of the system. $g(x) = 0$, $k(v) = Bv(t)$.

(c) Consider that $\phi(t)$ is the input, $g(x) = 0$, $k(v) = Bv(t)$, and derive the state equations of the system.

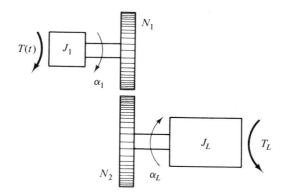

Figure 4P-13

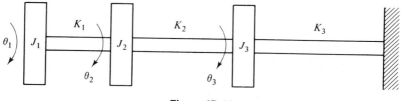

Figure 4P-14

4.13. Figure 4P-13 shows a gear-coupled mechanical system.
   (a) Find the optimum gear ratio n such that the load acceleration, α_L, is maximized.
   (b) Repeat part (a) when the load drag torque T_L is zero.

4.14. (a) Write the torque equations of the system in Fig. 4P-14 in the form

$$\ddot{\theta} + J^{-1}K\theta = 0$$

where θ is a 3×1 vector that contains all the displacement variables, θ_1, θ_2, and θ_3. J is the inertia matrix and K contains all the spring constants. Determine J and K.

   (b) Show that the torque equations can be expressed as a set of state equations of the form

$$\dot{x} = Ax$$

   where

$$A = \left[\begin{array}{c|c} 0 & I \\ \hline -J^{-1}K & 0 \end{array} \right]$$

   (c) Consider the following set of parameters with consistent units: $K_1 = 1000$, $K_2 = 3000$, $J_1 = 1$, $J_2 = 5$, $J_3 = 2$, and $K_3 = 1000$. Find the matrix A.

4.15. Figure 4P-15 shows the layout of the control of the unwind process of a cable reel with the object of maintaining constant linear cable velocity. Control is established by measuring the cable velocity, comparing it with a reference signal, and using the error to generate a control signal. A tachometer is used to sense the cable velocity. To maintain a constant linear cable velocity, the angular reel velocity $\dot{\theta}_R$ must increase as

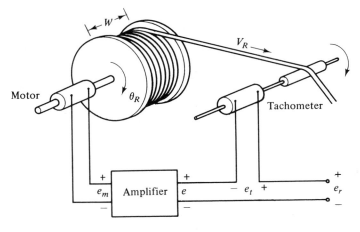

Figure 4P-15

the cable unwinds; that is, as the effective radius of the reel decreases. Let

$$D = \text{cable diameter} = 0.1 \text{ ft}$$
$$W = \text{width of reel} = 2 \text{ ft}$$
$$R_0 = \text{effective radius of reel (empty reel)} = 2 \text{ ft}$$
$$R_f = \text{effective radius of reel (full reel)} = 4 \text{ ft}$$
$$R = \text{effective radius of reel}$$
$$J_R = \text{moment of inertia of reel} = 18R^4 - 200 \text{ ft-lb-sec}^2$$
$$v_R = \text{linear speed of cable, ft/sec}$$
$$e_t = \text{output voltage of tachometer, volts}$$
$$T_m(t) = \text{motor torque, ft-lb}$$
$$e_m(t) = \text{motor input voltage, volts}$$
$$K = \text{amplifier gain}$$

Motor inertia and friction are negligible. The tachometer transfer function is

$$\frac{E_t(s)}{V_R(s)} = \frac{1}{1 + 0.5s}$$

and the motor transfer function is

$$\frac{T_m(s)}{E_m(s)} = \frac{50}{s + 1}$$

(a) Write an expression to describe the change of the radius of the reel R as a function of θ_R.

(b) Between layers of the cable, R and J_R are assumed to be constant, and the system is considered linear. Draw a block diagram for the system and indicate all the transfer functions. The input is e_r and the output is v_R.

(c) Derive the closed-loop transfer function $V_R(s)/E_r(s)$.

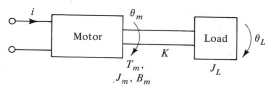

Figure 4P-16

4.16. Figure 4P-16 shows an inertial load driven by a motor. The following system variables and parameters are given:

Motor: i = motor current
K_i = motor torque constant
T_m = motor torque = $K_i i$
J_m = motor inertia
B_m = motor viscous frictional coefficient
θ_m = motor displacement

Load: K = torsional spring constant of shaft
J_L = load inertia
θ_L = load displacement

(a) Write the differential equations of the system.
(b) Draw a signal flow graph using $I(s)$, $s^2\Theta_m(s)$, $s\Theta_m(s)$, $\Theta_m(s)$, $s^2\Theta_L(s)$, $s\Theta_L(s)$, and $\Theta_L(s)$ as node variables.
(c) Derive the transfer function $\Theta_L(s)/I(s)$.

4.17. The linear model of a robot arm system driven by a motor is shown in Fig. 4P-17. The system parameters and variables are given as follows:

Motor: T_m = motor torque = $K_i i$
K_i = torque constant
i = armature current
J_m = motor inertia
B_m = motor viscous frictional coefficient
θ_m = motor shaft displacement

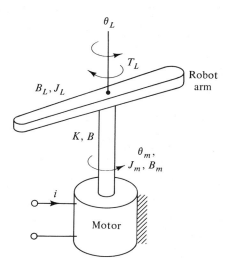

Figure 4P-17

Robot: J_L = inertia of arm
            T_L = disturbance torque on arm
            θ_L = displacement of robot arm
            K = torsional spring constant of shaft between motor and arm
            B = viscous damping coefficient of shaft between motor and arm
            B_L = viscous frictional coefficient of robot arm shaft

(a) Write the differential equations of the system with i and T_L as the inputs, and $\Theta_m(s)$ and $\Theta_L(s)$ as outputs.

(b) Draw a signal flow graph using $I(s)$, $T_L(s)$, $\Theta_M(s)$, and $\Theta_L(s)$ as node variables.

(c) Express the transfer function relations as

$$\begin{bmatrix} \Theta_m(s) \\ \Theta_L(s) \end{bmatrix} = \mathbf{G}(s) \begin{bmatrix} I(s) \\ -T_L(s) \end{bmatrix}$$

Find $\mathbf{G}(s)$.

4.18. Figure 4P-18 shows the schematic diagram of a ball-suspension control system. The steel ball is suspended in the air by the electromagnetic force generated by the electromagnet. The objective of the control is to keep the metal ball suspended at the nominal equilibrium position by controlling the current in the magnet with the voltage $e(t)$. The resistance of the coil is R, and the inductance is

$$L(y) = \frac{L}{y}$$

where L is a constant. The applied voltage $e(t)$ is a constant with amplitude E.

(a) Let E_{eq} be the nominal value of E. Find the nominal values of y and $\dot{y}$ at equilibrium.

(b) Define the state variables as

$$x_1 = i \qquad x_2 = y \qquad x_3 = \dot{y}$$

Find the nonlinear state equations in the form of

$$\dot{\mathbf{x}} = \mathbf{f}(\mathbf{x}, e)$$

(c) Linearize the state equations about the equilibrium point and express the linearized state equations as

$$\delta\dot{\mathbf{x}} = A^* \, \delta\mathbf{x} + B^* \, \delta e$$

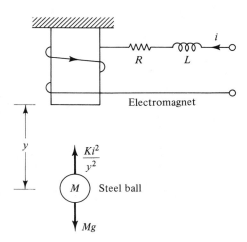

Figure 4P-18

The force generated by the electromagnet is Ki^2/y^2, where K is a proportional constant, and the gravitational force on the steel ball is Mg.

4.19. Figure 4P-19 shows the schematic diagram of a ball-suspension control system. The steel ball is suspended in the air by the electromagnetic force generated by the electromagnet. The objective of the control is to keep the metal ball suspended at the nominal position by controlling the current in the electromagnet. When the system is at the stable equilibrium point, any small perturbation of the ball position from its floating equilibrium position will cause the control to return the ball to the equilibrium position.

The free-body diagram of the system is given in Fig. 4P-19, where

$$M_1 = \text{mass of electromagnet} = 2$$
$$M_2 = \text{mass of steel ball} = 1$$
$$B = \text{viscous frictional coefficient of air} = 0.1$$
$$K = \text{proportional constant of electromagnet} = 1$$
$$g = \text{acceleration due to gravity} = 32.2$$

Let the stable equilibrium values of the variables i, y_1, and y_2 be I, Y_1, and Y_2, respectively.

(a) Given $Y_1 = 1$, find Y_2 and I.

(b) Define the state variables as

$$x_1 = y_1 \qquad x_2 = \dot{y}_1 \qquad x_3 = y_2 \qquad x_4 = \dot{y}_2$$

(i) Write the nonlinear state equations of the system in the form of $\dot{x} = f(x, i)$.

(ii) Find the state equations of the linearized system about the equilibrium state I, Y_1, and Y_2:

$$\delta \dot{\mathbf{x}} = \mathbf{A}^* \, \delta \mathbf{x} + \mathbf{B}^* \, \delta i$$

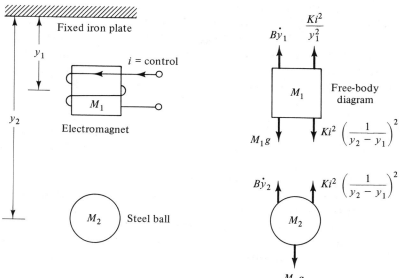

Figure 4P-19

chapter five

State-Variable Analysis of Linear Dynamic Systems

5.1 INTRODUCTION

In Chapter 2 the state equations of a dynamic system are defined simply as a set of first-order differential equations relating the state variables among themselves and to the inputs. In Chapter 3 the signal-flow-graph method is extended to state equations and the result is the *state diagram*. The concept and definition of state variables are formally presented in Chapter 4, together with several illustrative examples of how state equations are written for dynamic systems.

In contrast to the transfer function approach to the analysis and design of control systems, the state-variable method is regarded as modern, since it is the underlying force for optimal control. The basic characteristic of the state-variable formulation is that linear and nonlinear systems, time-invariant and time-varying systems, single-variable and multivariable systems can all be modeled in a unified manner. Transfer functions, on the other hand, are defined only for linear time-invariant systems.

The objective of this chapter is to introduce the fundamental concepts of state variables and state equations, so that the reader can gain a working knowledge of the subject for further studies when the state approach is used for optimal control design. Specifically, the closed-form solutions of linear time-invariant state equations are presented. Various transformations that may be used to facilitate the analysis and design of linear control systems in the state-variable domain are introduced. The relationship between the conventional transfer function approach and the state-variable approach is established, so that the analyst will be able to investigate a system problem with various alternative methods. Finally, the controllability and observability of linear systems are defined and their applications are investigated.

202

5.2 MATRIX REPRESENTATION OF STATE EQUATIONS

Let the n state equations of an nth-order dynamic system be represented as

$$\frac{dx_i(t)}{dt} = f_i\big[x_1(t), x_2(t), \ldots, x_n(t),$$

$$r_1(t), r_2(t), \ldots, r_p(t), w_1(t), w_2(t), \ldots, w_v(t)\big] \quad (5\text{-}1)$$

$i = 1, 2, \ldots, n$. The ith state variable is represented by $x_i(t)$, $r_j(t)$ denotes the jth input for $j = 1, 2, \ldots, p$, and $w_k(t)$ denotes the kth disturbance input, with $k = 1, 2, \ldots, v$.

Let the variables $c_1(t), c_2(t), \ldots, c_q(t)$ be the q output variables of the system. The output variables represent the link between the system and the outside world. In practice, not all the state variables are accessible, or, simply, measurable. However, an output variable must be measurable or accessible by some physical means. For instance, in an electric motor, such state variables as the winding current, rotor velocity, and displacement can be measured physically, and these variables all qualify as output variables. On the other hand, magnetic flux can also be regarded as a state variable in an electric motor; but it cannot be measured directly, and therefore it does not ordinarily qualify as an output variable.

In general, the output variables are functions of the state variables and the input variables. The *output equations* of a dynamic system can be expressed as

$$c_j(t) = q_j\big[x_1(t), x_2(t), \ldots, x_n(t), r_1(t), r_2(t), \ldots, r_p(t), w_1(t), w_2(t), \ldots, w_v(t)\big]$$
$$(5\text{-}2)$$

$j = 1, 2, \ldots, q$.

The set of n state equations in Eq. (5-1) and q output equations in Eq. (5-2) together form the so-called *dynamic equations*.

For ease of expression and manipulation it is convenient to represent the dynamic equations in vector-matrix form. Let us define the following column matrices, which are also called *vectors*:

$$\mathbf{x}(t) = \begin{bmatrix} x_1(t) \\ x_2(t) \\ \vdots \\ x_n(t) \end{bmatrix} \quad (n \times 1) \quad (5\text{-}3)$$

$$\mathbf{r}(t) = \begin{bmatrix} r_1(t) \\ r_2(t) \\ \vdots \\ r_p(t) \end{bmatrix} \quad (p \times 1) \quad (5\text{-}4)$$

$$\mathbf{c}(t) = \begin{bmatrix} c_1(t) \\ c_2(t) \\ \vdots \\ c_q(t) \end{bmatrix} \qquad (q \times 1) \tag{5-5}$$

$$\mathbf{w}(t) = \begin{bmatrix} w_1(t) \\ w_2(t) \\ \vdots \\ w_v(t) \end{bmatrix} \qquad (v \times 1) \tag{5-6}$$

The $n \times 1$ column matrix $\mathbf{x}(t)$ is called the *state vector*; $\mathbf{r}(t)$ is the $p \times 1$ *input vector*, $\mathbf{c}(t)$ is the $q \times 1$ *output vector*, and $\mathbf{w}(t)$ is the $v \times 1$ *disturbance vector*.

Using these vectors, the n state equations of Eq. (5-1) can be written

$$\frac{d\mathbf{x}(t)}{dt} = \mathbf{f}[\mathbf{x}(t), \mathbf{r}(t), \mathbf{w}(t)] \tag{5-7}$$

where $\mathbf{f}$ denotes an $n \times 1$ column matrix that contains the function $f_1, f_2, \ldots, f_n$, as elements. Similarly, the q output equations in Eq. (5-2) become

$$\mathbf{c}(t) = \mathbf{g}[\mathbf{x}(t), \mathbf{r}(t), \mathbf{w}(t)] \tag{5-8}$$

where $\mathbf{g}$ denotes a $q \times 1$ column matrix that contains the functions $g_1, g_2, \ldots, g_q$ as elements.

For a linear time-invariant system, the dynamic equations are written as

State equations: $\qquad \dfrac{d\mathbf{x}(t)}{dt} = \mathbf{A}\mathbf{x}(t) + \mathbf{B}\mathbf{r}(t) + \mathbf{F}\mathbf{w}(t) \tag{5-9}$

Output equations: $\qquad \mathbf{c}(t) = \mathbf{D}\mathbf{x}(t) + \mathbf{E}\mathbf{r}(t) + \mathbf{H}\mathbf{w}(t) \tag{5-10}$

where $\mathbf{A}$ is an $n \times n$ coefficient matrix with constant elements,

$$\mathbf{A} = \begin{bmatrix} a_{11} & a_{12} & \cdots & a_{1n} \\ a_{21} & a_{22} & \cdots & a_{2n} \\ \vdots & & & \vdots \\ a_{n1} & a_{n2} & \cdots & a_{nn} \end{bmatrix} \tag{5-11}$$

$\mathbf{B}$ is an $n \times p$ coefficient matrix with constant elements,

$$\mathbf{B} = \begin{bmatrix} b_{11} & b_{12} & \cdots & b_{1p} \\ b_{21} & b_{22} & \cdots & b_{2p} \\ \vdots & & & \vdots \\ b_{n1} & b_{n2} & \cdots & b_{np} \end{bmatrix} \tag{5-12}$$

D is a $q \times n$ coefficient matrix with constant elements,

$$\mathbf{D} = \begin{bmatrix} d_{11} & d_{12} & \cdots & d_{1n} \\ d_{21} & d_{22} & \cdots & d_{2n} \\ \vdots & & & \vdots \\ d_{q1} & d_{q2} & \cdots & d_{qn} \end{bmatrix} \tag{5-13}$$

E is a $q \times p$ coefficient matrix with constant elements,

$$\mathbf{E} = \begin{bmatrix} e_{11} & e_{12} & \cdots & e_{1p} \\ e_{21} & e_{22} & \cdots & e_{2p} \\ \vdots & & & \vdots \\ e_{q1} & e_{q2} & \cdots & e_{qp} \end{bmatrix} \tag{5-14}$$

F is an $n \times v$ coefficient matrix with constant elements,

$$\mathbf{F} = \begin{bmatrix} f_{11} & f_{12} & \cdots & f_{1v} \\ f_{21} & f_{22} & \cdots & f_{2v} \\ \vdots & & & \vdots \\ f_{n1} & f_{n2} & \cdots & f_{nv} \end{bmatrix} \tag{5-15}$$

and **H** is a $q \times v$ coefficient matrix with constant elements,

$$\mathbf{H} = \begin{bmatrix} h_{11} & h_{12} & \cdots & h_{1v} \\ h_{21} & h_{22} & \cdots & h_{2v} \\ \vdots & & & \vdots \\ h_{q1} & h_{q2} & \cdots & h_{qv} \end{bmatrix} \tag{5-16}$$

5.3 STATE TRANSITION MATRIX

Once the state equations of a linear time-invariant system are expressed in the form of Eq. (5-9), the next step often involves the solutions of these equations. The first term on the right-hand side of Eq. (5-9) is known as the homogeneous part of the state equation, and the last two terms represent the forcing functions $\mathbf{r}(t)$ and $\mathbf{w}(t)$.

The *state transition matrix* is defined as a matrix that satisfies the linear homogeneous state equation

$$\frac{d\mathbf{x}(t)}{dt} = \mathbf{A}\mathbf{x}(t) \tag{5-17}$$

Let $\boldsymbol{\phi}(t)$ be an $n \times n$ matrix that represents the state transition matrix; then it must satisfy the equation

$$\frac{d\boldsymbol{\phi}(t)}{dt} = \mathbf{A}\boldsymbol{\phi}(t) \tag{5-18}$$

Furthermore, let $\mathbf{x}(0)$ denote the initial state at $t = 0$; then $\boldsymbol{\phi}(t)$ is also defined by the matrix equation

$$\mathbf{x}(t) = \boldsymbol{\phi}(t)\mathbf{x}(0) \tag{5-19}$$

which is the solution of the homogeneous state equation for $t \geq 0$.

One way of determining $\boldsymbol{\phi}(t)$ is by taking the Laplace transform on both sides of Eq. (5-17); we have

$$s\mathbf{X}(s) - \mathbf{x}(0) = \mathbf{A}\mathbf{X}(s) \tag{5-20}$$

Solving for $\mathbf{X}(s)$ from the last equation, we get

$$\mathbf{X}(s) = (s\mathbf{I} - \mathbf{A})^{-1}\mathbf{x}(0) \tag{5-21}$$

where it is assumed that the matrix $(s\mathbf{I} - \mathbf{A})$ is nonsingular. Taking the inverse Laplace transform on both sides of the last equation yields

$$\mathbf{x}(t) = \mathcal{L}^{-1}\left[(s\mathbf{I} - \mathbf{A})^{-1}\right]\mathbf{x}(0) \qquad t \geq 0 \tag{5-22}$$

Comparing Eq. (5-19) with Eq. (5-22), the state transition matrix is identified to be

$$\boldsymbol{\phi}(t) = \mathcal{L}^{-1}\left[(s\mathbf{I} - \mathbf{A})^{-1}\right] \tag{5-23}$$

An alternative way of solving the homogeneous state equation is to assume a solution, as in the classical method of solving differential equations. We let the solution to Eq. (5-17) be

$$\mathbf{x}(t) = e^{\mathbf{A}t}\mathbf{x}(0) \tag{5-24}$$

for $t \geq 0$, where $e^{\mathbf{A}t}$ represents a power series of the matrix $\mathbf{A}t$ and

$$e^{\mathbf{A}t} = \mathbf{I} + \mathbf{A}t + \frac{1}{2!}\mathbf{A}^2 t^2 + \frac{1}{3!}\mathbf{A}^3 t^3 + \cdots \tag{5-25}[1]$$

It is easy to show that Eq. (5-24) is a solution of the homogeneous state equation, since, from Eq. (5-25),

$$\frac{de^{\mathbf{A}t}}{dt} = \mathbf{A}e^{\mathbf{A}t} \tag{5-26}$$

[1] It can be proved that this power series is uniformly convergent.

Therefore, in addition to Eq. (5-23), we have obtained another expression for the state transition matrix in

$$\phi(t) = e^{\mathbf{A}t} = \mathbf{I} + \mathbf{A}t + \frac{1}{2!}\mathbf{A}^2 t^2 + \frac{1}{3!}\mathbf{A}^3 t^3 + \cdots \tag{5-27}$$

Equation (5-27) can also be obtained directly from Eq. (5-23). This is left as an exercise for the reader (Problem 5.3).

■

Example 5-1

As a simple illustrative example of state variables, state equations, and the state transition matrix, let us consider the *RL* network shown in Fig. 5-1. The history of the network is completely specified by the initial current of the inductance, $i(0)$ at $t = 0$. Consider that at $t = 0$, a constant input voltage of magnitude E is applied to the network. The state equation of the network for $t \geq 0$ is written

$$\frac{di(t)}{dt} = -\frac{R}{L}i(t) + \frac{1}{L}e(t) \tag{5-28}$$

In this case the state equation is of the first order, or scalar. Comparing with Eq. (5-9), $i(t)$ is the state variable $x(t)$, $e(t)$ is the input $r(t)$, $w(t) = 0$, $A = -R/L$, $B = 1/L$, and $F = 0$. Taking the Laplace transform on both sides of the last equation, we get

$$sI(s) - i(0) = \frac{-R}{L}I(s) + \frac{E}{Ls} \tag{5-29}$$

Solving for $I(s)$ from Eq. (5-29) yields

$$I(s) = \frac{L}{R + Ls}i(0) + \frac{E}{s(R + Ls)} \tag{5-30}$$

The current $i(t)$ for $t \geq 0$ is obtained by taking the inverse Laplace transform on both sides of the last equation. We have

$$i(t) = e^{-Rt/L}i(0) + \frac{E}{R}(1 - e^{-Rt/L}) \tag{5-31}$$

Equation (5-31) represents the complete solution of the state equation with the current $i(t)$ as the state variable for the specified initial state $i(0)$ and the specific input $e(t)$ for $t \geq 0$. It is apparent that $i(t)$ satisfies the basic requirements as a state variable. This is not surprising since an inductor is an electric element that stores kinetic energy, and it is the energy-storage

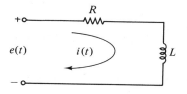

Figure 5-1 *RL* network.

capability that holds the information on the history of the system. Similarly, in general, the voltage across a capacitor also qualifies as a state variable.

The first term on the right-hand side of Eq. (5-31) is the solution of the homogeneous state equation, that is, the state equation with $e(t) = 0$. Thus, the state transition matrix of A, where $A = -R/L$, is

$$\phi(t) = e^{-Rt/L} \tag{5-32}$$

Apparently, $\phi(t)$ can also be obtained directly by substituting A into Eq. (5-27).

∎

Significance of the State Transition Matrix

Since the state transition matrix satisfies the homogeneous state equation, it represents the *free response* of the system. In other words, it governs the response that is excited by the initial conditions only. In view of Eqs. (5-23) and (5-27), the state transition matrix is dependent only upon the matrix **A**, and therefore is sometimes referred to as the *state transition matrix of* **A**. As the name implies, the state transition matrix $\phi(t)$ completely defines the transition of the states from the initial time $t = 0$ to any time t when the inputs are zero.

Properties of the State Transition Matrix

The state transition matrix $\phi(t)$ possesses the following properties:

1. $$\phi(0) = \mathbf{I} \qquad \text{the identity matrix} \tag{5-33}$$

   Proof. Equation (5-33) follows directly from Eq. (5-27) by setting $t = 0$.

2. $$\phi^{-1}(t) = \phi(-t) \tag{5-34}$$

   Proof. Postmultiplying both sides of Eq. (5-27) by $e^{-\mathbf{A}t}$, we get

$$\phi(t)e^{-\mathbf{A}t} = e^{\mathbf{A}t}e^{-\mathbf{A}t} = \mathbf{I} \tag{5-35}$$

Then premultiplying both sides of Eq. (5-35) by $\phi^{-1}(t)$, we get

$$e^{-\mathbf{A}t} = \phi^{-1}(t) \tag{5-36}$$

Thus,

$$\phi(-t) = \phi^{-1}(t) = e^{-\mathbf{A}t} \tag{5-37}$$

An interesting result from this property of $\phi(t)$ is that Eq. (5-24) can be rearranged to read

$$\mathbf{x}(0) = \phi(-t)\mathbf{x}(t) \tag{5-38}$$

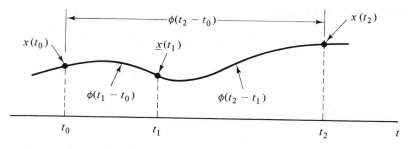

Figure 5-2 Property of the state transition matrix.

which means that the state transition process can be considered as bilateral in time. That is, the transition in time can take place in either direction.

3. $$\boldsymbol{\phi}(t_2 - t_1)\boldsymbol{\phi}(t_1 - t_0) = \boldsymbol{\phi}(t_2 - t_0) \qquad \text{for any } t_0, t_1, t_2 \qquad (5\text{-}39)$$

Proof.

$$\begin{aligned} \boldsymbol{\phi}(t_2 - t_1)\boldsymbol{\phi}(t_1 - t_0) &= e^{\mathbf{A}(t_2 - t_1)}e^{\mathbf{A}(t_1 - t_0)} \\ &= e^{\mathbf{A}(t_2 - t_0)} \\ &= \boldsymbol{\phi}(t_2 - t_0) \end{aligned} \qquad (5\text{-}40)$$

This property of the state transition matrix is important since it implies that a state transition process can be divided into a number of sequential transitions. Figure 5-2 illustrates that the transition from $t = t_0$ to $t = t_2$ is equal to the transition from t_0 to t_1, and then from t_1 to t_2. In general, of course, the transition process can be broken up into any number of parts.

Another way of proving Eq. (5-39) is to write

$$\mathbf{x}(t_2) = \boldsymbol{\phi}(t_2 - t_1)\mathbf{x}(t_1) \qquad (5\text{-}41)$$

$$\mathbf{x}(t_1) = \boldsymbol{\phi}(t_1 - t_0)\mathbf{x}(t_0) \qquad (5\text{-}42)$$

$$\mathbf{x}(t_2) = \boldsymbol{\phi}(t_2 - t_0)\mathbf{x}(t_0) \qquad (5\text{-}43)$$

The proper result is obtained by substituting Eq. (5-42) into Eq. (5-41) and comparing the result with Eq. (5-43).

4. $$[\boldsymbol{\phi}(t)]^k = \boldsymbol{\phi}(kt) \qquad \text{for } k = \text{integer} \qquad (5\text{-}44)$$

Proof.

$$\begin{aligned} [\boldsymbol{\phi}(t)]^k &= e^{\mathbf{A}t}e^{\mathbf{A}t} \cdots e^{\mathbf{A}t} \qquad (k \text{ terms}) \\ &= e^{k\mathbf{A}t} \\ &= \boldsymbol{\phi}(kt) \end{aligned} \qquad (5\text{-}45)$$

5.4 STATE TRANSITION EQUATION

The *state transition equation* is defined as the solution of the linear nonhomogeneous state equation. For example, Eq. (5-28) is a state equation of the RL network of Fig. 5-1. Then Eq. (5-31) is the state transition equation when the input voltage is constant of amplitude E for $t \geq 0$.

In general, the linear time-invariant state equation

$$\frac{d\mathbf{x}(t)}{dt} = \mathbf{A}\mathbf{x}(t) + \mathbf{B}\mathbf{r}(t) + \mathbf{F}\mathbf{w}(t) \qquad (5\text{-}46)$$

can be solved by using either the classical method of solving differential equations or the Laplace transform method. The Laplace transform method is presented in the following.

Taking the Laplace transform on both sides of Eq. (5-46), we have

$$s\mathbf{X}(s) - \mathbf{x}(0) = \mathbf{A}\mathbf{X}(s) + \mathbf{B}\mathbf{R}(s) + \mathbf{F}\mathbf{W}(s) \qquad (5\text{-}47)$$

where $\mathbf{x}(0)$ denotes the initial state vector evaluated at $t = 0$. Solving for $\mathbf{X}(s)$ in Eq. (5-47) yields

$$\mathbf{X}(s) = (s\mathbf{I} - \mathbf{A})^{-1}\mathbf{x}(0) + (s\mathbf{I} - \mathbf{A})^{-1}[\mathbf{B}\mathbf{R}(s) + \mathbf{F}\mathbf{W}(s)] \qquad (5\text{-}48)$$

The state transition equation of Eq. (5-46) is obtained by taking the inverse Laplace transform on both sides of Eq. (5-48).

$$\mathbf{x}(t) = \mathcal{L}^{-1}\big[(s\mathbf{I} - \mathbf{A})^{-1}\big]\mathbf{x}(0) + \mathcal{L}^{-1}\big\{(s\mathbf{I} - \mathbf{A})^{-1}[\mathbf{B}\mathbf{R}(s) + \mathbf{F}\mathbf{W}(s)]\big\} \qquad (5\text{-}49)$$

Using the definition of the state transition matrix of Eq. (5-23), and the convolution integral, Eq. (2-39), Eq. (5-49) is written

$$\mathbf{x}(t) = \boldsymbol{\phi}(t)\mathbf{x}(0) + \int_0^t \boldsymbol{\phi}(t - \tau)[\mathbf{B}\mathbf{r}(\tau) + \mathbf{F}\mathbf{w}(\tau)]\, d\tau \qquad t \geq 0 \qquad (5\text{-}50)$$

The state transition equation in Eq. (5-50) is useful only when the initial time is defined to be at $t = 0$. In the study of control systems, especially discrete-data control systems, it is often desirable to break up a state transition process into a sequence of transitions, so that a more flexible initial time must be chosen. Let the initial time be represented by t_0 and the corresponding initial state by $\mathbf{x}(t_0)$, and assume that the input $\mathbf{r}(t)$ and the disturbance $\mathbf{w}(t)$ are applied at $t \geq 0$.

We start with Eq. (5-50) by setting $t = t_0$, and solving for $\mathbf{x}(0)$, we get

$$\mathbf{x}(0) = \boldsymbol{\phi}(-t_0)\mathbf{x}(t_0) - \boldsymbol{\phi}(-t_0)\int_0^{t_0} \boldsymbol{\phi}(t_0 - \tau)[\mathbf{B}\mathbf{r}(\tau) + \mathbf{F}\mathbf{w}(\tau)]\, d\tau \qquad (5\text{-}51)$$

where the property on $\boldsymbol{\phi}(t)$ of Eq. (5-34) has been used.

Substituting Eq. (5-51) into Eq. (5-50) yields

$$\mathbf{x}(t) = \boldsymbol{\phi}(t)\boldsymbol{\phi}(-t_0)\mathbf{x}(t_0) - \boldsymbol{\phi}(t)\boldsymbol{\phi}(-t_0)\int_0^{t_0}\boldsymbol{\phi}(t_0 - \tau)[\mathbf{Br}(\tau) + \mathbf{Fw}(\tau)]\,d\tau$$

$$+ \int_0^t\boldsymbol{\phi}(t - \tau)[\mathbf{Br}(\tau) + \mathbf{Fw}(\tau)]\,d\tau \quad (5\text{-}52)$$

Now using the property of Eq. (5-39), and combining the last two integrals, Eq. (5-52) becomes

$$\mathbf{x}(t) = \boldsymbol{\phi}(t - t_0)\mathbf{x}(t_0) + \int_{t_0}^t\boldsymbol{\phi}(t - \tau)[\mathbf{Br}(\tau) + \mathbf{Fw}(\tau)]\,d\tau \quad (5\text{-}53)$$

It is apparent that Eq. (5-53) reverts to Eq. (5-50) when $t_0 = 0$.

Once the state transition equation is determined, the output vector can be expressed as a function of the initial state and the input vector simply by substituting $\mathbf{x}(t)$ from Eq. (5-53) into Eq. (5-10). Thus the output vector is written

$$\mathbf{c}(t) = \mathbf{D}\boldsymbol{\phi}(t - t_0)\mathbf{x}(t_0) + \int_{t_0}^t\mathbf{D}\boldsymbol{\phi}(t - \tau)[\mathbf{Br}(\tau) + \mathbf{Fw}(\tau)]\,d\tau + \mathbf{Er}(t) + \mathbf{Hw}(t)$$

$$(5\text{-}54)$$

The following example illustrates the application of the state transition equation.

∎

Example 5-2

Consider the state equation

$$\begin{bmatrix} \dfrac{dx_1(t)}{dt} \\[2ex] \dfrac{dx_2(t)}{dt} \end{bmatrix} = \begin{bmatrix} 0 & 1 \\ -2 & -3 \end{bmatrix}\begin{bmatrix} x_1(t) \\ x_2(t) \end{bmatrix} + \begin{bmatrix} 0 \\ 1 \end{bmatrix}r(t) \quad (5\text{-}55)$$

The problem is to determine the state vector $\mathbf{x}(t)$ for $t \geq 0$ when the input $r(t) = 1$ for $t \geq 0$; that is, $r(t) = u_s(t)$. The coefficient matrices are identified to be

$$\mathbf{A} = \begin{bmatrix} 0 & 1 \\ -2 & -3 \end{bmatrix} \quad \mathbf{B} = \begin{bmatrix} 0 \\ 1 \end{bmatrix} \quad \mathbf{F} = 0 \quad (5\text{-}56)$$

Therefore,

$$s\mathbf{I} - \mathbf{A} = \begin{bmatrix} s & 0 \\ 0 & s \end{bmatrix} - \begin{bmatrix} 0 & 1 \\ -2 & -3 \end{bmatrix} = \begin{bmatrix} s & -1 \\ 2 & s+3 \end{bmatrix} \quad (5\text{-}57)$$

The matrix of $(s\mathbf{I} - \mathbf{A})$ is

$$(s\mathbf{I} - \mathbf{A})^{-1} = \frac{1}{s^2 + 3s + 2}\begin{bmatrix} s+3 & 1 \\ -2 & s \end{bmatrix} \quad (5\text{-}58)$$

The state transition matrix of $\mathbf{A}$ is found by taking the inverse Laplace transform of the last equation. Thus,

$$\boldsymbol{\phi}(t) = \mathcal{L}^{-1}\left[(s\mathbf{I} - \mathbf{A})^{-1}\right] = \begin{bmatrix} 2e^{-t} - e^{-2t} & e^{-t} - e^{-2t} \\ -2e^{-t} + 2e^{-2t} & -e^{-t} + 2e^{-2t} \end{bmatrix} \tag{5-59}$$

The state transition equation for $t \geq 0$ is obtained by substituting Eq. (5-59), $\mathbf{B}$, and $r(t)$ into Eq. (5-50). We have

$$\mathbf{x}(t) = \begin{bmatrix} 2e^{-t} - e^{-2t} & e^{-t} - e^{-2t} \\ -2e^{-t} + e^{-2t} & -e^{-t} + 2e^{-2t} \end{bmatrix} \mathbf{x}(0)$$

$$+ \int_0^t \begin{bmatrix} 2e^{-(t-\tau)} - e^{-2(t-\tau)} & e^{-(t-\tau)} - e^{-2(t-\tau)} \\ -2e^{-(t-\tau)} + 2e^{-2(t-\tau)} & -e^{-(t-\tau)} + 2e^{-2(t-\tau)} \end{bmatrix} \begin{bmatrix} 0 \\ 1 \end{bmatrix} d\tau \tag{5-60}$$

or

$$\mathbf{x}(t) = \begin{bmatrix} 2e^{-t} - e^{-2t} & e^{-t} - e^{-2t} \\ -2e^{-t} + 2e^{-2t} & -e^{-t} + 2e^{-2t} \end{bmatrix} \mathbf{x}(0) + \begin{bmatrix} \frac{1}{2} - e^{-t} + \frac{1}{2}e^{-2t} \\ e^{-t} - e^{-2t} \end{bmatrix} \qquad t \geq 0 \tag{5-61}$$

As an alternative, the second term of the state transition equation can be obtained by taking the inverse Laplace transform of $(s\mathbf{I} - \mathbf{A})^{-1}\mathbf{B}R(s)$. Therefore,

$$\mathcal{L}^{-1}\left[(s\mathbf{I} - \mathbf{A})^{-1}\mathbf{B}R(s)\right] = \mathcal{L}^{-1} \frac{1}{s^2 + 3s + 2} \begin{bmatrix} s + 3 & 1 \\ -2 & s \end{bmatrix} \begin{bmatrix} 0 \\ 1 \end{bmatrix} \frac{1}{s}$$

$$= \mathcal{L}^{-1} \frac{1}{s^2 + 3s + 2} \begin{bmatrix} \frac{1}{s} \\ 1 \end{bmatrix} \tag{5-62}$$

$$= \begin{bmatrix} \frac{1}{2} - e^{-t} + \frac{1}{2}e^{-2t} \\ e^{-t} - e^{-2t} \end{bmatrix} \qquad t \geq 0$$

■

State Transition Equation Determined from the State Diagram

Equations (5-48) and (5-49) show that the Laplace transform method of solving the state equations requires the carrying out of the matrix inverse of $(s\mathbf{I} - \mathbf{A})$. We shall now show that the state diagram defined in Chapter 3 and the Mason's gain formula can be used to solve for the state transition equation.

The state transition equation in the Laplace transform domain is given by Eq. (5-48). Let the initial time be t_0; then Eq. (5-48) is written

$$\mathbf{X}(s) = (s\mathbf{I} - \mathbf{A})^{-1}\mathbf{x}(t_0) + (s\mathbf{I} - \mathbf{A})^{-1}[\mathbf{B}R(s) + \mathbf{F}W(s)] \qquad t \geq t_0 \tag{5-63}$$

Therefore, the last equation can be written directly from the state diagram by using the gain formula, with $X_i(s)$, $i = 1, 2, \ldots, n$, as the output nodes, and $x_i(t_0)$, $i = 1, 2, \ldots, n$, $R_j(s)$, $j = 1, 2, \ldots, p$, as the input nodes. The following example illustrates the state diagram method of finding the state transition equations for the system described in Example 5-2.

∎

Example 5-3

The state diagram for the system described by Eq. (5-55) is shown in Fig. 5-3 with t_0 as the initial time. The outputs of the integrators are assigned as state variables. Applying the gain formula to the state diagram in Fig. 5-3, with $X_1(s)$ and $X_2(s)$ as output nodes, $x_1(t_0)$, $x_2(t_0)$, and $R(s)$ as input nodes, we have

$$X_1(s) = \frac{s^{-1}(1 + 3s^{-1})}{\Delta} x_1(t_0) + \frac{s^{-2}}{\Delta} x_2(t_0) + \frac{s^{-2}}{\Delta} R(s) \qquad (5\text{-}64)$$

$$X_2(s) = \frac{-2s^{-2}}{\Delta} x_1(t_0) + \frac{s^{-1}}{\Delta} x_2(t_0) + \frac{s^{-1}}{\Delta} R(s) \qquad (5\text{-}65)$$

where

$$\Delta = 1 + 3s^{-1} + 2s^{-2} \qquad (5\text{-}66)$$

After simplification, Eqs. (5-64) and (5-65) are presented in matrix form:

$$\begin{bmatrix} X_1(s) \\ X_2(s) \end{bmatrix} = \frac{1}{(s+1)(s+2)} \begin{bmatrix} s+3 & 1 \\ -2 & s \end{bmatrix} \begin{bmatrix} x_1(t_0) \\ x_2(t_0) \end{bmatrix} + \begin{bmatrix} \dfrac{1}{(s+1)(s+2)} \\ \dfrac{s}{(s+1)(s+2)} \end{bmatrix} R(s) \quad (5\text{-}67)$$

The state transition equation for $t \ge t_0$ is obtained by taking the inverse Laplace transform on both sides of Eq. (5-67).

Consider that the input $r(t)$ is a unit step function applied at $t = t_0$. Then the following inverse Laplace transform relationships are identified:

$$\mathcal{L}^{-1}\left(\frac{1}{s}\right) = u_s(t - t_0) \qquad t \ge t_0 \qquad (5\text{-}68)$$

$$\mathcal{L}^{-1}\left(\frac{1}{s + a}\right) = e^{-a(t - t_0)} u_s(t - t_0) \qquad t \ge t_0 \qquad (5\text{-}69)$$

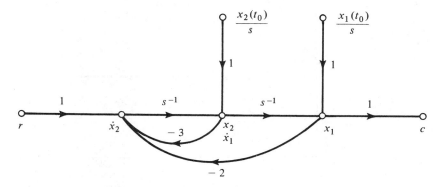

Figure 5-3 State diagram for Eq. (5-55).

The inverse Laplace transform of Eq. (5-67) is

$$\begin{bmatrix} x_1(t) \\ x_2(t) \end{bmatrix} = \begin{bmatrix} 2e^{-(t-t_0)} - e^{-2(t-t_0)} & e^{-(t-t_0)} - e^{-2(t-t_0)} \\ -2e^{-(t-t_0)} + 2e^{-2(t-t_0)} & -e^{-(t-t_0)} + 2e^{-2(t-t_0)} \end{bmatrix} \begin{bmatrix} x_1(t_0) \\ x_2(t_0) \end{bmatrix}$$

$$+ \begin{bmatrix} \frac{1}{2}u_s(t - t_0) - e^{-(t-t_0)} + \frac{1}{2}e^{-2(t-t_0)} \\ e^{-(t-t_0)} - e^{-2(t-t_0)} \end{bmatrix} \qquad t \geq t_0 \quad (5\text{-}70)$$

The reader should compare this result with that of Eq. (5-61), obtained for $t \geq 0$.

Example 5-4

In this example we illustrate the utilization of the state transition method to a system with input discontinuity. Let us consider that the input voltage to the RL network of Fig. 5-1 is as shown in Fig. 5-4. The state equation of the network is

$$\frac{di(t)}{dt} = -\frac{R}{L}i(t) + \frac{1}{L}e(t) \qquad (5\text{-}71)$$

Thus,

$$A = -\frac{R}{L} \qquad B = \frac{1}{L} \qquad F = 0 \qquad (5\text{-}72)$$

The state transition matrix is

$$\phi(t) = e^{-Rt/L} \qquad (5\text{-}73)$$

One approach to the problem of solving for $i(t)$ for $t \geq 0$ is to express the input voltage as

$$e(t) = Eu_s(t) + Eu_s(t - t_1) \qquad (5\text{-}74)$$

where $u_s(t)$ is the unit step function. The Laplace transform of $e(t)$ is

$$E(s) = \frac{E}{s}(1 + e^{-t_1 s}) \qquad (5\text{-}75)$$

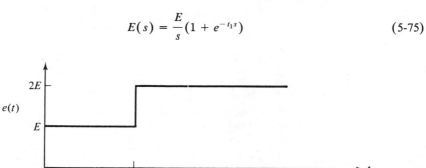

Figure 5-4 Input voltage waveform for the network in Fig. 5-1.

Then

$$(s\mathbf{I} - \mathbf{A})^{-1}\mathbf{BR}(s) = \frac{1}{s + R/L}\frac{1}{L}\frac{E}{s}(1 + e^{-t_1 s})$$

$$= \frac{E}{Rs[1 + (L/R)s]}(1 + e^{-t_1 s})$$

(5-76)

Substituting Eq. (5-76) into Eq. (5-49), the current for $t \geq 0$ is obtained:

$$i(t) = e^{-Rt/L}i(0)u_s(t) + \frac{E}{R}(1 - e^{-Rt/L})u_s(t) + \frac{E}{R}[1 - e^{-R(t-t_1)/L}]u_s(t - t_1)$$

(5-77)

Using the state transition approach we can divide the transition period in two parts: $t = 0$ to $t = t_1$, and $t = t_1$ to $t = \infty$. First for the time interval, $0 \leq t \leq t_1$, the input is

$$e(t) = Eu_s(t) \qquad 0 \leq t < t_1$$

(5-78)

Then

$$(s\mathbf{I} - \mathbf{A})^{-1}\mathbf{BR}(s) = \frac{1}{s + R/L}\frac{1}{L}\frac{E}{s}$$

$$= \frac{1}{Rs[1 + (L/R)s]}$$

(5-79)

Thus, the state transition equation for the time interval $0 \leq t \leq t_1$ is

$$i(t) = \left[e^{-Rt/L}i(0) + \frac{E}{R}(1 - e^{-Rt/L})\right]u_s(t)$$

(5-80)

Substituting $t = t_1$ into this equation, we get

$$i(t_1) = e^{-Rt_1/L}i(0) + \frac{E}{R}(1 - e^{-Rt_1/L})$$

(5-81)

The value of $i(t)$ at $t = t_1$ is now used as the initial state for the next transition period of $t_1 \leq t < \infty$. The magnitude of the input for this interval is $2E$. Therefore, the state transition equation for the second transition period is

$$i(t) = e^{-R(t-t_1)/L}i(t_1) + \frac{2E}{R}[1 - e^{-R(t-t_1)/L}] \qquad t \geq t_1$$

(5-82)

where $i(t_1)$ is given by Eq. (5-81).

This example illustrates two possible ways of solving a state transition problem. In the first approach, the transition is treated as one continuous process, whereas in the second, the transition period is divided into parts over which the input can be more easily represented. Although the first approach requires only one operation, the second method yields relatively simple results to the state transition equation, and it often presents computational advantages. Notice that in the second method the state at $t = t_1$ is used as the initial state for the next transition period, which begins at t_1.

5.5 RELATIONSHIP BETWEEN STATE EQUATIONS
AND HIGH-ORDER DIFFERENTIAL EQUATIONS

In preceding sections we defined the state equations and their solutions for linear time-invariant systems. In general, although it is always possible to write the state equations from the schematic diagram of a system, in practice the system may have been described by a high-order differential equation or transfer function. Therefore, it is necessary to investigate how state equations can be written directly from the differential equation or the transfer function. The relationship between a high-order differential equation and the state equations is discussed in this section.

Let us consider that a single-variable, linear time-invariant system is described by the following nth-order differential equation:

$$\frac{d^n c(t)}{dt^n} + a_n \frac{d^{n-1} c(t)}{dt^{n-1}} + a_{n-1} \frac{d^{n-2} c(t)}{dt^{n-2}} + \cdots + a_2 \frac{dc(t)}{dt} + a_1 c(t) = r(t)$$

$$(5\text{-}83)$$

where $c(t)$ is the output variable and $r(t)$ is the input.

The problem is to represent Eq. (5-83) by n state equations and an output equation. This simply involves the defining of the n state variables in terms of the output $c(t)$ and its derivatives. We have shown earlier that the state variables of a given system are not unique. Therefore, in general, we seek the most convenient way of assigning the state variables as long as the definition of state variables stated in Section 5.1 is met.

For the present case it is convenient to define the state variables as

$$x_1(t) = c(t)$$
$$x_2(t) = \frac{dc(t)}{dt}$$
$$\vdots$$
$$x_n(t) = \frac{d^{n-1} c(t)}{dt^{n-1}}$$

$$(5\text{-}84)$$

Then the state equations are

$$\frac{dx_1(t)}{dt} = x_2(t)$$
$$\frac{dx_2(t)}{dt} = x_3(t)$$
$$\vdots$$
$$\frac{dx_{n-1}(t)}{dt} = x_n(t)$$
$$\frac{dx_n(t)}{dt} = -a_1 x_1(t) - a_2 x_2(t) - \cdots - a_{n-1} x_{n-1}(t) - a_n x_n(t) + r(t)$$

$$(5\text{-}85)$$

where the last state equation is obtained by equating the highest-ordered derivative term to the rest of Eq. (5-83). The output equation is simply

$$c(t) = x_1(t) \tag{5-86}$$

In vector-matrix form, Eq. (5-85) is written

$$\frac{d\mathbf{x}(t)}{dt} = \mathbf{A}\mathbf{x}(t) + \mathbf{B}r(t) \tag{5-87}$$

where $\mathbf{x}(t)$ is the $n \times 1$ state vector and $r(t)$ is the scalar input. The coefficient matrices are

$$\mathbf{A} = \begin{bmatrix} 0 & 1 & 0 & 0 & 0 & \cdots & 0 \\ 0 & 0 & 1 & 0 & 0 & \cdots & 0 \\ 0 & 0 & 0 & 1 & 0 & \cdots & 0 \\ 0 & 0 & 0 & 0 & 0 & \cdots & 1 \\ -a_1 & -a_2 & -a_3 & -a_4 & -a_5 & \cdots & -a_n \end{bmatrix} \quad (n \times n) \tag{5-88}$$

$$\mathbf{B} = \begin{bmatrix} 0 \\ 0 \\ \vdots \\ 1 \end{bmatrix} \quad (n \times 1) \tag{5-89}$$

The output equation in vector-matrix form is

$$c(t) = \mathbf{D}\mathbf{x}(t) \tag{5-90}$$

where

$$\mathbf{D} = \begin{bmatrix} 1 & 0 & 0 & \cdots & 0 \end{bmatrix} \quad (1 \times n) \tag{5-91}$$

The state equation of Eq. (5-87) with the matrices $\mathbf{A}$ and $\mathbf{B}$ defined as in Eqs. (5-88) and (5-89), respectively, is called the *phase-variable canonical form* in the next section.

■

Example 5-5

Consider the differential equation

$$\frac{d^3c(t)}{dt^3} + 5\frac{d^2c(t)}{dt^2} + \frac{dc(t)}{dt} + 2c(t) = r(t) \tag{5-92}$$

Rearranging the last equation so that the highest-order derivative term is equated to the rest of the terms, we have

$$\frac{d^3c(t)}{dt^3} = -5\frac{d^2c(t)}{dt^2} - \frac{dc(t)}{dt} - 2c(t) + r(t) \tag{5-93}$$

The state variables are defined as

$$x_1(t) = c(t)$$

$$x_2(t) = \frac{dc(t)}{dt}$$ (5-94)

$$x_3(t) = \frac{d^2c(t)}{dt^2}$$

Then the state equations are represented by the vector-matrix equation of Eq. (5-87) with

$$\mathbf{A} = \begin{bmatrix} 0 & 1 & 0 \\ 0 & 0 & 1 \\ -2 & -1 & -5 \end{bmatrix}$$ (5-95)

and

$$\mathbf{B} = \begin{bmatrix} 0 \\ 0 \\ 1 \end{bmatrix}$$ (5-96)

The output equation is

$$c(t) = x_1(t)$$ (5-97)

∎

5.6 TRANSFORMATION TO PHASE-VARIABLE CANONICAL FORM

In general, when the coefficient matrices $\mathbf{A}$ and $\mathbf{B}$ are given by Eqs. (5-88) and (5-89), respectively, the state equation of Eq. (5-87) is called the *phase-variable canonical form*. It will be shown later that a linear time-invariant system which is representable in the phase-variable canonical form has certain unique properties with regard to controllability and pole-placement design through state feedback.

   Theorem 5-1.   *Let the state equation of a linear time-invariant system be given by*

$$\frac{d\mathbf{x}(t)}{dt} = \mathbf{A}\mathbf{x}(t) + \mathbf{B}r(t)$$ (5-98)

where $\mathbf{x}(t)$ is an $n \times 1$ state vector, $\mathbf{A}$ an $n \times n$ coefficient matrix, $\mathbf{B}$ an $n \times 1$ coefficient matrix, and $r(t)$ a scalar input. If the matrix

$$\mathbf{S} = \begin{bmatrix} \mathbf{B} & \mathbf{A}\mathbf{B} & \mathbf{A}^2\mathbf{B} & \cdots & \mathbf{A}^{n-1}\mathbf{B} \end{bmatrix}$$ (5-99)

is nonsingular, then there exists a nonsingular transformation

$$\mathbf{y}(t) = \mathbf{Q}\mathbf{x}(t)$$ (5-100)

or

$$\mathbf{x}(t) = \mathbf{Q}^{-1}\mathbf{y}(t) \tag{5-101}$$

which transforms Eq. (5-98) to the phase-variable canonical form

$$\dot{\mathbf{y}}(t) = \mathbf{A}_1\mathbf{y}(t) + \mathbf{B}_1 r(t) \tag{5-102}$$

where

$$\mathbf{A}_1 = \begin{bmatrix} 0 & 1 & 0 & 0 & \cdots & 0 \\ 0 & 0 & 1 & 0 & \cdots & 0 \\ 0 & 0 & 0 & 1 & \cdots & 0 \\ \cdots\cdots\cdots\cdots\cdots\cdots\cdots\cdots \\ 0 & 0 & 0 & 0 & \cdots & 1 \\ -a_1 & -a_2 & -a_3 & -a_4 & \cdots & -a_n \end{bmatrix} \tag{5-103}$$

and

$$\mathbf{B}_1 = \begin{bmatrix} 0 \\ 0 \\ \vdots \\ 1 \end{bmatrix} \tag{5-104}$$

The transforming matrix $\mathbf{Q}$ is given by

$$\mathbf{Q} = \begin{bmatrix} \mathbf{Q}_1 \\ \mathbf{Q}_1\mathbf{A} \\ \vdots \\ \mathbf{Q}_1\mathbf{A}^{n-1} \end{bmatrix} \tag{5-105}$$

where

$$\mathbf{Q}_1 = \begin{bmatrix} 0 & 0 & \cdots & 1 \end{bmatrix} \begin{bmatrix} \mathbf{B} & \mathbf{AB} & \mathbf{A}^2\mathbf{B} & \cdots & \mathbf{A}^{n-1}\mathbf{B} \end{bmatrix}^{-1} \tag{5-106}$$

Proof. Let

$$\mathbf{x}(t) = \begin{bmatrix} x_1(t) \\ x_2(t) \\ \vdots \\ x_n(t) \end{bmatrix} \tag{5-107}$$

$$\mathbf{y}(t) = \begin{bmatrix} y_1(t) \\ y_2(t) \\ \vdots \\ y_n(t) \end{bmatrix} \tag{5-108}$$

and

$$\mathbf{Q} = \begin{bmatrix} q_{11} & q_{12} & \cdots & q_{1n} \\ q_{21} & q_{22} & \cdots & q_{2n} \\ \cdots & \cdots & \cdots & \cdots \\ q_{n1} & q_{n2} & \cdots & q_{nn} \end{bmatrix} = \begin{bmatrix} \mathbf{Q}_1 \\ \mathbf{Q}_2 \\ \vdots \\ \mathbf{Q}_n \end{bmatrix} \tag{5-109}$$

where

$$\mathbf{Q}_i = [q_{i1} \quad q_{i2} \quad \cdots \quad q_{in}] \qquad i = 1, 2, \ldots, n \tag{5-110}$$

Then, from Eq. (5-100),

$$\begin{aligned} y_1(t) &= q_{11}x_1(t) + q_{12}x_2(t) + \cdots + q_{1n}x_n(t) \\ &= \mathbf{Q}_1\mathbf{x}(t) \end{aligned} \tag{5-111}$$

Taking the time derivative on both sides of the last equation and in view of Eqs. (5-102) and (5-103),

$$\dot{y}_1(t) = y_2(t) = \mathbf{Q}_1\dot{\mathbf{x}}(t) = \mathbf{Q}_1\mathbf{A}\mathbf{x}(t) + \mathbf{Q}_1\mathbf{B}r(t) \tag{5-112}$$

Since Eq. (5-100) states that $\mathbf{y}(t)$ is a function of $\mathbf{x}(t)$ only, in Eq. (5-112) $\mathbf{Q}_1\mathbf{B} = \mathbf{0}$. Therefore,

$$\dot{y}_1(t) = y_2(t) = \mathbf{Q}_1\mathbf{A}\mathbf{x}(t) \tag{5-113}$$

Taking the time derivative of the last equation once again leads to

$$\dot{y}_2(t) = y_3(t) = \mathbf{Q}_1\mathbf{A}^2\mathbf{x}(t) \tag{5-114}$$

with $\mathbf{Q}_1\mathbf{AB} = \mathbf{0}$.

Repeating the procedure leads to

$$\dot{y}_{n-1}(t) = y_n(t) = \mathbf{Q}_1\mathbf{A}^{n-1}\mathbf{x}(t) \tag{5-115}$$

with $\mathbf{Q}_1\mathbf{A}^{n-2}\mathbf{B} = 0$. Therefore, using Eq. (5-100), we have

$$\mathbf{y}(t) = \mathbf{Q}\mathbf{x}(t) = \begin{bmatrix} \mathbf{Q}_1 \\ \mathbf{Q}_1\mathbf{A} \\ \vdots \\ \mathbf{Q}_1\mathbf{A}^{n-1} \end{bmatrix} \mathbf{x}(t) \tag{5-116}$$

or

$$\mathbf{Q} = \begin{bmatrix} \mathbf{Q}_1 \\ \mathbf{Q}_1\mathbf{A} \\ \vdots \\ \mathbf{Q}_1\mathbf{A}^{n-1} \end{bmatrix} \tag{5-117}$$

and $\mathbf{Q}_1$ should satisfy the condition

$$\mathbf{Q}_1\mathbf{B} = \mathbf{Q}_1\mathbf{AB} = \cdots = \mathbf{Q}_1\mathbf{A}^{n-2}\mathbf{B} = \mathbf{0} \tag{5-118}$$

Now taking the derivative of Eq. (5-100) with respect to time, we get

$$\dot{\mathbf{y}}(t) = \mathbf{Q}\dot{\mathbf{x}}(t) = \mathbf{Q}\mathbf{A}\mathbf{x}(t) + \mathbf{Q}\mathbf{B}r(t) \tag{5-119}$$

Comparing Eq. (5-119) with Eq. (5-102), we obtain

$$\mathbf{A}_1 = \mathbf{Q}\mathbf{A}\mathbf{Q}^{-1} \tag{5-120}$$

and

$$\mathbf{B}_1 = \mathbf{Q}\mathbf{B} \tag{5-121}$$

Then, from Eq. (5-117),

$$\mathbf{QB} = \begin{bmatrix} \mathbf{Q}_1\mathbf{B} \\ \mathbf{Q}_1\mathbf{AB} \\ \vdots \\ \mathbf{Q}_1\mathbf{A}^{n-1}\mathbf{B} \end{bmatrix} = \begin{bmatrix} 0 \\ 0 \\ \vdots \\ 1 \end{bmatrix} \tag{5-122}$$

Since $\mathbf{Q}_1$ is an $1 \times n$ row matrix, Eq. (5-122) can be written

$$\mathbf{Q}_1[\mathbf{B} \quad \mathbf{AB} \quad \mathbf{A}^2\mathbf{B} \quad \cdots \quad \mathbf{A}^{n-1}\mathbf{B}] = [0 \quad 0 \quad \cdots \quad 1] \tag{5-123}$$

Thus, $\mathbf{Q}_1$ is obtained as

$$\begin{aligned} \mathbf{Q}_1 &= [0 \quad 0 \quad \cdots \quad 1][\mathbf{B} \quad \mathbf{AB} \quad \mathbf{A}^2\mathbf{B} \quad \cdots \quad \mathbf{A}^{n-1}\mathbf{B}]^{-1} \\ &= [0 \quad 0 \quad \cdots \quad 1]\mathbf{S}^{-1} \end{aligned} \tag{5-124}$$

if the matrix $\mathbf{S} = [\mathbf{B} \quad \mathbf{AB} \quad \mathbf{A}^2\mathbf{B} \quad \cdots \quad \mathbf{A}^{n-1}\mathbf{B}]$ is nonsingular. This is the condition of complete state controllability. Once $\mathbf{Q}_1$ is determined from Eq. (5-124), the transformation matrix $\mathbf{Q}$ is given by Eq. (5-117).

■

Example 5-6

Let a linear time-invariant system be described by Eq. (5-87) with

$$\mathbf{A} = \begin{bmatrix} 1 & -1 \\ 0 & -1 \end{bmatrix} \quad \mathbf{B} = \begin{bmatrix} 1 \\ 1 \end{bmatrix} \tag{5-125}$$

It is desired to transform the state equation into the phase-variable canonical form. Since the matrix

$$\mathbf{S} = [\mathbf{B} \quad \mathbf{AB}] = \begin{bmatrix} 1 & 0 \\ 1 & -1 \end{bmatrix} \tag{5-126}$$

is nonsingular, the system may be expressed in the phase-variable canonical form. Therefore, $\mathbf{Q}_1$ is obtained as a row matrix which contains the elements of the last row of $\mathbf{S}^{-1}$; that is,

$$\mathbf{Q}_1 = [1 \quad -1] \tag{5-127}$$

Using Eq. (5-117),

$$Q = \begin{bmatrix} Q_1 \\ Q_1 A \end{bmatrix} = \begin{bmatrix} 1 & -1 \\ 1 & 0 \end{bmatrix} \tag{5-128}$$

Thus,

$$A_1 = QAQ^{-1} = \begin{bmatrix} 0 & 1 \\ 1 & 0 \end{bmatrix} \tag{5-129}$$

$$B_1 = QB = \begin{bmatrix} 0 \\ 1 \end{bmatrix} \tag{5-130}$$

The method of defining state variables by inspection as described earlier with reference to Eq. (5-83) is inadequate when the right-hand side of the differential equation also includes the derivatives of $r(t)$. To illustrate the point we consider the following example.

Example 5-7

Given the differential equation

$$\frac{d^3c(t)}{dt^3} + 5\frac{d^2c(t)}{dt^2} + \frac{dc(t)}{dt} + 2c(t) = \frac{dr(t)}{dt} + 2r(t) \tag{5-131}$$

it is desired to represent the equation by three state equations. Since the right side of the state equations cannot include any derivatives of the input $r(t)$, it is necessary to include $r(t)$ when defining the state variables. Let us rewrite Eq. (5-131) as

$$\frac{d^3c(t)}{dt^3} - \frac{dr(t)}{dt} = -5\frac{d^2c(t)}{dt^2} - \frac{dc(t)}{dt} - 2c(t) + 2r(t) \tag{5-132}$$

The state variables are now defined as

$$x_1(t) = c(t)$$

$$x_2(t) = \frac{dc(t)}{dt} \tag{5-133}$$

$$x_3(t) = \frac{d^2c(t)}{dt^2} - r(t)$$

Using these last three equations and Eq. (5-132), the state equations are written

$$\frac{dx_1(t)}{dt} = x_2(t)$$

$$\frac{dx_2(t)}{dt} = x_3(t) + r(t) \tag{5-134}$$

$$\frac{dx_3(t)}{dt} = -2x_1(t) - x_2(t) - 5x_3(t) - 3r(t)$$

In general, it can be shown that for the nth-order differential equation

$$\frac{d^n c(t)}{dt^n} + a_n \frac{d^{n-1} c(t)}{dt^{n-1}} + \cdots + a_2 \frac{dc(t)}{dt} + a_1 c(t)$$

$$= b_{n+1} \frac{d^n r(t)}{dt^n} + b_n \frac{d^{n-1} r(t)}{dt^{n-1}} + \cdots + b_2 \frac{dr(t)}{dt} + b_1 r(t) \quad (5\text{-}135)$$

the state variables should be defined as

$$x_1(t) = c(t) - b_{n+1} r(t)$$

$$x_2(t) = \frac{dx_1(t)}{dt} - h_1 r(t)$$

$$x_3(t) = \frac{dx_2(t)}{dt} - h_2 r(t) \qquad (5\text{-}136)$$

$$\vdots$$

$$x_n(t) = \frac{dx_{n-1}(t)}{dt} - h_{n-1} r(t)$$

where

$$h_1 = b_n - a_n b_{n+1}$$
$$h_2 = (b_{n-1} - a_{n-1} b_{n+1}) - a_n h_1$$
$$h_3 = (b_{n-2} - a_{n-2} b_{n+1}) - a_{n-1} h_1 - a_n h_2 \qquad (5\text{-}137)$$
$$\vdots$$
$$h_n = (b_1 - a_1 b_{n+1}) - a_2 h_1 - a_3 h_2 - \cdots - a_{n-1} h_{n-2} - a_n h_{n-1}$$

Using Eqs. (5-136) and (5-137), we resolve the nth-order differential equation in Eq. (5-135) into the following n state equations:

$$\frac{dx_1(t)}{dt} = x_2(t) + h_1 r(t)$$

$$\frac{dx_2(t)}{dt} = x_3(t) + h_2 r(t)$$

$$\vdots \qquad (5\text{-}138)$$

$$\frac{dx_{n-1}(t)}{dt} = x_n(t) + h_{n-1} r(t)$$

$$\frac{dx_n(t)}{dt} = -a_1 x_1(t) - a_2 x_2(t) - \cdots - a_{n-1} x_{n-1}(t) - a_n x_n(t) + h_n r(t)$$

The output equation is obtained by rearranging the first equation of Eq. (5-136):

$$c(t) = x_1(t) + b_{n+1} r(t) \qquad (5\text{-}139)$$

Now if we apply these equations to the case of Example 5-7, we have

$$a_3 = 5 \qquad b_4 = 0 \qquad b_1 = 2$$
$$a_2 = 1 \qquad b_3 = 0$$
$$a_1 = 2 \qquad b_2 = 1$$
$$h_1 = b_3 - a_3 b_4 = 0$$
$$h_2 = (b_2 - a_2 b_4) - a_3 h_1 = 1$$
$$h_3 = (b_1 - a_1 b_4) - a_2 h_1 - a_3 h_2 = -3$$

When we substitute these parameters into Eqs. (5-136) and (5-137), we have the same results for the state variables and the state equations as obtained in Example 5-7.

The disadvantage with the method of Eqs. (5-136), (5-137), and (5-138) is that these equations are difficult and impractical to memorize. It is not expected that one will always have these equations available for reference. However, we shall later describe a more convenient method using the transfer function and the state diagram.

5.7 RELATIONSHIP BETWEEN STATE EQUATIONS AND TRANSFER FUNCTIONS

We have presented the methods of describing a linear time-invariant system by transfer functions and by dynamic equations. It is interesting to investigate the relationship between these two representations.

In Eq. (3-7) the transfer function of a linear single-variable system is defined in terms of the coefficients of the system's differential equation. Similarly, Eq. (3-15) gives the matrix transfer function relation for a multivariable system that has p inputs and q outputs. Now we investigate the transfer function matrix relation using the dynamic equation notation.

Consider that a linear time-invariant system is described by the dynamic equations

$$\frac{d\mathbf{x}(t)}{dt} = \mathbf{A}\mathbf{x}(t) + \mathbf{B}\mathbf{r}(t) + \mathbf{F}\mathbf{w}(t) \tag{5-140}$$

$$\mathbf{c}(t) = \mathbf{D}\mathbf{x}(t) + \mathbf{E}\mathbf{r}(t) + \mathbf{H}\mathbf{w}(t) \tag{5-141}$$

where $\mathbf{x}(t) = n \times 1$ state vector
$\mathbf{r}(t) = p \times 1$ input vector
$\mathbf{c}(t) = q \times 1$ output vector
$\mathbf{w}(t) = v \times 1$ disturbance vector

and $\mathbf{A}$, $\mathbf{B}$, $\mathbf{D}$, $\mathbf{E}$, $\mathbf{F}$, and $\mathbf{H}$ are matrices of appropriate dimensions.

Taking the Laplace transform on both sides of Eq. (5-140) and solving for $\mathbf{X}(s)$, we have

$$\mathbf{X}(s) = (s\mathbf{I} - \mathbf{A})^{-1}\mathbf{x}(0) + (s\mathbf{I} - \mathbf{A})^{-1}[\mathbf{BR}(s) + \mathbf{FW}(s)] \qquad (5\text{-}142)$$

The Laplace transform of Eq. (5-141) is

$$\mathbf{C}(s) = \mathbf{DX}(s) + \mathbf{ER}(s) + \mathbf{HW}(s) \qquad (5\text{-}143)$$

Substituting Eq. (5-142) into Eq. (5-143), we have

$$\mathbf{C}(s) = \mathbf{D}(s\mathbf{I} - \mathbf{A})^{-1}\mathbf{x}(0) + \mathbf{D}(s\mathbf{I} - \mathbf{A})^{-1}[\mathbf{BR}(s) + \mathbf{FW}(s)] + \mathbf{ER}(s) + \mathbf{HW}(s) \qquad (5\text{-}144)$$

Since the definition of transfer function requires that the initial conditions be set to zero, $\mathbf{x}(0) = \mathbf{0}$; thus Eq. (5-144) becomes

$$\mathbf{C}(s) = \Big[\mathbf{D}(s\mathbf{I} - \mathbf{A})^{-1}\mathbf{B} + \mathbf{E}\Big]\mathbf{R}(s) + \Big[\mathbf{D}(s\mathbf{I} - \mathbf{A})^{-1}\mathbf{F} + \mathbf{H}\Big]\mathbf{W}(s) \quad (5\text{-}145)$$

Since the system has two inputs in $r(t)$ and $w(t)$, we can only define the transfer function between one pair of input and output. Thus, the transfer functions are defined as:

$$w(t) = 0: \qquad \mathbf{G}(s) \; = \mathbf{D}(s\mathbf{I} - \mathbf{A})^{-1}\mathbf{B} + \mathbf{E} \qquad (5\text{-}146)$$

$$r(t) = 0: \qquad \mathbf{G}_w(s) = \mathbf{D}(s\mathbf{I} - \mathbf{A})^{-1}\mathbf{F} + \mathbf{H} \qquad (5\text{-}147)$$

where $\mathbf{G}(s)$ is a $q \times p$ transfer function matrix between $r(t)$ and $c(t)$, whereas $\mathbf{G}_w(s)$ is a $q \times v$ transfer function matrix between $w(t)$ and $c(t)$.

■

Example 5-8

Consider that a multivariable system is described by the differential equations

$$\frac{d^2 c_1}{dt^2} + 4\frac{dc_1}{dt} - 3c_2 = r_1 + 2w \qquad (5\text{-}148)$$

$$\frac{dc_2}{dt} + \frac{dc_1}{dt} + c_1 + 2c_2 = r_2 \qquad (5\text{-}149)$$

The state variables of the system are assigned as follows:

$$x_1 = c_1$$
$$x_2 = \frac{dc_1}{dt} \qquad (5\text{-}150)$$
$$x_3 = c_2$$

These state variables have been defined by mere inspection of the two differential equations, as no particular reasons for the definitions are given other than that these are the most convenient.

Now equating the first term of each of the equations of Eqs. (5-148) and (5-149) to the rest of the terms and using the state-variable relations of Eq. (5-150), we arrive at the following state equation and output equation in matrix form:

$$
\begin{bmatrix} \dfrac{dx_1}{dt} \\[2mm] \dfrac{dx_2}{dt} \\[2mm] \dfrac{dx_3}{dt} \end{bmatrix} = \begin{bmatrix} 0 & 1 & 0 \\ 0 & -4 & 3 \\ -1 & -1 & -2 \end{bmatrix} \begin{bmatrix} x_1 \\ x_2 \\ x_3 \end{bmatrix} + \begin{bmatrix} 0 & 0 \\ 1 & 0 \\ 0 & 1 \end{bmatrix} \begin{bmatrix} r_1 \\ r_2 \end{bmatrix} + \begin{bmatrix} 0 \\ 2 \\ 0 \end{bmatrix} w \tag{5-151}
$$

$$
\begin{bmatrix} c_1 \\ c_2 \end{bmatrix} = \begin{bmatrix} 1 & 0 & 0 \\ 0 & 0 & 1 \end{bmatrix} \begin{bmatrix} x_1 \\ x_2 \\ x_3 \end{bmatrix} = \mathbf{Dx} \tag{5-152}
$$

To determine the transfer function matrix of the system using the state-variable formulation, we substitute the $\mathbf{A}$, $\mathbf{B}$, $\mathbf{D}$, $\mathbf{E}$, and $\mathbf{F}$ matrices into Eq. (5-145). First, we form the matrix $(s\mathbf{I} - \mathbf{A})$,

$$
(s\mathbf{I} - \mathbf{A}) = \begin{bmatrix} s & -1 & 0 \\ 0 & s+4 & -3 \\ 1 & 1 & s+2 \end{bmatrix} \tag{5-153}
$$

The determinant of $(s\mathbf{I} - \mathbf{A})$ is

$$
|s\mathbf{I} - \mathbf{A}| = s^3 + 6s^2 + 11s + 3 \tag{5-154}
$$

Thus,

$$
(s\mathbf{I} - \mathbf{A})^{-1} = \frac{1}{|s\mathbf{I} - \mathbf{A}|} \begin{bmatrix} s^2 + 6s + 11 & s+2 & 3 \\ -3 & s(s+2) & 3s \\ -(s+4) & -(s+1) & s(s+4) \end{bmatrix} \tag{5-155}
$$

The transfer function matrix between $\mathbf{r}$ and $\mathbf{c}$ is

$$
\mathbf{G}(s) = \mathbf{D}(s\mathbf{I} - \mathbf{A})^{-1}\mathbf{B} = \frac{1}{s^3 + 6s^2 + 11s + 3} \begin{bmatrix} s+2 & 3 \\ -(s+1) & s(s+4) \end{bmatrix} \tag{5-156}
$$

and that between $\mathbf{w}$ and $\mathbf{c}$ is

$$
\mathbf{G}_w(s) = \mathbf{D}(s\mathbf{I} - \mathbf{A})^{-1}\mathbf{F} = \frac{1}{s^3 + 6s^2 + 11s + 3} \begin{bmatrix} 2(s+2) \\ -2(s+1) \end{bmatrix} \tag{5-157}
$$

Using the conventional approach, we take the Laplace transform on both sides of Eqs. (5-148) and (5-149) and assume zero initial conditions. The resulting transformed equations are written in matrix form as

$$\begin{bmatrix} s(s+4) & -3 \\ s+1 & s+2 \end{bmatrix} \begin{bmatrix} C_1(s) \\ C_2(s) \end{bmatrix} = \begin{bmatrix} R_1(s) \\ R_2(s) \end{bmatrix} + \begin{bmatrix} 2 \\ 0 \end{bmatrix} W(s) \qquad (5\text{-}158)$$

Solving for $\mathbf{C}(s)$ from Eq. (5-158), we obtain

$$\mathbf{C}(s) = \mathbf{G}(s)\mathbf{R}(s) + \mathbf{G}_w(s)\mathbf{W}(s) \qquad (5\text{-}159)$$

where

$$\mathbf{G}(s) = \begin{bmatrix} s(s+4) & -3 \\ s+1 & s+2 \end{bmatrix}^{-1} \qquad (5\text{-}160)$$

$$\mathbf{G}_w(s) = \begin{bmatrix} s(s+4) & -3 \\ s+1 & s+2 \end{bmatrix}^{-1} \begin{bmatrix} 2 \\ 0 \end{bmatrix} \qquad (5\text{-}161)$$

which are the same results as in Eqs. (5-156) and (5-157), respectively, when the matrix inverse is carried out.

■

5.8 CHARACTERISTIC EQUATION, EIGENVALUES, AND EIGENVECTORS

The characteristic equation plays an important part in the study of linear systems. It can be defined from the basis of the differential equation, the transfer function, or the state equations.

Consider that a linear time-invariant system is described by the differential equation

$$\frac{d^n c}{dt^n} + a_n \frac{d^{n-1}c}{dt^{n-1}} + a_{n-1} \frac{d^{n-2}c}{dt^{n-2}} + \cdots + a_2 \frac{dc}{dt} + a_1 c$$

$$= b_{n+1} \frac{d^n r}{dt^n} + b_n \frac{d^{n-1}r}{dt^{n-1}} + \cdots + b_2 \frac{dr}{dt} + b_1 r \qquad (5\text{-}162)$$

By defining the operator p as

$$p^k = \frac{d^k}{dt^k} \qquad k = 1, 2, \ldots, n$$

Eq. (5-162) is written

$$\left(p^n + a_n p^{n-1} + a_{n-1} p^{n-2} + \cdots + a_2 p + a_1 \right) c$$

$$= \left(b_{n+1} p^n + b_n p^{n-1} + \cdots + b_2 p + b_1 \right) r \qquad (5\text{-}163)$$

Then the characteristic equation of the system is defined as

$$s^n + a_n s^{n-1} + a_{n-1} s^{n-2} + \cdots + a_2 s + a_1 = 0 \qquad (5\text{-}164)$$

which is setting the homogeneous part of Eq. (5-162) to zero. Furthermore, the operator p is replaced by the Laplace transform variable s.

The transfer function of the system is

$$G(s) = \frac{C(s)}{R(s)} = \frac{b_{n+1} s^n + b_n s^{n-1} + \cdots + b_2 s + b_1}{s^n + a_n s^{n-1} + \cdots + a_2 s + a_1} \qquad (5\text{-}165)$$

Therefore, *the characteristic equation is obtained by equating the denominator of the transfer function to zero.*

From the state-variable approach, we can write Eq. (5-146) as

$$\begin{aligned} \mathbf{G}(s) &= \mathbf{D} \frac{\text{adj}(s\mathbf{I} - \mathbf{A})}{|s\mathbf{I} - \mathbf{A}|} \mathbf{B} + \mathbf{E} \\ &= \frac{\mathbf{D}[\text{adj}(s\mathbf{I} - \mathbf{A})]\mathbf{B} + |s\mathbf{I} - \mathbf{A}|\mathbf{E}}{|s\mathbf{I} - \mathbf{A}|} \end{aligned} \qquad (5\text{-}166)$$

Setting the denominator of the transfer function matrix $\mathbf{G}(s)$ to zero, we get the characteristic equation expressed as

$$|s\mathbf{I} - \mathbf{A}| = 0 \qquad (5\text{-}167)$$

which is an alternative form of Eq. (5-164).

Eigenvalues

The roots of the characteristic equation are often referred to as the eigenvalues of the matrix $\mathbf{A}$. It is interesting to note that if the state equations are represented in the phase-variable canonical form, the coefficients of the characteristic equation are readily given by the elements in the last row of the elements of the $\mathbf{A}$ matrix. That is, if $\mathbf{A}$ is of the form of Eq. (5-103), the characteristic equation is readily given by Eq. (5-164).

Another important property of the characteristic equation and the eigenvalues is that they are invariant under a nonsingular transformation. In other words, when the $\mathbf{A}$ matrix is transformed by a nonsingular transformation $\mathbf{x} = \mathbf{Py}$, so that

$$\hat{\mathbf{A}} = \mathbf{P}^{-1}\mathbf{A}\mathbf{P} \qquad (5\text{-}168)$$

then the characteristic equation and the eigenvalues of $\hat{\mathbf{A}}$ are identical to those of $\mathbf{A}$. This is proved by writing

$$\begin{aligned} s\mathbf{I} - \hat{\mathbf{A}} &= s\mathbf{I} - \mathbf{P}^{-1}\mathbf{A}\mathbf{P} \\ &= s\mathbf{P}^{-1}\mathbf{P} - \mathbf{P}^{-1}\mathbf{A}\mathbf{P} \end{aligned} \qquad (5\text{-}169)$$

The characteristic equation of $\hat{\mathbf{A}}$ is

$$
\begin{aligned}
|s\mathbf{I} - \hat{\mathbf{A}}| &= |s\mathbf{P}^{-1}\mathbf{P} - \mathbf{P}^{-1}\mathbf{AP}| \\
&= |\mathbf{P}^{-1}(s\mathbf{I} - \mathbf{A})\mathbf{P}|
\end{aligned}
\tag{5-170}
$$

Since the determinant of a product is equal to the product of the determinants, Eq. (5-170) becomes

$$
\begin{aligned}
|s\mathbf{I} - \hat{\mathbf{A}}| &= |\mathbf{P}^{-1}|\,|s\mathbf{I} - \mathbf{A}|\,|\mathbf{P}| \\
&= |s\mathbf{I} - \mathbf{A}|
\end{aligned}
\tag{5-171}
$$

Eigenvectors

The $n \times 1$ nonzero vector $\mathbf{p}_i$ that satisfies the matrix equation

$$
(\lambda_i \mathbf{I} - \mathbf{A})\mathbf{p}_i = \mathbf{0}
\tag{5-172}
$$

where λ_i is the ith eigenvalue of $\mathbf{A}$, is called the *eigenvector* of $\mathbf{A}$ associated with the eigenvalue λ_i. Illustrative examples of how the eigenvectors of a matrix are determined are given in the following section.

5.9 DIAGONALIZATION OF THE A MATRIX (SIMILARITY TRANSFORMATION)

One of the motivations for diagonalizing the **A** matrix is that if **A** is a diagonal matrix, the eigenvalues of **A**, $\lambda_1, \lambda_2, \ldots, \lambda_n$, all assumed to be distinct, are located on the main diagonal; then the state transition matrix $e^{\mathbf{A}t}$ will also be diagonal, with its nonzero elements given by $e^{\lambda_1 t}, e^{\lambda_2 t}, \ldots, e^{\lambda_n t}$. There are other reasons for wanting to diagonalize the **A** matrix, such as the controllability of a system (Section 5.12). We have to assume that all the eigenvalues of **A** are distinct, since, unless it is real and symmetric, **A** cannot always be diagonalized if it has multiple-order eigenvalues.

The problem can be stated as, given the linear system

$$
\dot{\mathbf{x}}(t) = \mathbf{A}\mathbf{x}(t) + \mathbf{B}\mathbf{u}(t)
\tag{5-173}
$$

where $\mathbf{x}(t)$ is an n-vector, $\mathbf{u}(t)$ an r-vector, and **A** has distinct eigenvalues $\lambda_1, \lambda_2, \ldots, \lambda_n$, it is desired to find a nonsingular matrix **P** such that the transformation

$$
\mathbf{x}(t) = \mathbf{P}\mathbf{y}(t)
\tag{5-174}
$$

transforms Eq. (5-173) into

$$
\dot{\mathbf{y}}(t) = \mathbf{\Lambda}\mathbf{y}(t) + \mathbf{\Gamma}\mathbf{u}(t)
\tag{5-175}
$$

with Λ given by the diagonal matrix

$$\Lambda = \begin{bmatrix} \lambda_1 & 0 & 0 & \cdots & 0 \\ 0 & \lambda_2 & 0 & \cdots & 0 \\ 0 & 0 & \lambda_3 & \cdots & 0 \\ \cdots\cdots\cdots\cdots\cdots\cdots\cdots \\ 0 & 0 & 0 & \cdots & \lambda_n \end{bmatrix} \quad (n \times n) \qquad (5\text{-}176)$$

This transformation is also known as the *similarity transformation*. The state equation of Eq. (5-175) is known as the *canonical form*.

Substituting Eq. (5-174) into Eq. (5-173) it is easy to see that

$$\Lambda = \mathbf{P}^{-1}\mathbf{A}\mathbf{P} \qquad (5\text{-}177)$$

and

$$\Gamma = \mathbf{P}^{-1}\mathbf{B} \quad (n \times r) \qquad (5\text{-}178)$$

In general, there are several methods of determining the matrix $\mathbf{P}$. We show in the following that $\mathbf{P}$ can be formed by use of the eigenvectors of $\mathbf{A}$; that is,

$$\mathbf{P} = [\mathbf{p}_1 \quad \mathbf{p}_2 \quad \mathbf{p}_3 \cdots \mathbf{p}_n] \qquad (5\text{-}179)$$

where $\mathbf{p}_i$ $(i = 1, 2, \ldots, n)$ denotes the eigenvector that is associated with the eigenvalue λ_i. This is proved by use of Eq. (5-172), which is written

$$\lambda_i \mathbf{p}_i = \mathbf{A}\mathbf{p}_i \quad i = 1, 2, \ldots, n \qquad (5\text{-}180)$$

Now forming the $n \times n$ matrix,

$$[\lambda_1 \mathbf{p}_1 \quad \lambda_2 \mathbf{p}_2 \quad \cdots \quad \lambda_n \mathbf{p}_n] = [\mathbf{A}\mathbf{p}_1 \quad \mathbf{A}\mathbf{p}_2 \cdots \mathbf{A}\mathbf{p}_n]$$
$$= \mathbf{A}[\mathbf{p}_1 \quad \mathbf{p}_2 \cdots \mathbf{p}_n] \qquad (5\text{-}181)$$

or

$$[\mathbf{p}_1 \quad \mathbf{p}_2 \cdots \mathbf{p}_n]\Lambda = \mathbf{A}[\mathbf{p}_1 \quad \mathbf{p}_2 \cdots \mathbf{p}_n] \qquad (5\text{-}182)$$

Therefore, if we let

$$\mathbf{P} = [\mathbf{p}_1 \quad \mathbf{p}_2 \cdots \mathbf{p}_n] \qquad (5\text{-}183)$$

Eq. (5-182) gives

$$\mathbf{P}\Lambda = \mathbf{A}\mathbf{P} \qquad (5\text{-}184)$$

or

$$\Lambda = \mathbf{P}^{-1}\mathbf{A}\mathbf{P} \qquad (5\text{-}185)$$

which is the desired transformation.

If the matrix **A** is of the phase-variable canonical form, it can be shown that the **P** matrix which diagonalizes **A** may be the Vandermonde matrix,

$$\mathbf{P} = \begin{bmatrix} 1 & 1 & \cdots & 1 \\ \lambda_1 & \lambda_2 & & \lambda_n \\ \lambda_1^2 & \lambda_2^2 & & \lambda_n^2 \\ \cdot & \cdot & \cdots & \cdot \\ \cdot & \cdot & \cdots & \cdot \\ \cdot & \cdot & \cdots & \cdot \\ \lambda_1^{n-1} & \lambda_2^{n-1} & & \lambda_n^{n-1} \end{bmatrix} \tag{5-186}$$

where $\lambda_1, \lambda_2, \ldots, \lambda_n$ are the eigenvalues of **A**.

Since it has been proven that **P** contains as its columns the eigenvectors of **A**, we shall show that the ith column of the matrix in Eq. (5-186) is the eigenvector of **A** that is associated with λ_i, $i = 1, 2, \ldots, n$.

Let

$$\mathbf{p}_i = \begin{bmatrix} p_{i1} \\ p_{i2} \\ \vdots \\ p_{in} \end{bmatrix} \tag{5-187}$$

be the ith eigenvector of **A**. Then

$$(\lambda_i \mathbf{I} - \mathbf{A})\mathbf{p}_i = \mathbf{0} \tag{5-188}$$

or

$$\begin{bmatrix} \lambda_i & -1 & 0 & 0 & \cdots & 0 \\ 0 & \lambda_i & -1 & 0 & \cdots & 0 \\ 0 & 0 & \lambda_i & -1 & \cdots & 0 \\ \cdot & \cdot & \cdot & & & \cdot \\ 0 & 0 & 0 & & \cdots & -1 \\ a_1 & a_2 & a_3 & a_4 & \cdots & \lambda_i + a_n \end{bmatrix} \begin{bmatrix} p_{i1} \\ p_{i2} \\ \vdots \\ p_{in} \end{bmatrix} = \mathbf{0} \tag{5-189}$$

This equation implies that

$$\lambda_i p_{i1} - p_{i2} = 0$$
$$\lambda_i p_{i2} - p_{i3} = 0$$
$$\vdots \tag{5-190}$$
$$\lambda_i p_{i,n-1} - p_{in} = 0$$
$$a_1 p_{i1} + a_2 p_{i2} + \cdots + (\lambda_i + a_n) p_{in} = 0$$

Now we arbitrarily let $p_{i1} = 1$. Then Eq. (5-190) gives

$$p_{i2} = \lambda_i$$
$$p_{i3} = \lambda_i^2$$
$$\vdots \tag{5-191}$$
$$p_{i,n-1} = \lambda_i^{n-2}$$
$$p_{in} = \lambda_i^{n-1}$$

which represent the elements of the ith column of the matrix in Eq. (5-186). Substitution of these elements of $\mathbf{p}_i$ into the last equation of Eq. (5-190) simply verifies that the characteristic equation is satisfied.

■

Example 5-9

Given the matrix

$$\mathbf{A} = \begin{bmatrix} 0 & 1 & 0 \\ 0 & 0 & 1 \\ -6 & -11 & -6 \end{bmatrix} \tag{5-192}$$

which is the phase-variable canonical form, the eigenvalues of $\mathbf{A}$ are $\lambda_1 = -1$, $\lambda_2 = -2$, $\lambda_3 = -3$. The similarity transformation may be carried out by use of the Vandermonde matrix of Eq. (5-186). Therefore,

$$\mathbf{P} = \begin{bmatrix} 1 & 1 & 1 \\ \lambda_1 & \lambda_2 & \lambda_3 \\ \lambda_1^2 & \lambda_2^2 & \lambda_3^2 \end{bmatrix} = \begin{bmatrix} 1 & 1 & 1 \\ -1 & -2 & -3 \\ 1 & 4 & 9 \end{bmatrix} \tag{5-193}$$

The canonical-form state equation is given by Eq. (5-175) with

$$\mathbf{\Lambda} = \mathbf{P}^{-1}\mathbf{A}\mathbf{P} = \begin{bmatrix} -1 & 0 & 0 \\ 0 & -2 & 0 \\ 0 & 0 & -3 \end{bmatrix} = \begin{bmatrix} \lambda_1 & 0 & 0 \\ 0 & \lambda_2 & 0 \\ 0 & 0 & \lambda_3 \end{bmatrix} \tag{5-194}$$

Example 5-10

Given the matrix

$$\mathbf{A} = \begin{bmatrix} 0 & 1 & -1 \\ -6 & -11 & 6 \\ -6 & -11 & 5 \end{bmatrix} \tag{5-195}$$

it can be shown that the eigenvalues of $\mathbf{A}$ are $\lambda_1 = -1$, $\lambda_2 = -2$, and $\lambda_3 = -3$. It is desired to find a nonsingular matrix $\mathbf{P}$ that will transform $\mathbf{A}$ into a diagonal matrix $\mathbf{\Lambda}$, such that $\mathbf{\Lambda} = \mathbf{P}^{-1}\mathbf{A}\mathbf{P}$.

We shall follow the guideline that $\mathbf{P}$ contains the eigenvectors of $\mathbf{A}$. Since $\mathbf{A}$ is not of the phase-variable canonical form, we cannot use the Vandermonde matrix.

Let the eigenvector associated with $\lambda_1 = -1$ be represented by

$$\mathbf{P}_1 = \begin{bmatrix} p_{11} \\ p_{21} \\ p_{31} \end{bmatrix} \tag{5-196}$$

Then $\mathbf{p}_1$ must satisfy

$$(\lambda_1 \mathbf{I} - \mathbf{A})\mathbf{p}_1 = \mathbf{0} \tag{5-197}$$

or

$$\begin{bmatrix} \lambda_1 & -1 & 1 \\ 6 & \lambda_1 + 11 & -6 \\ 6 & 11 & \lambda_1 - 5 \end{bmatrix} \begin{bmatrix} p_{11} \\ p_{21} \\ p_{31} \end{bmatrix} = \mathbf{0} \tag{5-198}$$

The last matrix equation leads to

$$-p_{11} - p_{21} + p_{31} = 0$$
$$6p_{11} + 10p_{21} - 6p_{31} = 0 \qquad (5\text{-}199)$$
$$6p_{11} + 11p_{21} - 6p_{31} = 0$$

from which we get $p_{21} = 0$ and $p_{11} = p_{31}$. Therefore, we can let $p_{11} = p_{31} = 1$, and get

$$\mathbf{p}_1 = \begin{bmatrix} 1 \\ 0 \\ 1 \end{bmatrix} \qquad (5\text{-}200)$$

For the eigenvector associated with $\lambda_2 = -2$, the following matrix equation must be satisfied:

$$\begin{bmatrix} \lambda_2 & -1 & 1 \\ 6 & \lambda_2 + 11 & -6 \\ 6 & 11 & \lambda_2 - 5 \end{bmatrix} \begin{bmatrix} p_{12} \\ p_{22} \\ p_{32} \end{bmatrix} = \mathbf{0} \qquad (5\text{-}201)$$

or

$$-2p_{12} - p_{22} + p_{32} = 0$$
$$6p_{12} + 9p_{22} - 6p_{32} = 0 \qquad (5\text{-}202)$$
$$6p_{12} + 11p_{22} - 7p_{32} = 0$$

In these three equations we let $p_{12} = 1$; then $p_{22} = 2$ and $p_{32} = 4$. Thus,

$$\mathbf{p}_2 = \begin{bmatrix} 1 \\ 2 \\ 4 \end{bmatrix} \qquad (5\text{-}203)$$

Finally, for the eigenvector $\mathbf{p}_3$, we have

$$\begin{bmatrix} \lambda_3 & -1 & 1 \\ 6 & \lambda_3 + 11 & -6 \\ 6 & 11 & \lambda_3 - 5 \end{bmatrix} \begin{bmatrix} p_{13} \\ p_{23} \\ p_{33} \end{bmatrix} = \mathbf{0} \qquad (5\text{-}204)$$

or

$$-3p_{13} - p_{23} + p_{33} = 0$$
$$6p_{13} + 8p_{23} - 6p_{33} = 0 \qquad (5\text{-}205)$$
$$6p_{13} + 11p_{23} - 8p_{33} = 0$$

Now if we arbitrarily let $p_{13} = 1$, the last three equations give $p_{23} = 6$ and $p_{33} = 9$. Therefore,

$$\mathbf{p}_3 = \begin{bmatrix} 1 \\ 6 \\ 9 \end{bmatrix} \qquad (5\text{-}206)$$

The matrix **P** is now given by

$$\mathbf{P} = [\mathbf{p}_1 \quad \mathbf{p}_2 \quad \mathbf{p}_3] = \begin{bmatrix} 1 & 1 & 1 \\ 0 & 2 & 6 \\ 1 & 4 & 9 \end{bmatrix} \tag{5-207}$$

It is easy to show that

$$\mathbf{\Lambda} = \mathbf{P}^{-1}\mathbf{A}\mathbf{P} = \begin{bmatrix} \lambda_1 & 0 & 0 \\ 0 & \lambda_2 & 0 \\ 0 & 0 & \lambda_3 \end{bmatrix} = \begin{bmatrix} -1 & 0 & 0 \\ 0 & -2 & 0 \\ 0 & 0 & -3 \end{bmatrix} \tag{5-208}$$

5.10 JORDAN CANONICAL FORM

In general when the **A** matrix has multiple-order eigenvalues, unless the matrix is symmetric and has real elements, it cannot be diagonalized. However, there exists a similarity transformation

$$\mathbf{\Lambda} = \mathbf{P}^{-1}\mathbf{A}\mathbf{P} \qquad (n \times n) \tag{5-209}$$

such that the matrix $\mathbf{\Lambda}$ is *almost* a diagonal matrix. The matrix $\mathbf{\Lambda}$ is called the *Jordan canonical form*. Typical Jordan canonical forms are shown in the following examples:

$$\mathbf{\Lambda} = \begin{bmatrix} \lambda_1 & 1 & 0 & 0 & 0 \\ 0 & \lambda_1 & 1 & 0 & 0 \\ 0 & 0 & \lambda_1 & 0 & 0 \\ 0 & 0 & 0 & \lambda_2 & 0 \\ 0 & 0 & 0 & 0 & \lambda_3 \end{bmatrix} \tag{5-210}$$

$$\mathbf{\Lambda} = \begin{bmatrix} \lambda_1 & 1 & 0 & 0 & 0 \\ 0 & \lambda_1 & 0 & 0 & 0 \\ 0 & 0 & \lambda_2 & 0 & 0 \\ 0 & 0 & 0 & \lambda_3 & 0 \\ 0 & 0 & 0 & 0 & \lambda_4 \end{bmatrix} \tag{5-211}$$

The Jordan canonical form generally has the following properties:

1. The elements on the main diagonal of $\mathbf{\Lambda}$ are the eigenvalues of the matrix.
2. All the elements below the main diagonal of $\mathbf{\Lambda}$ are zero.
3. Some of the elements immediately above the multiple-ordered eigenvalues on the main diagonal are 1s, such as the cases illustrated by Eqs. (5-210) and (5-211).

4. The 1s, together with the eigenvalues, form typical blocks which are called *Jordan blocks*. In Eqs. (5-210) and (5-211) the Jordan blocks are enclosed by dashed lines.

5. When the nonsymmetrical **A** matrix has multiple-order eigenvalues, its eigenvectors are not linearly independent. For an $n \times n$ **A**, there is only r $(r < n)$ linearly independent eigenvectors.

6. The number of Jordan blocks is equal to the number of independent eigenvectors, r. There is one and only one linearly independent eigenvector associated with each Jordan block.

7. The number of 1s above the main diagonal is equal to $n - r$.

The matrix **P** is determined with the following considerations. Let us assume that **A** has q distinct eigenvalues among n eigenvalues. In the first place, the eigenvectors that correspond to the first order eigenvalues are determined in the usual manner from

$$(\lambda_i \mathbf{I} - \mathbf{A})\mathbf{p}_i = \mathbf{0} \qquad (5\text{-}212)$$

where λ_i denotes the ith distinct eigenvalue, $i = 1, 2, \ldots, q$.

If λ_j is an mth-order eigenvalue, the corresponding mth-order Jordan block is of the form

$$\begin{bmatrix} \lambda_j & 1 & 0 & \cdots & 0 \\ 0 & \lambda_j & 1 & \cdots & 0 \\ \cdots & \cdots & \cdots & \lambda_j & 1 \\ 0 & 0 & \cdots & & \lambda_j \end{bmatrix} \qquad (m \times m) \qquad (5\text{-}123)$$

Then the following transformation must hold:

$$[\mathbf{p}_1 \quad \mathbf{p}_2 \quad \cdots \quad \mathbf{p}_m] \begin{bmatrix} \lambda_j & 1 & 0 & \cdots & 0 \\ 0 & \lambda_j & 1 & & 0 \\ \cdots & \cdots & \cdots & \lambda_j & 1 \\ 0 & 0 & \cdots & & \lambda_j \end{bmatrix} = \mathbf{A}[\mathbf{p}_1 \quad \mathbf{p}_2 \quad \cdots \quad \mathbf{p}_m]$$

$$(5\text{-}214)$$

or

$$\lambda_j \mathbf{p}_1 = \mathbf{A}\mathbf{p}_1$$
$$\mathbf{p}_1 + \lambda_j \mathbf{p}_2 = \mathbf{A}\mathbf{p}_2$$
$$\mathbf{p}_2 + \lambda_j \mathbf{p}_3 = \mathbf{A}\mathbf{p}_3 \qquad (5\text{-}215)$$
$$\vdots$$
$$\mathbf{p}_{m-1} + \lambda_j \mathbf{p}_m = \mathbf{A}\mathbf{p}_m$$

where $\mathbf{p}_1$ is the eigenvector associated with λ_j, and the remaining $m - 1$ vectors, $\mathbf{p}_2, \mathbf{p}_3, \ldots, \mathbf{p}_m$ are auxiliary vectors. Rearranging the vector equations in Eq. (5-215), the m vectors defined above can be determined from the following m vector equations:

$$\left(\lambda_j \mathbf{I} - \mathbf{A}\right)\mathbf{p}_1 = \mathbf{0}$$
$$\left(\lambda_j \mathbf{I} - \mathbf{A}\right)\mathbf{p}_2 = -\mathbf{p}_1$$
$$\left(\lambda_j \mathbf{I} - \mathbf{A}\right)\mathbf{p}_3 = -\mathbf{p}_2 \qquad (5\text{-}216)$$
$$\vdots$$
$$\left(\lambda_j \mathbf{I} - \mathbf{A}\right)\mathbf{p}_m = -\mathbf{p}_{m-1}$$

■

Example 5-11

Given the matrix

$$\mathbf{A} = \begin{bmatrix} 0 & 6 & -5 \\ 1 & 0 & 2 \\ 3 & 2 & 4 \end{bmatrix} \qquad (5\text{-}217)$$

the determinant of $\lambda \mathbf{I} - \mathbf{A}$ is

$$|\lambda \mathbf{I} - \mathbf{A}| = \begin{vmatrix} \lambda & -6 & 5 \\ -1 & \lambda & -2 \\ -3 & -2 & \lambda - 4 \end{vmatrix} = \lambda^3 - 4\lambda^2 + 5\lambda - 2$$
$$= (\lambda - 2)(\lambda - 1)^2 \qquad (5\text{-}218)$$

Therefore, $\mathbf{A}$ has a simple eigenvalue at $\lambda_1 = 2$ and a double eigenvalue at $\lambda_2 = 1$.

To find the Jordan canonical form of $\mathbf{A}$ involves the determination of the matrix $\mathbf{P}$ such that $\mathbf{\Lambda} = \mathbf{P}^{-1}\mathbf{A}\mathbf{P}$. The eigenvector that is associated with $\lambda_1 = 2$ is determined from

$$(\lambda_1 \mathbf{I} - \mathbf{A})\mathbf{p}_1 = \mathbf{0} \qquad (5\text{-}219)$$

Thus,

$$\begin{bmatrix} 2 & -6 & 5 \\ -1 & 2 & -2 \\ -3 & -2 & -2 \end{bmatrix} \begin{bmatrix} p_{11} \\ p_{21} \\ p_{31} \end{bmatrix} = \mathbf{0} \qquad (5\text{-}220)$$

Setting $p_{11} = 2$ arbitrarily, the last equation gives $p_{21} = -1$ and $p_{31} = -2$. Therefore,

$$\mathbf{p}_1 = \begin{bmatrix} 2 \\ -1 \\ -2 \end{bmatrix} \qquad (5\text{-}221)$$

For the eigenvector associated with the second-order eigenvalue, we turn to Eq. (5-216). We have

$$(\lambda_2 \mathbf{I} - \mathbf{A})\mathbf{p}_2 = \mathbf{0} \qquad (5\text{-}222)$$

and

$$(\lambda_2 \mathbf{I} - \mathbf{A})\mathbf{p}_3 = -\mathbf{p}_2 \qquad (5\text{-}223)$$

Equation (5-222) leads to

$$
\begin{bmatrix} 1 & -6 & 5 \\ -1 & 1 & -2 \\ -3 & -2 & -3 \end{bmatrix} \begin{bmatrix} p_{12} \\ p_{22} \\ p_{32} \end{bmatrix} = \mathbf{0}
\tag{5-224}
$$

Setting $p_{12} = 1$ arbitrarily, we have $p_{22} = -\frac{3}{7}$ and $p_{32} = -\frac{5}{7}$. Thus,

$$
\mathbf{p}_2 = \begin{bmatrix} 1 \\ -\frac{3}{7} \\ -\frac{5}{7} \end{bmatrix}
\tag{5-225}
$$

Equation (5-223), when expanded, gives

$$
\begin{bmatrix} 1 & -6 & -5 \\ -1 & 1 & -2 \\ -3 & -2 & -3 \end{bmatrix} \begin{bmatrix} p_{13} \\ p_{23} \\ p_{33} \end{bmatrix} = \begin{bmatrix} -1 \\ \frac{3}{7} \\ \frac{5}{7} \end{bmatrix}
\tag{5-226}
$$

from which we have

$$
\mathbf{p}_3 = \begin{bmatrix} p_{13} \\ p_{23} \\ p_{33} \end{bmatrix} = \begin{bmatrix} 1 \\ -\frac{22}{49} \\ -\frac{46}{49} \end{bmatrix}
\tag{5-227}
$$

Thus,

$$
\mathbf{P} = \begin{bmatrix} 2 & 1 & 1 \\ -1 & -\frac{3}{7} & -\frac{22}{49} \\ -2 & -\frac{5}{7} & -\frac{46}{49} \end{bmatrix}
\tag{5-228}
$$

The Jordan canonical form is now obtained as

$$
\Lambda = \mathbf{P}^{-1}\mathbf{A}\mathbf{P} = \left[\begin{array}{c|cc} 2 & 0 & 0 \\ \hline 0 & 1 & 1 \\ 0 & 0 & 1 \end{array}\right]
\tag{5-229}
$$

Note that in this case there are two Jordan blocks and there is one element of unity above the main diagonal.

5.11 DECOMPOSITION OF TRANSFER FUNCTIONS

Up to this point, various methods of characterizing a linear system have been presented. It will be useful to summarize briefly and gather thoughts at this point before proceeding to the main topics of this section.

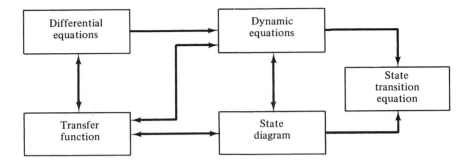

Figure 5-5 Block diagram showing the relationships among various methods of describing linear systems.

It has been shown that the starting point of the description of a linear system may be the system's differential equation, transfer function, or dynamic equations. It is demonstrated that all these methods are closely related. Further, the state diagram defined in Chapter 3 is shown to be a useful tool which not only can lead to the solutions of the state equations but also serves as a vehicle of translation from one type of description to the others. A block diagram is drawn as shown in Fig. 5-5 to illustrate the interrelationships between the various loops of describing a linear system. The block diagram shows that starting, for instance, with the differential equation of a system, one can get to the solution by use of the transfer function method or the state equation method. The block diagram also shows that the majority of the relationships are bilateral, so a great deal of flexibility exists between the methods.

One subject remains to be discussed. This involves the construction of the state diagram from the transfer function. In general, it is necessary to establish a better method than using Eqs. (5-135) through (5-138) in getting from a high-order differential equation to the state equations.

The process of going from the transfer function to the state diagram or the state equations is called the *decomposition* of the transfer function. In general, there are three basic ways of decomposing a transfer function: *direct decomposition*, *cascade decomposition*, and *parallel decomposition*. Each of these three schemes of decomposition has its own advantage and is best suited for a particular situation.

Direct Decomposition

The direct decomposition scheme is applied to a transfer function that is not in factored form. Without loss of generality, the method of direct composition can be described by the following transfer function:

$$\frac{C(s)}{R(s)} = \frac{a_0 s^2 + a_1 s + a_2}{b_0 s^2 + b_1 s + b_2} \tag{5-230}$$

The objective is to obtain the state diagram and the state equations. The following steps are outlined for the direct decomposition:

1. Alter the transfer function so that it has only negative powers of s. This is accomplished by multiplying the numerator and the denominator of the transfer function by the inverse of its highest power in s. For the transfer function of Eq. (5-230), we multiply the numerator and the denominator of $C(s)/R(s)$ by s^{-2}.

2. Multiply the numerator and the denominator of the transfer function by a dummy variable $X(s)$. Implementing steps 1 and 2, Eq. (5-230) becomes

$$\frac{C(s)}{R(s)} = \frac{a_0 + a_1 s^{-1} + a_2 s^{-2}}{b_0 + b_1 s^{-1} + b_2 s^{-2}} \frac{X(s)}{X(s)} \tag{5-231}$$

3. The numerators and the denominators on both sides of the transfer function resulting from steps 1 and 2 are equated to each other, respectively. From Eq. (5-231) this step results in

$$C(s) = (a_0 + a_1 s^{-1} + a_2 s^{-2}) X(s) \tag{5-232}$$

$$R(s) = (b_0 + b_1 s^{-1} + b_2 s^{-2}) X(s) \tag{5-233}$$

4. To construct a state diagram using these two equations, they must first be in the proper cause-and-effect relation. It is apparent that Eq. (5-232) already satisfies this prerequisite. However, Eq. (5-233) has the input on the left side and must be rearranged. Dividing both sides of Eq. (5-233) by b_0 and writing $X(s)$ in terms of the other terms, we have

$$X(s) = \frac{1}{b_0} R(s) - \frac{b_1}{b_0} s^{-1} X(s) - \frac{b_2}{b_0} s^{-2} X(s) \tag{5-234}$$

The state diagram is now drawn in Fig. 5-6 using the expressions in Eqs. (5-232) and (5-234). For simplicity, the initial states are not drawn on the diagram. As usual, the state variables are defined as the outputs of the integrators.

Following the method described in Section 5.10, the state equations are written directly from the state diagram:

$$\begin{bmatrix} \dfrac{dx_1}{dt} \\ \dfrac{dx_2}{dt} \end{bmatrix} = \begin{bmatrix} 0 & 1 \\ -\dfrac{b_2}{b_0} & -\dfrac{b_1}{b_0} \end{bmatrix} \begin{bmatrix} x_1 \\ x_2 \end{bmatrix} + \begin{bmatrix} 0 \\ \dfrac{1}{b_0} \end{bmatrix} r \tag{5-235}$$

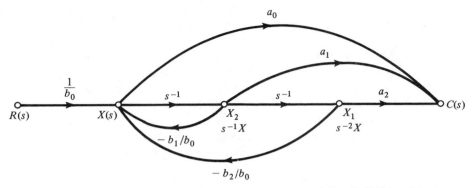

Figure 5-6 State diagram of the transfer function of Eq. (5-230) by direct decomposition.

The output equation is obtained from Fig. 5-6 by applying the gain formula with $c(t)$ as the output node and $x_1(t)$, $x_2(t)$, and $r(t)$ as the input nodes.

$$c = \left(a_2 - \frac{a_0 b_2}{b_0}\right)x_1 + \left(a_1 - \frac{a_0 b_1}{b_0}\right)x_2 + \frac{a_0}{b_0}r \tag{5-236}$$

Cascade Decomposition

Cascade decomposition may be applied to a transfer function that is in the factored form. Consider that the transfer function of Eq. (5-230) may be factored in the following form (of course, there are other possible combinations of factoring):

$$\frac{C(s)}{R(s)} = \frac{a_0}{b_0} \frac{s + z_1}{s + p_1} \frac{s + z_2}{s + p_2} \tag{5-237}$$

where z_1, z_2, p_1, and p_2 are real constants. Then it is possible to treat the functions as the product of two first-order transfer functions. The state diagram of each of the first-order transfer functions is realized by using the direct decomposition method. The complete state diagram is obtained by cascading the two first-order diagrams as shown in Fig. 5-7. As usual, the outputs of the integrators on the state diagram are

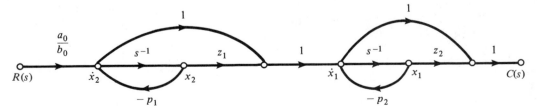

Figure 5-7 State diagram of the transfer function of Eq. (5-237) by cascade decomposition.

assigned as the state variables. The state equations are written in matrix form:

$$
\begin{bmatrix} \dfrac{dx_1}{dt} \\[2mm] \dfrac{dx_2}{dt} \end{bmatrix} = \begin{bmatrix} -p_2 & z_1 - p_1 \\[2mm] 0 & -p_1 \end{bmatrix} \begin{bmatrix} x_1 \\[2mm] x_2 \end{bmatrix} + \begin{bmatrix} \dfrac{a_0}{b_0} \\[2mm] \dfrac{a_0}{b_0} \end{bmatrix} r \tag{5-238}
$$

The output equation is

$$
c = (z_2 - p_2)x_1 + (z_1 - p_1)x_2 + \frac{a_0}{b_0} r \tag{5-239}
$$

The cascade decomposition has the advantage that the poles and zeros of the transfer function appear as isolated branch gains on the state diagram. This facilitates the study of the effects on the system when the poles and zeros are varied.

Parallel Decomposition

When the denominator of a transfer function is in factored form, it is possible to expand the transfer function by partial fractions. Consider that a second-order system is represented by the following transfer function:

$$
\frac{C(s)}{R(s)} = \frac{P(s)}{(s + p_1)(s + p_2)} \tag{5-240}
$$

where $P(s)$ is a polynomial of order less than 2. We assume that the poles p_1 and p_2 may be complex conjugate for analytical purposes, but it is difficult to implement complex coefficients on the computer.

In this case if p_1 and p_2 are equal it would not be possible to carry out a partial-fraction expansion of the transfer function of Eq. (5-240). With p_1 and p_2 being distinct, Eq. (5-240) is written

$$
\frac{C(s)}{R(s)} = \frac{K_1}{s + p_1} + \frac{K_2}{s + p_2} \tag{5-241}
$$

where K_1 and K_2 are constants.

The state diagram for the system is formed by the parallel combination of the state diagram representation of each of the first-order terms on the right side of Eq. (5-241), as shown in Fig. 5-8. The state equations of the system are written

$$
\begin{bmatrix} \dfrac{dx_1}{dt} \\[2mm] \dfrac{dx_2}{dt} \end{bmatrix} = \begin{bmatrix} -p_1 & 0 \\[2mm] 0 & -p_2 \end{bmatrix} \begin{bmatrix} x_1 \\[2mm] x_2 \end{bmatrix} = \begin{bmatrix} 1 \\[2mm] 1 \end{bmatrix} r \tag{5-242}
$$

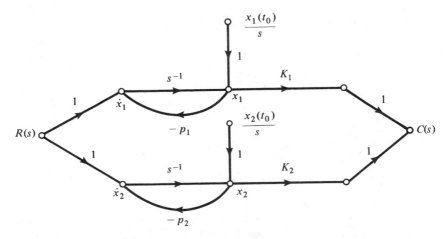

Figure 5-8 State diagram of the transfer function of Eq. (5-241) by parallel decomposition.

The output equation is

$$c = [K_1 \quad K_2]\begin{bmatrix} x_1 \\ x_2 \end{bmatrix} \tag{5-243}$$

One of the advantages of the parallel decomposition is that for transfer functions with simple poles, the resulting **A** matrix is always a diagonal matrix. Therefore, we can consider that parallel decomposition may be used for the diagonalization of the **A** matrix.

When a transfer function has multiple-order poles, care must be taken that the state diagram, as obtained through the parallel decomposition, contains a minimum number of integrators. To further clarify the point just made, consider the following transfer function and its partial-fraction expansion:

$$\frac{C(s)}{R(s)} = \frac{2s^2 + 6s + 5}{(s+1)^2(s+2)} = \frac{1}{(s+1)^2} + \frac{1}{s+1} + \frac{1}{s+2} \tag{5-244}$$

Note that the transfer function is of the third order, and although the total order of the terms on the right side of Eq. (5-244) is four, only three integrators should be used in the state diagram. The state diagram for the system is drawn as shown in Fig. 5-9. The minimum number of three integrators are used, with one integrator being shared by two channels. The state equations of the system are written

$$\begin{bmatrix} \dfrac{dx_1}{dt} \\[2mm] \dfrac{dx_2}{dt} \\[2mm] \dfrac{dx_3}{dt} \end{bmatrix} = \begin{bmatrix} -1 & 1 & 0 \\ 0 & -1 & 0 \\ 0 & 0 & -2 \end{bmatrix}\begin{bmatrix} x_1 \\ x_2 \\ x_3 \end{bmatrix} + \begin{bmatrix} 0 \\ 1 \\ 1 \end{bmatrix}r \tag{5-245}$$

Therefore, the **A** matrix is of the Jordan canonical form.

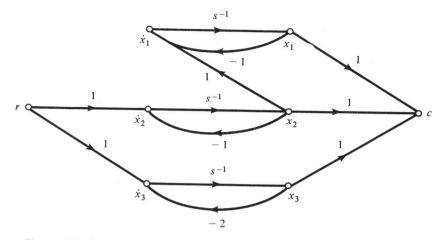

Figure 5-9 State diagram of the transfer function of Eq. (5-244) by parallel decomposition.

5.12 CONTROLLABILITY OF LINEAR SYSTEMS

The concepts of controllability and observability introduced first by Kalman play an important role in both theoretical and practical aspects of modern control theory. The conditions on controllability and observability often govern the existence of a solution to an optimal control problem. For instance, we shall show that the condition of controllability of a system is closely related to the existence of solutions of state feedback for the purpose of placing the eigenvalues of the system arbitrarily. The concept of observability relates to the condition of observing or estimating the state variables from the output variables, which are generally measurable.

One way of illustrating the motivation of investigating controllability and observability can be made by referring to the block diagrams shown in Fig. 5-10. Figure 5-10(a) shows a closed-loop system with the process dynamics described by

$$\dot{\mathbf{x}}(t) = \mathbf{A}\mathbf{x}(t) + \mathbf{B}\mathbf{u}(t) \tag{5-246}$$

The closed-loop system is formed by feeding back the state variables through a constant matrix $\mathbf{G}$. Thus,

$$\mathbf{u}(t) = -\mathbf{G}\mathbf{x}(t) + \mathbf{r}(t) \tag{5-247}$$

where $\mathbf{G}$ is a $p \times n$ feedback matrix with constant elements. The closed-loop system is thus described by

$$\dot{\mathbf{x}}(t) = (\mathbf{A} - \mathbf{B}\mathbf{G})\mathbf{x}(t) + \mathbf{B}\mathbf{r}(t) \tag{5-248}$$

The design objective in this case is to find the feedback matrix $\mathbf{G}$ such that the eigenvalues of $(\mathbf{A} - \mathbf{B}\mathbf{G})$, or of the closed-loop system, are at certain prescribed values. This problem is also known as the *pole-placement design* through state feedback. The word "pole" refers here to the poles of the closed-loop transfer function, which are the same as the eigenvalues of $\mathbf{A} - \mathbf{B}\mathbf{G}$.

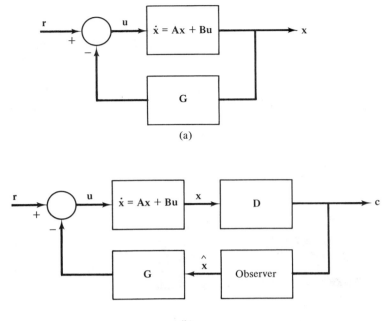

Figure 5-10 (a) Control system with state feedback. (b) Control system with observer and state feedback.

We shall show later that the existence of the solution to the pole-placement design through state feedback is directly based on the controllability of [**A**, **B**]. Thus, we can state that if the system of Eq. (5-246) is controllable, then there exists a constant feedback matrix **G** which allows the eigenvalues of **A** − **BG** to be arbitrarily placed.

Once the closed-loop system is designed, one has to deal with the practical problem of feeding back the state variables. There are two practical problems with this type of control. One is that the number of state variables, which is n, may be excessive, so that the cost of sensing each of these state variables for feedback may be prohibitive. Another problem is that not all the state variables are accessible directly from the system. In reality, only the output variables $c_1, c_2, \ldots, c_p$ are guaranteed to be accessible. Figure 5-10(b) shows the block diagram of a closed-loop system with an observer that estimates the state vector from the output vector **c**. The observed or the estimated state vector is designated as $\hat{\mathbf{x}}$, which is then used to generate the control **u** through the feedback matrix **G**. The condition that such an observer exists for the system of Eq. (5-246), together with the output equation $\mathbf{c}(t) = \mathbf{D}\mathbf{x}(t)$, is called the observability of the system.

General Concept of Controllability

The concept of controllability can be stated with reference to the block diagram of Fig. 5-11. The process G is said to be completely controllable if *every* state variable

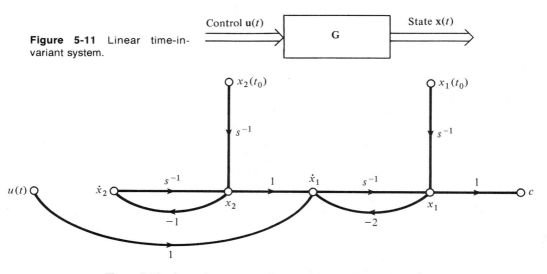

Figure 5-11 Linear time-invariant system.

Figure 5-12 State diagram of a system that is not state controllable.

of **G** can be affected or controlled to reach a certain objective in finite time by some unconstrained control $\mathbf{u}(t)$. Intuitively, it is simple to understand that if any one of the state variables is independent of the control $\mathbf{u}(t)$, there would be no way of driving this particular state variable to a desired state in finite time by means of a control effort. Therefore, this particular state is said to be uncontrollable, and as long as there is at least one uncontrollable state, the system is said to be not completely controllable, or simply uncontrollable.

As a simple example of an uncontrollable system, Fig. 5-12 illustrates the state diagram of a linear system with two state variables. Since the control $u(t)$ affects only the state $x_1(t)$, $x_2(t)$ is uncontrollable. In other words, it would be impossible to drive $x_2(t)$ from an initial state $x_2(t_0)$ to a desired state $x_2(t_f)$ in a finite time interval $t_f - t_0$ by a control $u(t)$. Therefore, the entire system is said to be uncontrollable.

The concept of controllability given above refers to the states and is sometimes referred to as the *state controllability*. Controllability can also be defined for the outputs of a system, so there is a difference between state controllability and output controllability.

Definition of Controllability (State Controllability)

Consider that a linear time-invariant system is described by the following dynamic equations:

$$\dot{\mathbf{x}}(t) = \mathbf{A}\mathbf{x}(t) + \mathbf{B}\mathbf{u}(t) \tag{5-249}$$

$$\mathbf{c}(t) = \mathbf{D}\mathbf{x}(t) + \mathbf{E}\mathbf{u}(t) \tag{5-250}$$

where $\mathbf{x}(t) = n \times 1$ state vector
$\mathbf{u}(t) = r \times 1$ input vector
$\mathbf{c}(t) = p \times 1$ output vector
$\mathbf{A} = n \times n$ coefficient matrix
$\mathbf{B} = n \times r$ coefficient matrix
$\mathbf{D} = p \times n$ coefficient matrix
$\mathbf{E} = p \times r$ coefficient matrix

The state $\mathbf{x}(t)$ is said to be controllable at $t = t_0$ if there exists a piecewise continuous input $\mathbf{u}(t)$ that will drive the state to any final state $\mathbf{x}(t_f)$ for a finite time $(t_f - t_0) \geq 0$. If every state $\mathbf{x}(t_0)$ of the system is controllable in a finite time interval, the system is said to be completely state controllable or simply state controllable.

The following theorem shows that the condition of controllability depends on the coefficient matrices $\mathbf{A}$ and $\mathbf{B}$ of the system. The theorem also gives one way of testing state controllability.

Theorem 5-2. *For the system described by the state equation of Eq. (5-249) to be completely state controllable, it is necessary and sufficient that the following $n \times nr$ matrix has a rank of n:*

$$\mathbf{S} = [\mathbf{B} \quad \mathbf{AB} \quad \mathbf{A}^2\mathbf{B} \cdots \mathbf{A}^{n-1}\mathbf{B}] \qquad (5\text{-}251)$$

Since the matrices $\mathbf{A}$ and $\mathbf{B}$ are involved, sometimes we say that the pair $[\mathbf{A}, \mathbf{B}]$ is controllable, which implies that $\mathbf{S}$ is of rank n.

The proof of this theorem is given in any standard textbook on optimal control systems. The idea is to start with the state transition equation of Eq. (5-53) and then proceed to show that Eq. (5-251) must be satisfied in order that all the states are accessible by the input control.

Although the criterion of state controllability given by Theorem 5-2 is quite straightforward, it is not very easy to implement for multiple-input systems. Even with $r = 2$, there are $2n$ columns in $\mathbf{S}$, and there would be a large number of possible combinations of $n \times n$ matrices. A practical way may be to use one column of $\mathbf{B}$ at a time, each time giving an $n \times n$ matrix for $\mathbf{S}$. However, failure to find an $\mathbf{S}$ with a rank for n this way does not mean that the system is uncontrollable, until all the columns of $\mathbf{B}$ are used. An easier way would be to form the matrix $\mathbf{SS}'$, which is $n \times n$; then if $\mathbf{SS}'$ is nonsingular, $\mathbf{S}$ has rank n.

■

Example 5-12

Consider the system shown in Fig. 5-12, which was reasoned earlier to be uncontrollable. Let us investigate the same problem using the condition of Eq. (5-251).

The state equations of the system are written, from Fig. 5-12,

$$\begin{bmatrix} \dfrac{dx_1(t)}{dt} \\[2mm] \dfrac{dx_2(t)}{dt} \end{bmatrix} = \begin{bmatrix} -2 & 1 \\ 0 & -1 \end{bmatrix} \begin{bmatrix} x_1(t) \\ x_2(t) \end{bmatrix} + \begin{bmatrix} 1 \\ 0 \end{bmatrix} u(t) \qquad (5\text{-}252)$$

Therefore, from Eq. (5-251),

$$\mathbf{S} = [\mathbf{B} \quad \mathbf{AB}] = \begin{bmatrix} 1 & -2 \\ 0 & 0 \end{bmatrix} \tag{5-253}$$

which is singular, and the system is not state controllable.

Example 5-13

Determine the state controllability of the system described by the state equation

$$\begin{bmatrix} \dfrac{dx_1(t)}{dt} \\ \dfrac{dx_2(t)}{dt} \end{bmatrix} = \begin{bmatrix} 0 & 1 \\ -1 & 0 \end{bmatrix} \begin{bmatrix} x_1(t) \\ x_2(t) \end{bmatrix} + \begin{bmatrix} 0 \\ 1 \end{bmatrix} u(t) \tag{5-254}$$

From Eq. (5-251),

$$\mathbf{S} = [\mathbf{B} \quad \mathbf{AB}] = \begin{bmatrix} 0 & 1 \\ 1 & 0 \end{bmatrix} \tag{5-255}$$

which is nonsingular. Therefore, the system is completely state controllable.

■

Alternative Tests on Controllability

Theorem 5-2 and Eq. (5-251) give a simple and practical method of testing the state controllability of a linear time-invariant system. There are several other alternative methods of testing controllability, and some of these may be more convenient to apply under certain conditions.

Theorem 5-3. *Consider that a linear time-invariant system is described by the state equation*

$$\dot{\mathbf{x}}(t) = \mathbf{Ax}(t) + \mathbf{Bu}(t) \tag{5-256}$$

If the eigenvalues of **A** *are distinct and are denoted by* λ_i, $i = 1, 2, \ldots, n$, *then there exists an nth-order nonsingular matrix* **P** *which transforms* **A** *into a diagonal matrix* Λ *that is given by Eqs. (5-176) and (5-177). The new state vector is*

$$\mathbf{y} = \mathbf{P}^{-1}\mathbf{x} \tag{5-257}$$

and the new state equation is

$$\dot{\mathbf{y}} = \Lambda\mathbf{y} + \Gamma\mathbf{u} \tag{5-258}$$

where

$$\Gamma = \mathbf{P}^{-1}\mathbf{B} \tag{5-259}$$

Then the system in Eq. (5-256) or (5-258) is completely state controllable if Γ *has no rows that are all zeros.*

The proof of this theorem follows directly from the definition of state controllability. Since the state equations in Eq. (5-258) are completely decoupled from each other, the only way the states are controllable is that each state is controlled directly by at least one input. Therefore, if any one row of Γ contains all zeros, this means that the corresponding state is not controlled by any one of the inputs.

If the system of Eq. (5-258) is completely controllable, it implies that the following matrix is of rank n.

$$\mathbf{S}_1 = [\Gamma \quad \Lambda\Gamma \quad \Lambda^2\Gamma \quad \cdots \quad \Lambda^{n-1}\Gamma] \qquad (5\text{-}260)$$

or using Eqs. (5-177) and (5-260), we have

$$\mathbf{S}_1 = \mathbf{P}^{-1}[\mathbf{B} \quad \mathbf{AB} \quad \mathbf{A}^2\mathbf{B} \quad \cdots \quad \mathbf{A}^{n-1}\mathbf{B}]$$
$$= \mathbf{P}^{-1}\mathbf{S} \qquad (5\text{-}261)$$

Since $\mathbf{P}$ is nonsingular, the rank of $\mathbf{S}$ must be the same as that of $\mathbf{S}_1$. Thus, if the system in Eq. (5-258) is controllable, so is that of Eq. (5-256).

It should be noted that the prerequisite on distinct eigenvalues precedes the condition of diagonalization of $\mathbf{A}$. In other words, all square matrices with distinct eigenvalues can be diagonalized. However, certain matrices with multiple-order eigenvalues can also be diagonalized. The natural question is: Does the alternative definition apply to a system with multiple-order eigenvalues but whose $\mathbf{A}$ matrix can be diagonalized? The answer is *no*. We must not lose sight of the original definition on state controllability that any state $\mathbf{x}(t_0)$ is brought to any state $\mathbf{x}(t_f)$ in finite time. Thus the question of independent control must enter the picture. In other words, consider that we have two states which are uncoupled and are related by the following state equations:

$$\frac{dx_1(t)}{dt} = ax_1(t) + b_1u(t) \qquad (5\text{-}262)$$

$$\frac{dx_2(t)}{dt} = ax_2(t) + b_2u(t) \qquad (5\text{-}263)$$

This system is apparently uncontrollable, since

$$\mathbf{S} = [\mathbf{B} \quad \mathbf{AB}] = \begin{bmatrix} b_1 & ab_1 \\ b_2 & ab_2 \end{bmatrix} \qquad (5\text{-}264)$$

is singular. Therefore, just because $\mathbf{A}$ is diagonal, and $\mathbf{B}$ has no rows which are zeros does not mean that the system is controllable. The reason in this case is that $\mathbf{A}$ has multiple-order eigenvalues.

When **A** has multiple-order eigenvalues and cannot be diagonalized, there is a nonsingular matrix **P** which transforms **A** into a Jordan canonical form $\Lambda = \mathbf{P}^{-1}\mathbf{AP}$. *The condition of state controllability is that all the elements of* $\Gamma = \mathbf{P}^{-1}\mathbf{B}$ *that correspond to the last row of each Jordan block are nonzero.* The reason behind this is that the last row of each Jordan block corresponds to a state equation that is completely uncoupled from the other state equations.

The elements in the other rows of Γ need not all be nonzero, since the corresponding states are all coupled. For instance, if the matrix **A** has four eigenvalues, $\lambda_1, \lambda_1, \lambda_1, \lambda_2$, three of which are equal, then there is a nonsingular **P** which transforms **A** into the Jordan canonical form:

$$\Lambda = \mathbf{P}^{-1}\mathbf{AP} = \begin{bmatrix} \lambda_1 & 1 & 0 & 0 \\ 0 & \lambda_1 & 1 & 0 \\ 0 & 0 & \lambda_1 & 0 \\ \hline 0 & 0 & 0 & \lambda_2 \end{bmatrix} \tag{5-265}$$

Then the condition given above becomes self-explanatory.

∎

Example 5-14

Consider the system of Example 5-11. The **A** and **B** matrices are, respectively,

$$\mathbf{A} = \begin{bmatrix} -2 & 1 \\ 0 & -1 \end{bmatrix} \qquad \mathbf{B} = \begin{bmatrix} 1 \\ 0 \end{bmatrix}$$

Let us check the controllability of the system by checking the rows of the matrix Γ. It can be shown that **A** is diagonalized by the matrix

$$\mathbf{P} = \begin{bmatrix} 1 & 1 \\ 0 & 1 \end{bmatrix}$$

Therefore,

$$\Gamma = \mathbf{P}^{-1}\mathbf{B} = \begin{bmatrix} 1 & -1 \\ 0 & 1 \end{bmatrix}\begin{bmatrix} 1 \\ 0 \end{bmatrix} = \begin{bmatrix} 1 \\ 0 \end{bmatrix} \tag{5-266}$$

The transformed state equation is

$$\dot{\mathbf{y}}(t) = \begin{bmatrix} -2 & 0 \\ 0 & -1 \end{bmatrix}\mathbf{y}(t) + \begin{bmatrix} 1 \\ 0 \end{bmatrix}u(t) \tag{5-267}$$

Since the second row of Γ is zero, the state variable $y_2(t)$, or $x_2(t)$, is uncontrollable, and the system is uncontrollable.

Example 5-15

Consider that a third-order system has the coefficient matrices

$$\mathbf{A} = \begin{bmatrix} 1 & 2 & -1 \\ 0 & 1 & 0 \\ 1 & -4 & 3 \end{bmatrix} \qquad \mathbf{B} = \begin{bmatrix} 0 \\ 0 \\ 1 \end{bmatrix}$$

Then

$$S = [B \quad AB \quad A^2 B] = \begin{bmatrix} 0 & -1 & -4 \\ 0 & 0 & 0 \\ 1 & 3 & 8 \end{bmatrix} \qquad (5\text{-}268)$$

Since S is singular, the system is not state controllable.

Using the alternative method, the eigenvalues of A are found to be $\lambda_1 = 2$, $\lambda_2 = 2$, and $\lambda_3 = 1$. The Jordan canonical form of A is obtained with

$$P = \begin{bmatrix} 1 & 0 & 0 \\ 0 & 0 & 1 \\ -1 & 1 & 2 \end{bmatrix} \qquad (5\text{-}269)$$

Then

$$\Lambda = P^{-1}AP = \begin{bmatrix} 2 & 1 & 0 \\ 0 & 2 & 0 \\ 0 & 0 & 1 \end{bmatrix} \qquad (5\text{-}270)$$

$$\Gamma = P^{-1}B = \begin{bmatrix} 0 \\ -1 \\ 0 \end{bmatrix} \qquad (5\text{-}271)$$

Since the last row of Γ is zero, the state variable y_3 is uncontrollable. Since $x_2 = y_3$, this corresponds to x_2 being uncontrollable.

Example 5-16

Consider a linear system whose input-output relationship is described by the differential equation

$$\frac{d^2c(t)}{dt^2} + 2\frac{dc(t)}{dt} + c(t) = \frac{du(t)}{dt} + u(t) \qquad (5\text{-}272)$$

We shall show that the state controllability of the system depends upon how the state variables are defined.

Let the state variables be defined as

$$x_1 = c$$
$$x_2 = \dot{c} - u$$

The state equations of the system are expressed in matrix form as

$$\begin{bmatrix} \dot{x}_1 \\ \dot{x}_2 \end{bmatrix} = \begin{bmatrix} 0 & 1 \\ -1 & -2 \end{bmatrix} \begin{bmatrix} x_1 \\ x_2 \end{bmatrix} + \begin{bmatrix} 1 \\ -1 \end{bmatrix} u \qquad (5\text{-}273)$$

The output equation is

$$c = x_1$$

The state controllability matrix is

$$S = [B \quad AB] = \begin{bmatrix} 1 & -1 \\ -1 & 1 \end{bmatrix} \qquad (5\text{-}274)$$

which is singular. The system is *not state controllable*.

Now let us define the state variables of the system in a different way. By the method of direct decomposition, the state equations are written in matrix form:

$$\begin{bmatrix} \dot{x}_1 \\ \dot{x}_2 \end{bmatrix} = \begin{bmatrix} 0 & 1 \\ -1 & -2 \end{bmatrix} \begin{bmatrix} x_1 \\ x_2 \end{bmatrix} + \begin{bmatrix} 0 \\ 1 \end{bmatrix} u \tag{5-275}$$

The output equation is

$$c = x_1 + x_2$$

The system is now *completely state controllable* since

$$\mathbf{S} = [\mathbf{B} \quad \mathbf{AB}] = \begin{bmatrix} 0 & 1 \\ 1 & -2 \end{bmatrix} \tag{5-276}$$

which is nonsingular.

5.13 OBSERVABILITY OF LINEAR SYSTEMS

The concept of observability is quite similar to that of controllability. Essentially, a system is completely observable if every state variable of the system affects some of the outputs. In other words, it is often desirable to obtain information on the state variables from measurements of the outputs and the inputs. If any one of the states cannot be observed from the measurements of the outputs, the state is said to be unobservable, and the system is not completely observable, or is simply unobservable. Figure 5-13 shows the state diagram of a linear system in which the state x_2 is not connected to the output c in any way. Once we have measured c, we can observe the state x_1, since $x_1 = c$. However, the state x_2 cannot be observed from the information on c. Thus, the system is described as not completely observable, or simply unobservable.

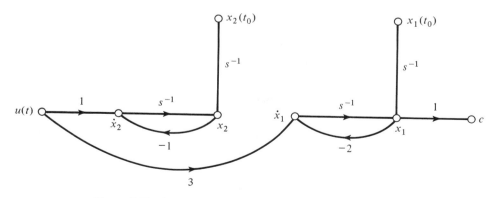

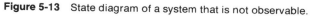

Figure 5-13 State diagram of a system that is not observable.

Definition of Observability. *Given a linear time-invariant system that is described by the dynamic equations of Eqs. (5-249) and (5-250) the state $\mathbf{x}(t_0)$ is said to be observable if given any input $\mathbf{u}(t)$, there exists a finite time $t_f \geq t_0$ such that the knowledge of $\mathbf{u}(t)$ for $t_0 \leq t < t_f$; the matrices $\mathbf{A}, \mathbf{B}, \mathbf{D},$ and $\mathbf{E}$; and the output $\mathbf{c}(t)$ for $t_0 \leq t < t_f$ are sufficient to determine $\mathbf{x}(t_0)$. If every state of the system is observable for a finite t_f, we say that the system is completely observable, or simply observable.*

The following theorem shows that the condition of observability depends on the coefficient matrices $\mathbf{A}$ and $\mathbf{D}$ of the system. The theorem also gives one method of testing observability.

Theorem 5-4. *For the system described by the dynamic equation of Eqs. (5-249) and (5-250) to be completely observable, it is necessary and sufficient that the following $n \times np$ matrix has a rank of n:*

$$\mathbf{V} = \begin{bmatrix} \mathbf{D}' & \mathbf{A}'\mathbf{D}' & (\mathbf{A}')^2\mathbf{D}' & \cdots & (\mathbf{A}')^{n-1}\mathbf{D}' \end{bmatrix} \qquad (5\text{-}277)$$

The condition is also referred to as the pair $[\mathbf{A}, \mathbf{D}]$ being observable. In particular, if the system has only one output, $\mathbf{D}$ is an $1 \times n$ matrix; $\mathbf{V}$ of Eq. (5-277) is an $n \times n$ square matrix. Then the system is completely observable if $\mathbf{V}$ is nonsingular.

The proof of this theorem is not given here, but it is based on the principle that Eq. (5-277) must be satisfied so that $\mathbf{x}(t_0)$ can be uniquely determined from the output vector $\mathbf{c}(t)$.

■

Example 5-17

Consider the system shown in Fig. 5-13, which was earlier defined to be unobservable. The dynamic equations of the system are written directly from the state diagram.

$$\begin{bmatrix} \dot{x}_1 \\ \dot{x}_2 \end{bmatrix} = \begin{bmatrix} -2 & 0 \\ 0 & -1 \end{bmatrix} \begin{bmatrix} x_1 \\ x_2 \end{bmatrix} + \begin{bmatrix} 3 \\ 1 \end{bmatrix} u \qquad (5\text{-}278)$$

$$c = \begin{bmatrix} 1 & 0 \end{bmatrix} \begin{bmatrix} x_1 \\ x_2 \end{bmatrix} \qquad (5\text{-}279)$$

Therefore,

$$\mathbf{D} = \begin{bmatrix} 1 & 0 \end{bmatrix} \qquad \mathbf{D}' = \begin{bmatrix} 1 \\ 0 \end{bmatrix}$$

$$\mathbf{A}'\mathbf{D}' = \begin{bmatrix} -2 & 0 \\ 0 & -1 \end{bmatrix} \begin{bmatrix} 1 \\ 0 \end{bmatrix} = \begin{bmatrix} -2 \\ 0 \end{bmatrix}$$

and, from Eq. (5-277),

$$\mathbf{V} = \begin{bmatrix} \mathbf{D}' & \mathbf{A}'\mathbf{D}' \end{bmatrix} = \begin{bmatrix} 1 & -2 \\ 0 & 0 \end{bmatrix} \qquad (5\text{-}280)$$

Since $\mathbf{V}$ is singular, the system is unobservable.

Example 5-18

Let us consider the system described by the differential equation of Eq. (5-272), Example 5-16. In Example 5-16 we have shown that state controllability of a system depends on how the state variables are defined. We shall now show that the observability also depends on the definition of the state variables. Let the dynamic equations of the system be defined as in Eq. (5-273),

$$\mathbf{A} = \begin{bmatrix} 0 & 1 \\ -1 & -2 \end{bmatrix} \quad \mathbf{D} = [1 \ \ 0]$$

Then

$$\mathbf{V} = [\mathbf{D}' \ \ \mathbf{A}'\mathbf{D}'] = \begin{bmatrix} 1 & 0 \\ 0 & 1 \end{bmatrix} \tag{5-281}$$

and thus the system is completely observable.

Let the dynamic equations of the system be given by Eq. (5-275). Then

$$\mathbf{A} = \begin{bmatrix} 0 & 1 \\ -1 & -2 \end{bmatrix} \quad \mathbf{D} = [1 \ \ 1]$$

Then

$$\mathbf{V} = [\mathbf{D}' \ \ \mathbf{A}'\mathbf{D}'] = \begin{bmatrix} 1 & -1 \\ 1 & -1 \end{bmatrix}$$

which is singular. Thus the system is unobservable, and we have shown that given the input-output relation of a linear system, the observability of the system depends on how the state variables are defined. It should be noted that for the system of Eq. (5-272), one method of state variable assignment, Eq. (5-273), yields a system that is observable but not state controllable. On the other hand, if the dynamic equations of Eq. (5-275) are used, the system is completely state controllable but not observable. There are definite reasons behind these results, and we investigate these phenomena further in the following discussions.

■

Alternative Tests of Observability.   If the matrix **A** has distinct eigenvalues, it can be diagonalized as in Eq. (5-258). The new state variable is

$$\mathbf{y} = \mathbf{P}^{-1}\mathbf{x} \tag{5-282}$$

The new dynamic equations are

$$\dot{\mathbf{y}} = \mathbf{\Lambda}\mathbf{y} + \mathbf{\Gamma}\mathbf{u} \tag{5-283}$$

$$\mathbf{c} = \mathbf{F}\mathbf{y} + \mathbf{E}\mathbf{u} \tag{5-284}$$

where

$$\mathbf{F} = \mathbf{D}\mathbf{P} \tag{5-285}$$

Then the system is completely observable if **F** *has no zero columns.*

The reason behind the above condition is that if the jth $(j = 1, 2, \ldots, n)$ column of $\mathbf{F}$ contains all zeros, the state variable y_j will not appear in Eq. (5-284) and is not related to the output $\mathbf{c}(t)$. Therefore, y_j will be unobservable. In general, the states that correspond to zero columns of $\mathbf{F}$ are said to be unobservable, and the rest of the state variables are observable.

∎

Example 5-19

Consider the system of Example 5-17 which was found to be unobservable. Since the A matrix, as shown in Eq. (5-278), is already a diagonal matrix, the alternative condition of observability stated above requires that the matrix $\mathbf{D} = [1 \quad 0]$ must not contain any zero columns. Since the second column of $\mathbf{D}$ is indeed zero, the state x_2 is unobservable, and the system is unobservable.

∎

5.14 INVARIANT THEOREMS ON CONTROLLABILITY AND OBSERVABILITY

We now introduce several theorems that relate to the effects of nonsingular transformations on controllability and observability. The effects on controllability and observability due to state and output feedbacks will also be investigated.

Theorem 5-5. *Invariant Theorem on Nonsingular Transformation: Controllability.* Given the nth-order system

$$\dot{\mathbf{x}}(t) = \mathbf{A}\mathbf{x}(t) + \mathbf{B}\mathbf{u}(t) \tag{5-286}$$

where the pair $[\mathbf{A}, \mathbf{B}]$ is completely controllable. The transformation $\mathbf{x}(t) = \mathbf{P}\mathbf{y}(t)$, where $\mathbf{P}$ is nonsingular, transforms the system of Eq. (5-286) to

$$\dot{\mathbf{y}}(t) = \boldsymbol{\Lambda}\mathbf{y}(t) + \boldsymbol{\Gamma}\mathbf{u}(t) \tag{5-287}$$

where $\quad \boldsymbol{\Lambda} = \mathbf{P}^{-1}\mathbf{A}\mathbf{P}$
$\quad\quad\quad \boldsymbol{\Gamma} = \mathbf{P}^{-1}\mathbf{B}$

Then, the pair $[\boldsymbol{\Lambda}, \boldsymbol{\Gamma}]$ is also controllable. Conversely, if $[\mathbf{A}, \mathbf{B}]$ is uncontrollable, then $[\boldsymbol{\Lambda}, \boldsymbol{\Gamma}]$ is also uncontrollable.

Proof. The proof of this theorem follows directly from Theorem 5-2 using Eq. (5-251).

Theorem 5-5 can be easily extended to the phase-variable canonical-form transformation. Thus, the property of state controllability is invariant if the system of Eq. (5-286) is transformed into the phase-variable canonical form by Eq. (5-100). On the other hand, it is interesting to note that since the condition of the existence of a phase-variable canonical-form transformation is that the matrix $\mathbf{S}$ of Eq. (5-99) or Eq. (5-251) is of rank n, we state that any system that can be transformed into or is already in the phase-variable canonical form is completely state controllable.

Theorem 5-6. *Invariant Theorem on Nonsingular Transformation*: *Observability*. Given the nth-order system

$$\dot{\mathbf{x}}(t) = \mathbf{A}\mathbf{x}(t) + \mathbf{B}\mathbf{u}(t) \tag{5-288}$$

$$\mathbf{c}(t) = \mathbf{D}\mathbf{x}(t) \tag{5-289}$$

where the pair $[\mathbf{A}, \mathbf{D}]$ is completely observable. The transformation $\mathbf{x}(t) = \mathbf{P}\mathbf{y}(t)$, where $\mathbf{P}$ is nonsingular, transforms the system equation to

$$\dot{\mathbf{y}}(t) = \mathbf{\Lambda}\mathbf{y}(t) + \mathbf{\Gamma}\mathbf{u}(t) \tag{5-290}$$

and

$$\mathbf{c}(t) = \mathbf{D}\mathbf{P}\mathbf{y}(t) \tag{5-291}$$

Then the pair $[\mathbf{\Lambda}, \mathbf{DP}]$ is also observable. On the other hand, if $[\mathbf{A}, \mathbf{D}]$ is unobservable, the same is true for $[\mathbf{\Lambda}, \mathbf{DP}]$.

Proof. The proof of this theorem is similar to that of Theorem 5-5 and is not detailed here.

Theorem 5-7. *Theorem on Controllability of Closed-Loop Systems with State Feedback*. If the system

$$\dot{\mathbf{x}}(t) = \mathbf{A}\mathbf{x}(t) + \mathbf{B}\mathbf{u}(t) \tag{5-292}$$

is completely state controllable, then the closed-loop system obtained through state feedback,

$$\mathbf{u}(t) = \mathbf{r}(t) - \mathbf{G}\mathbf{x}(t) \tag{5-293}$$

so that the state equation becomes

$$\dot{\mathbf{x}}(t) = (\mathbf{A} - \mathbf{BG})\mathbf{x}(t) + \mathbf{B}\mathbf{r}(t) \tag{5-294}$$

is also completely controllable. On the other hand, if the pair $[\mathbf{A}, \mathbf{B}]$ is uncontrollable, then there is no $\mathbf{G}$ that will make the pair $[\mathbf{A} - \mathbf{BG}, \mathbf{B}]$ controllable. In other words, if an open-loop system is uncontrollable, it cannot be made controllable through state feedback.

Proof. By controllable of the pair $[\mathbf{A}, \mathbf{B}]$ we mean that there exists a control $\mathbf{u}(t)$ over the interval $[t_0, t_f]$ such that the initial state $\mathbf{x}(t_0)$ is driven to the final state $\mathbf{x}(t_f)$ over the finite time interval $t_f - t_0$. We can write Eq. (5-293) as

$$\mathbf{r}(t) = \mathbf{u}(t) + \mathbf{G}\mathbf{x}(t) \tag{5-295}$$

which is the control of the closed-loop system. Thus, if $u(t)$ exists which can drive $x(t_0)$ to $x(t_f)$ in finite time, Eq. (5-295) implies that $r(t)$ also exists, and the closed-loop system is also controllable.

Conversely, if the pair $[A, B]$ is uncontrollable, which means that no $u(t)$ exists which will drive any $x(t_0)$ to any $x(t_f)$ in finite time, then we cannot find an input $r(t)$ that will do the same to $x(t)$, since otherwise we can set $u(t)$ as in Eq. (5-293) to control the open-loop system.

Theorem 5-8. *Theorem on Observability of Closed-Loop Systems with State Feedback.* If the system described by Eqs. (5-288) and (5-289) is controllable and observable, then state feedback of the form of Eq. (5-293) could destroy observability. In other words, the observability of open-loop and closed-loop systems due to state feedback are unrelated.

The following example will illustrate the relation between observability and state feedback.

∎

Example 5-20

For the system described by Eqs. (5-288) and (5-289), let

$$A = \begin{bmatrix} 0 & 1 \\ -2 & -3 \end{bmatrix} \quad B = \begin{bmatrix} 1 \\ 1 \end{bmatrix} \quad D = [1 \quad 2]$$

We can show that the pair $[A, B]$ is controllable and the pair $[A, D]$ is observable.

Let the state feedback be defined by

$$u(t) = r(t) - Gx(t) \tag{5-296}$$

where

$$G = [g_1 \quad g_2] \tag{5-297}$$

Then the closed-loop system is described by the state equation

$$\dot{x}(t) = (A - BG)x(t) + Br(t) \tag{5-298}$$

$$A - BG = \begin{bmatrix} -g_1 & 1 - g_2 \\ -2 - g_1 & -3 - g_2 \end{bmatrix} \tag{5-299}$$

The observability matrix of the closed-loop system is

$$V = [D' \quad (A - BG)'D'] = \begin{bmatrix} 1 & -3g_1 - 4 \\ 2 & -3g_2 - 5 \end{bmatrix} \tag{5-300}$$

The determinant of V is

$$|V| = 6g_1 - 3g_2 + 3 \tag{5-301}$$

Therefore, if g_1 and g_2 are chosen so that $|\mathbf{V}| = 0$, the closed-loop system would be unobservable.

■

5.15 RELATIONSHIP AMONG CONTROLLABILITY, OBSERVABILITY, AND TRANSFER FUNCTIONS

In the classical analysis of control systems, transfer functions are often used for the modeling of linear time-invariant systems. Although controllability and observability are concepts of modern control theory, they are closely related to the properties of the transfer function.

Let us focus our attention on the system considered in Examples 5-16 and 5-18. It was demonstrated in these two examples that the system is either not state controllable or not observable, depending on the ways the state variables are defined. These phenomena can be explained by referring to the transfer function of the system, which is obtained from Eq. (5-272). We have

$$\frac{C(s)}{U(s)} = \frac{s+1}{s^2 + 2s + 1} = \frac{s+1}{(s+1)^2} = \frac{1}{s+1} \tag{5-302}$$

which has an identical pole and zero at $s = -1$. The following theorem gives the relationship between controllability and observability and the pole-zero cancellation of a transfer function.

Theorem 5-9. *If the input–output transfer function of a linear system has pole–zero cancellation, the system will be either not state controllable or unobservable, depending on how the state variables are defined. If the input–output transfer function of a linear system does not have pole–zero cancellation, the system can always be represented by dynamic equations as a completely controllable and observable system.*

Proof. Consider that an nth-order system with a single input and single output and distinct eigenvalues is represented by the dynamic equations

$$\dot{\mathbf{x}}(t) = \mathbf{A}\mathbf{x}(t) + \mathbf{B}u(t) \tag{5-303}$$

$$c(t) = \mathbf{D}\mathbf{x}(t) \tag{5-304}$$

Let the $\mathbf{A}$ matrix be diagonalized by an $n \times n$ Vandermonde matrix $\mathbf{P}$,

$$\mathbf{P} = \begin{bmatrix} 1 & 1 & 1 & \cdots & 1 \\ \lambda_1 & \lambda_2 & \lambda_3 & \cdots & \lambda_n \\ \lambda_1^2 & \lambda_2^2 & \lambda_3^2 & \cdots & \lambda_n^2 \\ \cdots & \cdots & \cdots & \cdots & \cdots \\ \lambda_1^{n-1} & \lambda_2^{n-1} & \lambda_3^{n-1} & \cdots & \lambda_n^{n-1} \end{bmatrix} \tag{5-305}$$

The new state equation in canonical form is

$$\dot{\mathbf{y}}(t) = \mathbf{\Lambda}\mathbf{y}(t) + \mathbf{\Gamma}u(t) \tag{5-306}$$

where $\mathbf{\Lambda} = \mathbf{P}^{-1}\mathbf{A}\mathbf{P}$. The output equation is transformed into

$$c(t) = \mathbf{F}\mathbf{y}(t) \tag{5-307}$$

where $\mathbf{F} = \mathbf{D}\mathbf{P}$. The state vectors $\mathbf{x}(t)$ and $\mathbf{y}(t)$ are related by

$$\mathbf{x}(t) = \mathbf{P}\mathbf{y}(t) \tag{5-308}$$

Since $\mathbf{\Lambda}$ is a diagonal matrix, the ith equation of Eq. (5-306) is

$$\dot{y}_i(t) = \lambda_i y_i(t) + \gamma_i u(t) \tag{5-309}$$

where λ_i is the ith eigenvalue of $\mathbf{A}$ and γ_i is the ith element of $\mathbf{\Gamma}$, where $\mathbf{\Gamma}$ is an $n \times 1$ matrix in the present case. Taking the Laplace transform on both sides of Eq. (5-309) and assuming zero initial conditions, we obtain the transfer function relation between $Y_i(s)$ and $U(s)$ as

$$Y_i(s) = \frac{\gamma_i}{s - \lambda_i} U(s) \tag{5-310}$$

The Laplace transform of Eq. (5-307) is

$$C(s) = \mathbf{F}\mathbf{Y}(s) = \mathbf{D}\mathbf{P}\mathbf{Y}(s) \tag{5-311}$$

Now if it is assumed that

$$\mathbf{D} = [d_1 \quad d_2 \quad \cdots \quad d_n] \tag{5-312}$$

then

$$\mathbf{F} = \mathbf{D}\mathbf{P} = [f_1 \quad f_2 \quad \cdots \quad f_n] \tag{5-313}$$

where

$$f_i = d_i + d_2\lambda_i + \cdots + d_n\lambda_i^{n-1} \tag{5-314}$$

for $i = 1, 2, \ldots, n$. Equation (5-311) is written as

$$\begin{aligned}
C(s) &= [f_1 \quad f_2 \quad \cdots \quad f_n]\mathbf{Y}(s) \\
&= [f_1 \quad f_2 \quad \cdots \quad f_n] \begin{bmatrix} \dfrac{\gamma_1}{s - \lambda_1} \\[2mm] \dfrac{\gamma_2}{s - \lambda_2} \\[1mm] \vdots \\[1mm] \dfrac{\gamma_n}{s - \lambda_n} \end{bmatrix} U(s) \\
&= \sum_{i=1}^{n} \frac{f_i \gamma_i}{s - \lambda_i} U(s)
\end{aligned} \tag{5-315}$$

For the nth-order system with distinct eigenvalues, let us assume that the input-output transfer function is of the form

$$\frac{C(s)}{U(s)} = \frac{K(s - a_1)(s - a_2) \cdots (s - a_m)}{(s - \lambda_1)(s - \lambda_2) \cdots (s - \lambda_n)} \quad n > m \qquad (5\text{-}316)$$

which is expanded by partial fraction into

$$\frac{C(s)}{U(s)} = \sum_{i=1}^{n} \frac{\sigma_i}{s - \lambda_i} \qquad (5\text{-}317)$$

where σ_i denotes the residue of $C(s)/U(s)$ at $s = \lambda_i$.

It was established earlier that for the system described by Eq. (5-306) to be state controllable, all the rows of Γ must be nonzero; that is, $\gamma_i \neq 0$ for $i = 1, 2, \ldots, n$. If $C(s)/U(s)$ has one or more pairs of identical pole and zero, for instance in Eq. (5-316), $a_1 = \lambda_1$, then in Eq. (5-317) $\sigma_1 = 0$. Comparing Eq. (5-315) with Eq. (5-317), we see that in general

$$\sigma_i = f_i \gamma_i \qquad (5\text{-}318)$$

Therefore, when $\sigma_i = 0$, γ_i will be zero if $f_i \neq 0$, and the state y_i is uncontrollable.

For observability, it was established earlier that $\mathbf{F}$ must not have columns containing zeros. Or, in the present case, $f_i \neq 0$ for $i = 1, 2, \ldots, n$. However, from Eq. (5-318),

$$f_i = \frac{\sigma_i}{\gamma_i} \qquad (5\text{-}319)$$

When the transfer function has an identical pair of pole and zero at $a_i = \lambda_i$, $\sigma_i = 0$. Thus, from Eq. (5-319), $f_i = 0$ if $\gamma_i \neq 0$.

5.16 STATE EQUATIONS OF LINEAR DISCRETE-DATA SYSTEMS

Similar to the continuous-data systems case, a modern way of modeling a discrete-data system is by means of discrete state equations. As described earlier, when dealing with discrete-data systems, we often encounter two different situations. The first one is that the components of the system are continuous-data elements, but the signals at certain points of the system are discrete or discontinuous with respect to time, because of the sample-and-hold operations. In this case the components of the system are still described by differential equations, but because of the discrete data, a set of difference equations may be generated from the original differential equations. The second situation involves systems that are completely discrete with respect to time in the sense that they receive and send out discrete data only, such as in the case of a digital controller or digital computer. Under this condition, the system dynamics should be described by difference equations.

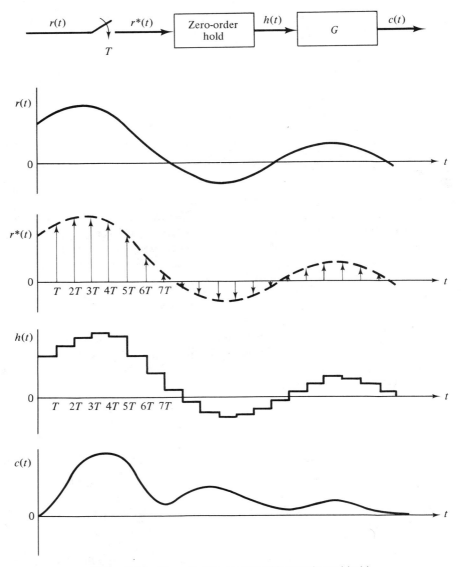

Figure 5-14   Discrete-data system with sample-and-hold.

Let us consider the open-loop discrete-data control system with a sample-and-hold device, as shown in Fig. 5-14. Typical signals that appear at various points in the system are also shown in the figure. The output signal, $c(t)$, ordinarily is a continuous-data signal. The output of the sample-and-hold, $h(t)$, is a train of steps. Therefore, we can write

$$h(kT) = r(kT) \qquad k = 0, 1, 2, \ldots \tag{5-320}$$

Now we let the linear process G be described by the state equation and output equation:

$$\frac{d\mathbf{x}(t)}{dt} = \mathbf{A}\mathbf{x}(t) + \mathbf{B}h(t) \tag{5-321}$$

$$c(t) = \mathbf{D}\mathbf{x}(t) + \mathbf{E}h(t) \tag{5-322}$$

where $\mathbf{x}(t)$ is the state vector and $h(t)$ and $c(t)$ are the scalar input and output signals, respectively. The matrices $\mathbf{A}$, $\mathbf{B}$, $\mathbf{D}$, and $\mathbf{E}$ are coefficient matrices which have been defined earlier. Using Eq. (5-53), the state transition equation of the system is written

$$\mathbf{x}(t) = \boldsymbol{\phi}(t - t_0)\mathbf{x}(t_0) + \int_{t_0}^{t} \boldsymbol{\phi}(t - \tau)\mathbf{B}h(\tau)\,d\tau \tag{5-323}$$

for $t \geq t_0$.

If we are interested only in the responses at the sampling instants, just as in the case of the z-transform solution, we let $t = (k + 1)T$ and $t_0 = kT$. Then Eq. (5-323) becomes

$$\mathbf{x}[(k + 1)T] = \boldsymbol{\phi}(T)\mathbf{x}(kT) + \int_{kT}^{(k+1)T} \boldsymbol{\phi}[(k + 1)T - \tau]\mathbf{B}h(\tau)\,d\tau \tag{5-324}$$

where $\boldsymbol{\phi}(t)$ is the state transition matrix as defined in Section 5.3.

Since $h(t)$ is piecewise constant, that is, $h(kT) = r(kT)$ for $kT \leq t < (k + 1)T$, the input function $h(\tau)$ in Eq. (5-324) can be taken outside the integral sign. Equation (5-324) is written

$$\mathbf{x}[(k + 1)T] = \boldsymbol{\phi}(T)\mathbf{x}(kT) + \int_{kT}^{(k+1)T} \boldsymbol{\phi}[(k + 1)T - \tau]\mathbf{B}\,d\tau\, r(kT) \tag{5-325}$$

or

$$\mathbf{x}[(k + 1)T] = \boldsymbol{\phi}(T)\mathbf{x}(kT) + \boldsymbol{\theta}(T)r(kT) \tag{5-326}$$

where

$$\boldsymbol{\theta}(T) = \int_{kT}^{(k+1)T} \boldsymbol{\phi}[(k + 1)T - \tau]\mathbf{B}\,d\tau \tag{5-327}$$

Equation (5-326) is of the form of a linear difference equation in vector-matrix form. Since it represents a set of first-order difference equations, it is referred to as the *vector-matrix discrete state equation*.

The discrete state equation in Eq. (5-326) can be solved by means of a simple recursion procedure. Setting $k = 0, 1, 2, \ldots$ in Eq. (5-326), we find that the following equations result:

$$k = 0: \qquad \mathbf{x}(T) = \boldsymbol{\phi}(T)\mathbf{x}(0) + \boldsymbol{\theta}(T)r(0) \qquad (5\text{-}328)$$

$$k = 1: \qquad \mathbf{x}(2T) = \boldsymbol{\phi}(T)\mathbf{x}(T) + \boldsymbol{\theta}(T)r(T) \qquad (5\text{-}329)$$

$$k = 2: \qquad \mathbf{x}(3T) = \boldsymbol{\phi}(T)\mathbf{x}(2T) + \boldsymbol{\theta}(T)r(2T) \qquad (5\text{-}330)$$

$$\vdots \qquad\qquad \vdots$$

$$k = k - 1: \qquad \mathbf{x}(kT) = \boldsymbol{\phi}(T)\mathbf{x}[(k-1)T] + \boldsymbol{\theta}(T)r[(k-1)T] \quad (5\text{-}331)$$

Substituting Eq. (5-328) into Eq. (5-329), and then Eq. (5-329) into Eq. (5-330), ..., and so on, we obtain the following solution for Eq. (5-326):

$$\mathbf{x}(kT) = \boldsymbol{\phi}^k(T)\mathbf{x}(0) + \sum_{i=0}^{k-1} \boldsymbol{\phi}^{k-i-1}(T)\boldsymbol{\theta}(T)r(iT) \qquad (5\text{-}332)$$

Equation (5-332) is defined as the *discrete state transition equation* of the discrete-data system. It is interesting to note that Eq. (5-332) is analogous to its continuous counterpart in Eq. (5-50). In fact, the state transition equation of Eq. (5-50) describes the state of the system of Fig. 5-14 with or without sampling. The discrete state transition equation of Eq. (5-332) is more restricted in that it describes the state only at $t = kT$ ($k = 0, 1, 2, \ldots$), and only if the system has a sample-and-hold device such as in Fig. 5-14.

With kT considered as the initial time, a discrete state transition equation similar to that of Eq. (5-53) can be obtained as

$$\mathbf{x}[(k + N)T] = \boldsymbol{\phi}^N(T)\mathbf{x}(kT) + \sum_{i=0}^{N-1} \boldsymbol{\phi}^{N-i-1}(T)\boldsymbol{\theta}(T)r[(k+i)T] \quad (5\text{-}333)$$

where N is a positive integer. The derivation of Eq. (5-333) is left as an exercise for the reader.

The output of the system of Fig. 5-14 at the sampling instants is obtained by substituting $t = kT$ and Eq. (5-332) into Eq. (5-322), yielding

$$c(kT) = \mathbf{D}\mathbf{x}(kT) + \mathbf{E}h(kT)$$

$$= \mathbf{D}\boldsymbol{\phi}^k(T)\mathbf{x}(0) + \mathbf{D}\sum_{i=0}^{k-1} \boldsymbol{\phi}^{k-i-1}(T)\boldsymbol{\theta}(T)r(iT) + \mathbf{E}h(kT) \quad (5\text{-}334)$$

An important advantage of the state-variable method over the z-transform method is that it can be modified easily to describe the states and the output between sampling instants. In Eq. (5-323) if we let $t = (k + \Delta)T$, where $0 < \Delta \le 1$ and $t_0 = kT$, we get

$$\mathbf{x}[(k + \Delta)T] = \boldsymbol{\phi}(\Delta T)\mathbf{x}(kT) + \int_{kT}^{(k+\Delta)T} \boldsymbol{\phi}[(k+\Delta)T - \tau]\mathbf{B}\,d\tau\,r(kT)$$

$$= \boldsymbol{\phi}(\Delta T)\mathbf{x}(kT) + \boldsymbol{\theta}(\Delta T)r(kT) \qquad (5\text{-}335)$$

By varying the value of Δ between 0 and 1, the information between the sampling instants is completely described by Eq. (5-335).

One of the interesting properties of the state transition matrix $\phi(t)$ is that

$$\phi^k(T) = \phi(kT) \tag{5-336}$$

which is proved as follows.

Using the homogeneous solution of the state equation of Eq. (5-321), we have

$$\mathbf{x}(t) = \phi(t - t_0)\mathbf{x}(t_0) \tag{5-337}$$

Let $t = kT$ and $t_0 = 0$; the last equation becomes

$$\mathbf{x}(kT) = \phi(kT)\mathbf{x}(0) \tag{5-338}$$

Also, by the recursive procedure with $t = (k + 1)T$ and $t_0 = kT$, $k = 0, 1, 2, \ldots,$ Eq. (5-337) leads to

$$\mathbf{x}(kT) = \phi^k(T)\mathbf{x}(0) \tag{5-339}$$

Comparison of Eqs. (5-338) and (5-339) gives the identity in Eq. (5-336).

In view of the relation of Eq. (5-336), the discrete state transition equations of Eqs. (5-332) and (5-333) are written

$$\mathbf{x}(kT) = \phi(kT)\mathbf{x}(0) + \sum_{i=0}^{k-1} \phi[(k - i - 1)T]\theta(T)r(iT) \tag{5-340}$$

$$\mathbf{x}[(k + N)T] = \phi(NT)\mathbf{x}(kT) + \sum_{i=0}^{N-1} \phi[(N - i - 1)T]\theta(T)r[(k + i)T] \tag{5-341}$$

respectively. These two equations can be modified to represent systems with multiple inputs simply by changing the input r into a vector $\mathbf{r}$.

When a linear system has only discrete data throughout the system, its dynamics can be described by a set of discrete state equations

$$\mathbf{x}[(k + 1)T] = \mathbf{A}\mathbf{x}(kT) + \mathbf{B}r(kT) \tag{5-342}$$

and output equations

$$\mathbf{c}(kT) = \mathbf{D}\mathbf{x}(kT) + \mathbf{E}r(kT) \tag{5-343}$$

where $\mathbf{A}$, $\mathbf{B}$, $\mathbf{D}$, and $\mathbf{E}$ are coefficient matrices of the appropriate dimensions. Notice that Eq. (5-342) is basically of the same form as Eq. (5-326). The only difference in the two situations is the starting point of the system representation. In the case of Eq. (5-326), the starting point is the continuous-data state equations of Eq. (5-321); $\phi(T)$ and $\theta(T)$ are determined from the $\mathbf{A}$ and $\mathbf{B}$ matrices of Eq. (5-321). In the case of Eq. (5-342), the equation itself represents an outright description of the discrete-data system, which has only discrete signals.

The solution of Eq. (5-342) follows directly from that of Eq. (5-326). Therefore, the discrete state transition equation of Eq. (5-342) is written

$$\mathbf{x}(kT) = \mathbf{A}^k\mathbf{x}(0) + \sum_{i=0}^{k-1} \mathbf{A}^{k-i-1}\mathbf{Br}(iT) \tag{5-344}$$

where

$$\mathbf{A}^k = \underbrace{\mathbf{AAAA\ldots A}}_{k} \tag{5-345}$$

5.17 z-TRANSFORM SOLUTION OF DISCRETE STATE EQUATIONS

The discrete state equation in vector-matrix form,

$$\mathbf{x}[(k+1)T] = \mathbf{Ax}(kT) + \mathbf{Br}(kT) \tag{5-346}$$

can be solved by means of the z-transform method. Taking the z-transform on both sides of Eq. (5-346) yields

$$z\mathbf{X}(z) - z\mathbf{x}(0) = \mathbf{AX}(z) + \mathbf{BR}(z) \tag{5-347}$$

Solving for $\mathbf{X}(z)$ from the last equation gives

$$\mathbf{X}(z) = (z\mathbf{I} - \mathbf{A})^{-1}z\mathbf{x}(0) + (z\mathbf{I} - \mathbf{A})^{-1}\mathbf{BR}(z) \tag{5-348}$$

The inverse z-transform of the last equation is

$$\mathbf{x}(kT) = \mathfrak{z}^{-1}\left[(z\mathbf{I} - \mathbf{A})^{-1}z\right]\mathbf{x}(0) + \mathfrak{z}^{-1}\left[(z\mathbf{I} - \mathbf{A})^{-1}\mathbf{BR}(z)\right] \tag{5-349}$$

In order to carry out the inverse z-transform operation of the last equation, we write the z-transform of $\mathbf{A}^k$ as

$$\mathfrak{z}(\mathbf{A}^k) = \sum_{k=0}^{\infty} \mathbf{A}^k z^{-k} = \mathbf{I} + \mathbf{A}z^{-1} + \mathbf{A}^2 z^{-2} + \cdots \tag{5-350}$$

Premultiplying both sides of the last equation by $\mathbf{A}z^{-1}$ and subtracting the result from the last equation, we get

$$(\mathbf{I} - \mathbf{A}z^{-1})\mathfrak{z}(\mathbf{A}^k) = \mathbf{I} \tag{5-351}$$

Therefore, solving for $\mathfrak{z}(\mathbf{A}^k)$ from the last equation yields

$$\mathfrak{z}(\mathbf{A}^k) = (\mathbf{I} - \mathbf{A}z^{-1})^{-1} = (z\mathbf{I} - \mathbf{A})^{-1}z \tag{5-352}$$

or

$$\mathbf{A}^k = \mathfrak{z}^{-1}\left[(z\mathbf{I} - \mathbf{A})^{-1}z\right] \tag{5-353}$$

Equation (5-353) also represents a way of finding $\mathbf{A}^k$ by using the *z*-transform method. Similarly, we can prove that

$$\mathfrak{z}^{-1}\left[(z\mathbf{I} - \mathbf{A})^{-1}\mathbf{B}\mathbf{R}(z)\right] = \sum_{i=0}^{k-1} \mathbf{A}^{k-i-1}\mathbf{B}\mathbf{r}(iT) \tag{5-354}$$

Now we substitute Eqs. (5-353) and (5-354) into Eq. (5-349) and we have the solution for $\mathbf{x}(kT)$ as

$$\mathbf{x}(kT) = \mathbf{A}^k\mathbf{x}(0) + \sum_{i=0}^{k-1} \mathbf{A}^{k-i-1}\mathbf{B}\mathbf{r}(iT) \tag{5-355}$$

which is identical to the expression in Eq. (5-344).

Once a discrete-data system is represented by the dynamic equations of Eqs. (5-342) and (5-343), the transfer function relation of the system can be expressed in terms of the coefficient matrices.

Setting the initial state $\mathbf{x}(0)$ to zero, Eq. (5-348) gives

$$\mathbf{X}(z) = (z\mathbf{I} - \mathbf{A})^{-1}\mathbf{B}\mathbf{R}(z) \tag{5-356}$$

When this equation is substituted into the *z*-transformed version of Eq. (5-343), we have

$$\mathbf{C}(z) = \left[\mathbf{D}(z\mathbf{I} - \mathbf{A})^{-1}\mathbf{B} + \mathbf{E}\right]\mathbf{R}(z) \tag{5-357}$$

Thus the transfer function matrix of the system is

$$\mathbf{G}(z) = \mathbf{D}(z\mathbf{I} - \mathbf{A})^{-1}\mathbf{B} + \mathbf{E} \tag{5-358}$$

This equation can be written

$$\mathbf{G}(z) = \frac{\mathbf{D}[\mathrm{adj}(z\mathbf{I} - \mathbf{A})]\mathbf{B} + |z\mathbf{I} - \mathbf{A}|\mathbf{E}}{|z\mathbf{I} - \mathbf{A}|} \tag{5-359}$$

The characteristic equation of the system is defined as

$$|z\mathbf{I} - \mathbf{A}| = 0 \tag{5-360}$$

In general, a linear time-invariant discrete-data system with one input and one output can be described by the following linear difference equation with constant coefficients:

$$
\begin{aligned}
c[(k + n)T] &+ a_1 c[(k + n - 1)T] + a_2 c[(k + n - 2)T] \\
&+ \cdots + a_{n-1} c[(k + 1)T] + a_n c(kT) \\
&= b_0 r[(k + m)T] + b_1 r[(k + m - 1)T] \\
&+ \cdots + b_{m-1} r[(k + 1)T] + b_m r(kT) \quad n \geq m
\end{aligned} \tag{5-361}
$$

Taking the z-transform on both sides of this equation and rearranging terms, the transfer function of the system is written

$$
\frac{C(z)}{R(z)} = \frac{b_0 z^m + b_1 z^{m-1} + \cdots + b_{m-1} z + b_m}{z^n + a_1 z^{n-1} + \cdots + a_{n-1} z + a_n} \tag{5-362}
$$

The characteristic equation is defined as

$$
z^n + a_1 z^{n-1} + \cdots + a_{n-1} z + a_n = 0 \tag{5-363}
$$

∎

Example 5-21

Consider that a discrete-data system is described by the difference equation

$$
c(k + 2) + 5c(k + 1) + 3c(k) = r(k + 1) + 2r(k) \tag{5-364}
$$

Taking the z-transform on both sides of the last equation and assuming zero initial conditions yields

$$
z^2 C(z) + 5zC(z) + 3C(z) = zR(z) + 2R(z) \tag{5-365}
$$

From the last equation the transfer function of the system is easily written

$$
\frac{C(z)}{R(z)} = \frac{z + 2}{z^2 + 5z + 3} \tag{5-366}
$$

The characteristic equation is obtained by setting the denominator polynomial of the transfer function to zero,

$$
z^2 + 5z + 3 = 0 \tag{5-367}
$$

The state variables of the system are arbitrarily defined as

$$
x_1(k) = c(k) \tag{5-368}
$$
$$
x_2(k) = x_1(k + 1) - r(k) \tag{5-369}
$$

Substitution of the last two relations into the original difference equation of Eq. (5-364) gives the two state equations of the system as

$$x_1(k + 1) = x_2(k) + r(k) \tag{5-370}$$

$$x_2(k + 1) = -3x_1(k) - 5x_2(k) - 3r(k) \tag{5-371}$$

from which we have the **A** matrix of the system,

$$\mathbf{A} = \begin{bmatrix} 0 & 1 \\ -3 & -5 \end{bmatrix} \tag{5-372}$$

The same characteristic equation as in Eq. (5-367) is obtained by using $|z\mathbf{I} - \mathbf{A}| = 0$.

■

5.18 STATE DIAGRAMS FOR DISCRETE-DATA SYSTEMS

When a discrete-data system is described by difference equations or discrete state equations, a discrete state diagram may be constructed for the system. Similar to the relations between the analog computer diagram and the state diagram for a continuous-data system, the elements of a discrete state diagram resemble the computing elements of a digital computer. Some of the operations of a digital computer are multiplication by a constant, addition of several machine variables, time delay, or shifting. The mathematical descriptions of these basic digital computations and their corresponding z-transform expressions are as follows:

1. *Multiplication by a constant:*

$$x_2(kT) = ax_1(kT) \tag{5-373}$$

$$X_2(z) = aX_1(z) \tag{5-374}$$

2. *Summing:*

$$x_2(kT) = x_0(kT) + x_1(kT) \tag{5-375}$$

$$X_2(z) = X_0(z) + X_1(z) \tag{5-376}$$

3. *Shifting or time delay:*

$$x_2(kT) = x_1[(k + 1)T] \tag{5-377}$$

$$X_2(z) = zX_1(z) - zx_1(0) \tag{5-378}$$

or

$$X_1(z) = z^{-1}X_2(z) + x_1(0) \tag{5-379}$$

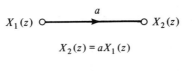

$$X_2(z) = aX_1(z)$$

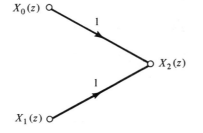

$$X_2(z) = X_0(z) + X_1(z)$$

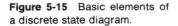

$$X_1(z) = z^{-1}X_2(z) + x_1(0)$$

Figure 5-15 Basic elements of a discrete state diagram.

The state diagram representations of these operations are illustrated in Fig. 5-15. The initial time $t = 0$ in Eq. (5-379) can be generalized to $t = t_0$. Then Eq. (5-379) is written

$$X_1(z) = z^{-1}X_2(z) + x_1(t_0) \tag{5-380}$$

which represents the discrete-time state transition for time greater than or equal to t_0.

∎

Example 5-22

Consider again the difference equation in Eq. (5-364), which is

$$c(k + 2) + 5c(k + 1) + 3c(k) = r(k + 1) + 2r(k) \tag{5-381}$$

One way of constructing the discrete state diagram for the system is to use the state equations. In this case the state equations are available in Eqs. (5-370) and (5-371) and these are repeated here:

$$x_1(k + 1) = x_2(k) + r(k) \tag{5-382}$$
$$x_2(k + 1) = -3x_1(k) - 5x_2(k) - 3r(k) \tag{5-383}$$

Using essentially the same principle as for the state diagrams for continuous-data systems,

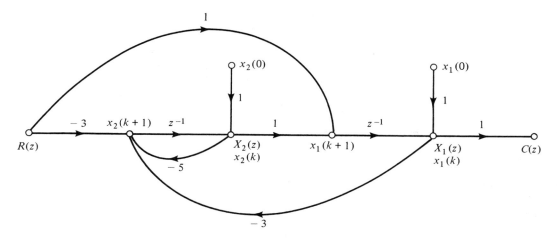

Figure 5-16 Discrete state diagram of the system described by the difference equation of Eq. (5-381) or by the state equations of Eqs. (5-382) and (5-383).

the state diagram for Eqs. (5-382) and (5-383) is constructed in Fig. 5-16. The time delay unit z^{-1} is used to relate $x_1(k + 1)$ to $x_1(k)$. The state variables will always appear as outputs of the delay units on the state diagram.

As an alternative, the state diagram can also be drawn directly from the difference equation by means of the decomposition schemes. The decomposition of a discrete transfer function will be discussed in the following section, after we have demonstrated some of the practical applications of the discrete state diagram.

The state transition equation of the system can be obtained directly from the state diagram using the gain formula. Referring to $X_1(z)$ and $X_2(z)$ as the output notes and to $x_1(0)$, $x_2(0)$, and $R(z)$ as input nodes in Fig. 5-16, the state transition equations are written in the following vector-matrix form:

$$\begin{bmatrix} X_1(z) \\ X_2(z) \end{bmatrix} = \frac{1}{\Delta} \begin{bmatrix} 1 + 5z^{-1} & z^{-1} \\ -3z^{-1} & 1 \end{bmatrix} \begin{bmatrix} x_1(0) \\ x_2(0) \end{bmatrix} + \frac{1}{\Delta} \begin{bmatrix} z^{-1}(1 + 5z^{-1}) - 3z^{-2} \\ -3z^{-1} - 3z^{-2} \end{bmatrix} R(z) \quad (5\text{-}384)$$

where

$$\Delta = 1 + 5z^{-1} + 3z^{-2} \quad (5\text{-}385)$$

The same transfer function between $R(z)$ and $C(z)$ as in Eq. (5-366) can be obtained directly from the state diagram by applying the gain formula between these two nodes.

■

Decomposition of Discrete Transfer Functions

The three schemes of decomposition discussed earlier for continuous-data systems can be applied to transfer functions of discrete-data systems without the need of modification. As an illustrative example, the following transfer function is decomposed by the three methods, and the corresponding state diagrams are shown in Fig.

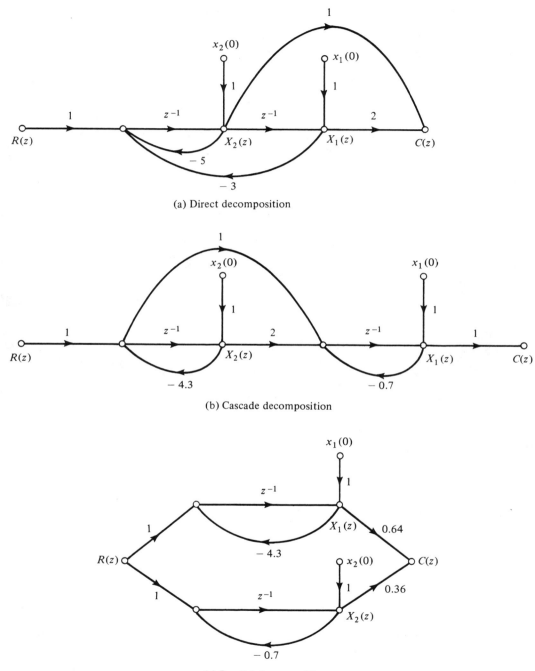

(a) Direct decomposition

(b) Cascade decomposition

(c) Parallel decomposition

Figure 5-17 State diagrams of the transfer function $C(z) / R(z) = (z + 2) / (z^2 + 5z + 3)$ by the three methods of decomposition. (a) Direct decomposition. (b) Cascade decomposition. (c) Parallel decomposition.

5-17:

$$\frac{C(z)}{R(z)} = \frac{z + 2}{z^2 + 5z + 3} \qquad (5\text{-}386)$$

Equation (5-386) is used for direct decomposition after the numerator and the denominator are both multiplied by z^{-2}. For cascade decomposition, the transfer function is first written in factored form as

$$\frac{C(z)}{R(z)} = \frac{z + 2}{(z + 4.3)(z + 0.7)} \qquad (5\text{-}387)$$

For the parallel decomposition, the transfer function is first fractioned by partial fraction into the following form:

$$\frac{C(z)}{R(z)} = \frac{0.64}{z + 4.3} + \frac{0.36}{z + 0.7} \qquad (5\text{-}388)$$

5.19 STATE DIAGRAMS FOR SAMPLED-DATA SYSTEMS

When a discrete-data system has continuous-data as well as discrete-data elements, with the two types of elements separated by sample-and-hold devices, a special treatment of the state diagram is necessary if a description of the continuous-data states is desired for all times.

Let us first establish the state diagram of the zero-order hold. Consider that the input of the zero-order hold is denoted by $e*(t)$ which is a train of impulses, and the output by $h(t)$. Since the zero-order hold simply holds the magnitude of the input impulse at the sampling instant until the next input comes along, the signal $h(t)$ is a sequence of steps. The input–output relation in the Laplace domain is written

$$H(s) = \frac{1 - e^{-Ts}}{s} E*(s) \qquad (5\text{-}389)$$

In the time domain, the relation is simply

$$h(t) = e(kT+) \qquad (5\text{-}390)$$

for $kT \le t < (k + 1)T$.

In the state diagram notation, we need the relation between $H(s)$ and $e(kT+)$. For this purpose we take the Laplace transform on both sides of Eq. (5-390) to give

$$H(s) = \frac{e(kT+)}{s} \qquad (5\text{-}391)$$

Figure 5-18 State diagram representation of the zero-order hold.

for $kT \leq t < (k + 1)T$. The state diagram representation of the zero-order hold is shown in Fig. 5-18. As an illustrative example on how the state diagram of a sampled-data system is constructed, let us consider the system shown in Fig. 5-19. We shall demonstrate the various available ways of modeling the input–output relations of the system.

First, the Laplace transform of the output of the system is written

$$C(s) = \frac{1 - e^{-Ts}}{s} \frac{1}{s + 1} E^*(s) \tag{5-392}$$

Taking the z-transform on both sides of the last equation yields

$$C(z) = \frac{1 - e^{-T}}{z - e^{-T}} E(z) \tag{5-393}$$

Given information on the input $e(t)$ or $e^*(t)$, Eq. (5-393) gives the output response at the sampling instants.

A state diagram can be drawn from Eq. (5-393) using the decomposition technique. Figure 5-20 illustrates the discrete state diagram of the system through decomposition. The discrete dynamic equations of the system are written directly from this state diagram:

$$x_1[(k + 1)T] = e^{-T}x_1(kT) + (1 - e^{-T})e(kT +) \tag{5-394}$$

$$c(kT) = x_1(kT) \tag{5-395}$$

Therefore, the output response of the system can also be obtained by solving the difference equation of Eq. (5-394).

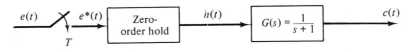

Figure 5-19 Sampled-data system.

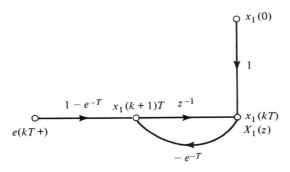

Figure 5-20 Discrete state diagram of the system in Fig. 5-19.

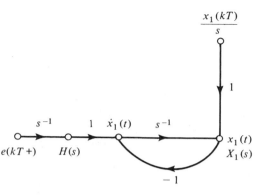

Figure 5-21 State diagram for the system of Fig. 5-19 for the time interval $kT \le t \le (k+1)T$.

If the response of the output $c(t)$ is desired for all t, we may construct the state diagram shown in Fig. 5-21. This state diagram is obtained by cascading the state diagram representations of the zero-order hold and the process $G(s)$. To determine $c(t)$, which is also $x_1(t)$, we must first obtain $X_1(s)$ by applying the gain formula to the state diagram of Fig. 5-21. We have

$$X_1(s) = \frac{s^{-2}}{1 + s^{-1}} e(kT+) + \frac{s^{-1}}{1 + s^{-1}} x_1(kT) \qquad (5\text{-}396)$$

for $kT \le t \le (k+1)T$. Taking the inverse Laplace transform of the last equation gives

$$x_1(t) = \left[1 - e^{-(t-kT)}\right] e(kT+) + e^{-(t-kT)} x_1(kT) \qquad (5\text{-}397)$$

$kT \le t \le (k+1)T$. It is interesting to note that in Eq. (5-397) t is valid for one sampling period, whereas the result in Eq. (5-394) gives information on $x_1(t)$ only at the sampling instants. It is easy to see that if we let $t = (k+1)T$ in Eq. (5-397), the latter becomes Eq. (5-394).

REFERENCES

State Variables and State Equations

1. L. A. ZADEH, "An Introduction to State Space Techniques," Workshop on Techniques for Control Systems, *Proceedings, Joint Automatic Control Conference*, Boulder, Colo., 1962.

2. B. C. KUO, *Linear Networks and Systems*, McGraw-Hill Book Company, New York, 1967.

3. D. W. WIBERG, *Theory and Problems of State Space and Linear Systems* (Schaum's Outline Series), McGraw-Hill Book Company, New York, 1971.

State Transition Matrix

4. R. B. KIRCHNER, "An Explicit Formula for e^{At}," *Amer. Math. Monthly*, Vol. 74, pp. 1200, 1204, 1967.

5. W. EVERLING, "On the Evaluation of e^{At} by Power Series," *Proc. IEEE*, Vol. 55, p. 413, Mar. 1967.

6. T. A. BICKART, "Matrix Exponential: Approximation by Truncated Power Series," *Proc. IEEE*, Vol. 56, pp. 872–873, May 1968.

7. T. M. APOSTOL, "Some Explicit Formulas for the Exponential Matrix e^{At}," *Amer. Math. Monthly*, Vol. 76, pp. 289–292, 1969.

8. M. VIDYASAGAR, "A Novel Method of Evaluating e^{At} in Closed Form," *IEEE Trans. Automatic Control*, Vol. AC-15, pp. 600–601, Oct. 1970.

9. C. G. CULLEN, "Remarks on Computing e^{At}," *IEEE Trans. Automatic Control*, Vol. AC-16, pp. 94–95, Feb. 1971.

10. J. C. JOHNSON and C. L. PHILLIPS, "An Algorithm for the Computation of the Integral of the State Transition Matrix," *IEEE Trans. Automatic Control*, Vol. AC-16, pp. 204–205, Apr. 1971.

11. M. HEALEY, "Study of Methods of Computing Transition Matrices," *Proc. IEE*, Vol. 120, No. 8, pp. 905–912, Aug. 1973.

PROBLEMS

5.1. Write state equations for the electric networks in Fig. 5P-1.

5.2. The following differential equations represent linear time-invariant systems. Write the dynamic equations (state equations and output equations) in vector-matrix form.

(a) $\dfrac{d^2c(t)}{dt^2} + 4\dfrac{dc(t)}{dt} + 3c(t) = r(t)$

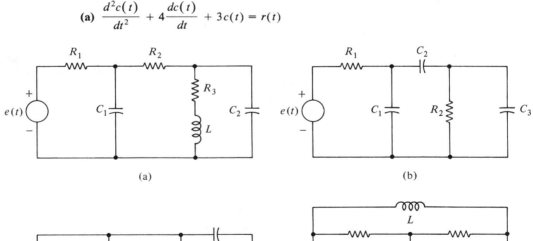

(a)   (b)

(c)   (d)

Figure 5P-1

(b) $3\dfrac{d^3c(t)}{dt^3} + 5\dfrac{d^2c(t)}{dt^2} + 2\dfrac{dc(t)}{dt} + c(t) = 2r(t)$

(c) $\dfrac{d^3c(t)}{dt^3} + 3\dfrac{d^2c(t)}{dt^2} + 2c(t) + \displaystyle\int_0^t c(\tau)\,d\tau = r(t)$

(d) $\dfrac{d^3c(t)}{dt^3} + \dfrac{d^2c(t)}{dt^2} + 3\dfrac{dc(t)}{dt} = 2\dfrac{dr(t)}{dt} + r(t)$

5.3. Using Eq. (5-23), show that

$$\phi(t) = I + At + \frac{1}{2!}A^2 t^2 + \frac{1}{3!}A^3 t^3 + \cdots$$

5.4. The state equations of a linear time-invariant system are represented by

$$\dot{x}(t) = Ax(t) + Bu(t)$$

Find the state transition matrix $\phi(t)$ of A for the following cases.

(a) $A = \begin{bmatrix} 0 & 1 \\ -2 & -3 \end{bmatrix}$ $B = \begin{bmatrix} 0 & 1 \\ 0 & 0 \end{bmatrix}$

(b) $A = \begin{bmatrix} 0 & 1 \\ -1 & -3 \end{bmatrix}$ $B = \begin{bmatrix} 0 \\ 1 \end{bmatrix}$

(c) $A = \begin{bmatrix} -4 & 0 \\ 0 & -4 \end{bmatrix}$ $B = \begin{bmatrix} 1 \\ 1 \end{bmatrix}$

(d) $A = \begin{bmatrix} 4 & 0 \\ 0 & -4 \end{bmatrix}$ $B = \begin{bmatrix} 0 \\ 1 \end{bmatrix}$

(e) $A = \begin{bmatrix} 0 & 1 \\ 1 & 0 \end{bmatrix}$ $B = \begin{bmatrix} 0 \\ 1 \end{bmatrix}$

(f) $A = \begin{bmatrix} -2 & 0 & 0 \\ 0 & -1 & 1 \\ 0 & 0 & -1 \end{bmatrix}$ $B = \begin{bmatrix} 0 \\ 1 \\ 0 \end{bmatrix}$

(g) $A = \begin{bmatrix} -2 & 1 & 0 \\ 0 & -2 & 1 \\ 0 & 0 & -2 \end{bmatrix}$ $B = \begin{bmatrix} 0 \\ 0 \\ 1 \end{bmatrix}$

Does the state transition matrix $\phi(t)$ depend on the matrix B? Find the characteristic equations of these systems. Does the characteristic equation depend on B?

5.5. Find the state transition equation for each of the systems described in Problem 5.4 for $t \geq 0$. Assume that $x(0)$ is given and the components of $u(t)$ are all unit-step functions.

5.6. Given the state equation

$$\dot{x}(t) = Ax(t) + Bu(t)$$

$$A = \begin{bmatrix} 0 & 1 & 0 \\ 1 & 2 & 0 \\ -1 & 0 & 1 \end{bmatrix} \quad B = \begin{bmatrix} 0 \\ 1 \\ 1 \end{bmatrix}$$

Find the transformation $y = Qx$ such that the system

$$\dot{y}(t) = A_1 y(t) + B_1 u(t)$$

is in the phase-variable canonical form.

5.7. Explain why the following state equations cannot be transformed into phase-variable canonical forms.

$$\dot{x} = Ax(t) + Bu(t)$$

(a) $\quad A = \begin{bmatrix} -1 & 0 \\ 0 & -2 \end{bmatrix} \quad B = \begin{bmatrix} 0 \\ 1 \end{bmatrix}$

(b) $\quad A = \begin{bmatrix} 1 & 2 \\ 1 & 1 \end{bmatrix} \quad B = \begin{bmatrix} 2 \\ 2 \end{bmatrix}$

(c) $\quad A = \begin{bmatrix} -2 & 1 & 0 \\ 0 & -2 & 0 \\ -1 & -2 & -3 \end{bmatrix} \quad B = \begin{bmatrix} 1 \\ 0 \\ 1 \end{bmatrix}$

5.8. Given a system described by the dynamic equations

$$\dot{x}(t) = Ax(t) + Bu(t)$$
$$c(t) = Dx(t) + Eu(t)$$

(a) $\quad A = \begin{bmatrix} 0 & 1 & 0 \\ 0 & 0 & 1 \\ -3 & -2 & -1 \end{bmatrix} \quad B = \begin{bmatrix} 0 \\ 0 \\ 1 \end{bmatrix} \quad D = \begin{bmatrix} 1 & 0 & 0 \end{bmatrix} \quad E = 0$

(b) $\quad A = \begin{bmatrix} 1 & -1 \\ 0 & 1 \end{bmatrix} \quad B = \begin{bmatrix} 0 \\ 1 \end{bmatrix} \quad D = \begin{bmatrix} 1 & 1 \end{bmatrix} \quad E = 0$

(c) $\quad A = \begin{bmatrix} 0 & 1 & 0 \\ 0 & 0 & 1 \\ 0 & -1 & -2 \end{bmatrix} \quad B = \begin{bmatrix} 0 \\ 0 \\ 1 \end{bmatrix} \quad D = \begin{bmatrix} 1 & 1 & 0 \end{bmatrix} \quad E = 0$

Find (1) the eigenvalues of A, (2) the transfer function relation between $X(s)$ and $U(s)$, (3) the transfer function $C(s)/U(s)$.

5.9. Given the state equation and output equation

$$\dot{x}(t) = Ax(t) + Bu(t)$$
$$c(t) = Dx(t)$$

where

$$A = \begin{bmatrix} 0 & 1 & 0 \\ 0 & 0 & 1 \\ -3 & -2 & -1 \end{bmatrix} \quad B = \begin{bmatrix} 0 \\ 0 \\ 1 \end{bmatrix} \quad D = \begin{bmatrix} 1 & 1 & 0 \end{bmatrix}$$

Find the matrices A_1 and B_1 so that the state equations are written as

$$\dot{y}(t) = A_1 y(t) + B_1 u(t)$$

where

$$y(t) = \begin{bmatrix} x_1(t) \\ c(t) \\ \dot{c}(t) \end{bmatrix}$$

5.10. The differential equations of a motor control system are given by

$$T = J\frac{d^2\theta_c}{dt^2} + B\frac{d\theta_c}{dt} + K\theta_c$$

$$v = Ri + L\frac{di}{dt} + K_b\frac{d\theta_c}{dt}$$

The algebraic equations relating some of the variables of the system are:

$$T = K_i i \qquad\qquad v = Ke$$

$$e = K_s(\theta_r - \theta_c) \qquad T = \text{motor torque}$$

where i = motor current, v = voltage applied to motor, θ_c = motor output displacement, e = error signal, θ_r = reference input, L = motor inductance, R = motor resistance, K_b = motor back-emf constant, J = motor-load inertia, K_a = amplifier gain, K_s = gain of error detector, B = viscous friction coefficient, and K = spring constant.
(a) Assign state variables and write the state equations for the system in the following form:

$$\dot{\mathbf{x}}(t) = \mathbf{A}\mathbf{x}(t) + \mathbf{B}u(t)$$

where

$$\mathbf{x}(t) = \begin{bmatrix} x_1(t) \\ x_2(t) \\ x_3(t) \end{bmatrix} \qquad u(t) = \theta_r(t)$$

Write the output equation in the form $c(t) = \mathbf{D}\mathbf{x}(t) + Eu(t)$, where $c(t) = \theta_c(t)$.
(b) Write the transfer functions

$$G(s) = \frac{\theta_c(s)}{E(s)} \qquad \text{and} \qquad M(s) = \frac{\theta_c(s)}{\theta_r(s)}$$

5.11. The state equation of a linear time-invariant system is represented by

$$\dot{\mathbf{x}}(t) = \mathbf{A}\mathbf{x}(t) + \mathbf{B}u(t)$$

Given that

$$\mathbf{A} = \begin{bmatrix} 0 & 1 \\ -1 & 0 \end{bmatrix}$$

find the state transition matrix $\phi(t) = e^{\mathbf{A}t}$ using the following methods:
(a) Infinite series expansion of $e^{\mathbf{A}t}$ and express it in a closed form.
(b) The inverse Laplace transform of $(s\mathbf{I} - \mathbf{A})^{-1}$.

5.12. The schematic diagram of a feedback control system using a dc motor is shown in Fig. 5P-12. The torque developed by the motor is $T = K_i i_a$. The constants of the system

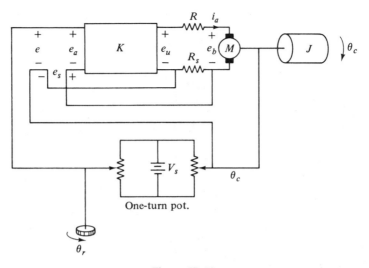

Figure 5P-12

are:

$$K_s = \frac{V_s}{2\pi} \qquad V_s = 2\pi \quad \text{volts}$$

$$K = 9 \qquad R = 0.1 \quad R_s = 0.15 \text{ ohms}$$

$$K_b = 1 \qquad K_i = 1 \quad J = 0.01$$

Assume that all the units are consistent.

(a) Write state equations of the system in vector-matrix form. Let θ_r be the input; $\theta_c = x_1$ and $\dot{\theta}_c = x_2$. Show that the matrices **A** and **B** are in phase-variable canonical form.

(b) Let $\theta_r(t)$ be a unit-step-function input. Find $\mathbf{x}(t)$ in terms of $\mathbf{x}(0)$, the initial state. Use the Laplace transform table.

(c) Find the characteristic equation of **A** and the eigenvalues of **A**.

(d) Comment on the purpose of the "feedback" resistor R_s.

5.13. For the mechanical system shown in Fig. 5P-13, write the differential equation of the system. Express the equation in state equation form. Find the characteristic equation of

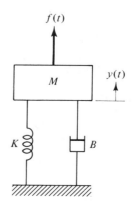

Figure 5P-13

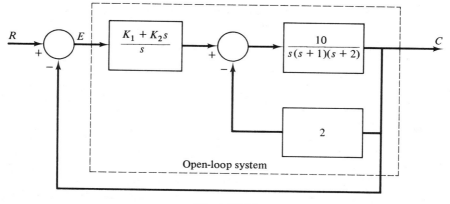

Figure 5P-14

A. Assume that the system is already in equilibrium so that the gravitational force is not included.

5.14. The block diagram of a feedback control system is shown in Fig. 5P-14.
   (a) Find the open-loop transfer function $C(s)/E(s)$ and the closed-loop transfer function $C(s)/R(s)$.
   (b) Write the dynamic equations in the form of

$$\dot{\mathbf{x}}(t) = \mathbf{Ax}(t) + \mathbf{B}r(t)$$
$$c(t) = \mathbf{Dx}(t) + \mathbf{E}r(t)$$

   Find **A**, **B**, **D**, and **E** in terms of the system parameters.
   (c) If the input $r(t)$ is a unit step function, and assuming that the closed-loop system is stable, what is the steady-state value of the output $c(t)$?

5.15. For the linear time-invariant system whose state equations have the coefficient matrices given by Eqs. (5-88) and (5-89) (phase-variable canonical form), show that

$$\mathrm{adj}(s\mathbf{I} - \mathbf{A})\mathbf{B} = \begin{bmatrix} 1 \\ s \\ s^2 \\ \vdots \\ s^{n-1} \end{bmatrix}$$

and the characteristic equation of **A** is

$$s^n + a_n s^{n-1} + a_{n-1} s^{n-2} + \cdots + a_2 s + a_1 = 0$$

5.16. A closed-loop control system is described by

$$\dot{\mathbf{x}}(t) = \mathbf{Ax}(t) + \mathbf{Bu}(t)$$
$$\mathbf{u}(t) = -\mathbf{Gx}(t)$$

where $\mathbf{x}(t) = n \times 1$ state vector, $\mathbf{u}(t) = r \times 1$ input vector, **A** is $n \times n$, **B** is $n \times r$, **G** is the $r \times n$ feedback matrix.

(a) Show that the roots of the characteristic equation of the closed-loop system are the eigenvalues of $A - BG$.

(b) Let

$$A = \begin{bmatrix} 0 & 1 & 0 & 0 \\ 0 & 0 & 1 & 0 \\ 0 & 0 & 0 & 1 \\ 0 & -2 & -5 & -10 \end{bmatrix} \qquad B = \begin{bmatrix} 0 \\ 0 \\ 0 \\ 1 \end{bmatrix} \qquad G = \begin{bmatrix} g_1 & g_2 & g_3 & g_4 \end{bmatrix}$$

where the elements of G are constants. Find the characteristic equation of the closed-loop system. Determine the elements of G so that the eigenvalues of $A - BG$ are at -1, -2, $-1 + j1$, and $-1 - j1$. Can all the eigenvalues of $A - BG$ be arbitrarily assigned for this system?

5.17. A linear time-invariant system is described by the following differential equation:

$$\frac{d^2c(t)}{dt^2} + 2\frac{dc(t)}{dt} + c(t) = r(t)$$

(a) Find the state transition matrix $\phi(t)$.

(b) Let $c(0) = 1$, $\dot{c}(0) = 0$, and $r(t) = u_s(t)$, the unit step function; find the state transition equations for the system.

(c) Determine the characteristic equation of the system and the eigenvalues.

5.18. A linear multivariable system is described by the following set of differential equations:

$$\frac{d^2c_1(t)}{dt^2} + \frac{dc_1(t)}{dt} + 2c_1(t) - 2c_2(t) = r_1(t)$$

$$\frac{d^2c_2(t)}{dt^2} - c_1(t) + c_2(t) = r_2(t)$$

(a) Write the state equations of the system in vector-matrix form. Write the output equation in vector-matrix form.

(b) Find the transfer function relation between the outputs and the inputs of the system.

5.19. Given the state transition equation $\dot{x}(t) = Ax(t)$, where

$$A = \begin{bmatrix} \sigma & -\omega \\ \omega & \sigma \end{bmatrix} \qquad \sigma \text{ and } \omega \text{ are real numbers}$$

(a) Find the state transition matrix of A, $\phi(t)$.

(b) Find the eigenvalues of A.

(c) Find the eigenvectors of the eigenvalues.

5.20. **(a)** Show that the transfer functions $C(s)/U(s)$ of the two systems shown in Figs. 5P-20(a) and 5P-20(b) are the same.

(b) Write the dynamic equations of the system in Fig. 5P-20(a) as

$$\dot{x}(t) = A_1 x(t) + B_1 u_1(t)$$

$$c_1(t) = D_1 x(t)$$

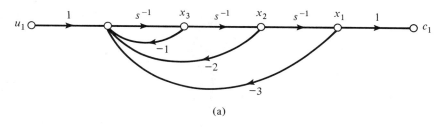

(a)

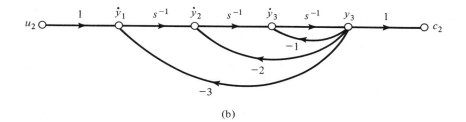

(b)

Figure 5P-20

and those of the system in Fig. 5P-20(b) as

$$\dot{\mathbf{y}}(t) = \mathbf{A}_2 \mathbf{y}(t) + \mathbf{B}_2 u_2(t)$$
$$c_2(t) = \mathbf{D}_2 y(t)$$

Find $\mathbf{A}_1$, $\mathbf{B}_1$, $\mathbf{D}_1$, $\mathbf{A}_2$, $\mathbf{B}_2$, and $\mathbf{D}_2$. Show that $\mathbf{A}_2 = \mathbf{A}'_1$.

5.21. Given the state equations of a linear system as

$$\dot{\mathbf{x}}(t) = \mathbf{A}\mathbf{x}(t) + \mathbf{B}u(t)$$

where

$$\mathbf{A} = \begin{bmatrix} 0 & 1 & 0 \\ 0 & 0 & 1 \\ 0 & -2 & -3 \end{bmatrix} \quad \mathbf{B} = \begin{bmatrix} 0 \\ 0 \\ 1 \end{bmatrix}$$

Find the transformation $\mathbf{x}(t) = \mathbf{P}\mathbf{y}(t)$ that will transform $\mathbf{A}$ into a diagonal matrix $\Lambda = \text{diag}[\lambda_1, \lambda_2, \lambda_3]$, where λ_1, λ_2 and λ_3 are the eigenvalues of $\mathbf{A}$.

5.22. Given a linear system with the state equations described by

$$\dot{\mathbf{x}}(t) = \mathbf{A}\mathbf{x}(t) + \mathbf{B}u(t)$$

where

$$\mathbf{A} = \begin{bmatrix} 0 & 1 & 0 \\ 0 & 0 & 1 \\ -25 & -35 & -11 \end{bmatrix} \quad \mathbf{B} = \begin{bmatrix} 0 \\ 0 \\ 1 \end{bmatrix}$$

The eigenvalues are $\lambda_1 = -1$, $\lambda_2 = \lambda_3 = -5$. Find the transformation $\mathbf{x}(t) = \mathbf{P}\mathbf{y}(t)$ so

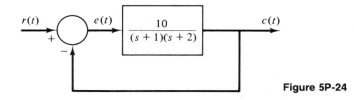

Figure 5P-24

that $\mathbf{A}$ is transformed to the Jordan canonical form. The transformed state equations are

$$\dot{\mathbf{y}}(t) = \mathbf{\Lambda}\mathbf{y}(t) + \mathbf{\Gamma}u(t)$$

Find $\mathbf{\Lambda}$ and $\mathbf{\Gamma}$.

5.23. Draw state diagrams for the following systems:
(a) $\dot{\mathbf{x}}(t) = \mathbf{A}\mathbf{x}(t) + \mathbf{B}u(t)$

$$\mathbf{A} = \begin{bmatrix} -3 & 1 & 0 \\ -1 & 1 & 0 \\ -4 & -3 & -2 \end{bmatrix} \qquad \mathbf{B} = \begin{bmatrix} 0 \\ 0 \\ 1 \end{bmatrix}$$

(b) $\dot{\mathbf{x}}(t) = \mathbf{A}\mathbf{x}(t) + \mathbf{B}u(t)$. Same $\mathbf{A}$ as in part (a) but with

$$\mathbf{B} = \begin{bmatrix} 0 & 1 \\ 1 & 0 \\ 1 & 0 \end{bmatrix}$$

5.24. The block diagram of a feedback control system is shown in Fig. 5P-24.
(a) Write the dynamic equations of the system in vector-matrix form.
(b) Draw a state diagram for the system.
(c) Find the state transition equations for the system. Express the equations in vector-matrix form. The initial states are represented by $\mathbf{x}(t_0)$ and the input $r(t)$ is a unit step function, $u_s(t - t_0)$, which is applied at $t = t_0$.

5.25. The block diagram of a linearized idle-speed engine control system is shown in Fig. 5P-25. The system is linearized about a nominal operating point, so that all the

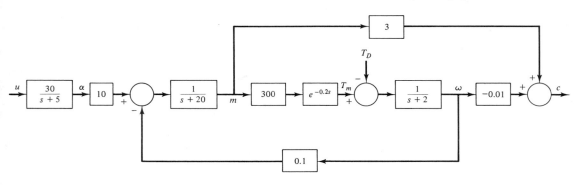

Figure 5P-25

variables represent linearized perturbed quantities.

$$m = \text{average air mass in manifold}$$
$$T_m = \text{engine torque}$$
$$T_D = \text{load disturbance torque}$$
$$c = \text{manifold pressure}$$
$$\omega = \text{engine speed}$$
$$u = \text{input voltage to throttle actuator}$$
$$\alpha = \text{throttle angle}$$

The time delay in the engine model may be approximated by

$$e^{-0.2s} \cong \frac{1 - 0.1s}{1 + 0.1s}$$

(a) Draw a state diagram for the system by decomposing each block individually. Write the state equations in the form of

$$\dot{\mathbf{x}}(t) = \mathbf{A}\mathbf{x}(t) + \mathbf{B}\begin{bmatrix} u(t) \\ T_D(t) \end{bmatrix}$$

(b) For feedback purposes, it is desirable to define the state vector as

$$\mathbf{y}(t) = \begin{bmatrix} x_1(t) \\ \dot{x}_1(t) \\ c(t) \\ \dot{c}(t) \end{bmatrix}$$

where $x_1(t) = \omega(t)$ from part (a). Show that the new state equations can be approximated (neglecting the small coefficients) by

$$\dot{\mathbf{y}} = \begin{bmatrix} \dot{x}_1 \\ \ddot{x}_1 \\ \dot{c} \\ \ddot{c} \end{bmatrix} = \begin{bmatrix} 0 & 1 & 0 & 0 \\ -14.4 & -12 & 998 & -100 \\ 0 & 0 & 0 & 1 \\ -3.4 & -0.18 & -110 & -23.8 \end{bmatrix} \begin{bmatrix} x_1 \\ \dot{x}_1 \\ c \\ \dot{c} \end{bmatrix} + \begin{bmatrix} 0 & 0 \\ 0 & -10 \\ 0 & 0 \\ 900 & 0 \end{bmatrix} \begin{bmatrix} u \\ T_D \end{bmatrix}$$

Assume that the inputs u and T_D are constants.

5.26. Draw state diagrams for the following transfer functions by means of direct decomposition.

(a) $G(s) = \dfrac{10}{s^3 + 10s^2 + 5s + 10}$

(b) $G(s) = \dfrac{5(s + 2)}{s^2(s + 3)(s + 5)}$

Write the state equations from the state diagrams and show that the equations are in the phase-variable canonical form.

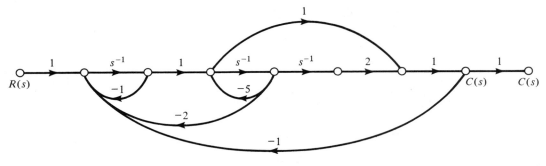

Figure 5P-29

5.27. Draw state diagrams for the following systems by means of parallel decomposition.

(a) $G(s) = \dfrac{5(s+2)}{s^2(s+3)(s+5)}$

(b) $G(s) = \dfrac{10(s+1)}{s(s+2)(s+3)}$

(c) $\dfrac{d^2c(t)}{dt^2} + 5\dfrac{dc(t)}{dt} + 4c(t) = 2\dfrac{dr(t)}{dt} + r(t)$

Make certain that the state diagrams contain a minimum number of integrators. Write the state equations from the state diagrams and show that the states are decoupled from each other.

5.28. Draw state diagrams for the systems in Problem 5.27 by means of cascade decomposition.

5.29. The state diagram of a linear system is shown in Fig. 5P-29.
  (a) Assign the state variables and write the dynamic equations of the system.
  (b) Determine the closed-loop transfer function $C(s)/R(s)$.

5.30. The state diagram of a linear system is shown in Fig. 5P-30.
  (a) Assign state variables on the state diagram; create additional artificial nodes if necessary as long as the system is not altered.
  (b) Write the dynamic equations for the system.

5.31. Given the state equation, $\dot{x}(t) = Ax(t)$, where

$$A = \begin{bmatrix} -2 & 1 & 0 \\ 0 & -2 & 1 \\ 0 & 0 & -2 \end{bmatrix}$$

  (a) Find the eigenvalues of A.
  (b) Determine the state transition matrix $\phi(t)$.

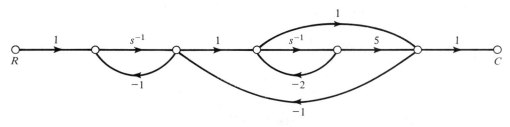

Figure 5P-30

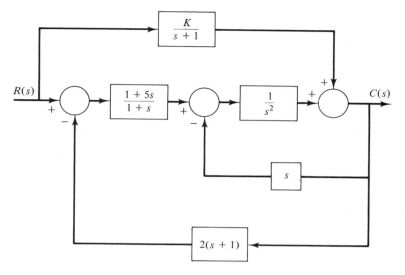

Figure 5P-33

5.32. Given the linear system

$$\dot{\mathbf{x}}(t) = \mathbf{A}\mathbf{x}(t) + \mathbf{B}\mathbf{u}(t)$$

where $\mathbf{u}(t)$ is generated by state feedback, $\mathbf{u}(t) = -\mathbf{G}\mathbf{x}(t)$. The state transition matrix of the closed-loop system is

$$\phi(t) = e^{(\mathbf{A} - \mathbf{B}\mathbf{G})t} = \mathcal{L}^{-1}\left[\left(s\mathbf{I} - \mathbf{A} + \check{\mathbf{B}}\mathbf{G}\right)^{-1}\right]$$

Is the following relation valid?

$$e^{(\mathbf{A} - \mathbf{B}\mathbf{G})t} = e^{\mathbf{A}t}e^{-\mathbf{B}\mathbf{G}t}$$

where $\quad e^{\mathbf{A}t} = \mathcal{L}^{-1}[(s\mathbf{I} - \mathbf{A})^{-1}]$

$\quad e^{-\mathbf{B}\mathbf{G}t} = \mathcal{L}^{-1}[(s\mathbf{I} + \mathbf{B}\mathbf{G})^{-1}]$

Explain your conclusions.

5.33. The block diagram of a control system is shown in Fig. 5P-33.
  (a) Determine the transfer function $C(s)/R(s)$.
  (b) Determine what value (or values) of K must be avoided if the system is to be both completely state controllable and observable.
  (c) For $K = 1$, find the characteristic equation and the eigenvalues of the system.
  (d) For $K = 1$, write the state equations of the system in phase-variable canonical form.
  (e) Find a nonsingular matrix $\mathbf{P}$ that will diagonalize the $\mathbf{A}$ matrix found in part (d) to given $\mathbf{\Lambda} = \mathbf{P}^{-1}\mathbf{A}\mathbf{P} = \text{diag}(\lambda_i)$, where λ_i are the eigenvalues of $\mathbf{A}$.
  (f) Write the state transition matrix $\phi(t) = e^{\mathbf{\Lambda}t}$, where $\mathbf{\Lambda}$ is found in part (e).
  (g) Draw a state diagram for the system with $K = 1$ and using a minimum number of integrators.
  (h) Draw a state diagram of the system with a minimum number of integrators and K equal to the value found in part (b).

5.34. A considerable amount of effort is being spent by automobile manufacturers to meet the exhaust emission performance standards of the various governmental agencies. Modern automotive power-plant systems consist of an internal combustion engine

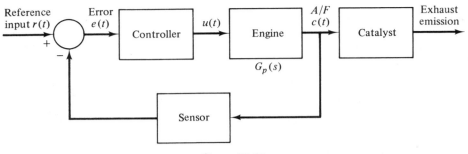

Figure 5P-34

which has an internal cleanup device called a catalytic converter. Such a system requires control of such variables as engine air–fuel ratio (A/F), ignition spark timing, exhaust gas recirculation, and injection air. The control system problem considered in this exercise deals with the control of the air–fuel ratio A/F. In general, depending on fuel composition and other factors, a typical stoichiometric A/F is $14.7 : 1$, that is, 14.7 g of air to each gram of fuel. An A/F greater or less than stoichiometry will cause high hydrocarbons, carbon monoxide, and oxides of nitrogen in the tailpipe emission.

The control system whose block diagram is shown in Fig. 5P-34 is devised to control the air–fuel ratio so that a desired output variable is maintained for a given command signal. Refer to the block diagram in Fig. 5P-34, which shows that the sensor senses the composition of the exhaust-gas mixture entering the catalytic converter. The electronic controller detects the difference or the error between the command and the sensor signals and computes the control signal necessary to achieve the desired exhaust-gas composition. The output variable $c(t)$ denotes the effective air–fuel ratio A/F. The transfer function of the engine is given by

$$\frac{C(s)}{U(s)} = \frac{e^{-T_d s}}{1 + \tau s} = G_p(s)$$

where T_d is the time delay (0.25 sec). The time constant τ is 0.25 sec. The time delay term, $e^{-T_d s}$, can be approximated by a truncated series,

$$e^{-T_d s} = \frac{1}{e^{T_d s}} = \frac{1}{1 + T_d s + T_d^2 s^2/2! + \cdots} \cong \frac{1}{1 + T_d s + T_d^2 s^2/2!}$$

The transfer function of the sensor can be regarded as 1.

(a) Using the approximation for $e^{-T_d s}$ given above, find the expression for

$$G_p(s) = \frac{C(s)}{U(s)}$$

Decompose this transfer function and express the system between C and U by the following state equations:

$$\dot{\mathbf{x}} = \mathbf{A}\mathbf{x} + \mathbf{B}u$$

where $\mathbf{A}$ and $\mathbf{B}$ are in phase-variable canonical form. Find $\mathbf{A}$ and $\mathbf{B}$.

(b) Assuming that there is no controller in the system (i.e., $u = e$), find the characteristic equation of the closed-loop system and the eigenvalues.

(c) Define the state feedback control as

$$u = -\mathbf{G}\mathbf{x} + r$$

where r is the reference input and $\mathbf{G}$ is the feedback matrix, of the form

$$\mathbf{G} = [\, g_1 \quad g_2 \quad g_3 \,]$$

Find the elements of $\mathbf{G}$ so that the eigenvalues of the closed-loop system are at $-8, -1 + j1, -1 - j1$. With this controller, find the steady-state value of $c(t)$ when the input $r(t)$ is a unit step function.

(d) If it is required that the steady-state value of $c(t)$ be identical to the input unit step (zero steady-state error), find the elements of the feedback matrix $\mathbf{G}$ so that the two complex eigenvalues are at $-1 + j1$ and $-1 - j1$. Find the eigenvalues of the closed-loop system.

5.35. The schematic diagram in Fig. 5P-35(a) shows a permanent-magnet dc motor coupled to a viscous-inertia damper. A mechanical damper such as the viscous-inertia type shown here is sometimes used in practice as a simple and economical way of stabilizing a control system.

The differential equations and the other algebraic equations of the system are given below. You should check and make sure you know how to arrive at these equations yourself.

$$v_a(t) = Ke(t)$$
$$v_a(t) = R_a i_a(t) + v_b(t)$$
$$v_b(t) = K_b \omega(t)$$
$$T_m(t) = J\dot{\omega}(t) + K_D[\omega(t) - \omega_D(t)]$$
$$T_m(t) = K_i i_a(t)$$
$$K_D[\omega(t) - \omega_D(t)] = J_r \dot{\omega}_D(t)$$

In these equations,

$\quad e(t) =$ input signal to the open-loop dc motor control system

$\quad K =$ gain of controller and amplifier $=$ constant

$\quad v_a(t) =$ input voltage to dc motor armature

$\quad i_a(t) =$ armature current of dc motor

$\quad R_a =$ armature resistance (the inductance is very small and is neglected)

$\quad \omega(t) =$ angular velocity of dc motor shaft, which is directly coupled to the damper housing

$\quad T_m(t) =$ torque developed by dc motor

$\quad K_i =$ torque constant of dc motor

$\quad K_b =$ back-emf constant of dc motor

$\quad v_b(t) =$ back emf of dc motor

$\quad \omega_D(t) =$ angular velocity of damper rotor

$\quad K_D =$ viscous damping coefficient of damper

$\quad J = J_H + J_M =$ housing inertia of damper $+$ motor rotor inertia (since they are directly coupled)

$\quad J_R =$ rotor inertia of damper

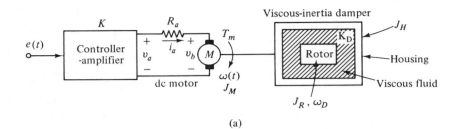

(a)

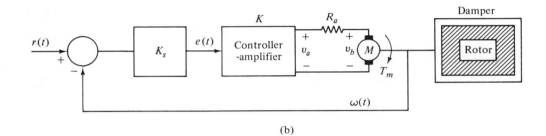

(b)

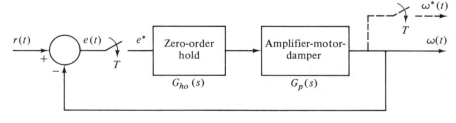

(c)

(d)

Figure 5P-35

(a) Let the state variables be defined as

$$x_1(t) = \omega(t) \quad \text{and} \quad x_2(t) = \omega_D(t)$$

You should understand why these variables are selected as state variables. Write the state equations in the form

$$\dot{\mathbf{x}}(t) = \mathbf{A}\mathbf{x}(t) + \mathbf{B}u(t)$$

where $\mathbf{x}(t)$ denotes the 2×1 state vector

$$\mathbf{x}(t) = \begin{bmatrix} x_1(t) \\ x_2(t) \end{bmatrix}$$

$u(t) = e(t) =$ input variable. Find the $\mathbf{A}$ and $\mathbf{B}$ matrices. Let $\omega(t)$ be the output variable, $c(t)$. Express the output equation in the form of

$$c(t) = \mathbf{D}\mathbf{x}(t) + \mathbf{E}u(t)$$

Find the matrices $\mathbf{D}$ and $\mathbf{E}$.

(b) Draw state diagram for the open-loop system described above. The state diagram should be constructed from the state equations and the output equations, and only the input, the output, the state variables, and the derivatives of the state variables should be found on the state diagram. Omit the initial states for now.

(c) Derive the transfer function $\Omega(s)/E(s)$ for the system.

(d) Find the characteristic equation of the system.

(e) Let $K_b = K_i = 1$, $R_a = 1$, $K = 10$, $K_D = 1$, $J = J_R = 0.05$. Assume that all the units of the system parameters are consistent. Find the roots of the characteristic equation.

(f) The system shown in Fig. 5P-35(a) is an open-loop system and thus is not very useful for control purposes. Figure 5P-35(b) shows the closed-loop system using the components of Fig. 5P-35(a). In this case

$$e(t) = K_s[r(t) - \omega(t)]$$

where K_s is the gain of the error detector and $r(t)$ is the input command. Now find the transfer function of the closed-loop system $\Omega(s)/R(s)$. Find the characteristic equation of the closed-loop system and its roots when $K_s = 1$ and all the other constants are as given above. Find the unit step response of the closed-loop system.

(g) The open-loop dc motor system is now incorporated in a digital feedback control system. The block diagram of the system is shown in Fig. 5P-35(c). In this case, the microprocessor takes the information from the encoder and computes for the velocity information. This generates the sequence of numbers $\omega(T)$, $\omega(2T), \ldots, \omega(kT), \ldots$. The microprocessor then generates the error signal $e(kT) = r(kT) - \omega(kT)$. The digital control system is modeled by the block diagram shown in Fig. 5P-35(d). Find the open-loop transfer function $\Omega(z)/E(z)$ with T (sampling period) $= 0.1$ sec. Use the parameter values given earlier.

(h) Find the closed-loop transfer function $\Omega(z)/R(z)$ for the digital control system. Find the characteristic equation and its roots. Locate these roots in the z-plane.

(i) Find the response of $\omega(kT)$ when the input is a unit step function. Repeat part (h) for $T = 0.01$ sec and $T = 0.001$ sec. Again find the unit step response of $\omega(kT)$.

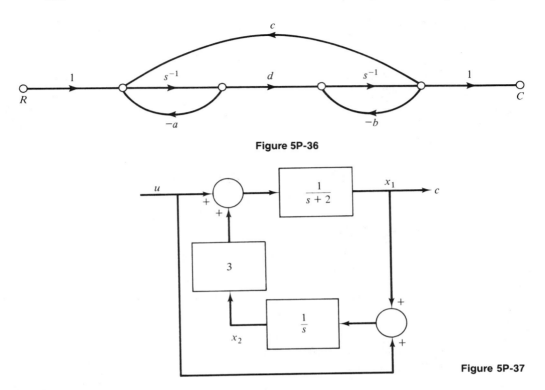

Figure 5P-36

Figure 5P-37

5.36. Determine the state controllability of the system shown in Fig. 5P-36.
  (a) $a = 1$, $b = 2$, $c = 2$, and $d = 1$.
  (b) Are there any nonzero values for a, b, c, and d such that the system is not completely controllable?

5.37. Figure 5P-37 shows the block diagram of a feedback control system. Determine the state controllability and observability of the system by the following methods.
  (a) Conditions on the **A**, **B**, **D**, and **E** matrices.
  (b) Condition on the pole–zero cancellation of the transfer function.
  (c) Transform the state equations with x_1 and x_2 as state variables into canonical form and check the coupling of states.

5.38. The transfer function of a linear control system is given by

$$\frac{C(s)}{R(s)} = \frac{s + a}{s^3 + 7s^2 + 14s + 8}$$

  (a) Determine the value of **a** so that the system is either uncontrollable or unobservable.
  (b) Define the state variables so that one of them is uncontrollable.
  (c) Define the state variables so that one of them is unobservable.

5.39. Consider the system described by the state equation

$$\dot{x}(t) = \mathbf{A}x(t) + \mathbf{B}u(t)$$

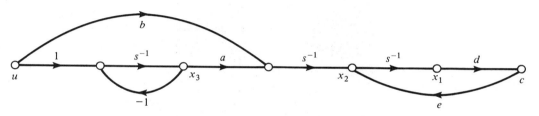

Figure 5P-41

where

$$\mathbf{A} = \begin{bmatrix} 0 & 1 \\ -1 & a \end{bmatrix} \qquad \mathbf{B} = \begin{bmatrix} 1 \\ b \end{bmatrix}$$

Find the region in the *a*-versus-*b* plane such that the system is completely controllable.

5.40. Determine the condition on b_1, b_2, d_1, and d_2 so that the following system is completely state controllable and observable.

$$\dot{\mathbf{x}}(t) = \mathbf{A}\mathbf{x}(t) + \mathbf{B}u(t) \qquad c(t) = \mathbf{D}\mathbf{x}(t)$$

$$\mathbf{A} = \begin{bmatrix} 1 & 1 \\ 0 & 1 \end{bmatrix} \qquad \mathbf{B} = \begin{bmatrix} b_1 \\ b_2 \end{bmatrix} \qquad \mathbf{D} = [\, d_1 \quad d_2 \,]$$

5.41. The state diagram of a linear controlled process is shown in Fig. 5P-41.
   (a) Determine the relation between the parameters *a* and *b* that should be avoided in order that the system be completely controllable. Find the eigenvalues of the process.
   (b) Let $a = b = 1$ [so you know these cannot be the answers to part (a)]. State feedback in the form $u = -\mathbf{G}\mathbf{x}$, where $\mathbf{G} = [g_1, g_2, g_3]$ and **x** is the state vector, is applied to the process. Find the feedback matrix **G** so that the eigenvalues of the closed-loop system are at -1, -1, and -1.

5.42. The schematic diagram shown in Fig. 5P-42 represents a control system whose purpose is to hold the level of the liquid in the tank at a desired fixed level. The liquid level is

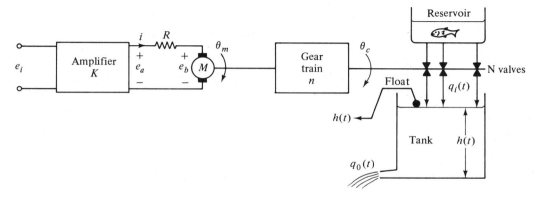

Figure 5P-42

controlled by a float whose position $h(t)$ is monitored. The control signal of the open-loop system is $e_i(t)$. The system parameters and equations are as follows:

DC motor:

Armature resistance	$R = 10$ ohms
Armature inductance	$L =$ negligible
Torque constant	$K_i = 10$ oz-in./A
Back-emf constant	$K_b = 0.075$ V/rad/sec
Rotor inertia	$J_m = 0.005$ oz-in.-sec^2
Load inertia	$J_L = 10$ oz-in.-sec^2
Load and motor friction	negligible
Gear ratio	$n = n_1/n_2 = 1/100$
Amplifier gain	$K = 50$

Motor equations:

$$e_a = Ri + e_b \qquad\qquad e_b = K_b\omega_m = K_b\dot{\theta}_m$$

$$\omega_m = d\theta_m/dt \qquad\qquad T_m = K_i i = J\dot{\omega}_m$$

$$\theta_c = n_1\theta_m/n_2 = n\theta_m \qquad\qquad J = J_m + n^2 J_L$$

Dynamics of tank: area $A = 50$ ft^2.

There are N valves connected to the tank from the reservoir, $N = 10$. All the valves have the same characteristics, and are controlled simultaneously by θ_c. The equations that govern the volume of flow are:

$$q_i(t) = K_I N\theta_c(t) \qquad K_I = 10 \text{ ft}^3/\text{sec-rad}$$

$$q_0(t) = K_o h(t) \qquad K_o = 50 \text{ ft}^3/\text{sec}$$

$$h(t) = \frac{\text{volume of tank}}{\text{area of tank}} = \frac{1}{A}\int[q_i(t) - q_o(t)]\, dt$$

(a) Draw a state diagram for the open-loop system using $x_1 = h$, $x_2 = \theta_m$, and $x_3 = \dot{\theta}_m$ as state variables.
Write the state equations in the form of $\dot{\mathbf{x}} = \mathbf{A}\mathbf{x} + \mathbf{B}e_i$.
(b) Find the characteristic equation and the eigenvalues of the $\mathbf{A}$ matrix.
(c) Show that the open-loop system is completely controllable; that is, the pair, $[\mathbf{A}, \mathbf{B}]$ is controllable.
(d) Let us apply state feedback control, $e_i = -\mathbf{G}\mathbf{x} + r$, where $\mathbf{G}$ is the feedback matrix and r is the reference input. Find $\mathbf{G}$ such that the eigenvalues of $\mathbf{A} - \mathbf{B}\mathbf{G}$ are at $-10 + j10$, $-10 - j10$, and -100.
(e) Let all the initial states be zero. Set $r(t) = Ru_s(t)$ ft, where $u_s(t)$ is the unit step function. Find the value of R so that the final level is 100 ft as time goes to infinity.
(f) For reason of economy, only one of the three state variables is measured and fed back for control purposes. The output equation is $c = \mathbf{D}\mathbf{x}$, where $\mathbf{D}$ can be one of

the following forms:

(1) $\mathbf{D} = [1 \quad 0 \quad 0]$
(2) $\mathbf{D} = [0 \quad 1 \quad 0]$
(3) $\mathbf{D} = [0 \quad 0 \quad 1]$

Determine which case (or cases) corresponds to a completely observable system.

5.43. Figure 5P-43 illustrates a well-known "broom-balancing" problem in control systems. The objective of the control system is to maintain the "broom" in the upright position by means of the force $u(t)$ applied to the car as shown. Any small displacement $\theta(t)$ encountered by the broom should be "damped out" quickly.

    The dynamic equations of the broom–car system are nonlinear. The standard practice is to write these nonlinear differential equations and then linearize them about a nominal solution and input. You are not concerned with the linearization problem here, although you should refer to Section 4.8 so that you can appreciate how the linearized equations are determined. After linearization, the small-signal model of the broom–car system is described by the following set of equations;

$$\dot{x}_1 = x_2$$
$$\dot{x}_2 = 2x_1 + u$$
$$\dot{x}_3 = x_4$$
$$\dot{x}_4 = 2x_1 - u$$

where $x_1 = \theta$, $x_2 = \dot{\theta}$, $x_3 = y$, $x_4 = \dot{y}$; y is the linear displacement of the car.

(a) Write the state equations in vector-matrix form

$$\dot{\mathbf{x}}(t) = \mathbf{A}\mathbf{x}(t) + \mathbf{B}u(t)$$

(b) Find the eigenvalues of $\mathbf{A}$.
(c) Find the eigenvectors associated with eigenvalues of $\mathbf{A}$.
(d) Transform the state equations into phase-variable canonical form.
(e) Find the similarity transformation $\mathbf{P}$ which transforms $\mathbf{A}$ into Jordan canonical form,

$$\mathbf{\Lambda} = \mathbf{P}^{-1}\mathbf{A}\mathbf{P} \qquad \mathbf{\Gamma} = \mathbf{P}^{-1}\mathbf{B}$$

Find $\mathbf{\Lambda}$ and $\mathbf{\Gamma}$.

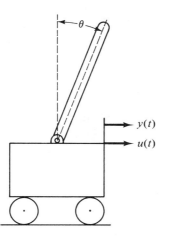

Figure 5P-43

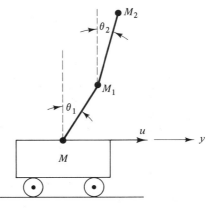

Figure 5P-44

(f) Determine the controllability of the pair $(\mathbf{A}, \mathbf{B})$.

(g) For reason of economy, only one of the four state variables is going to be measured. Therefore, the output equation $c = \mathbf{D}x$ can assume the form with one of the following four matrices for $\mathbf{D}$:

(1) $\mathbf{D} = \begin{bmatrix} 1 & 0 & 0 & 0 \end{bmatrix}$
(2) $\mathbf{D} = \begin{bmatrix} 0 & 1 & 0 & 0 \end{bmatrix}$
(3) $\mathbf{D} = \begin{bmatrix} 0 & 0 & 1 & 0 \end{bmatrix}$
(4) $\mathbf{D} = \begin{bmatrix} 0 & 0 & 0 & 1 \end{bmatrix}$

Determine which case (or cases) corresponds to an observable system.

5.44. The double inverted pendulum shown in Fig. 5P-44 is approximated by the following linear model.

$$\dot{\mathbf{x}}(t) = \mathbf{A}\mathbf{x}(t) + \mathbf{B}u(t)$$

where

$$\mathbf{x}(t) = \begin{bmatrix} \theta_1(t) \\ \dot{\theta}_1(t) \\ \theta_2(t) \\ \dot{\theta}_2(t) \\ y(t) \\ \dot{y}(t) \end{bmatrix}$$

$$\mathbf{A} = \begin{bmatrix} 0 & 1 & 0 & 0 & 0 & 0 \\ 16 & 0 & -8 & 0 & 0 & 0 \\ 0 & 0 & 0 & 1 & 0 & 0 \\ -16 & 0 & 16 & 0 & 0 & 0 \\ 0 & 0 & 0 & 0 & 0 & 1 \\ 0 & 0 & 0 & 0 & 0 & 0 \end{bmatrix} \qquad \mathbf{B} = \begin{bmatrix} 0 \\ -1 \\ 0 \\ 0 \\ 0 \\ 1 \end{bmatrix}$$

Determine the controllability of the states.

5.45. The block diagram of a simplified control system for the Large Space Telescope (LST) is shown in Fig. 5P-45. For simulation and control purposes, it would be desirable to represent the system by state equations and a state diagram.

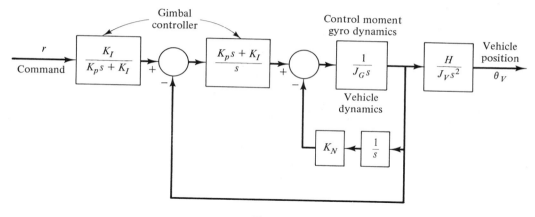

Figure 5P-45

(a) Draw a state diagram for the system and write the state equations in vector-matrix form. Note that you should use a minimum number of state variables.

(b) Find the characteristic equation of the system.

5.46. The state diagram shown in Fig. 5P-46 represents two subsystems connected in cascade.

(a) Determine the controllability and observability of the overall system.

(b) Consider that output feedback is applied by feeding back c_2 to u_2; that is,

$$u_2 = -gc_2$$

where g is a real constant. Determine how the value of g affects the controllability and observability of the system.

5.47. Given the system

$$\dot{\mathbf{x}}(t) = \mathbf{A}\mathbf{x}(t) + \mathbf{B}u(t)$$
$$c(t) = \mathbf{D}\mathbf{x}(t)$$

where

$$\mathbf{A} = \begin{bmatrix} 0 & 1 \\ -1 & -3 \end{bmatrix} \quad \mathbf{B} = \begin{bmatrix} 1 \\ 2 \end{bmatrix} \quad \mathbf{D} = \begin{bmatrix} 1 & 1 \end{bmatrix}$$

(a) Determine the state controllability and observability of the system.

(b) Let $u = -\mathbf{G}\mathbf{x}$, where $\mathbf{G} = [g_1, g_2]$. Determine if and how controllability and observability of the closed-loop system are affected by the elements of $\mathbf{G}$.

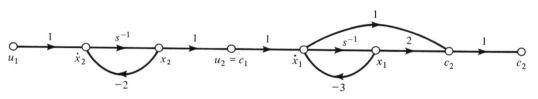

Figure 5P-46

chapter six

Time-Domain Analysis
of Control Systems

6.1 INTRODUCTION

Since time is used as an independent variable in most control systems, it is usually of interest to evaluate the state and output responses with respect to time, or simply, the time response. In the analysis problem, a reference input signal is applied to a system, and the performance of the system is evaluated by studying the system response in the time domain. For instance, if the objective of the control system is to have the output variable track the input signal, starting at some initial condition, it is necessary to compare the input and the output response as functions of time. Therefore, in most control system problems the final evaluation of the performance of the system is based on the time responses.

The time response of a control system is usually divided into two parts: the *transient response* and the *steady-state* response. Let $c(t)$ denote a time response; then, in general, it may be written as

$$c(t) = c_t(t) + c_{ss}(t) \tag{6-1}$$

where $c_t(t)$ = transient response
        $c_{ss}(t)$ = steady-state response

The definition of steady state has not been entirely standardized in the studies of systems and network theory. In network analysis, it is sometimes useful to define steady state as a condition when the response has reached a constant value with respect to the independent variable. In control system studies, however, it is more appropriate to define steady state as the fixed response when time reaches infinity. Therefore, a sine wave is considered as a steady-state response because its behavior is fixed for any time interval, as when time approaches infinity. Similarly, the ramp function $c(t) = t$ is a steady-state response, although it increases with time.

Transient response is defined as the part of the response that goes to zero as time becomes very large. Therefore, $c_t(t)$ has the property

$$\lim_{t \to \infty} c_t(t) = 0 \qquad (6\text{-}2)$$

It can also be stated that the steady-state response is that part of the response which remains after the transient has died out.

All control systems exhibit transient phenomenon to some extent before a steady state is reached. Since inertia, mass, and inductance cannot be avoided entirely in physical systems, the responses of a typical control system cannot follow sudden changes in the input instantaneously, and transients are usually observed. Therefore, the control of the transient response is necessarily important, as it is a significant part of the dynamic behavior of the system; and the deviation between the output response and the input or the desired response, before the steady state is reached, must be closely watched.

The steady-state response of a control system is also very important, since when compared with the input, it gives an indication of the final accuracy of the system. If the steady-state response of the output does not agree with the steady state of the input exactly, the system is said to have a *steady-state error*.

The study of a control system in the time domain essentially involves the evaluation of the transient and the steady-state responses of the system. In the design problem, specifications are usually given in terms of the transient and steady-state performances, and controllers are designed so that the specifications are all met by the designed system.

6.2 TYPICAL TEST SIGNALS FOR THE TIME RESPONSE OF CONTROL SYSTEMS

Unlike many electrical circuits and communication systems, the input excitations to many practical control systems are not known ahead of time. In many cases, the actual inputs of a control system may vary in random fashions with respect to time. For instance, in a radar tracking system, the position and speed of the target to be tracked may vary in an unpredictable manner, so that they cannot be expressed deterministically by a mathematical expression. This poses a problem for the designer, since it is difficult to design the control system so that it will perform satisfactorily to any input signal. For the purposes of analysis and design, it is necessary to assume some basic types of input functions so that the performance of a system can be evaluated with respect to these test signals. By selecting these basic test signals properly, not only the mathematical treatment of the problem is systematized, but the responses due to these inputs allow the prediction of the system's performance to other more complex inputs. In a design problem, performance criteria may be specified with respect to these test signals so that a system may be designed to meet the criteria.

When the response of a linear time-invariant system is analyzed in the frequency domain, a sinusoidal input with variable frequency is used. When the input frequency is swept from zero to beyond the significant range of the system characteristics, curves in terms of the amplitude ratio and phase between input and output are drawn as functions of frequency. It is possible to predict the time-domain behavior of the system from its frequency-domain characteristics.

To facilitate the time-domain analysis, the following deterministic test signals are often used.

Step Input Function. The step input function represents an instantaneous change in the reference input variable. For example, if the input is the angular position of a mechanical shaft, the step input represents the sudden rotation of the shaft. The mathematical representation of a step function is

$$r(t) = \begin{cases} R & t > 0 \\ 0 & t < 0 \end{cases} \tag{6-3}$$

where R is a constant. Or

$$r(t) = Ru_s(t) \tag{6-4}$$

where $u_s(t)$ is the unit step function. The step function is not defined at $t = 0$. The step function as a function of time is shown in Fig. 6-1(a).

Ramp Input Function. In the case of the ramp function, the signal is considered to have a constant change in value with respect to time. Mathematically, a ramp function is represented by

$$r(t) = \begin{cases} Rt & t \geq 0 \\ 0 & t < 0 \end{cases} \tag{6-5}$$

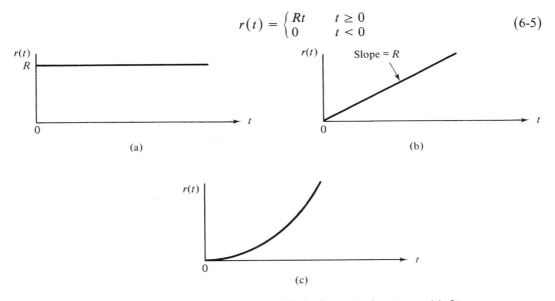

(a)   (b)

(c)

Figure 6-1 Basic time-domain test signals for control systems. (a) Step function input, $r(t) = Ru_s(t)$. (b) Ramp function input, $r(t) = Rtu_s(t)$. (c) Parabolic function input, $r(t) = Rt^2u_s(t)$.

or simply

$$r(t) = Rtu_s(t) \tag{6-6}$$

The ramp function is shown in Fig. 6-1(b). If the input variable is of the form of the angular displacement of a shaft, the ramp input represents the constant-speed rotation of the shaft.

Parabolic Input Function. The mathematical representation of a parabolic input function is

$$r(t) = \begin{cases} Rt^2 & t \geq 0 \\ 0 & t < 0 \end{cases} \tag{6-7}$$

or simply

$$r(t) = Rt^2 u_s(t) \tag{6-8}$$

The graphical representation of the parabolic function is shown in Fig. 6-1(c).

These test signals all have the common feature that they are simple to describe mathematically, and from the step function to the parabolic function they become progressively faster with respect to time. The step function is very useful as a test signal since its initial instantaneous jump in amplitude reveals a great deal about a system's quickness to respond. Also, since the step function has, in principle, a wide band of frequencies in its spectrum, as a result of the jump discontinuity, as a test signal it is equivalent to the application of numerous sinusoidal signals with a wide range of frequencies.

The ramp function has the ability to test how the system would respond to a signal that changes linearly with time. A parabolic function is one degree faster than a ramp function. In practice, we seldom find it necessary to use a test signal faster than a parabolic function. This is because, as we shall show later, to track or follow a high-order input, the system is necessarily of high order, which may mean that stability problems will be encountered.

6.3 TIME-DOMAIN PERFORMANCE OF CONTROL SYSTEMS — THE STEADY-STATE ERROR

It was mentioned in preceding sections that the steady-state error is a measure of system accuracy when a specific type of input is applied to a control system. In a practical system, because of friction, other imperfections, and the nature of the system, the steady state of the output response seldom agrees exactly with the reference input. Therefore, steady-state errors in control systems are almost unavoidable, and in a design problem one of the objectives is to keep the error to a minimum, or below a certain tolerable value.

In practice, the type of error and the relative tolerance of errors found in control systems could vary over a wide range. For instance, in a velocity control system, the steady-state value of the difference between the actual velocity and the desired velocity of the system is an error in velocity. Numerous control systems are devised for the purpose of controlling position. In this case the difference between the actual controlled position and the desired position is a position error.

The accuracy requirement on control systems depends to a great extent on the control objectives of the system. For instance, if the controlled variable is the position of an elevator, then the steady-state error can be tolerable if it is kept under a fraction of an inch. Similarly, in the guidance control of a missile carrying an explosive warhead, it is necessary only to guide the missile to the vicinity of the target, although a direct hit would be ideal. On the other hand, the error requirements on certain control systems can be extremely stringent. For instance, the pointing accuracy on the control of the Large Space Telescope (a space shuttle with a large telescope on board) is measured in microradians.

Before embarking on the analytical study of the steady-state error, it is useful to investigate the contributing factors to steady-state errors in control systems.

Steady-State Error Caused by Nonlinear Elements

Most of the steady-state errors found in control systems are attributed to some nonlinear characteristics such as nonlinear friction or dead zone. For instance, if an amplifier used in a control system has the input-output characteristics shown in Fig. 6-2, then, when the input signal magnitude is less than the dead zone, the output of the amplifier would be zero, and the control would not be able to correct the error. Similarly, Fig. 6-3 shows the input-output characteristics of a relay that may cause steady-state errors related to the size of the dead zone.

Digital components used in control systems, such as a microprocessor, output signals that can take on only discrete or quantized levels. This property is illustrated by the quantization characteristics shown in Fig. 6-4. When the input to the quantizer is within $\pm q/2$, again the output is zero, and the system may generate an error whose magnitude is related to $\pm q/2$. This type of error is also known as the *quantization error*.

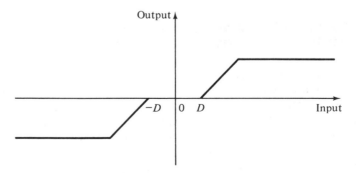

Figure 6-2 Typical input-output characteristics of an amplifier with a dead zone.

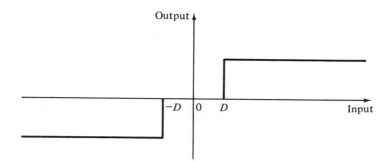

Figure 6-3 Typical input-output characteristics of a relay with a dead zone.

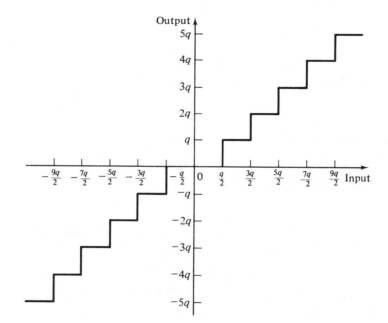

Figure 6-4 Typical input-output characteristics of a quantizer.

When the control of physical objects is involved, friction is almost always unavoidable. We shall show in the following that Coulomb friction is a common cause of steady-state errors in control systems.

Let us consider that the torque generated by a step motor is related to the rotor position of the motor as shown in Fig. 6-5. The point O designates a stable equilibrium point. The torque on either side of the point O represents a restoring torque that tends to return the rotor to the equilibrium point when some angular displacement disturbance takes place. When there is no friction, the step motor with the torque-position relation shown in Fig. 6-5 is supposed to have a zero steady-state error. However, if the rotor of the motor sees a Coulomb friction torque T_f, then the

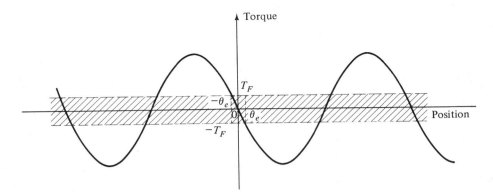

Figure 6-5 Torque-angle curve of a step motor with Coulomb friction.

motor torque must first overcome this frictional torque before producing any motion. Thus, as the motor torque falls below T_f as the rotor position approaches the stable equilibrium point, it may stop at any position inside the shaded band shown in Fig. 6-5 (i.e., $\pm\theta_e$).

Although it is relatively simple to comprehend the effects of nonlinearities or errors and to establish maximum upper bounds on the error magnitudes, it is difficult to establish general or closed-form solutions for nonlinear systems. Usually, exact and detailed analysis of errors in nonlinear control systems can be carried out only by computer simulations.

Steady-State Error of Linear Systems

Next we investigate the properties of steady-state errors in linear control systems. In general, the steady-state errors of linear control systems depend on the input and the type of the system.

In a control system, if the reference input $r(t)$ and the controlled output $c(t)$ are dimensionally the same, for example, a voltage controlling a voltage, a position controlling a position, and so on, and these signals are at the same level, the error signal is simply

$$e(t) = r(t) - c(t) \tag{6-9}$$

However, sometimes it may be impossible or inconvenient to provide a reference input that is at the same level or even of the same dimension as the controlled variable. For instance, it may be necessary to use a low-voltage source for the control of the output of a high-voltage power source; for a velocity-control system it is more practical to use a voltage source or position input to control the velocity of the output shaft. Under these conditions, the error signal cannot be defined simply as the difference between the reference input and the controlled output, and Eq. (6-9) becomes meaningless. The input and the output signals must be of the same dimension and at the same level before subtraction. Therefore, a nonunity element, $H(s)$, is usually incorporated in the feedback path, as shown in

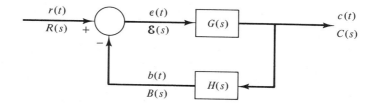

Figure 6-6 Nonunity feedback control system.

Fig. 6-6. The error of this nonunity-feedback control system is defined as

$$\epsilon(t) = r(t) - b(t) \tag{6-10}$$

or

$$\mathcal{E}(s) = R(s) - B(s) = R(s) - H(s)C(s) \tag{6-11}$$

For example, if a 10-V reference is used to regulate a 100-V voltage supply, H is a constant and is equal to 0.1. When the output voltage is exactly 100 V, the error signal is

$$\epsilon(t) = 10 - 0.1 \cdot 100 = 0 \tag{6-12}$$

As another example, let us consider that the system shown in Fig. 6-6 is a velocity-control system in that the input $r(t)$ is used as a reference to control the output velocity of the system. Let $c(t)$ denote the output displacement. Then, we need a device such as a tachometer in the feedback path, so that $H(s) = K_t s$. Thus, the error in velocity is defined as

$$\epsilon(t) = r(t) - b(t)$$

$$= r(t) - K_t \frac{dc(t)}{dt} \tag{6-13}$$

The error becomes zero when the output velocity $dc(t)/dt$ is equal to $r(t)/K_t$.

The steady-state error of a feedback control system is defined as the error when time reaches infinity; that is,

$$\text{steady-state error} = e_{ss} = \lim_{t \to \infty} \epsilon(t) \tag{6-14}$$

With reference to Fig. 6-6 the Laplace-transformed error function is

$$\mathcal{E}(s) = \frac{R(s)}{1 + G(s)H(s)} \tag{6-15}$$

By use of the final-value theorem, the steady-state error of the system is

$$e_{ss} = \lim_{t \to \infty} \epsilon(t) = \lim_{s \to 0} s\mathcal{E}(s) \tag{6-16}$$

where $s\mathscr{E}(s)$ is to have no poles that lie on the imaginary axis and in the right half of the s-plane. Substituting Eq. (6-15) into Eq. (6-16), we have

$$e_{ss} = \lim_{s \to 0} \frac{sR(s)}{1 + G(s)H(s)} \qquad (6\text{-}17)$$

which shows that the steady-state error depends on the reference input $R(s)$ and the loop transfer function $G(s)H(s)$.

Let us first establish the *type* of control system by referring to the form of $G(s)H(s)$. In general, $G(s)H(s)$ may be written

$$G(s)H(s) = \frac{K(1 + T_1 s)(1 + T_2 s) \cdots (1 + T_m s)}{s^j(1 + T_a s)(1 + T_b s) \cdots (1 + T_n s)} \qquad (6\text{-}18)$$

where K and all the Ts are constants. The *type* of feedback control system refers to the *order* of the pole of $G(s)H(s)$ at $s = 0$. Therefore, the system that is described by the $G(s)H(s)$ of Eq. (6-18) is of type j, where $j = 0, 1, 2, \ldots$. The values of m, n, and the Ts are not important to the system type and do not affect the value of the steady-state error. For instance, a feedback control system with

$$G(s)H(s) = \frac{K(1 + 0.5s)}{s(1 + s)(1 + 2s)} \qquad (6\text{-}19)$$

is of type 1, since $j = 1$.

Now let us consider the effects of the types of inputs on the steady-state error. We shall consider only the step, ramp, and parabolic inputs.

Steady-State Error Due to a Step Input.   If the reference input to the control system of Fig. 6-6 is a step input of magnitude R, the Laplace transform of the input is R/s. Equation (6-17) becomes

$$e_{ss} = \lim_{s \to 0} \frac{sR(s)}{1 + G(s)H(s)} = \lim_{s \to 0} \frac{R}{1 + G(s)H(s)} = \frac{R}{1 + \lim\limits_{s \to 0} G(s)H(s)} \qquad (6\text{-}20)$$

For convenience we define

$$K_p = \lim_{s \to 0} G(s)H(s) \qquad (6\text{-}21)$$

where K_p is the *step error constant*. Then Eq. (6-20) is written

$$e_{ss} = \frac{R}{1 + K_p} \qquad (6\text{-}22)$$

We see that for e_{ss} to be zero, when the input is a step function, K_p must be infinite. If $G(s)H(s)$ is described by Eq. (6-18), we see that for K_p to be infinite, j must be at least equal to unity; that is, $G(s)H(s)$ must have at least one pure integration. Therefore, we can summarize the steady-state error due to a step input as follows:

$$\text{type 0 system:} \qquad e_{ss} = \frac{R}{1 + K_p} = \text{constant}$$

$$\text{type 1 (or higher) system:} \quad e_{ss} = 0$$

Steady-State Error Due to a Ramp Input. If the input to the control system of Fig. 6-6 is

$$r(t) = Rtu_s(t) \tag{6-23}$$

where R is a constant, the Laplace transform of $r(t)$ is

$$R(s) = \frac{R}{s^2} \tag{6-24}$$

Substituting Eq. (6-24) into Eq. (6-17), we have

$$e_{ss} = \lim_{s \to 0} \frac{R}{s + sG(s)H(s)} = \frac{R}{\lim_{s \to 0} sG(s)H(s)} \tag{6-25}$$

If we define

$$K_v = \lim_{s \to 0} sG(s)H(s) = \text{ramp error constant} \tag{6-26}$$

Eq. (6-25) reads

$$e_{ss} = \frac{R}{K_v} \tag{6-27}$$

which is the steady-state error when the input is a ramp function. A typical e_{ss} due to a ramp input is shown in Fig. 6-7.

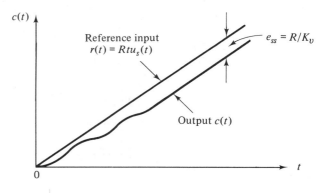

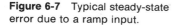

Figure 6-7 Typical steady-state error due to a ramp input.

Equation (6-27) shows that for e_{ss} to be zero when the input is a ramp function, K_v must be infinite. Using Eqs. (6-18) and (6-26), we obtain

$$K_v = \lim_{s \to 0} sG(s)H(s) = \lim_{s \to 0} \frac{K}{s^{j-1}} \qquad j = 0, 1, 2, \ldots \qquad (6\text{-}28)$$

Therefore, in order for K_v to be infinite, j must be at least equal to 2, or the system must be of type 2 or higher. The following conclusions may be stated with regard to the steady-state error of a system with ramp input:

type 0 system: $e_{ss} = \infty$

type 1 system: $e_{ss} = \dfrac{R}{K_v} = \text{constant}$

type 2 (or higher) system: $e_{ss} = 0$

Steady-State Error Due to a Parabolic Input.   If the input is described by

$$r(t) = \frac{Rt^2}{2} u_s(t) \qquad (6\text{-}29)$$

the Laplace transform of $r(t)$ is

$$R(s) = \frac{R}{s^3} \qquad (6\text{-}30)$$

The steady-state error of the system of Fig. 6-6 is

$$e_{ss} = \frac{R}{\lim\limits_{s \to 0} s^2 G(s) H(s)} \qquad (6\text{-}31)$$

Defining the *parabolic error constant* as

$$K_a = \lim_{s \to 0} s^2 G(s) H(s) \qquad (6\text{-}32)$$

the steady-state error is

$$e_{ss} = \frac{R}{K_a} \qquad (6\text{-}33)$$

The following conclusions can be made with regard to the steady-state error of a system with parabolic input:

type 0 system: $e_{ss} = \infty$

type 1 system: $e_{ss} = \infty$

type 2 system: $e_{ss} = \dfrac{R}{K_a} = \text{constant}$

type 3 (or higher) system: $e_{ss} = 0$

Table 6-1   Summary of the Steady-State Errors Due to Step, Ramp, and Parabolic Inputs

Type of System j	K_p	K_v	K_a	Step Input, $e_{ss} = \dfrac{R}{1 + K_p}$	Ramp Input, $e_{ss} = \dfrac{R}{K_v}$	Parabolic Input, $e_{ss} = \dfrac{R}{K_a}$
0	K	0	0	$e_{ss} = \dfrac{R}{1 + K}$	$e_{ss} = \infty$	$e_{ss} = \infty$
1	∞	K	0	$e_{ss} = 0$	$e_{ss} = \dfrac{R}{K}$	$e_{ss} = \infty$
2	∞	∞	K	$e_{ss} = 0$	$e_{ss} = 0$	$e_{ss} = \dfrac{R}{K}$
3	∞	∞	∞	$e_{ss} = 0$	$e_{ss} = 0$	$e_{ss} = 0$

As a summary of the error analysis, the relations among the error constants, the types of the system, and the input types are tabulated in Table 6-1. The transfer function of Eq. (6-18) is used as a reference.

It should be noted that the step, ramp, and parabolic error constants are significant in the error analysis only when the input signal is a step function, a ramp function, and a parabolic function, respectively.

It should be noted further that the steady-state error analysis in this section is conducted by applying the final-value theorem to the error function, which is defined as the difference between the actual output and the desired output signal. In certain cases the error signal may be defined as the difference between the output and the reference input, whether or not the feedback element is unity. For instance, one may define the error signal for the system of Fig. 6-6 as

$$\epsilon(t) = r(t) - c(t) \tag{6-34}$$

Then

$$\mathcal{E}(s) = \frac{1 + G(s)[H(s) - 1]}{1 + G(s)H(s)} R(s) \tag{6-35}$$

and

$$e_{ss} = \lim_{s \to 0} s \frac{1 + G(s)[H(s) - 1]}{1 + G(s)H(s)} R(s) \tag{6-36}$$

It should be kept in mind that since the steady-state error analysis discussed here relies on the use of the final-value theorem, it is important to first check to see if $sE(s)$ has any poles on the $j\omega$ axis or in the right half of the s-plane.

One of the disadvantages of the error constants is, of course, that they do not give information on the steady-state error when inputs are other than the three basic types mentioned. Another difficulty is that when the steady-state error is a function of time, the error constants give only an answer of infinity, and do not provide any information on how the error varies with time. We shall present the error series in the following section, which gives a more general representation of the steady-state error.

Error Series

In this section, the error-constant concept is generalized to include inputs of almost any arbitrary function of time. We start with the transformed error function of Eq. (6-15),

$$\mathcal{E}(s) = \frac{R(s)}{1 + G(s)H(s)} \tag{6-37}$$

or of Eq. (6-35), as the case may be.

Using the principle of the convolution integral as discussed in Section 3.2, the error signal $\epsilon(t)$ may be written

$$\epsilon(t) = \int_{-\infty}^{t} w_e(\tau) r(t - \tau) \, d\tau \tag{6-38}$$

where $w_e(\tau)$ is the inverse Laplace transform of

$$W_e(s) = \frac{1}{1 + G(s)H(s)} \tag{6-39}$$

which is known as the *error transfer function*.

If the first n derivatives of $r(t)$ exist for all values of t, the function $r(t - \tau)$ can be expanded into a Taylor series; that is,

$$r(t - \tau) = r(t) - \tau \dot{r}(t) + \frac{\tau^2}{2!} \ddot{r}(t) - \frac{\tau^3}{3!} \dddot{r}(t) + \cdots \tag{6-40}$$

where $\dot{r}(t)$ represents the first derivative of $r(t)$ with respect to time.

Since $r(t)$ is considered to be zero for negative time, the limit of the convolution integral of Eq. (6-38) may be taken from 0 to t. Substituting Eq. (6-40) into Eq. (6-38), we have

$$\epsilon(t) = \int_0^t w_e(\tau) \left[r(t) - \tau \dot{r}(t) + \frac{\tau^2}{2!} \ddot{r}(t) - \frac{\tau^3}{3!} \dddot{r}(t) + \cdots \right] d\tau$$

$$= r(t) \int_0^t w_e(\tau) \, d\tau - \dot{r}(t) \int_0^t \tau w_e(\tau) \, d\tau + \ddot{r}(t) \int_0^t \frac{\tau^2}{2!} w_e(\tau) \, d\tau - \cdots \tag{6-41}$$

As before, the steady-state error is obtained by taking the limit of $\epsilon(t)$ as t approaches infinity; thus

$$e_{ss} = \lim_{t \to \infty} \epsilon(t) = \lim_{t \to \infty} \epsilon_s(t) \tag{6-42}$$

where $\epsilon_s(t)$ denotes the steady-state part of $\epsilon(t)$ and is given by

$$\epsilon_s(t) = r_s(t)\int_0^\infty w_e(\tau)\,d\tau - \dot{r}_s(t)\int_0^\infty \tau w_e(\tau)\,d\tau + \ddot{r}_s(t)\int_0^\infty \frac{\tau^2}{2!}w_e(\tau)\,d\tau$$

$$-\dddot{r}_s(t)\int_0^\infty \frac{\tau^3}{3!}w_e(\tau)\,d\tau + \cdots \tag{6-43}$$

and $r_s(t)$ denotes the steady-state part of $r(t)$.

Let us define

$$C_0 = \int_0^\infty w_e(\tau)\,d\tau$$

$$C_1 = -\int_0^\infty \tau w_e(\tau)\,d\tau$$

$$C_2 = \int_0^\infty \tau^2 w_e(\tau)\,d\tau \tag{6-44}$$

$$\vdots$$

$$C_n = (-1)^n \int_0^\infty \tau^n w_e(\tau)\,d\tau$$

Equation (6-43) is written

$$\epsilon_s(t) = C_0 r_s(t) + C_1\dot{r}_s(t) + \frac{C_2}{2!}\ddot{r}_s(t) + \cdots + \frac{C_n}{n!}r_s^{(n)}(t) + \cdots \tag{6-45}$$

which is called the *error series*, and the coefficients, $C_0, C_1, C_2, \ldots, C_n$ are defined as the *generalized error coefficients*, or simply as the *error coefficients*.

The error coefficients may be readily evaluated directly from the error transfer function, $W_e(s)$. Since $W_e(s)$ and $w_e(\tau)$ are related through the Laplace transform, we have

$$W_e(s) = \int_0^\infty w_e(\tau)e^{-\tau s}\,d\tau \tag{6-46}$$

Taking the limit on both sides of Eq. (6-46) as s approaches zero, we have

$$\lim_{s\to 0} W_e(s) = \lim_{s\to 0} \int_0^\infty w_e(\tau)e^{-\tau s}\,d\tau \tag{6-47}$$

The derivative of $W_e(s)$ of Eq. (6-46) with respect to s gives

$$\frac{dW_e(s)}{ds} = -\int_0^\infty \tau w_e(\tau)e^{-\tau s}\,d\tau \tag{6-48}$$

from which we get

$$C_1 = \lim_{s\to 0} \frac{dW_e(s)}{ds} \tag{6-49}$$

The rest of the error coefficients are obtained in a similar fashion by taking successive differentiation of Eq. (6-46) with respect to s. Therefore,

$$C_2 = \lim_{s \to 0} \frac{d^2 W_e(s)}{ds^2} \tag{6-50}$$

$$C_3 = \lim_{s \to 0} \frac{d^3 W_e(s)}{ds^3} \tag{6-51}$$

$$\vdots$$

$$C_n = \lim_{s \to 0} \frac{d^n W_e(s)}{ds^n} \tag{6-52}$$

The following examples illustrate the general application of the error series and its advantages over the error constants.

■

Example 6-1

In this illustrative example the steady-state error of a feedback control system will be evaluated by use of the error series and the error coefficients. Consider a unity feedback control system with the open-loop transfer function given as

$$G(s) = \frac{K}{s + 1} \tag{6-53}$$

Since the system is of type 0, the error constants are $K_p = K$, $K_v = 0$, and $K_a = 0$. Thus, the steady-state errors of the system due to the three basic types of inputs are as follows:

unit step input, $u_s(t)$: $e_{ss} = \dfrac{1}{1 + K}$

unit ramp input, $tu_s(t)$: $e_{ss} = \infty$

unit parabolic input, $t^2 u_s(t)$: $e_{ss} = \infty$

Notice that when the input is either a ramp or a parabolic function, the steady-state error is infinite in magnitude, since it apparently increases with time. It is apparent that the error constants fail to indicate the exact manner in which the steady-state function increases with time. Therefore, ordinarily, if the steady-state response of this system due to a ramp or parabolic input is desired, the differential equation of the system must be solved. We now show that the steady-state response of the system can actually be determined from the error series.

Using Eq. (6-39), we have for this system

$$W_e(s) = \frac{1}{1 + G(s)} = \frac{s + 1}{s + K + 1} \tag{6-54}$$

The error coefficients are evaluated as

$$C_0 = \lim_{s \to 0} W_e(s) = \frac{1}{K+1} \tag{6-55}$$

$$C_1 = \lim_{s \to 0} \frac{dW_e(s)}{ds} = \frac{K}{(1+K)^2} \tag{6-56}$$

$$C_2 = \lim_{s \to 0} \frac{d^2 W_e(s)}{ds^2} = \frac{-2K}{(1+K)^3} \tag{6-57}$$

Although higher-order coefficients can be obtained, they will become less significant as their values will be increasingly smaller. The error series is written

$$e_s(t) = \frac{1}{1+K} r_s(t) + \frac{K}{(1+K)^2} \dot{r}_s(t) + \frac{-K}{(1+K)^3} \ddot{r}_s(t) + \cdots \tag{6-58}$$

Now let us consider the three basic types of inputs.

1. When the input signal is a unit step function, $r_s(t) = u_s(t)$, and all derivatives of $r_s(t)$ are zero. The error series gives

$$e_s(t) = \frac{1}{1+K} \tag{6-59}$$

   which agrees with the result given by the error-constant method.

2. When the input signal is a unit ramp function, $r_s(t) = tu_s(t)$, $\dot{r}_s(t) = u_s(t)$, and all higher-order derivatives of $\dot{r}_s(t)$ are zero. Therefore, the error series is

$$e_s(t) = \left[\frac{1}{1+K} t + \frac{K}{(1+K)^2} \right] u_s(t) \tag{6-60}$$

   which indicates that the steady-state error increases linearly with time. The error-constant method simply yields the result that the steady-state error is infinite but fails to give details of the time dependence.

3. For a parabolic input, $r_s(t) = (t^2/2)u_s(t)$, $\dot{r}_s(t) = tu_s(t)$, $\ddot{r}_s(t) = u_s(t)$, and all higher derivatives are zero. The error series becomes

$$e_s(t) = \left[\frac{1}{1+K} \frac{t^2}{2} + \frac{K}{(1+K)^2} t - \frac{K}{(1+K)^3} \right] u_s(t) \tag{6-61}$$

   In this case the error increases in magnitude as the second power of t.

4. Consider that the input signal is represented by a polynomial of t and an exponential term,

$$r(t) = \left[a_0 + a_1 t + \frac{a_2 t^2}{2} + e^{-a_3 t} \right] u_s(t) \tag{6-62}$$

where a_0, a_1, a_2, and a_3 are constants. Then,

$$r_s(t) = \left[a_0 + a_1 t + \frac{a_2 t^2}{2} \right] u_s(t) \tag{6-63}$$

$$\dot{r}_s(t) = (a_1 + a_2 t) u_s(t) \tag{6-64}$$

$$\ddot{r}_s(t) = a_2 u_s(t) \tag{6-65}$$

In this case the error series becomes

$$e_s(t) = \frac{1}{1+K} r_s(t) + \frac{K}{(1+K)^2} \dot{r}_s(t) - \frac{K}{(1+K)^3} \ddot{r}_s(t) \tag{6-66}$$

Example 6-2

In this example we shall consider a situation in which the error constant is totally inadequate in providing a solution to the steady-state error. Let us consider that the input to the system described in Example 6-1 is a sinusoid,

$$r(t) = \sin \omega_0 t \tag{6-67}$$

where $\omega_0 = 2$. Then

$$\begin{aligned}
r_s(t) &= \sin \omega_0 t \\
\dot{r}_s(t) &= \omega_0 \cos \omega_0 t \\
\ddot{r}_s(t) &= -\omega_0^2 \sin \omega_0 t \\
\dddot{r}_s(t) &= -\omega_0^3 \cos \omega_0 t
\end{aligned} \tag{6-68}$$

The error series can be written

$$e_s(t) = \left[C_0 - \frac{C_2}{2!} \omega_0^2 + \frac{C_4}{4!} \omega_0^4 - \cdots \right] \sin \omega_0 t + \left[C_1 \omega_0 - \frac{C_3}{3!} \omega_0^3 + \cdots \right] \cos \omega_0 t \tag{6-69}$$

Because of the sinusoidal input, the error series is now an infinite series. The convergence of the series is important in arriving at a meaningful answer to the steady-state error. It is clear that the convergence of the error series depends on the value of ω_0 and K. Let us assign the value of K to be 100. Then

$$C_0 = \frac{1}{1+K} = 0.0099$$

$$C_1 = \frac{K}{(1+K)^2} = 0.0098$$

$$C_2 = -\frac{2K}{(1+K)^3} = -0.000194 \tag{6-70}$$

$$C_3 = \frac{6K}{(1+K)^5} = 5.65 \times 10^{-8}$$

Thus, using only the first four error coefficients, Eq. (6-69) becomes

$$e_s(t) \cong \left[0.0099 + \frac{0.000194}{2} \cdot 4\right] \sin 2t + 0.0196 \cos 2t$$

$$= 0.01029 \sin 2t + 0.0196 \cos 2t \qquad (6\text{-}71)$$

or

$$e_s(t) \cong 0.02215 \sin(2t + 62.3°) \qquad (6\text{-}72)$$

Therefore, the steady-state error in this case is also a sinusoid, as given by Eq. (6-72).

∎

6.4 TIME-DOMAIN PERFORMANCE OF CONTROL SYSTEMS — TRANSIENT RESPONSE

The transient portion of the time response is that part which goes to zero as time becomes large. Of course, the transient response has significance only when a stable system is referred to, since for an unstable system the response does not diminish and is out of control.

The transient performance of a control system is usually characterized by the use of a unit step input. Typical performance criteria that are used to characterize the transient response to a unit step input include overshoot, delay time, rise time, and settling time. Figure 6-8 illustrates a typical unit step response of a linear control system. The above-mentioned criteria are defined with respect to the step response:

1. *Maximum overshoot.* The maximum overshoot is defined as the largest deviation of the output over the step input during the transient state. The amount of maximum overshoot is also used as a measure of the relative stability of the system. The maximum overshoot is often represented as a percentage of the final value of the step response; that is,

$$\text{percent maximum overshoot} = \frac{\text{maximum overshoot}}{\text{final value}} \times 100\% \qquad (6\text{-}73)$$

2. *Delay time.* The delay time T_d is defined as the time required for the step response to reach 50 percent of its final value.

3. *Rise time.* The rise time T_r is defined as the time required for the step response to rise from 10 percent to 90 percent of its final value. Sometimes an alternative measure is to represent the rise time as a reciprocal of the slope of the step response at the instant that the response is equal to 50 percent of its final value.

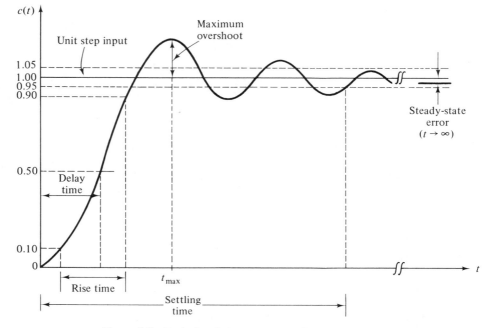

Figure 6-8   Typical unit step response of a control system.

4. *Settling time.* The settling time T_s is defined as the time required for the step response to decrease and stay within a specified percentage of its final value. A frequently used figure is 5 percent.

The four quantities defined above give a direct measure of the transient characteristics of the step response. These quantities are relatively easy to measure when a step response is already plotted. However, analytically these quantities are difficult to determine except for the simple cases.

6.5 TRANSIENT RESPONSE OF A SECOND-ORDER SYSTEM

Although true second-order control systems are rare in practice, their analysis generally helps to form a basis for the understanding of design and analysis techniques.

Consider that a second-order feedback control system is represented by the state diagram of Fig. 6-9. The state equations are written

$$\begin{bmatrix} \dot{x}_1(t) \\ \dot{x}_2(t) \end{bmatrix} = \begin{bmatrix} 0 & 1 \\ -\omega_n^2 & -2\zeta\omega_n \end{bmatrix}\begin{bmatrix} x_1(t) \\ x_2(t) \end{bmatrix} + \begin{bmatrix} 0 \\ 1 \end{bmatrix} r(t) \tag{6-74}$$

where ζ and ω_n are constants.

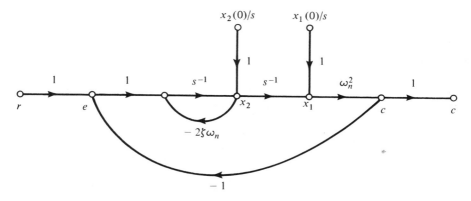

Figure 6-9 State diagram of a second-order feedback control system.

The output equation is

$$c(t) = \omega_n^2 x_1(t) \tag{6-75}$$

Applying the gain formula to the state diagram of Fig. 6-9, the state transition equations are written

$$\begin{bmatrix} X_1(s) \\ X_2(s) \end{bmatrix} = \frac{1}{\Delta} \begin{bmatrix} s + 2\zeta\omega_n & 1 \\ -\omega_n^2 & s \end{bmatrix} \begin{bmatrix} x_1(0) \\ x_2(0) \end{bmatrix} + \frac{1}{\Delta} \begin{bmatrix} 1 \\ s \end{bmatrix} R(s) \tag{6-76}$$

where

$$\Delta = s^2 + 2\zeta\omega_n s + \omega_n^2 \tag{6-77}$$

The inverse Laplace transform of Eq. (6-76) is carried out with the help of the Laplace transform table. For a unit step function input, we have

$$\begin{bmatrix} x_1(t) \\ x_2(t) \end{bmatrix} = \frac{e^{-\zeta\omega_n t}}{\sqrt{1-\zeta^2}} \begin{bmatrix} \sin\left(\omega_n\sqrt{1-\zeta^2}\,t + \psi\right) & \dfrac{1}{\omega_n}\sin\omega_n\sqrt{1-\zeta^2}\,t \\ -\omega_n\sin\omega_n\sqrt{1-\zeta^2}\,t & \sin\left(\omega_n\sqrt{1-\zeta^2}\,t + \phi\right) \end{bmatrix} \begin{bmatrix} x_1(0) \\ x_2(0) \end{bmatrix}$$

$$+ \begin{bmatrix} \dfrac{1}{\omega_n^2}\left\{1 + \dfrac{1}{\sqrt{1-\zeta^2}}e^{-\zeta\omega_n t}\sin\left[\omega_n\sqrt{1-\zeta^2}\,t - \phi\right]\right\} \\ \dfrac{1}{\omega_n\sqrt{1-\zeta^2}}e^{-\zeta\omega_n t}\sin\omega_n\sqrt{1-\zeta^2}\,t \end{bmatrix} \qquad t \geq 0 \quad (6\text{-}78)$$

where

$$\psi = \tan^{-1} \frac{\sqrt{1-\zeta^2}}{\zeta} \tag{6-79}$$

$$\phi = \tan^{-1} \frac{\sqrt{1-\zeta^2}}{-\zeta} \tag{6-80}$$

Although Eq. (6-78) gives the complete solution of the state variable in terms of the initial states and the unit step input, it is a rather formidable-looking expression, especially in view of the fact that the system is only of the second order. However, the analysis of control systems does not rely completely on the evaluation of the complete state and output responses. The development of linear control theory allows the study of control system performance by use of the transfer function and the characteristic equation. We shall show that a great deal can be learned about the system's behavior by studying the location of the roots of the characteristic equation.

The closed-loop transfer function of the system is determined from Fig. 6-9.

$$\frac{C(s)}{R(s)} = \frac{\omega_n^2}{s^2 + 2\zeta\omega_n s + \omega_n^2} \tag{6-81}$$

The characteristic equation of the system is obtained by setting Eq. (6-77) to zero; that is,

$$\Delta = s^2 + 2\zeta\omega_n s + \omega_n^2 = 0 \tag{6-82}$$

For a unit step function input, $R(s) = 1/s$, the output response of the system is determined by taking the inverse Laplace transform of

$$C(s) = \frac{\omega_n^2}{s(s^2 + 2\zeta\omega_n s + \omega_n^2)} \tag{6-83}$$

Or, $c(t)$ is determined by use of Eqs. (6-75) and (6-78) with zero initial states:

$$c(t) = 1 + \frac{e^{-\zeta\omega_n t}}{\sqrt{1-\zeta^2}} \sin\left[\omega_n\sqrt{1-\zeta^2}\, t - \tan^{-1}\frac{\sqrt{1-\zeta^2}}{-\zeta}\right] \qquad t \geq 0 \tag{6-84}$$

It is interesting to study the relationship between the roots of the characteristic equation and the behavior of the step response $c(t)$. The two roots of Eq. (6-82) are

$$s_1, s_2 = -\zeta\omega_n \pm j\omega_n\sqrt{1-\zeta^2}$$

$$= -\alpha \pm j\omega \tag{6-85}$$

The physical significance of the constant ζ, ω_n, α, and ω is now described as follows. As seen from Eq. (6-85), $\alpha = \zeta\omega_n$, and α appears as the constant that is multiplied to t in the exponential term of Eq. (6-84). Therefore, α controls the rate of rise and decay of the time response. In other words, α controls the "damping" of the system and is called the *damping constant* or the *damping factor*. The inverse of α, $1/\alpha$, is proportional to the time constant of the system.

When the two roots of the characteristic equation are real and identical we call the system *critically damped*. From Eq. (6-85) we see that *critical damping occurs when $\zeta = 1$*. Under this condition the damping factor is simply $\alpha = \omega_n$. Therefore, we can regard ζ as the *damping ratio*, which is the ratio between the actual damping factor and the damping factor when the damping is critical.

ω_n is defined as the *natural undamped frequency*. As seen from Eq. (6-85), when the damping is zero, $\zeta = 0$, the roots of the characteristic equation are imaginary, and Eq. (6-84) shows that the step response is purely sinusoidal. Therefore, ω_n corresponds to the frequency of the undamped sinusoid.

Equation (6-85) shows that

$$\omega = \omega_n\sqrt{1 - \zeta^2} \tag{6-86}$$

However, since unless $\zeta = 0$, the response of Eq. (6-84) is not a periodic function. Therefore, strictly, ω is not a frequency. For the purpose of reference ω is sometimes defined as *conditional frequency*.

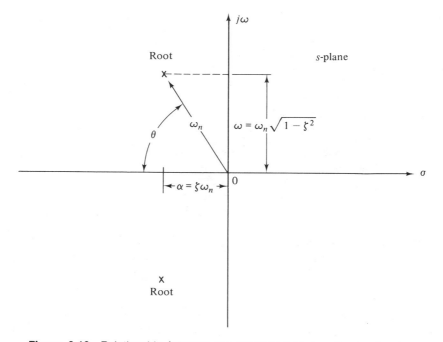

Figure 6-10 Relationship between the characteristic equation roots of a second-order system and α, ζ, ω_n, and ω.

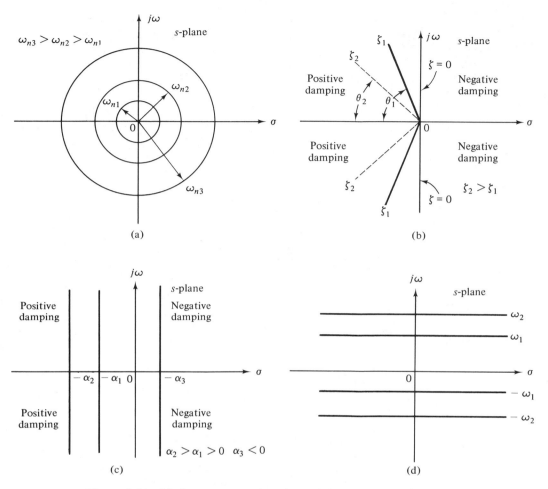

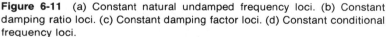

Figure 6-11 (a) Constant natural undamped frequency loci. (b) Constant damping ratio loci. (c) Constant damping factor loci. (d) Constant conditional frequency loci.

Figure 6-10 illustrates the relationship between the location of the characteristic equation roots and α, ζ, ω_n, and ω. For the complex-conjugate roots shown, ω_n is the radial distance from the roots to the origin of the s-plane. The damping factor α is the real part of the roots; the conditional frequency is the imaginary part of the roots, and the damping ratio ζ is equal to the cosine of the angle between the radial line to the roots and the negative real axis; that is,

$$\zeta = \cos \theta \qquad (6\text{-}87)$$

Figure 6-11 shows the constant-ω_n loci, the constant-ζ loci, the constant-α loci, and the constant-ω loci. Note that the left-half of the s-plane corresponds to positive damping (i.e., the damping factor or ratio is positive), and the right-half of the s-plane corresponds to negative damping. The imaginary axis corresponds to

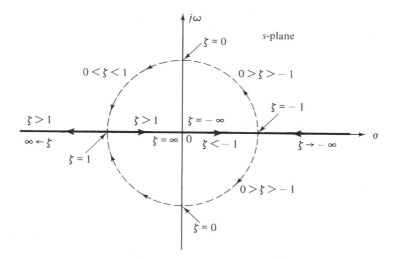

Figure 6-12 Locus of roots of Eq. (6-82) when ω_n is held constant while the damping ratio is varied from $-\infty$ to $+\infty$.

zero damping ($\alpha = 0$, $\zeta = 0$). As shown by Eq. (6-84), when the damping is positive, the step response will settle to its constant final value because of the negative exponent of $e^{-\zeta\omega_n t}$. Negative damping will correspond to a response that grows without bound, and zero damping gives rise to a sustained sinusoidal oscillation. These last two cases are defined as *unstable* for linear systems. Therefore, we have demonstrated that the location of the characteristic equation roots plays a great part in the dynamic behavior of the transient response of the system.

The effect of the characteristic-equation roots on the damping of the second-order system is further illustrated by Figs. 6-12 and 6-13. In Fig. 6-12 ω_n is held constant while the damping ratio ζ is varied from $-\infty$ to $+\infty$. The following classification of the system dynamics with respect to the value of ζ is given:

$0 < \zeta < 1$:	$s_1, s_2 = -\zeta\omega_n \pm j\omega_n\sqrt{1 - \zeta^2}$	underdamped case
$\zeta = 1$:	$s_1, s_2 = -\omega_n$	critically damped case
$\zeta > 1$:	$s_1, s_2 = -\zeta\omega_n \pm \omega_n\sqrt{\zeta^2 - 1}$	overdamped case
$\zeta = 0$:	$s_1, s_2 = \pm j\omega_n$	undamped case
$\zeta < 0$:	$s_1, s_2 = -\zeta\omega_n \pm j\omega_n\sqrt{1 - \zeta^2}$	negatively damped case

Figure 6-13 illustrates typical step responses that correspond to the various root locations.

In practical applications only stable systems are of interest. Therefore, the cases when ζ is positive are of particular interest. In Fig. 6-14 is plotted the variation of the unit step response described by Eq. (6-84) as a function of the normalized time $\omega_n t$, for various values of the damping ratio ζ. It is seen that the

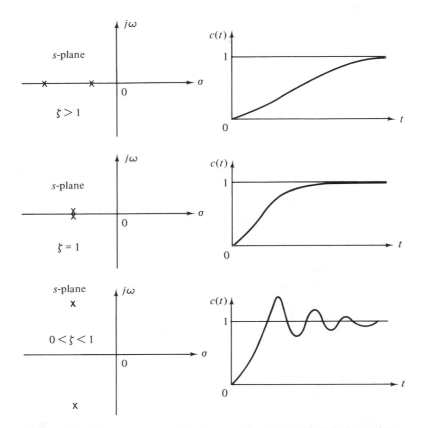

Figure 6-13 Response comparison for various root locations in the s-plane.

response becomes more oscillatory as ζ decreases in value. When $\zeta \geq 1$ there is no overshoot in the step response; that is, the output never exceeds the value of the reference input.

The exact relation between the damping ratio and the amount of overshoot can be obtained by taking the derivative of Eq. (6-84) and setting the result to zero. Thus,

$$\frac{dc(t)}{dt} = -\frac{\zeta \omega_n e^{-\zeta \omega_n t}}{\sqrt{1 - \zeta^2}} \sin(\omega t - \phi)$$

$$+ \frac{e^{-\zeta \omega_n t}}{\sqrt{1 - \zeta^2}} \omega_n \sqrt{1 - \zeta^2} \cos(\omega t - \phi) \qquad t \geq 0 \quad (6\text{-}88)$$

where

$$\phi = \tan^{-1} \frac{\sqrt{1 - \zeta^2}}{-\zeta} \qquad (6\text{-}89)$$

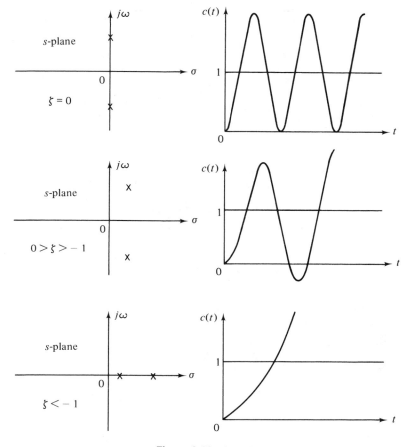

Figure 6-13 (cont.)

Equation (6-88) is simplified to

$$\frac{dc(t)}{dt} = \frac{\omega_n}{\sqrt{1 - \zeta^2}} e^{-\zeta\omega_n t} \sin \omega_n \sqrt{1 - \zeta^2}\, t \qquad t \geq 0 \qquad (6\text{-}90)$$

Therefore, setting Eq. (6-90) to zero, we have $t = \infty$ and

$$\omega_n \sqrt{1 - \zeta^2}\, t = n\pi \qquad n = 0, 1, 2, \ldots \qquad (6\text{-}91)$$

or

$$t = \frac{n\pi}{\omega_n \sqrt{1 - \zeta^2}} \qquad (6\text{-}92)$$

The first maximum value of the step response $c(t)$ occurs at $n = 1$. Therefore, the

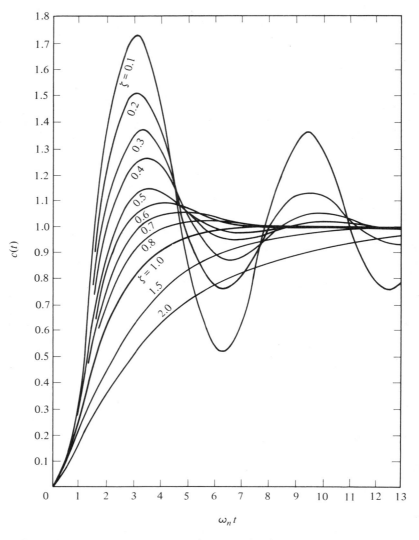

Figure 6-14 Transient response of a second-order system to a unit step function input.

time at which the maximum overshoot occurs is given by

$$t_{max} = \frac{\pi}{\omega_n \sqrt{1 - \zeta^2}} \qquad (6\text{-}93)$$

In general, for all odd values of n, that is, $n = 1, 3, 5, \ldots$, Eq. (6-92) gives the times at which the overshoots occur. For all even values of n, Eq. (6-92) gives the times at which the undershoots occur, as shown in Fig. 6-15. It is interesting to note

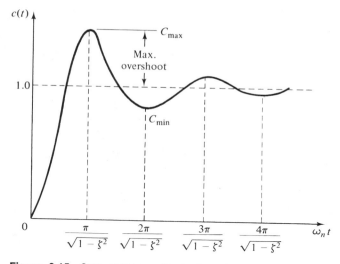

Figure 6-15 Step response illustrating that the maxima and minima occur at periodic intervals.

that, although the maxima and the minima of the response occur at periodic intervals, the response is a damped sinusoid and is not a periodic function.

The magnitudes of the overshoots and the undershoots can be obtained by substituting Eq. (6-92) into Eq. (6-84). Thus,

$$c(t)|_{\text{max or min}} = 1 + \frac{e^{-n\pi\zeta/\sqrt{1-\zeta^2}}}{\sqrt{1-\zeta^2}} \sin\left(n\pi - \tan^{-1}\frac{\sqrt{1-\zeta^2}}{-\zeta}\right) \qquad n = 1, 2, 3, \ldots$$

$$(6\text{-}94)$$

or

$$c(t)|_{\text{max or min}} = 1 + (-1)^{n-1} e^{-n\pi\zeta/\sqrt{1-\zeta^2}} \qquad (6\text{-}95)$$

The maximum overshoot is obtained by letting $n = 1$ in Eq. (6-95). Therefore,

$$\text{maximum overshoot} = c_{\text{max}} - 1 = e^{-\pi\zeta/\sqrt{1-\zeta^2}} \qquad (6\text{-}96)$$

and

$$\text{percent maximum overshoot} = 100 e^{-\pi\zeta/\sqrt{1-\zeta^2}} \qquad (6\text{-}97)$$

Note that for the second-order system, the maximum overshoot of the step response is only a function of the damping ratio. The relationship between the percent maximum overshoot and damping ratio for the second-order system is shown in Fig. 6-16.

From Eqs. (6-93) and (6-94) it is seen that for the second-order system under consideration, the maximum overshoot and the time at which it occurs are all exactly expressed in terms of ζ and ω_n. For the delay time, rise time, and settling time, however, the relationships are not so simple. It would be difficult to determine the exact expressions for these quantities. For instance, for the delay time, we would

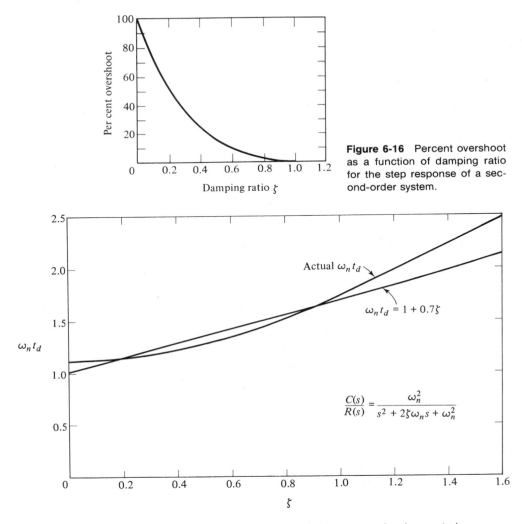

Figure 6-16 Percent overshoot as a function of damping ratio for the step response of a second-order system.

Figure 6-17 Normalized delay time versus ζ for a second-order control system.

have to set $c(t) = 0.5$ in Eq. (6-84) and solve for t. An easier way would be to plot $\omega_n t_d$ versus ζ as shown in Fig. 6-17. Then, over the range of $0 < \zeta < 1.0$ is it possible to approximate the curve by a straight line,

$$\omega_n t_d \cong 1 + 0.7\zeta \tag{6-98}$$

Thus, the delay time

$$t_d \cong \frac{1 + 0.7\zeta}{\omega_n} \tag{6-99}$$

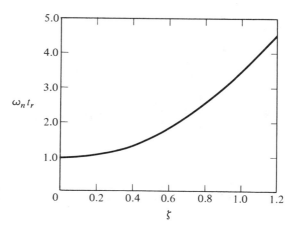

Figure 6-18 Normalized rise time versus ζ for a second-order system.

For a wider range of ζ, a second-order equation should be used. Then

$$t_d \cong \frac{1 + 0.6\zeta + 0.15\zeta^2}{\omega_n} \qquad (6\text{-}100)$$

For the rise time t_r, which is the time for the step response to reach from 10 to 90 percent of its final value, the exact values can again be obtained directly from the responses of Fig. 6-14. The plot of $\omega_n t_r$ versus ζ is shown in Fig. 6-18. In this case the rise time versus ζ relation can again be approximated by a straight line over a limited range of ζ. Therefore,

$$t_r \cong \frac{0.8 + 2.5\zeta}{\omega_n} \qquad 0 < \zeta < 1 \qquad (6\text{-}101)$$

A better approximation may be obtained by using a second-order equation; then

$$t_r \cong \frac{1 + 1.1\zeta + 1.4\zeta^2}{\omega_n} \qquad (6\text{-}102)$$

From the definition of settling time, it is clear that the expression for the settling time is the most difficult to determine. However, we can obtain an approximation for the case of $0 < \zeta < 1$ by using the envelope of the damped sinusoid, as shown in Fig. 6-19.

From the figure it is clear that the same result is obtained with the approximation whether the upper envelope or the lower envelope is used. Therefore,

$$c(t) = 1 + \frac{e^{-\zeta \omega_n t_s}}{\sqrt{1 - \zeta^2}} = 1.05 \qquad (6\text{-}103)$$

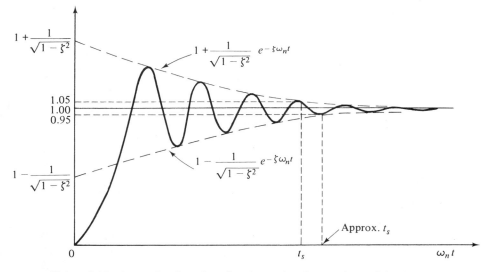

Figure 6-19 Approximation of settling time using the envelope of the decaying step response of a second order system ($0 < \zeta < 1$).

Solving for $\omega_n t_s$ from the last equation, we have

$$\omega_n t_s = -\frac{1}{\zeta}\ln\left[0.05\sqrt{1-\zeta^2}\right] \tag{6-104}$$

For small values of ζ, Eq. (6-104) is simplified to

$$\omega_n t_s \cong \frac{3}{\zeta} \tag{6-105}$$

or

$$t_s \cong \frac{3}{\zeta\omega_n} \qquad 0 < \zeta < 1 \tag{6-106}$$

Now reviewing the relationships for the delay time, rise time, and settling time, it is seen that small values of ζ would yield short rise time and short delay time. However, a fast settling time requires a large value for ζ. Therefore, a compromise in the value of ζ should be made when all these criteria are to be satisfactorily met in a design problem. Together with the consideration on maximum overshoot, a generally accepted range of damping ratio for satisfactory all-around performance is between 0.5 and 0.8.

6.6 TIME-DOMAIN ANALYSIS OF A PRINTWHEEL CONTROL SYSTEM

In this section we study the time-domain performance of a control system whose objective is to control the position of the printwheel of a word processor or terminal printer. Figure 6-20 shows a typical word processor, and Fig. 6-21 shows the printwheel, the dc motor that controls the printwheel, and the incremental encoder assembly for speed and position feedback. The printwheel, which is nicknamed the "daisywheel," typically has 96 character positions. The control of the printwheel requires that the proper print character be placed in front of the hammer for hard-copy printing. The printwheel is mounted directly on the motor shaft, and rotation can be made in either direction.

The block diagram of the printwheel control system is shown in Fig. 6-22. When a command to print a certain character is given, either through the keyboard or the computer communication link, the control system first interprets the total distance and the direction to be traveled, and then commands the dc motor to drive

Figure 6-20   A word processor.

Figure 6-21 A "daisywheel" print element.

the printwheel to the correct position. Such a control system usually consists of two modes of operation: the velocity mode and the position mode. When the command to travel a certain number of character positions is given by the set point, the velocity mode is first activated, and the motor-printwheel system is driven to follow a specific velocity profile, which is stored in the microprocessor controller. After the load is driven to the vicinity of the desired position, in the region between $-a$ and 0 or a and 0, as shown in Fig. 6-22, depending on the direction of rotation, the system is switched to the position-control mode. The main purpose of the position mode is to null the position error, or drive the printwheel to point 0 as fast as possible without prolonged or excessive oscillations. The encoder used in this case is an incremental encoder (Section 4.5, Figs. 4-40 and 4-45). As shown in Fig. 6-22, the gain of the encoder for position control is determined by using the slope of the encoder output waveform at the equilibrium point. It should be noted that the velocity-control problem is not treated in the present discussion.

The following system parameters are given:

Gain of encoder	$K_s = 1 \text{ V/rad}$
Gain of power amplifier	$K = \text{variable}$
Resistance of armature of motor	$R_a = 5 \text{ ohms}$
Inductance of armature of motor	$L_a = \text{negligible}$
Torque constant of motor	$K_i = 3 \text{ oz-in./A}$
Back emf constant of motor	$K_b = 0.02125 \text{ V/rad/sec}$
Inertia of motor rotor	$J_m = 3 \times 10^{-4} \text{ oz-in.-sec}^2$
Inertia of printwheel	$J_L = 12 \times 10^{-4} \text{ oz-in.-sec}^2$
Viscous friction coefficient of motor shaft	$B_m = 0.03 \text{ oz-in.-sec.}$
Viscous friction coefficient of load shaft	$B_L = 0.03 \text{ oz-in.-sec}$

The first step in the analysis of the positional control system is to write the equations of the system. This is done by proceeding from the input to the output following through the cause-and-effect relationship.

1. *Encoder-error computation*:

$$\theta_e(t) = \theta_r(t) - \theta_c(t) \qquad (6\text{-}107)$$

$$e(t) = K_s\theta_e(t) \qquad (6\text{-}108)$$

We have assumed that in the position-control mode the microprocessor simply compares the encoder output with the reference set point and sends out an error signal in proportion to the difference between the two signals.

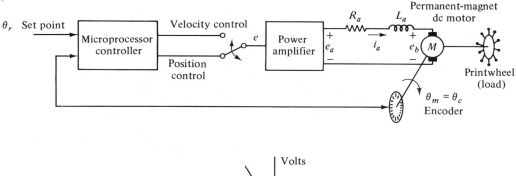

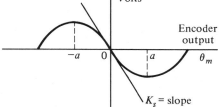

Figure 6-22 Printwheel control system and the encoder characteristics.

2. *Power amplifier*:

$$e_a(t) = Ke(t) \qquad (6\text{-}109)$$

3. *Permanent-magnet dc motor*:

$$L_a \frac{di_a(t)}{dt} = -R_a i_a(t) + e_a(t) - e_b(t) \qquad (6\text{-}110)$$

$$e_b(t) = K_b \omega_m(t) \qquad (6\text{-}111)$$

$$T_m(t) = K_i i_a(t) \qquad (6\text{-}112)$$

$$J \frac{d\omega_m(t)}{dt} = -B\omega_m(t) + T_m(t) \qquad (6\text{-}113)$$

where J and B are the total inertia and viscous frictional coefficient seen by the motor, respectively. In the present case, since the load is directly coupled to the motor shaft,

$$J = J_m + J_L = 15 \times 10^{-4} \text{ oz-in.-sec}^2 \qquad (6\text{-}114)$$

$$B = B_m + B_L = 0.06 \text{ oz-in.-sec} \qquad (6\text{-}115)$$

4. *Mechanical output*:

$$\frac{d\theta_m(t)}{dt} = \omega_m(t) \qquad (6\text{-}116)$$

$$\theta_c(t) = \theta_m(t) \qquad (6\text{-}117)$$

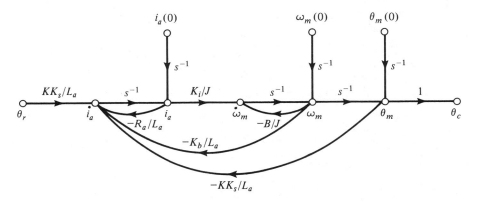

Figure 6-23 State diagram of the position-control mode of the printwheel control system.

Using Eqs. (6-107) through (6-116), the state equations of the system are written in vector-matrix form:

$$
\begin{bmatrix} \dfrac{di_a(t)}{dt} \\[3mm] \dfrac{d\omega_m(t)}{dt} \\[3mm] \dfrac{d\theta_m(t)}{dt} \end{bmatrix} = \begin{bmatrix} -\dfrac{R_a}{L_a} & -\dfrac{K_b}{L_a} & -\dfrac{KK_s}{L_a} \\[3mm] \dfrac{K_i}{J} & -\dfrac{B}{J} & 0 \\[3mm] 0 & 1 & 0 \end{bmatrix} \begin{bmatrix} i_a(t) \\[3mm] \omega_m(t) \\[3mm] \theta_m(t) \end{bmatrix} + \begin{bmatrix} \dfrac{KK_s}{L_a} \\[3mm] 0 \\[3mm] 0 \end{bmatrix} \theta_r(t) \quad (6\text{-}118)
$$

The output equation is readily given by Eq. (6-117).

The state diagram of the system is drawn as shown in Fig. 6-23. The closed-loop transfer function of the system is determined from Fig. 6-23.

$$
\frac{\Theta_c(s)}{\Theta_r(s)} = \frac{K_s KK_i}{R_a Bs(1 + \tau_a s)(1 + \tau s) + K_b K_i s + K_s KK_i} \quad (6\text{-}119)
$$

where

$$
\tau_a = \frac{L_a}{R_a} \cong 0 \quad (6\text{-}120)
$$

$$
\tau = \frac{J}{B} = \frac{15 \times 10^{-4}}{0.06} = 0.025 \text{ sec} \quad (6\text{-}121)
$$

The open-loop transfer function of the system is

$$
G(s) = \frac{\Theta_c(s)}{\Theta_e(s)} = \frac{K_s KK_i}{R_a Bs(1 + \tau_a s)(1 + \tau s) + K_b K_i s} \quad (6\text{-}122)
$$

The state transition equations of the system can be obtained from the state diagram by use of the gain formula, in the usual fashion. However, the main objective of this problem is to demonstrate the behavior of the time response of the positional control system with respect to the system parameters, and it is sufficient to work with the transfer functions.

Since $\tau_a \cong 0$, Eq. (6-119) is simplified to

$$\frac{\Theta_c(s)}{\Theta_r(s)} = \frac{K_s KK_i}{R_a J s^2 + (K_b K_i + R_a B)s + K_s KK_i} \tag{6-123}$$

This transfer function is of the second order; thus, it can be written in the standard form of Eq. (6-81). The natural undamped frequency of the system is

$$\omega_n = \pm \sqrt{\frac{K_s KK_i}{R_a J}} \tag{6-124}$$

The damping ratio is

$$\zeta = \frac{K_b K_i + R_a B}{2 R_a J \omega_n} = \frac{K_b K_i + R_a B}{2\sqrt{K_s KK_i R_a J}} \tag{6-125}$$

Substituting the system parameters into Eq. (6-123), the closed-loop transfer function becomes

$$\frac{\Theta_c(s)}{\Theta_r(s)} = \frac{400K}{s^2 + 48.5s + 400K} \tag{6-126}$$

and

$$\omega_n = \pm 20\sqrt{K} \qquad \text{rad/sec} \tag{6-127}$$

$$\zeta = \frac{1.2125}{\sqrt{K}} \tag{6-128}$$

Thus, we see that the natural undamped frequency is proportional to $\sqrt{K}$, whereas the damping ratio is inversely proportional to $\sqrt{K}$. The characteristic equation of the system is obtained by equating the denominator of Eq. (6-126) to zero,

$$s^2 + 48.5s + 400K = 0 \tag{6-129}$$

Time Response to a Unit Step Input

Although the main objective of the printwheel positional control system is to bring the printwheel to the desired equilibrium position once it has reached to within $\frac{1}{4}$ of a cycle of the encoder output, which corresponds to $\frac{1}{4}$ of the distance between the character positions on the printwheel, for time-domain analysis it is informative to analyze the system behavior by applying a unit step input with zero initial conditions. This way it would be possible to characterize the performance of the system in terms of the maximum overshoot and some of the other measures, such as rise time and delay time, if necessary. We shall show that in general, the damping ratio ζ and the natural undamped frequency ω_n are adequate to describe this second-order system. In fact, for a second-order system the maximum overshoot is directly related to ζ through Eq. (6-96).

Let the reference input be a unit step function, $\theta_r(t) = u_s(t)$ rad; then $\Theta_r(s) = 1/s$. The output of the system, with zero initial conditions, is

$$\theta_c(t) = \mathcal{L}^{-1}\left[\frac{400K}{s(s^2 + 48.5s + 400K)}\right] \tag{6-130}$$

The inverse Laplace transform on the right-hand side of Eq. (6-130) is carried out using the Laplace transform table in Appendix B. Thus,

$$\theta_c(t) = 1 + \frac{1}{\sqrt{1 - \zeta^2}} e^{-\zeta\omega_n t} \sin\left(\omega_n\sqrt{1 - \zeta^2}\, t - \phi\right) \tag{6-131}$$

where

$$\phi = \tan^{-1}\frac{\sqrt{1 - \zeta^2}}{-\zeta} \tag{6-132}$$

and ω_n and ζ are given in Eqs. (6-127) and (6-128), respectively.

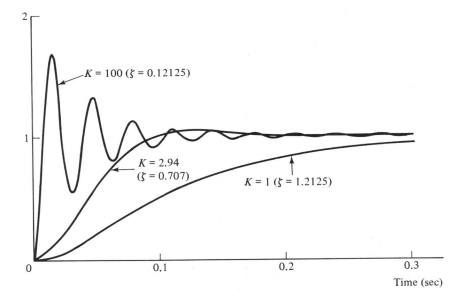

Figure 6-24 Unit step responses of the positional control system shown in Fig. 6-22.

Table 6-2   Comparison of the Performance of the Second-Order Positional Control System When the Gain K Varies

Gain K	ζ	ω_n	Maximum Overshoot	T_d	T_r	T_s	t_{max}
1	1.2125	20	0	0.1	0.215	0.115	—
2.94	0.707	34.3	0.043	0.044	0.065	0.086	0.13
100	0.12125	200	0.681	0.005	0.006	0.31	0.016

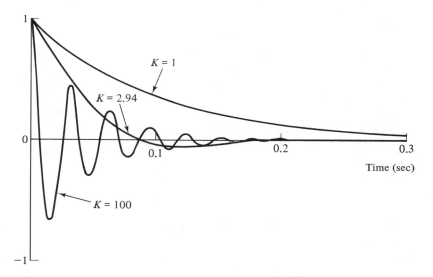

Figure 6-25   Free responses of the position-control system shown in Fig. 6-22 with zero input and nonzero initial conditions.

The output responses of the system are plotted in Fig. 6-24 for three typical values of K. Table 6-2 gives a comparison of the unit step responses for the three values of K used.

When $K = 100$, $\zeta = 0.12125$, the system is underdamped, and the overshoot is very high, although the rise time and delay time are very short. The settling time is 0.115 sec. When the value of K is set at a very low value, $K = 1$, the damping ratio is 1.2125 (i.e., the system is overdamped). The step response approaches the final value without any oscillations and overshoot. The rise time, delay time, and settling time are all very long. When K is set at the value of 2.94, the damping ratio is 0.707, and we see that the overshoot is only 4.3 percent. The rise time is moderate, but the settling time is 0.086 sec, which is actually less than that with $K = 100$.

Figure 6-25 illustrates the responses of $\theta_c(t)$ when $\theta_r(t) = 0$, $\theta_c(0) = 1$, and $\dot{\theta}_c(0) = -10$, with $K = 1$, 2.94, and 100. These responses are more realistic in the illustration of the time-domain performance of the position-control system. In this case, the initial time $t = 0$ represents the instant the velocity mode is switched off and the position mode is switched on, and the printwheel is driven to the final equilibrium point from the nonzero initial conditions. The important point to make here is that the unit step responses of Fig. 6-24 completely predicted the zero-input responses of Fig. 6-25, and thus the importance of the unit step responses of a linear system is demonstrated.

Since the problem illustrated here is not a design problem, only the amplifier gain K is variable. It seems that, from the time-response standpoint, the system performance with $\zeta = 0.707$ or $K = 2.94$ is the best possible when K is the only design parameter. If the rise time and settling time are to be reduced while maintaining a relatively small overshoot, the microprocessor would have to provide some dynamic operations of the error signal. In other words, a dynamic controller is needed in the control loop.

It is important to point out that in practice it would be time consuming and costly to evaluate the unit step response for each change in the system parameter for either analysis or design purposes. Indeed, one of the main objectives of studying control systems theory using either the conventional or the modern approach is to establish methods so that the total reliance on computer simulation can be avoided or reduced. The motivation behind this discussion is to show that the performance of a control system can be predicted by investigating the roots of the characteristic equation of the system. For the characteristic equation of Eq. (6-129), the roots are

$$s_1 = -24.25 + \sqrt{588.06 - 400K} \tag{6-133}$$

$$s_2 = -24.25 - \sqrt{588.06 - 400K} \tag{6-134}$$

For $K = 1$, 2.94, and 100, the roots of the characteristic equation are tabulated as follows:

K	s_1	s_2
1	-10.54	-37.96
2.94	$-24.25 + j24.25$	$-24.25 - j24.25$
100	$-24.25 + j198.5$	$-24.25 - j198.5$

These roots are plotted in the s-plane as shown in Fig. 6-26. The heavy lines shown in Fig. 6-24 represent the trajectories of the two characteristic equation roots when K varies between $-\infty$ and ∞. In general, these trajectories are called the *root loci* of Eq. (6-129), and are used extensively for the analysis and design of linear control systems.

From Eqs. (6-133) and (6-134) we can see that the two roots are real and negative for values of K between 0 and 1.47. This means that the system is overdamped and the step response will have no overshoot for this range of K. For values of K greater than 1.47, the roots are complex conjugate with the real parts of the roots equal to -24.25. This means that for K greater than 1.47, the system is underdamped, and the damping factor is always equal to 24.25 sec^{-1}, independent of K. The root loci also clearly show that as K increases beyond 1.47, the natural undamped frequency will increase with K, since for the second-order complex roots the distance from the origin to the roots is ω_n. When K is negative, one of the roots is positive, which corresponds to a time response that increases monotonically with

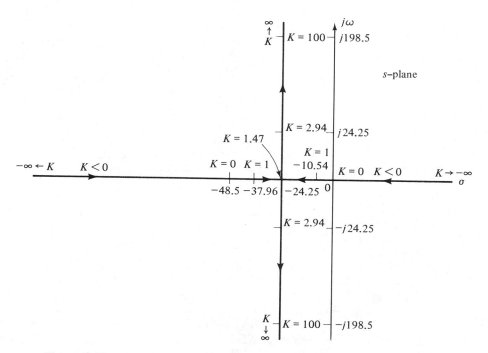

Figure 6-26 Root loci of the characteristic equation in Eq. (6-129) as K varies.

time, and the system is said to be unstable. The dynamic characteristics of the transient response as determined from the root loci of Fig. 6-26 are summarized as follows:

Amplifier Gain	Characteristic Equation Roots	Systems Dynamics
$0 < K < 1.47$	Two negative distinct real roots	Overdamped ($\zeta > 1$)
$K = 1.47$	Two negative equal real roots	Critically damped ($\zeta = 1$)
$1.47 < K < \infty$	Two complex-conjugate roots with negative real roots	Underdamped ($\zeta < 1$)
$-\infty < K < 0$	Two distinct real roots, one positive and one negative	Unstable system ($\zeta < 0$)

Since the position-control system is type 1, the steady-state error of the system is zero for all positive values of A, when the input is a step function. In other words, for a step input, the step error constant, K_p, is to be used.

From Eq. (6-122) the open-loop transfer function of the system is written

$$G(s) = \frac{400K}{s(s + 48.5)} \tag{6-135}$$

Substituting $G(s)$ into Eq. (6-21), with $H(s) = 1$, we have

$$K_p = \lim_{s \to 0} \frac{400K}{s(s + 48.5)} = \infty \qquad (6\text{-}136)$$

Therefore, the steady-state error of the position-control system due to a step input, as given by Eq. (6-22), is zero. The unit step responses of Fig. 6-24 verify this result.

The zero-steady-state condition is achieved because in the system model only the viscous friction is considered. In the practical case, Coulomb friction is almost always present, so that the steady-state positioning accuracy of the system can never be perfect.

Time Response to a Unit Ramp Input

When the printwheel control system is in the velocity-control mode, the objective is to control the acceleration and deceleration of the load so that when the print character reaches the vicinity of the desired position the velocity will not be excessive, or the printwheel position may overshoot the target. Once the total distance to be traveled is determined, there is a preferred velocity profile to be followed by the printwheel. Figure 6-27 illustrates several typical velocity profiles for various distances to be traversed. For short distances, the system may simply follow an acceleration and deceleration profile. For long distances, the velocity profile may consist of a constant-speed portion. The acceleration and deceleration characteristics of the system can be realized by applying ramp inputs to the system; that is, $\theta_r(t) = Rtu_s(t)$. The output response is described by

$$\theta_c(t) = \mathcal{L}^{-1} \left[\frac{400K}{s^2(s^2 + 48.5s + 400K)} \right] \qquad (6\text{-}137)$$

From the Laplace transform table, Eq. (6-137) is written

$$\theta_c(t) = t - \frac{2\zeta}{\omega_n} + \frac{1}{\omega_n\sqrt{1 - \zeta^2}} e^{-\zeta\omega_n t} \sin\left[\omega_n\sqrt{1 - \zeta^2}\, t - \phi \right] \qquad t \geq 0 \quad (6\text{-}138)$$

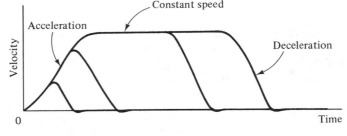

Figure 6-27 Typical velocity profiles.

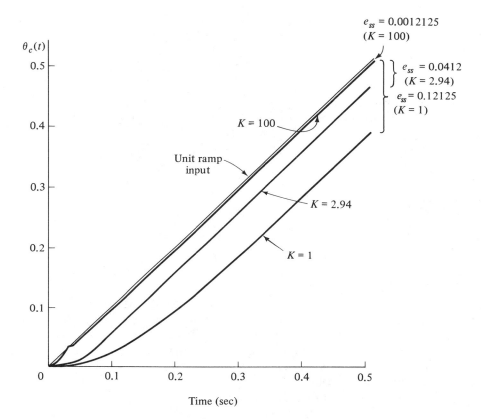

Figure 6-28 Unit ramp responses of the position-control system shown in Figure 6-22.

where

$$\phi = 2 \tan^{-1} \frac{\sqrt{1 - \zeta^2}}{-\zeta} \qquad (6\text{-}139)$$

$\zeta = 1.2125/\sqrt{K}$ and $\omega_n = 20\sqrt{K}$.

The ramp responses for $K = 1$, 2.94, and 100 are shown in Fig. 6-28. Notice that in this case the steady-state error of the system is not equal to zero. The last term in Eq. (6-138) represents the transient response, which is entirely characterized by the parameters ζ and ω_n. The steady-state response of the system due to a unit ramp input is

$$\lim_{t \to \infty} \theta_c(t) = \lim_{t \to \infty} \left(t - \frac{2\zeta}{\omega_n} \right) \qquad (6\text{-}140)$$

Thus, the steady-state response of the system due to a unit ramp input is

$$e_{ss} = \frac{2\zeta}{\omega_n} = \frac{0.12125}{K} \tag{6-141}$$

which is a constant.

A simpler method of determining the steady-state error to a ramp input is to use the ramp error constant K_v. From Eq. (6-26),

$$K_v = \lim_{s \to 0} sG(s) = \lim_{s \to 0} \frac{400K}{s + 48.5} = 8.247K \tag{6-142}$$

Therefore, from Eq. (6-27),

$$e_{ss} = \frac{1}{K_v} = \frac{0.12125}{K} \tag{6-143}$$

which agrees with the result of Eq. (6-141).

The result in Eq. (6-143) shows that the steady-state error is inversely proportional to the magnitude of K. For $K = 2.94$, which corresponds to a damping ratio of 0.707, the steady-state error is 0.04124 rad, or more appropriately, 4.124 percent of the ramp input magnitude. Apparently, if we attempt to improve the steady-state accuracy of the system by increasing the value of K, the transient step response will become more oscillatory. This phenomenon is rather typical in all control systems. For higher-order systems, if the loop gain of the system is too high, the system can become unstable.

Time Response of a Third-Order System

It was shown in the preceding sections that if the armature inductance L_a of the dc motor is neglected, the control system is of the second order and is stable for all positive values of K. Suppose now that L_a is 0.005 H in the system in Fig. 6-22, and the other system parameters are unchanged. The armature time constant τ_a is now 0.001 sec. The open-loop transfer function given in Eq. (6-122) is now

$$G(s) = \frac{\Theta_c(s)}{\Theta_e(s)} = \frac{400,000K}{s(s^2 + 1040s + 48,500)}$$

$$= \frac{400,000K}{s(s + 49)(s + 991)} \tag{6-144}$$

and the closed-loop transfer function becomes

$$\frac{\Theta_c(s)}{\Theta_r(s)} = \frac{400,000K}{s^3 + 1040s^2 + 48,500s + 400,000K} \tag{6-145}$$

The system is now of third order, and the characteristic equation is

$$s^3 + 1040s^2 + 48,500s + 400,000K = 0 \qquad (6\text{-}146)$$

For $K = 1$, the closed-loop transfer function of Eq. (6-145) becomes

$$\frac{\Theta_c(s)}{\Theta_r(s)} = \frac{400,000}{(s + 10.66)(s + 37.85)(s + 991.5)} \qquad (6\text{-}147)$$

The characteristic equation has three negative real roots at $s = -10.66$, -37.85, and -991.5. We see that the first two roots are very close to the roots of the characteristic equation of the second-order system when $L_a = 0$. The third root is relatively far to the left on the real axis in the s-plane.

For a unit step function input, the output response for $K = 1$ is obtained by use of Eq. (6-147).

$$\theta_c(t) = 1 - 1.407e^{-10.66t} + 0.408e^{-37.85t} - 0.00043e^{-991.5t} \qquad (6\text{-}148)$$

We see that the output response is dominated by the two roots at $s = -10.66$ and -37.85. The root at -991.5 gives rise to the last term in Eq. (6-148), which decays to zero very rapidly, and furthermore, the magnitude of the starting value at $t = 0$ is very small. Thus, we can conclude that, in general, the contribution of roots that lie relatively far to the left in the s-plane to the time response will be small. The roots that are closer to the imaginary axis, which dominate the transient response, are generally called the "dominant roots."

For $K = 2.94$, the closed-loop transfer function of Eq. (6-145) becomes

$$\frac{\Theta_c(s)}{\Theta_r(s)} = \frac{1,176,000}{(s + 992.32)(s^2 + 47.68s + 1185.24)} \qquad (6\text{-}149)$$

In this case the characteristic equation has two complex-conjugate roots at $s = -23.84 - j24.83$ and $-23.84 + j24.83$, and a real root at $s = -992.3$. The transient response is again controlled by the complex-conjugate roots, and the real root at -992.3 will have a negligible effect on the transient response, since 992.3 is much larger than the real parts of the complex roots, 23.84. However, we *cannot* simply throw out the term $(s + 992.32)$ in Eq. (6-149), since it still effects the steady-state response. The proper way to approximate the third-order system by a second-order system is to write Eq. (6-149) as

$$\frac{\Theta_c(s)}{\Theta_r(s)} = \frac{1185.2}{(s^2 + 47.68s + 1185.2)(1 + 0.001s)}$$

$$\cong \frac{1185.2}{s^2 + 47.68s + 1185.2} \qquad (6\text{-}150)$$

Then the unit step response is described by Eqs. (6-131) and (6-132), where $\omega_n = 34.43$ rad/sec and $\zeta = 0.6925$. It should be noted that these are not the true natural frequency and damping ratio of the system, since the system is of the third

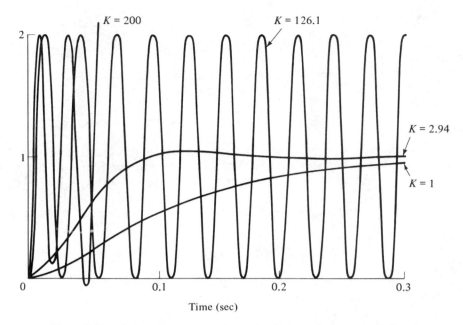

Figure 6-29 Unit step responses of the third-order position-control system.

order. Realistically, we refer to these as the *relative natural frequency* and the *relative damping ratio*, which are determined from the dominant roots.

When $K = 126.1$, the closed-loop transfer function of Eq. (6-145) becomes

$$\frac{\Theta_c(s)}{\Theta_r(s)} = \frac{50,440,000}{(s + 1040)(s^2 + 48,500)} \tag{6-151}$$

The roots of the characteristic equations are at $s = -1040$, $-j220.23$, and $j220.23$. The unit step response of the system is

$$\theta_c(t) = 1 - 0.043e^{-1040t} - 0.9783 \sin(220.23t + 78°) \tag{6-152}$$

Thus, the steady-state response is an undamped sinusoid with a frequency of 220.23 rad/sec. The system is said to be on the verge of instability. When K becomes greater than 126.1, the two complex-conjugate roots of the characteristic equation will have positive real parts, the sinusoidal component of the time response will increase with time, and the system becomes unstable. Thus, we see that the third-order system is capable of being unstable; that is, the output response to a bounded input will increase with time if the gain K exceeds 126.1. Earlier we found that the second-order system is always stable for all finite positive values of K. Figure 6-29 shows the step responses of the third-order system for $K = 1$, 2.94, 126.1, and 200. The responses for $K = 1$ and $K = 2.94$ are very close to those of the second-order system with the same values of K. This is because one of the roots of

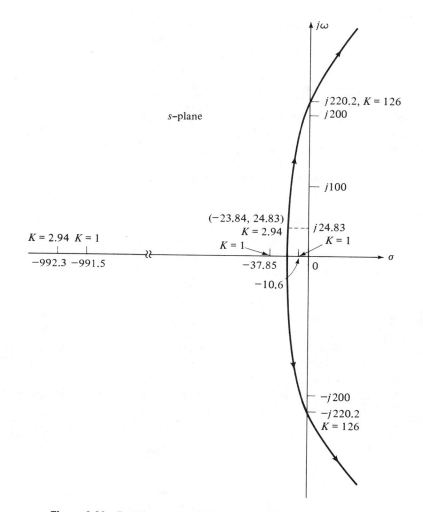

Figure 6-30 Root loci of the third-order position-control system.

the third-order system is far to the left in the s-plane, so that the transient responses can be approximated by the two dominant roots.

Figure 6-30 illustrates the root loci of the third-order system, or of Eq. (6-146), as K varies. The loci clearly show that when K is greater than 126.1, the two complex-conjugate roots are in the right half of the s-plane.

6.7 EFFECTS OF ADDING POLES AND ZEROS TO TRANSFER FUNCTIONS

The illustrative example given in the preceding section reveals some of the important properties of the time response of typical second- and third-order closed-loop systems. Specifically, the effects on the transient response relative to the location of the roots of the characteristic equation are demonstrated. However, in practice, successful design of a control system cannot depend only on choosing values of the system parameters so that the characteristic equation roots are properly placed. Addition of poles and zeros and/or cancellation of undesirable poles and zeros of the transfer function often are necessary to achieve the designed time-domain performance.

In this section we shall show that the addition of poles and zeros to open-loop and closed-loop transfer functions has varying effects on the transient response of the closed-loop system.

Addition of a Pole to the Open-Loop Transfer Function

For the printwheel control system discussed in Sec. 6.6, when the motor inductance is neglected, the system is of the second order, and the open-loop transfer function is given by Eq. (6-135). When the motor inductance is restored, the system is of the third order, and the open-loop transfer functions is given in Eq. (6-144). Comparing the two transfer functions in Eqs. (6-135) and (6-144), we see that the effect of considering the inductance is equivalent to adding a pole at $s = -991$ to the open-loop transfer function of Eq. (6-135), although the pole at -48.5 is shifted slightly to -49, and the forward-path gain is also increased. The apparent effect of adding a pole to the open-loop transfer function is that the third-order system can now become unstable if the value of the amplifier gain K exceeds 126.1. As shown by the root loci of Figs. 6-26 and 6-30, the open-loop pole at $s = -991$ essentially "pushes" and "bends" the root loci of the second-order system toward the right-half

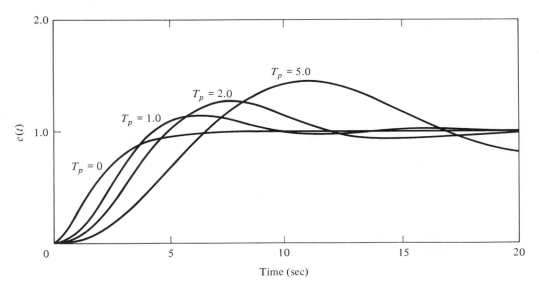

Figure 6-31   Unit step responses of the closed-loop system with

$$M(s) = \frac{1}{T_p s^3 + (1 + 2T_p)s^2 + 2s + 1}, \quad T_p = 0, 1, 2 \text{ and } 5.$$

s-plane. Actually, owing to the specific value of the inductance chosen, the additional pole of the third-order system is far to the left of the pole at $s = -49$, so that its effect is small except when the value K is relatively large.

To study the effect of the addition of a pole, and its relative location, to an open-loop transfer function, consider the transfer function

$$G(s) = \frac{\omega_n^2}{s(s + 2\zeta\omega_n)(1 + T_p s)} \tag{6-153}$$

The pole at $s = -1/T_p$ is considered to be added to an otherwise second-order transfer function. The transfer function of the closed-loop system with unity feedback is

$$M(s) = \frac{C(s)}{R(s)} = \frac{G(s)}{1 + G(s)} = \frac{\omega_n^2}{T_p s^3 + (1 + 2\zeta\omega_n T_p)s^2 + 2\zeta\omega_n s + \omega_n^2} \tag{6-154}$$

Figure 6-31 illustrates the unit step responses of the closed-loop system when $\omega_n = 1$, $\zeta = 1$, and $T_p = 0, 1, 2,$ and 5, respectively. These responses again show that the addition of a pole to the open-loop transfer function generally has the effect of increasing the peak overshoot of the closed-loop step response. As the value of T_p increases, the pole, $-1/T_p$, moves closer to the origin in the s-plane, and the peak overshoot increases. However, these responses also show that the pole increases the rise time of the step response. This is not surprising, since the additional pole has

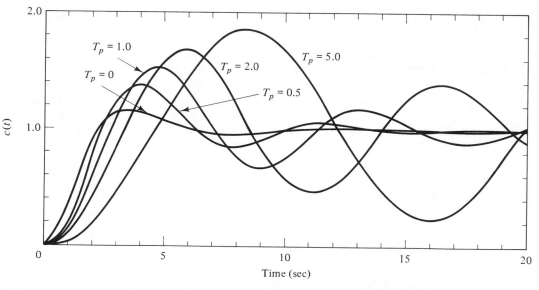

Figure 6-32 Unit step responses of the closed-loop system with
$$M(s) = \frac{1}{T_p s^3 + (1 + T_p)s^2 + s + 1}, \quad T_p = 0, 0.5, 1, 2, \text{ and } 5.$$

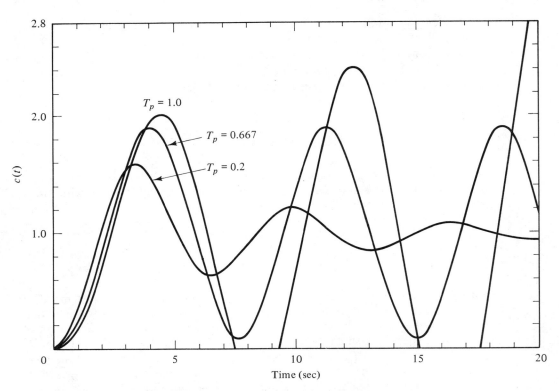

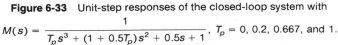

Figure 6-33   Unit-step responses of the closed-loop system with

$$M(s) = \frac{1}{T_p s^3 + (1 + 0.5T_p)s^2 + 0.5s + 1}, \; T_p = 0, 0.2, 0.667, \text{ and } 1.$$

the effect of reducing the bandwidth (see Chapter 9) of the system, thus cutting out the high-frequency components of the signal.

The same conclusions can be drawn from the unit step responses of Fig. 6-32, which are obtained with $\omega_n = 1$, $\zeta = 0.5$, and $T_p = 0, 0.5, 1, 2,$ and 5. It can be shown that for the $G(s)$ given in Eq. (6-153), with $\omega_n = 1$, $\zeta = 0.5$, and $\zeta = 1.0$, the closed-loop system is stable for all positive values of T_p.

Figure 6-33 illustrates the unit step responses of the closed-loop system when $\omega_n = 1$, $\zeta = 0.25$, and $T_p = 0, 0.2, 0.667,$ and 1.0, respectively. In this case, when T_p is greater than 0.667, the amplitude of the unit step response increases with time, and the system is said to be unstable.

Addition of a Pole to the Closed-Loop Transfer Function

Since the poles of the closed-loop transfer function are the roots of the characteristic equation, they control the transient response of the system directly. Consider the

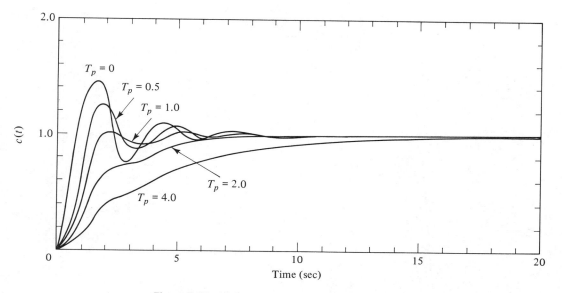

Figure 6-34 Unit step responses of closed-loop system with

$$M(s) = \frac{1}{(s^2 + s + 1)(1 + T_p s)}, \, T_p = 0, 0.5, 1, 2, \text{ and } 4.$$

closed-loop transfer function

$$M(s) = \frac{C(s)}{R(s)} = \frac{\omega_n^2}{\left(s^2 + 2\zeta\omega_n s + \omega_n^2\right)(1 + T_p s)} \tag{6-155}$$

where the term $(1 + T_p s)$ is added to an otherwise second-order system. Figure 6-34 illustrates the unit step responses of the system with $\omega_n = 1$, $\zeta = 0.5$, and $T_p = 0$, 0.5, 1, 2, and 4, respectively. As the pole at $s = -1/T_p$ is moved toward the origin in the s-plane, the rise time increases, and the maximum overshoot decreases. Thus, as far as the overshoot is concerned, adding a pole to the closed-loop transfer function has just the opposite effect to that of adding a pole to the open-loop transfer function.

Addition of a Zero to the Closed-Loop Transfer Function

Figure 6-35 shows the unit step responses of the closed-loop system with the transfer function

$$M(s) = \frac{C(s)}{R(s)} = \frac{\omega_n^2(1 + T_z s)}{\left(s^2 + 2\zeta\omega_n s + \omega_n^2\right)} \tag{6-156}$$

where $\omega_n = 1$, $\zeta = 0.5$, and $T_z = 0$, 1, 3, 6, and 10, respectively. In this case, we see

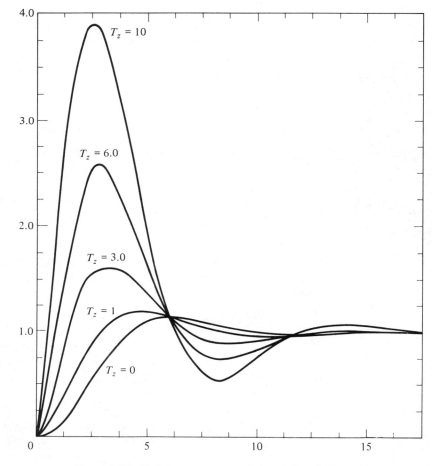

Figure 6-35 Unit step responses of closed-loop system with

$$M(s) = \frac{1 + T_z s}{s^2 + s + 1}, \; T_z = 0, 1, 3, 6, \text{ and } 10.$$

that adding a zero at $s = -1/T_z$ to the closed-loop transfer function decreases the rise time and increases the maximum overshoot of the system.

We can analyze the general case by writing Eq. (6-156) as

$$M(s) = \frac{C(s)}{R(s)} = \frac{\omega_n^2}{s^2 + 2\zeta\omega_n s + \omega_n^2} + \frac{T_z s \omega_n^2}{s^2 + 2\zeta\omega_n s + \omega_n^2} \qquad (6\text{-}157)$$

For a unit step input, let the output response that corresponds to the first term of the right-hand side of Eq. (6-157) be $c_1(t)$. Then, the total output is written

$$c(t) = c_1(t) + T_z \frac{dc_1(t)}{dt} \qquad (6\text{-}158)$$

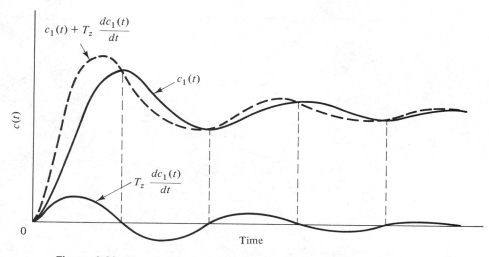

Figure 6-36 Time responses showing the effect of adding a zero to the closed-loop transfer function.

Figure 6-36 shows why the addition of the zero at $s = -1/T_z$ reduces the rise time and increases the maximum overshoot of the step response, according to Eq. (6-158). In fact, as T_z increases to infinity, the maximum overshoot also becomes infinitely large.

Addition of a Zero to the Open-Loop Transfer Function

Let us consider that a zero at $s = -1/T_z$ is added to an open-loop transfer function, so that

$$G(s) = \frac{6(1 + T_z s)}{s(s + 1)(s + 2)} \qquad (6\text{-}159)$$

The closed-loop system transfer function is

$$M(s) = \frac{C(s)}{R(s)} = \frac{6(1 + T_z s)}{s^3 + 3s^2 + (2 + 6T_z)s + 6} \qquad (6\text{-}160)$$

The difference between this case and that of adding a zero to the closed-loop transfer function is that in the present case, the term $(1 + T_z s)$ also appears in the numerator of the closed-loop transfer function, but the denominator of $M(s)$ also contains T_z. The term $(1 + T_z s)$ in the numerator of $M(s)$ increases the maximum overshoot, but T_z appears in the coefficient of the s term in the denominator, which has the effect of improving the damping, or reducing the maximum overshoot. Figure 6-37 illustrates the unit step responses when $T_z = 0$, 0.2, 0.5, 2, 5, and 10, respectively. Notice that when $T_z = 0$, the closed-loop system is on the verge of

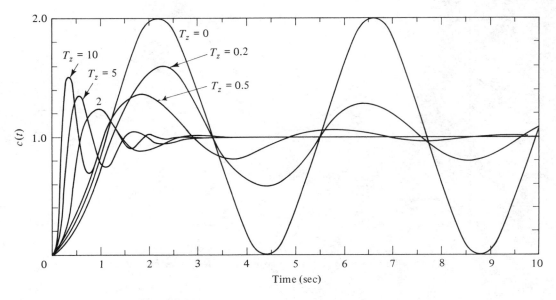

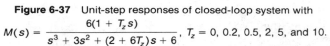

Figure 6-37   Unit-step responses of closed-loop system with

$$M(s) = \frac{6(1 + T_z s)}{s^3 + 3s^2 + (2 + 6T_z)s + 6}, \; T_z = 0, 0.2, 0.5, 2, 5, \text{ and } 10.$$

becoming unstable. When $T_z = 0.2$ and 0.5, the maximum overshoots are reduced, owing mainly to the improved damping. As T_z increases beyond 2, although the damping effect is still improved, the $(1 + T_z s)$ term in the numerator becomes more dominant, so that the maximum overshoot actually becomes greater as T_z is increased further.

An important finding from these discussions is that while the characteristic equation roots are generally used to study the relative damping and stability of linear control systems, the zeros of the transfer function should not be overlooked in their effects on the overshoot and rise time of the step response.

6.8 DOMINANT POLES OF TRANSFER FUNCTIONS

From the discussions and illustrative examples given in the preceding sections, it becomes apparent that the relative locations of the poles of a transfer function in the s-plane have a great deal to do with the transient response of the system. For analysis and design purposes, it is important to sort out the poles that have a dominant effect on the transient response and call these the "dominant poles."

Since most control systems found in practice are of the high order, it would be useful to establish guidelines on the approximation of high-order systems by lower-order systems insofar as the transient response is concerned. In design, we can use the dominant poles to control the dynamic performance of the system, whereas the "insignificant" poles are used for the purpose of ensuring that the controller can be physically realized.

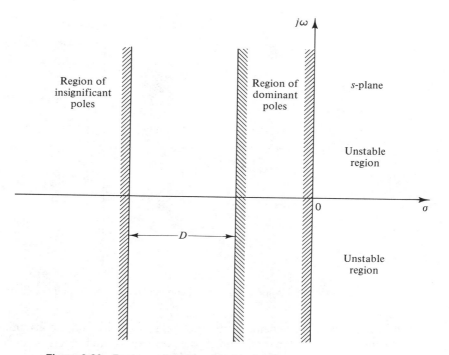

Figure 6-38 Regions of dominant and insignificant poles in the s-plane.

For all practical purposes we can qualitatively sectionalize the s-plane into regions in which the dominant and less dominant poles can lie, as shown in Fig. 6-38. We purposefully do not assign specific values to the coordinates, since these are all relative to a given system.

The poles that are close to the imaginary axis in the left-half s-plane give rise to transient responses that will decay relatively slowly, whereas the poles that are far away from the axis (relative to the dominant poles) correspond to fast decaying-time responses.

The distance D between the dominant region and the least-significant region shown in Fig. 6-38 will be subject to discussion. The question is: "How large a pole is considered to be really large?" It has been recognized in practice and in the literature that if the magnitude of the real part of a pole is at least 5 to 10 times that of a dominant pole or a pair of complex dominant poles, then the pole can be regarded as insignificant insofar as the transient response is concerned.

We shall show that the guideline in general depends on the specific transfer function and is not an exact science, unless one carries out the detailed steps to justify the approximation. On the other hand, in design, we cannot simply place the insignificant poles arbitrarily far to the left in the s-plane, or these will correspond to unrealistic physical parameters when the pencil-and-paper design is to be implemented by physical components.

We must also point out that the regions shown in Fig. 6-38 are there merely for the definitions of "dominant" and "insignificant" poles. For design purposes,

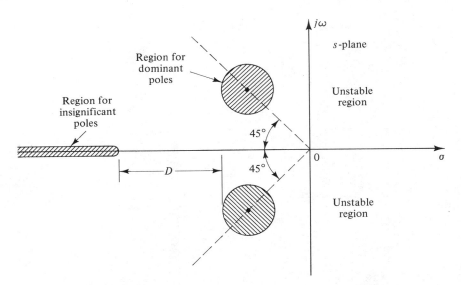

Figure 6-39   Regions of dominant and insignificant poles in the *s*-plane.

such as in the pole-placement design, the dominant poles and the insignificant poles, as selected by the designer, should most likely be located in the shaded regions in Fig. 6-39. Again, we do not show any absolute values of the coordinates, except that it is assumed that the desired region of the dominant poles is centered around the $\zeta = 0.707$ line.

We shall use the following examples to illustrate the effects of the insignificant poles.

Real Dominant Pole or Poles — Closed-Loop System

Consider the closed-loop transfer function

$$\frac{C(s)}{R(s)} = \frac{2a}{(s+2)(s+a)} \tag{6-161}$$

which has two real poles. Figure 6-40 shows the unit step responses of the system with $a = 8, 10, 20, 40,$ and ∞, respectively. When $a = \infty$, the pole at $-a$ is moved to $-\infty$, and the transfer function becomes

$$\frac{C(s)}{R(s)} = \frac{2}{s+2} \tag{6-162}$$

The problem is to investigate the accuracy of approximating Eq. (6-161) by Eq. (6-162) when the value of a becomes large relative to the pole at -2, which is considered to be dominant. The following tabulation shows the maximum error

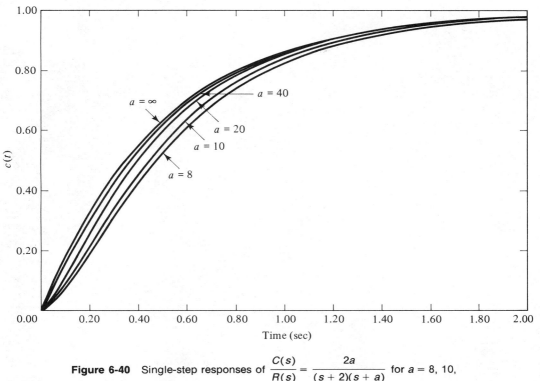

Figure 6-40 Single-step responses of $\dfrac{C(s)}{R(s)} = \dfrac{2a}{(s+2)(s+a)}$ for $a = 8$, 10, 20, 40 and ∞.

between the unit step responses of Eq. (6-161) and Eq. (6-162) for various values of a. Notice that when $a = 20$, which is 10 times the magnitude of the dominant pole at -2, the maximum error between the responses is 0.0762, or 7.62 percent of the final value of $c(t)$.

a	Maximum Error	Percent Maximum Error
8	0.1565	15.65
10	0.1337	13.37
20	0.0762	7.62
40	0.0421	4.21

Thus, when the dominant roots are real, the criterion on the insignificant poles is simply based on the dominant pole that is farthest to the left in the s-plane.

Complex-Conjugate Dominant Poles — Closed-Loop Systems

When the dominant poles are complex, the guidelines on whether any insignificant poles can be neglected strictly depends on both the real and the imaginary parts of the dominant poles.

Let us consider the following transfer function:

$$\frac{C(s)}{R(s)} = \frac{ac}{(s + a)(s^2 + bs + c)} \tag{6-163}$$

where a, b, and c are real positive constants. The values of b and c are such that the roots of the quadratic equation are complex conjugate. For a unit step input, the output transform $C(s)$ is written

$$C(s) = \frac{K_1}{s} + \frac{K_2}{s + a} + \frac{K_3 + K_4 s}{s^2 + bs + c} \tag{6-164}$$

where

$$K_1 = 1$$

$$K_2 = \frac{-c}{a^2 - ab + c}$$

$$K_3 = \frac{-a^2 + ab}{a^2 - ab + c} \tag{6-165}$$

$$K_4 = \frac{a(-ab + b^2 - c)}{a^2 - ab + c}$$

We see that the magnitude of K_2, which is the coefficient of the term $s + a$ in Eq. (6-164), depends on the values of a, b, and c. Thus, the significance of the pole at $-a$ depends not just on the real part of the complex dominant poles, but also on the imaginary parts.

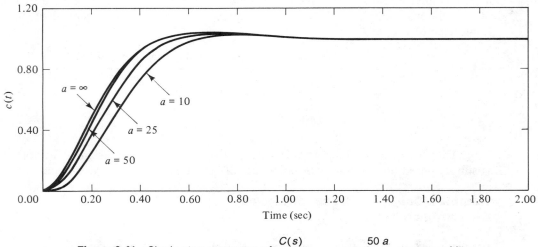

Figure 6-41 Single-step responses of $\dfrac{C(s)}{R(s)} = \dfrac{50\,a}{(s + a)(s^2 + 10s + 50)}$ with $a = 10, 25, 50,$ and ∞.

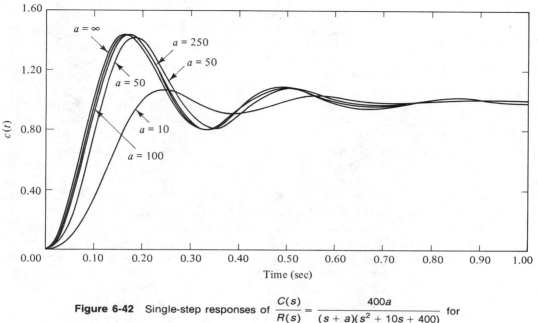

Figure 6-42 Single-step responses of $\dfrac{C(s)}{R(s)} = \dfrac{400a}{(s + a)(s^2 + 10s + 400)}$ for $a = 10, 50, 100, 250,$ and ∞.

Figure 6-41 shows the unit step responses of Eq. (6-163) with $b = 10$, $c = 50$, and $a = 10, 25, 50,$ and ∞. In this case, the two complex-conjugate poles have a real part of -5, but the relative damping ratio is 0.707, so that the maximum overshoot is less than 4.32 percent.

As shown in Fig. 6-41, when $a = 50$, which is 10 times the magnitude of the real parts of the complex dominant poles, and 5.77 times the distance from the origin to the poles, the second-order system with the transfer function

$$\frac{C(s)}{R(s)} = \frac{50}{s^2 + 10s + 50} \tag{6-166}$$

is a very good approximation of the third-order system. In fact, the maximum overshoot of the system, which is slightly underdamped, is not greatly affected by the value of a.

Figure 6-42 shows the unit-step responses of Eq. (6-163) with $b = 10$, $c = 400$, and $a = 10, 50, 100, 250,$ and ∞. In this case, the real part of the complex poles is -5 and the distance from the origin to the poles is 22.36. We see that when a is greater than or equal to 100, the second-order system

$$\frac{C(s)}{R(s)} = \frac{400}{s^2 + 10s + 400} \tag{6-167}$$

is a very good approximation of the third-order system. Thus, we see that when the dominant poles are complex, the proper criterion for the determination of the insignificant poles should be referenced to the distance from the origin to the complex poles, rather than just the real parts of the poles.

Open-Loop Systems

In the analysis and design of control systems we often work directly with the open-loop transfer function. Thus, the effects of neglecting the "insignificant" open-loop poles on the performance of closed-loop systems should be studied.

   Although the general meanings of the dominant poles are the same for open-loop and closed-loop transfer functions, we must keep in mind that generally, as we move an open-loop pole toward the imaginary axis in the s-plane, the rise time and the overshoot (if any) of the step response of the closed-loop system are increased. The effect of moving a pole of a closed-loop transfer function toward the imaginary axis generally slows down the step response and reduces the overshoot (Figs. 6-40 to 6-42). Furthermore, additional care must be taken in "throwing away" the insignificant poles of open-loop transfer functions because they have negligible effects on the transient response of the closed-loop system. For example, the open-loop transfer function in Eq. (6-144) has a pole at -991, which is 20 times the pole at -49. However, the justification on whether the pole at -991 can be neglected from the closed-loop system standpoint depends on the value of K. The root locus diagram in Fig. 6-26 shows that if we approximate the third-order transfer function Eq. (6-144) by

$$G(s) = \frac{403.63K}{s(s + 49)} \tag{6-168}$$

the second-order closed-loop system is always stable for all positive values of K. The root locus diagram in Fig. 6-30 clearly shows that for relatively small values of K, the second-order system is not a bad approximation to the third-order one, but as K increases, the two closed-loop complex roots approach the $j\omega$ axis, and the system becomes unstable for $K > 126$.

   The conclusions from the discussions given above are that the approximation of high-order systems by lower-order ones by casting away the insignificant poles is not an exact science, and that for complex dominant poles the distance from the origin to the poles should be considered rather than just the real part. When working with open-loop transfer functions, there is still the loop gain K that has to be dealt with.

The Proper Way of Neglecting the Insignificant Poles

Once we have found that it is justifiable to cast away a certain insignificant pole or poles of a transfer function, analytically, this must be handled properly. Consider the open-loop transfer function

$$G(s) = \frac{K}{s(s + a)(s^2 + bs + c)} \tag{6-169}$$

If the value of a is relatively large so that the pole at $s = -a$ can be neglected from the transient standpoint, we should first write Eq. (6-169) as

$$G(s) = \frac{K}{as(1 + s/a)(s^2 + bs + c)} \tag{6-170}$$

Then we can reason that $|s/a| \ll 1$ when the real and imaginary components of s are much smaller than a, owing to the dominant nature of the complex poles. Equation (6-170) can be approximated by

$$G(s) \cong \frac{K}{as(s^2 + bs + c)} \tag{6-171}$$

Another way of explaining the procedure is that the ramp error constant of Eq. (6-170) and Eq. (6-171) is

$$K_v = \lim_{s \to 0} sG(s) = \frac{K}{ac} \tag{6-172}$$

Thus, when we neglect the insignificant pole(s), the steady-state response of the closed-loop systems should not be affected.

6.9 STABILITY OF CONTROL SYSTEMS — INTRODUCTION

In preceding sections we have learned that the transient response of a linear time-invariant control system is governed by the roots of the characteristic equation. Basically, the design of linear feedback control systems may be regarded as a problem of arranging the location of the characteristic equation roots such that the corresponding system will perform according to the prescribed specifications.

Among the many forms of performance specifications used in the design of control systems, the most important requirement is that the system be stable at all times. Generally speaking, stability is used to distinguish two classes of systems: *useful* and *useless*; that is, from a practical standpoint, we consider that a stable system may be useful, whereas an unstable system is useless.

When all types of systems are considered—linear, nonlinear, time-invariant, time-varying—the definition of stability can be given in many different forms. In these sections we shall deal only with the stability of linear time-invariant systems.

For analysis and design purposes we may classify stability into *absolute stability* and *relative stability*. Absolute stability refers to the condition of stable or unstable; it is a *yes* or *no* condition. Once the system is found to be stable, it is of interest to determine how stable it is, and this degree of stability is a measure of relative stability. Parameters such as the overshoot and damping ratio used in relation to the transient response in preceding sections often indicate the relative stability of linear time-invariant systems in the time domain.

From the illustrative examples of preceding sections we may summarize the relation between the transient response and the characteristic equation roots as follows:

1. When all the roots of the characteristic equation are found in the left half of the s-plane, the system responses due to the initial conditions will decrease to zero as time approaches infinity.

2. If one or more pairs of simple roots are located on the imaginary axis of the s-plane, but there are no roots in the right half of the s-plane, the responses due to initial conditions will be undamped sinusoidal oscillations.

3. If one or more roots are found in the right half of the s-plane, the responses will increase in magnitude as time increases.

In linear system theory, the last two categories are usually defined as *unstable* conditions. Note that the responses referred to in the foregoing conditions are due to initial conditions only, and they are often called the *zero-input responses*.

6.10 STABILITY, CHARACTERISTIC EQUATION, AND THE STATE TRANSITION MATRIX

We can show from a more rigorous approach that the zero-input stability of a linear time-invariant system is determined by the location of the roots of the characteristic equation or the behavior of the state transition matrix $\phi(t)$.

Let a linear time-invariant system be described by the state equation

$$\dot{\mathbf{x}}(t) = \mathbf{A}\mathbf{x}(t) + \mathbf{B}\mathbf{u}(t) \qquad (6\text{-}173)$$

where $\mathbf{x}(t)$ is the state vector and $\mathbf{u}(t)$ the input vector. For zero input, $\mathbf{x}(t) = \mathbf{0}$ satisfies the homogeneous state equation $\dot{\mathbf{x}}(t) = \mathbf{A}\mathbf{x}(t)$ and is defined as the *equilibrium state* of the system. The zero-input stability is defined as follows: *If the zero-input response* $\mathbf{x}(t)$, *subject to finite initial state* $\mathbf{x}(t_0)$, *returns to the equilibrium state* $\mathbf{x}(t) = \mathbf{0}$ *as t approaches infinity, the system is said to be stable; otherwise, the system is unstable. This type of stability is also known as the asymptotic stability.*

In a more mathematical manner, the foregoing definition may be stated: *A linear time-invariant system is said to be stable* (*zero input*) *if for any finite initial state* $\mathbf{x}(t_0)$ *there is a positive number M* [*which depends on* $\mathbf{x}(t_0)$] *such that*

$$(1) \qquad\qquad \|\mathbf{x}(t)\| < M \qquad \text{For all } t \geq t_0 \qquad\qquad (6\text{-}174)$$

and

$$(2) \qquad\qquad \lim_{t \to \infty} \|\mathbf{x}(t)\| = 0 \qquad\qquad (6\text{-}175)$$

where $\|\mathbf{x}(t)\|$ *represents the norm*[1] *of the state vector* $\mathbf{x}(t)$, *or*

$$\|\mathbf{x}(t)\| = \left[\sum_{i=1}^{n} x_i^2(t) \right]^{1/2} \qquad\qquad (6\text{-}176)$$

[1] The norm of a function $\mathbf{x}(t)$, $\|\mathbf{x}(t)\|$, is defined as a real number which is the greatest lower bound of the set of numbers N such that $\|\mathbf{x}(t)\| \leq N$.

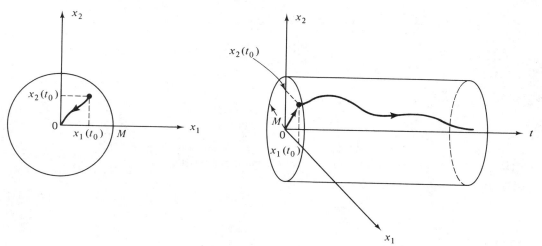

Figure 6-43 Stability concept illustrated in the state space.

The condition stated in Eq. (6-174) implies that the transition of state for any $t > t_0$ as represented by the norm of the vector $\mathbf{x}(t)$ must be bounded. Equation (6-175) states that the system must reach its equilibrium state as time approaches infinity.

The interpretation of the stability criterion in the state space is illustrated by the second-order case shown in Fig. 6-43. The state trajectory represents the transition of $\mathbf{x}(t)$ for $t > t_0$ from a finite initial state $\mathbf{x}(t_0)$. As shown in the figure, $\mathbf{x}(t_0)$ is represented by a point that is the tip of the vector obtained from the vector sum $x_1(t_0)$ and $x_2(t_0)$. A cylinder with radius M forms the upper bound for the trajectory points for all $t > t_0$, and as t approaches infinity, the system reaches its equilibrium state $\mathbf{x}(t) = \mathbf{0}$.

Next we shall show that the definition of stability of linear time-invariant systems given above leads to the same conclusion on the condition of the roots of the characteristic equation. For the zero-input condition, the state transition equation of the system is

$$\mathbf{x}(t) = \boldsymbol{\phi}(t - t_0)\mathbf{x}(t_0) \tag{6-177}$$

where $\boldsymbol{\phi}(t - t_0)$ is the state transition matrix.

Taking the norm on both sides of Eq. (6-177) gives

$$\|\mathbf{x}(t)\| = \|\boldsymbol{\phi}(t - t_0)\mathbf{x}(t_0)\| \tag{6-178}$$

An important property of the norm of a vector is

$$\|\mathbf{x}(t)\| \le \|\boldsymbol{\phi}(t - t_0)\|\|\mathbf{x}(t_0)\| \tag{6-179}$$

which is analogous to the relation between lengths of vectors.

Then the condition in Eq. (6-174) requires that $\|\boldsymbol{\phi}(t - t_0)\|\|\mathbf{x}(t_0)\|$ be finite. Thus, if $\|\mathbf{x}(t_0)\|$ is finite as postulated, $\|\boldsymbol{\phi}(t - t_0)\|$ must also be finite for $t > t_0$.

Similarly, Eq. (6-175) leads to the condition that

$$\lim_{t \to \infty} \| \boldsymbol{\phi}(t - t_0) \| = 0 \tag{6-180}$$

or

$$\lim_{t \to \infty} \boldsymbol{\phi}_{ij}(t - t_0) = 0 \tag{6-181}$$

$i, j = 1, 2, \ldots, n$, where $\boldsymbol{\phi}_{ij}(t - t_0)$ is the ijth element of $\boldsymbol{\phi}(t - t_0)$.

In Eq. (5-23) the state transition matrix is written

$$\boldsymbol{\phi}(t) = \mathcal{L}^{-1}\left[(s\mathbf{I} - \mathbf{A})^{-1}\right] \tag{6-182}$$

or

$$\boldsymbol{\phi}(t) = \mathcal{L}^{-1}\left[\frac{\text{adj}(s\mathbf{I} - \mathbf{A})}{|s\mathbf{I} - \mathbf{A}|}\right] \tag{6-183}$$

Since $|s\mathbf{I} - \mathbf{A}| = 0$ is the characteristic equation of the system, Eq. (6-171) implies that the time response of $\boldsymbol{\phi}(t)$ is governed by the roots of the characteristic equation. Thus the condition in Eq. (6-180) requires that the roots of the characteristic equation must all have negative real parts.

Stability of Linear Time-Invariant Systems with Inputs

Although the stability criterion for linear time-invariant systems given in the preceding section is for the zero-input condition, we can show that, in general, the stability condition for this class of systems is *independent* of the inputs. An alternative way of defining stability of linear time-invariant systems is as follows: *A system is stable if its output is bounded for any bounded input.*

In other words, let $c(t)$ be the output and $r(t)$ the input of a linear system with a single input–output. If

$$|r(t)| \le N < \infty \text{ for } t \ge t_0 \tag{6-184}$$

then

$$|c(t)| \le M < \infty \text{ for } t \ge t_0 \tag{6-185}$$

However, there are a few exceptions to the foregoing definition. A differentiator gives rise to an impulse response at $t = t_0$ when it is subjected to a unit step function input. In this case the amplitude of the input is bounded, but the amplitude of the output is not, since an impulse is known to have an infinite amplitude. Also, when a unit step function is applied to a perfect integrator, the output is an unbounded ramp function. However, since a differentiator and an integrator are all useful systems, they are defined as stable systems, and are exceptions to the stability condition defined above.

We shall show that the bounded input–bounded output definition of stability again leads to the requirement that the roots of the characteristic equation be located in the left half of the s-plane.

Let us express the input–output relation of a linear system by a convolution integral

$$c(t) = \int_0^\infty r(t - \tau)g(\tau)\,d\tau \qquad (6\text{-}186)$$

where $g(\tau)$ is the impulse response of the system.

Taking the absolute value on both sides of Eq. (6-186), we get

$$|c(t)| = \left| \int_0^\infty r(t - \tau)g(\tau)\,d\tau \right| \qquad (6\text{-}187)$$

Since the absolute value of an integral is not greater than the integral of the absolute value of the integrand, Eq. (6-187) is written

$$|c(t)| \le \int_0^\infty |r(t - \tau)|\,|g(\tau)|\,d\tau \qquad (6\text{-}188)$$

Now if $r(t)$ is a bounded signal, then, from Eq. (6-185),

$$|c(t)| \le \int_0^\infty N|g(\tau)|\,d\tau = N\int_0^\infty |g(\tau)|\,d\tau \qquad (6\text{-}189)$$

Therefore, if $c(t)$ is to be a bounded output,

$$N\int_0^\infty |g(\tau)|\,d\tau \le M < \infty \qquad (6\text{-}190)$$

or

$$\int_0^\infty |g(\tau)|\,d\tau \le P < \infty \qquad (6\text{-}191)$$

A physical interpretation of Eq. (6-191) is that the area under the absolute-value curve of the impulse response $g(t)$, evaluated from $t = 0$ to $t = \infty$, must be finite.

We shall now show that the requirement on the impulse response for stability can be linked to the restrictions on the characteristic equation roots. By definition, the transfer function $G(s)$ of the system and the impulse response $g(t)$ are related through the Laplace transform integral

$$G(s) = \int_0^\infty g(t)e^{-st}\,dt \qquad (6\text{-}192)$$

Taking the absolute value on the left side of Eq. (6-192) gives

$$|G(s)| \le \int_0^\infty |g(t)||e^{-st}| \, dt \qquad (6\text{-}193)$$

The roots of the characteristic equation are the poles of $G(s)$, and when s takes on these values, $|G(s)| = \infty$. Also, $s = \sigma + j\omega$; the absolute value of e^{-st} is $|e^{-\sigma t}|$. Equation (6-193) becomes

$$\infty \le \int_0^\infty |g(t)||e^{-\sigma t}| \, dt \qquad (6\text{-}194)$$

If one or more roots of the characteristic equation are in the right half or on the imaginary axis of the s-plane, $\sigma \ge 0$, and thus $|e^{-\sigma t}| \le N = 1$. Thus, Eq. (6-194) is written

$$\infty \le \int_0^\infty N|g(t)| \, dt = \int_0^\infty |g(t)| \, dt \qquad (6\text{-}195)$$

for $\text{Re}(s) = \sigma \ge 0$.

Since Eq. (6-195) contradicts the stability criterion given in Eq. (6-191), we conclude that *for the system to be stable, the roots of the characteristic equation must all lie inside the left half of the s-plane.*

The discussions conducted in the preceding sections lead to the conclusions that the stability of linear time-invariant systems can be determined by checking whether any roots of the characteristic equation are in the right half or on the imaginary axis of the s-plane. The regions of stability and instability in the s-plane are illustrated in Fig. 6-44. The imaginary axis, excluding the origin, is included in the unstable region.

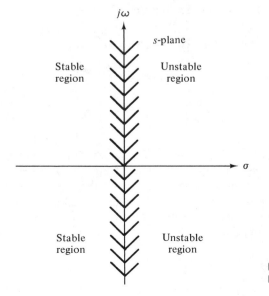

Figure 6-44 Stable and unstable regions in the s-plane.

6.11 METHODS OF DETERMINING STABILITY OF LINEAR CONTROL SYSTEMS

Although the stability of linear time-invariant systems may be checked by investigating the impulse response, the state transition matrix, or by finding the roots of the characteristic equation, these criteria are difficult to implement in practice. For instance, the impulse response is obtained by taking the inverse Laplace transform of the transfer function, which is not always a simple task; a similar process is required to evaluate the state transition matrix $\phi(t)$. The solving of the roots of a high-order polynomial can only be carried out by a digital computer. In practice, therefore, the stability analysis of a linear system is seldom carried out by working with the impulse response or the state transition matrix, or even by finding the exact location of the roots of the characteristic equation. In general, we are interested in algorithms that are straightforward to apply and which can provide answers to stability or instability, without excessive computations. The methods outlined below are frequently used for the stability studies of linear time-invariant systems.

1. *Routh–Hurwitz criterion*: an algebraic method that provides information on the absolute stability of a linear time-invariant system. The criterion tests whether any roots of the characteristic equation lie in the right half of the s-plane. The number of roots that lie on the imaginary axis and in the right half of the s-plane are also indicated.

2. *Nyquist criterion*: a semigraphical method that gives information on the difference between the number of poles and zeros of the closed-loop transfer function by observing the behavior of the Nyquist plot of the loop transfer function. The poles of the closed-loop transfer function are the roots of the characteristic equation. This method requires that we know the relative location of the zeros of the closed-loop transfer function.

3. *Root locus plot*: represents a diagram of loci of the characteristic equation roots when a certain system parameter varies. When the root loci lie in the right half of the s-plane, the closed-loop system is unstable.

4. *Bode diagram*: the Bode plot of the loop transfer function $G(s)H(s)$ may be used to determine the stability of the closed-loop system. However, the method can be used only if $G(s)H(s)$ has no poles and zeros in the right-half s-plane.

5. *Lyapunov's stability criterion*: a method of determining the stability of nonlinear systems, although it can also be applied to linear systems. The stability of the system is determined by checking on the properties of the *Lyapunov function* of the system.

6.12 ROUTH–HURWITZ CRITERION [4–8]

The Routh–Hurwitz criterion represents a method of determining the location of zeros of a polynomial with constant real coefficients with respect to the left half and the right half of the s-plane, without actually solving for the zeros.

It would seem that with all the root-solving programs available on digital computers, there is no need for the Routh-Hurwitz criterion. However, for systems with variable parameter(s), the criterion represents a convenient way of determining the range(s) of the parameter(s) for stability. For high-order and complex systems with variable parameter(s), it may still be less cumbersome to use the root locus method (Chapter 7) to determine the stability. The designer simply has to make judgment on the choice of the best tool in a given situation.

Consider that the characteristic equation of a linear time-invariant system is of the form

$$F(s) = a_0 s^n + a_1 s^{n-1} + a_2 s^{n-2} + \cdots + a_{n-1} s + a_n = 0 \qquad (6\text{-}196)$$

where all the coefficients are real numbers.

In order that there be no roots of the last equation with positive real parts, it is necessary but not sufficient that[2]

1. All the coefficients of the polynomial have the same sign.
2. None of the coefficients vanishes.

These two necessary conditions can easily be checked by inspection. However, these conditions are not sufficient; it is quite possible that a polynomial with all its coefficients nonzero and of the same sign still have zeros in the right half of the s-plane.

The necessary and sufficient condition that all roots of Eq. (6-196) lie in the left half of the s-plane is that the polynomial's Hurwitz determinants, D_k, $k = 1, 2, \ldots, n$, must all be positive.

---

[2] From the basic laws of algebra, the following relations are observed for the polynomial in Eq. (6-196)

$$\frac{a_1}{a_0} = -\Sigma \text{ all roots}$$

$$\frac{a_2}{a_0} = \Sigma \text{ products of the roots taken 2 at a time}$$

$$\frac{a_3}{a_0} = -\Sigma \text{ products of the roots taken 3 at a time}$$

$$\vdots$$

$$\frac{a_n}{a_0} = (-1)^n \text{ products of all roots}$$

All the ratios must be positive and nonzero unless at least one of the roots has a positive real part.

The Hurwitz determinants of Eq. (6-196) are given by

$$D_1 = a_1 \quad D_2 = \begin{vmatrix} a_1 & a_3 \\ a_0 & a_2 \end{vmatrix} \quad D_3 = \begin{vmatrix} a_1 & a_3 & a_5 \\ a_0 & a_2 & a_4 \\ 0 & a_1 & a_3 \end{vmatrix}$$

$$\cdots D_n = \begin{vmatrix} a_1 & a_3 & a_5 & \cdots & a_{2n-1} \\ a_0 & a_2 & a_4 & \cdots & a_{2n-2} \\ 0 & a_1 & a_3 & \cdots & a_{2n-3} \\ 0 & a_0 & a_2 & \cdots & a_{2n-4} \\ 0 & 0 & a_1 & \cdots & a_{2n-5} \\ \cdots & \cdots & \cdots & \cdots & \cdots \\ 0 & 0 & 0 & \cdots & a_n \end{vmatrix} \tag{6-197}$$

where the coefficients with indices larger than n or with negative indices are replaced by zeros.

At first glance the application of the Hurwitz determinants may seem to be formidable for high-order polynomials, because of the labor involved in evaluating the determinants in Eq. (6-197). Fortunately, the rule was simplified by Routh into a tabulation, so one does not have to work with the determinants of Eq. (6-197).

The first step in the simplification of the Routh-Hurwitz criterion is to arrange the polynomial coefficients into two rows. The first row consists of the first, third, fifth, ... coefficients, and the second row consists of the second, the fourth, sixth, ... coefficients, as shown in the following tabulation:

$$\begin{matrix} a_0 & a_2 & a_4 & a_6 & a_8 & \cdots \\ a_1 & a_3 & a_5 & a_7 & a_9 & \cdots \end{matrix}$$

The next step is to form the following array of numbers by the indicated operations (the example shown is for a sixth-order system):

s^6	a_0	a_2	a_4	a_6
s^5	a_1	a_3	a_5	0
s^4	$\dfrac{a_1 a_2 - a_0 a_3}{a_1} = A$	$\dfrac{a_1 a_4 - a_0 a_5}{a_1} = B$	$\dfrac{a_1 a_6 - a_0 \times 0}{a_1} = a_6$	0
s^3	$\dfrac{A a_3 - a_1 B}{A} = C$	$\dfrac{A a_5 - a_1 a_6}{A} = D$	$\dfrac{A \times 0 - a_1 \times 0}{A} = 0$	0
s^2	$\dfrac{BC - AD}{C} = E$	$\dfrac{C a_6 - A \times 0}{C} = a_6$	$\dfrac{C \times 0 - A \times 0}{C} = 0$	0
s^1	$\dfrac{ED - C a_6}{E} = F$	0	0	0
s^0	$\dfrac{F a_6 - E \times 0}{F} = a_6$	0	0	0

The array of numbers and operations given above is known as the *Routh tabulation* or the *Routh array*. The column of ss on the left side is used for identification purpose.

Once the Routh tabulation has been completed, the last step in the Routh–Hurwitz criterion is to investigate the signs of the numbers in the first column of the tabulation. The following conclusions are drawn: *The roots of the polynomials are all in the left half of the s-plane if all the elements of the first column of the Routh tabulation are of the same sign. If there are changes of signs in the elements of the first column, the number of sign changes indicates the number of roots with positive real parts.*

The reason for the foregoing conclusion is simple, based on the requirements of the Hurwitz determinants. The relations between the elements in the first column of the Routh tabulation and the Hurwitz determinants are:

$$s^6 \qquad a_0 = a_0$$

$$s^5 \qquad a_1 = D_1$$

$$s^4 \qquad A = \frac{D_2}{D_1}$$

$$s^3 \qquad C = \frac{D_3}{D_2}$$

$$s^2 \qquad E = \frac{D_4}{D_3}$$

$$s^1 \qquad F = \frac{D_5}{D_4}$$

$$s^0 \qquad a_6 = \frac{D_6}{D_5}$$

Therefore, if all the Hurwitz determinants are positive, the elements in the first column would also be of the same sign.

The following two examples illustrate the application of the Routh–Hurwitz criterion to simple problems.

■

Example 6-3

Consider the equation

$$(s - 2)(s + 1)(s - 3) = s^3 - 4s^2 + s + 6 = 0 \qquad (6\text{-}198)$$

which has two negative coefficients. Thus, from the necessary condition, we know without applying the Routh–Hurwitz test that the equation has at least one root with positive real parts. But for the purpose of illustrating the Routh–Hurwitz criterion, the Routh tabulation is formed as follows:

$$s^3 \qquad\qquad 1 \qquad\qquad 1$$

Sign change

$$s^2 \qquad\qquad -4 \qquad\qquad 6$$

Sign change

$$s^1 \qquad \frac{(-4)(1) - (6)(1)}{-4} = 2.5 \qquad 0$$

$$s^0 \qquad \frac{(2.5)(6) - (-4)(0)}{2.5} = 6 \qquad 0$$

Since there are two sign changes in the first column of the tabulation, the polynomial has two roots located in the right half of the s-plane. This agrees with the known result, as Eq. (6-198) clearly shows that the two right-half plane roots are at $s = 2$ and $s = 3$.

Example 6-4

Consider the equation

$$2s^4 + s^3 + 3s^2 + 5s + 10 = 0 \qquad\qquad (6\text{-}199)$$

Since the equation has no missing terms and the coefficients are all of the same sign, it satisfies the necessary condition for not having roots in the right half of the s-plane or on the imaginary axis. However, the sufficient condition must still be checked. The Routh tabulation is made as follows:

$$s^4 \qquad\qquad 2 \qquad\qquad 3 \qquad 10$$

$$s^3 \qquad\qquad 1 \qquad\qquad 5 \qquad 0$$

Sign change

$$s^2 \qquad \frac{(1)(3) - (2)(5)}{1} = -7 \qquad 10 \qquad 0$$

Sign change

$$s^1 \qquad \frac{(-7)(5) - (1)(10)}{-7} = 6.43 \qquad 0 \qquad 0$$

$$s^0 \qquad\qquad 10$$

Since there are two changes in sign in the first column, the equation has two roots in the right half of the s-plane.

∎

Special Cases

The two illustrative examples given above are designed so that the Routh–Hurwitz criterion can be carried out without any complications. However, depending upon the equation to be tested, the following difficulties may occur occasionally when carrying out the Routh test:

1. The first element in any one row of the Routh tabulation is zero, but the other elements are not.
2. The elements in one row of the Routh tabulation are all zero.

In the first case, if a zero appears in the first position of a row, the elements in the next row will all become infinite, and the Routh test breaks down. This situation may be corrected by multiplying the equation by the factor $(s + a)$, where a is any number,[3] and then carry on the usual tabulation.

■

Example 6-5

Consider the equation

$$(s - 1)^2(s + 2) = s^3 - 3s + 2 = 0 \qquad (6\text{-}200)$$

Since the coefficient of the s^2 term is zero, we know from the necessary condition that at least one root of the equation is located in the right half of the s-plane. To determine how many of the roots are in the right-half plane, we carry out the Routh tabulation as follows:

s^3	1	-3
s^2	0	2
s^1	∞	

Because of the zero in the first element of the second row, the first element of the third row is infinite. To correct this situation, we multiply both sides of Eq. (6-200) by the factor $(s + a)$, where a is an arbitrary number. The simplest number that enters one's mind is 1. However, for reasons that will become apparent later, we do not choose a to be 1 or 2. Let $a = 3$; then Eq. (6-200) becomes

$$(s - 1)^2(s + 2)(s + 3) = s^4 + 3s^3 - 3s^2 - 7s + 6 = 0 \qquad (6\text{-}201)$$

The Routh tabulation of Eq. (6-201) is

s^4	1	-3	6
s^3	3	-7	0

Sign change

s^2	$\dfrac{-9 + 7}{3} = -\dfrac{2}{3}$	6	0

Sign change

s^1	$\dfrac{\left(-\frac{2}{3}\right)(-7) - 18}{-\frac{2}{3}} = 20$	0
s^0	6	

[3] If one chooses to use a negative number for a, the $(s + a)$ term will contribute a root in the right half of the s-plane, and this root must be taken into account when interpreting the Routh tabulation.

Since there are two changes in sign in the first column of the Routh tabulation, the equation has two roots in the right half of the *s*-plane.

■

As an alternative to the remedy of the situation described above, we may replace the zero element in the Routh tabulation by an arbitrary small positive number ϵ and then proceed with the Routh test.[4] For instance, for the equation given in Eq. (6-200), we may replace the zero element in the second row of the Routh tabulation by ϵ; then we have

$$
\begin{array}{ccc}
s^3 & 1 & -3 \\
s^2 & \epsilon & 2 \\
\end{array}
$$

Sign change

$$
\begin{array}{ccc}
s^1 & \dfrac{-3\epsilon - 2}{\epsilon} & 0 \\
\end{array}
$$

Sign change

$$
\begin{array}{cc}
s^0 & 2 \\
\end{array}
$$

Since ϵ is postulated to be a small positive number, $(-3\epsilon - 2)/\epsilon$ approaches $-2/\epsilon$, which is a negative number; thus the first column of the last tabulation has two sign changes. This agrees with the result obtained earlier. On the other hand, we may assume ϵ to be negative, and we can easily verify that the number of sign changes is still two, but they are between the first three rows.

In the second special case, when all the elements in one row of the Routh tabulation are zeros, it indicates that one or more of the following conditions may exist:

1. Pairs of real roots with opposite signs.
2. Pairs of imaginary roots.
3. Pairs of complex-conjugate roots forming symmetry about the origin of the *s*-plane.

The equation that is formed by using the coefficients of the row just above the row of zeros is called the auxiliary equation. The order of the auxiliary equation is always even, and it indicates the number of root pairs that are equal in magnitude but opposite in sign. For instance, if the auxiliary equation is of the second order, there are two equal and opposite roots. For a fourth-order auxiliary equation, there must be two pairs of equal and opposite roots. *All these roots of equal magnitude can be obtained by solving the auxiliary equation.*

When a row of zeros appears in the Routh tabulation, again the test breaks down. The test may be carried on by performing the following remedies:

1. Take the derivative of the auxiliary equation with respect to *s*.

---

[4] The ϵ-method may not give correct results if the polynomial has pure imaginary roots [9, 10].

2. Replace the row of zeros with the coefficients of the resultant equation obtained by taking the derivative of the auxiliary equation.

3. Carry on the Routh test in the usual manner with the newly formed tabulation.

∎

Example 6-6

Consider the same equation, Eq. (6-200), which is used in Example 6-5. In multiplying this equation by a factor $(s + a)$, logically the first number that comes into one's mind would be $a = 1$. Multiplying both sides of Eq. (6-200) by $(s + 1)$, we have

$$(s - 1)^2(s + 2)(s + 1) = s^4 + s^3 - 3s^2 - s + 2 = 0 \qquad (6\text{-}202)$$

The Routh tabulation is made as follows:

s^4	1	-3	2
s^3	1	-1	0
s^2	$\dfrac{(1)(-3) - (1)(-1)}{1} = -2$	2	
s^1	$\dfrac{2 - 2}{-2} = 0$	0	

Since the s^1 row contains all zeros, the Routh test terminates prematurely. The difficulty in this case is due to the multiplication of the original equation, which already has a root at $s = 1$, by the factor $(s + 1)$. This makes the new equation fit special case (2). To remedy this situation, we form the auxiliary equation using the coefficients contained in the s^2 row, that is, the row preceding the row of zeros. Thus the auxiliary equation is written

$$A(s) = -2s^2 + 2 = 0 \qquad (6\text{-}203)$$

Taking the derivative of $A(s)$ with respect to s gives

$$\frac{dA(s)}{ds} = -4s \qquad (6\text{-}204)$$

Now, the row of zeros in the Routh tabulation is replaced by the coefficients of Eq. (6-204), and the new tabulation reads as follows:

	s^4	1	-3	2
	s^3	1	-1	
Sign change				
	s^2	-2	2	(coefficients of auxiliary equations)
	s^1	-4	0	[coefficients of $dA(s)/ds$]
Sign change				
	s^0	2	0	

Since the preceding tabulation has two sign changes, the equation of Eq. (6-200) has two roots in the right-half plane. By solving the roots of the auxiliary equation in Eq. (6-204), we have

$$s^2 = 1 \quad \text{or} \quad s = \pm 1$$

These are also the roots of the equation in Eq. (6-203). It should be remembered that the roots of the auxiliary equation are also roots of the original equation, which is under the Routh test.

Example 6-7

In this example we shall consider equations with imaginary roots. Consider

$$(s + 2)(s - 2)(s + j)(s - j)(s^2 + s + 1) = s^6 + s^5 - 2s^4 - 3s^3 - 7s^2 - 4s - 4 = 0$$

$$(6\text{-}205)$$

which is known to have two pairs of equal roots with opposite signs at $s = \pm 2$ and $s = \pm j$. The Routh tabulation is

s^6	1	-2	-7	-4
s^5	1	-3	-4	
s^4	1	-3	-4	
s^3	0	0	0	

Since a row of zeros appears prematurely, we form the auxiliary equation using the coefficients of the s^4 row. The auxiliary equation is

$$A(s) = s^4 - 3s^2 - 4 = 0 \qquad (6\text{-}206)$$

The derivative of $A(s)$ with respect to s is

$$\frac{dA(s)}{ds} = 4s^3 - 6s = 0$$

from which the coefficients 4 and -6 are used to replace the row of zeros in the Routh tabulation. The new Routh tabulation is

s^6	1	-2	-7	-4
s^5	1	-3	-4	
s^4	1	-3	-4	
s^3	4	-6	0	$\left[\text{coefficients of } \dfrac{dA(s)}{ds} \right]$

Sign change

s^2	-1.5	-4	0
s^1	-16.7	0	
s^0	-4	0	

Since there is one change in sign in the first column of the new Routh tabulation, Eq. (6-205) has one root in the right half of the s-plane. The equal, but opposite roots that caused the all-zero row to occur are solved from the auxiliary equation. From Eq. (6-206) these roots are found to be

$$s = +2, -2, +j, \text{ and } -j.$$

∎

A frequent use of the Routh–Hurwitz criterion is for a quick check of the stability and the simple design of a linear feedback control system. For example, the third-order printwheel control system treated in Section 6.6 has the characteristic equation in Eq. (6-146),

$$s^3 + 1040s^2 + 48,500s + 400,000K = 0 \qquad (6\text{-}207)$$

The root loci of this equation are sketched as shown in Fig. 6-30. The Routh–Hurwitz criterion may be used to determine the critical value of K for stability. The Routh tabulation of Eq. (6-195) is

s^3	1	48,500
s^2	1040	$400,000K$
s^1	$\dfrac{50,440,000 - 400,000K}{1040}$	0
s^0	$400,000K$	

For the system to be stable, or, for all the roots of Eq. (6-207) to be in the left-half s-plane, all the coefficients in the first column of the Routh tabulation must have the same sign. This leads to the following conditions:

$$\frac{50,440,000 - 400,000K}{1040} > 0 \qquad (6\text{-}208)$$

$$400,000K > 0 \qquad (6\text{-}209)$$

From the inequality of Eq. (6-208), we have

$$K < 126.1$$

and the condition in Eq. (6-209) gives

$$K > 0$$

Therefore, the condition of asymptotic stability of the overall system is

$$0 < K < 126.1 \qquad (6\text{-}210)$$

If we let $K = 126.1$, this corresponds to Eq. (6-207) having two roots on the $j\omega$ axis. To find these roots, we substitute $K = 126.1$ in the auxiliary equation, which is obtained from the Routh tabulation by using the coefficients of the s^2 row. Thus,

$$A(s) = 1040s^2 + 50,440,000 = 0 \qquad (6\text{-}211)$$

which has roots at $s = \pm j220.23$. Notice that in Fig. 6-30, these are the points where the root loci cross the imaginary axis in the s-plane, and the corresponding value of K at these points is also the critical value for stability.

∎

Example 6-8

Consider the characteristics equation of a certain closed-loop control system,

$$s^3 + 3Ks^2 + (K + 2)s + 4 = 0 \qquad (6\text{-}212)$$

It is desired to determine the range of K so that the system is stable. The Routh tabulation of Eq. (6-212) is

s^3	1	$(K + 2)$
s^2	$3K$	4
s^1	$\dfrac{3K(K + 2) - 4}{3K}$	0
s^0	4	

From the s^2 row, the condition of stability is

$$K > 0$$

and from the s^1 row, the condition of stability is

$$3K^2 + 6K - 4 > 0 \qquad \text{or} \qquad K < -2.528 \qquad \text{or} \qquad K > 0.528$$

When the conditions of $K > 0$ and $K > 0.528$ are compared, it is apparent that the latter limitation is the more stringent one. Thus, for the closed-loop system to be stable, K must satisfy

$$K > 0.528$$

The requirement of $K < -2.528$ is disregarded since K cannot be negative.

∎

It should be reiterated that the Routh–Hurwitz criterion is valid only if the characteristic equation is algebraic and all the coefficients are real. If any one of the coefficients of the characteristic equation is a complex number, or if the equation contains exponential functions of s, such as in the case of a system with time delays, the Routh–Hurwitz criterion cannot be applied.

Another limitation of the Routh–Hurwitz criterion is that it is generally used to determine only the absolute stability of the system, and no information on relative stability can be obtained. If a system is found to be stable by the Routh test, one still does not know how good the system is—in other words, how closely the roots of the characteristic equation roots are located to the imaginary axis of the s-plane. However, it is possible to apply a transformation to the characteristic equation so that the Routh–Hurwitz criterion may be applied to test certain types of relative stability of a control system (see Problem 6.20).

REFERENCES

Time-Domain Analysis

1. O. L. R. JACOBS, "The Damping Ratio of an Optimal Control System," *IEEE Trans. Automatic Control*, Vol. AC-10, pp. 473–476, Oct. 1965.

2. G. A. JONES, "On the Step Response of a Class of Third-Order Linear Systems," *IEEE Trans. Automatic Control*, Vol. AC-12, p. 341, June 1967.

3. R. A. MONZINGO, "On Approximating the Step Response of a Third-Order Linear System by a Second-Order Linear System," *IEEE Trans. Automatic Control*, Vol. AC-13, p. 739, Dec. 1968.

Routh – Hurwitz Criterion

4. E. J. ROUTH, *Dynamics of a System of Rigid Bodies*, Chap. 6, Part II, Macmillan & Co. Ltd., London, 1905.

5. N. N. PURI and C. N. WEYGANDT, "Second Method of Liapunov and Routh's Canonical Form," *J. Franklin Inst.*, Vol. 76, pp. 365–384, Nov. 1963.

6. G. V. S. S. RAJU, "The Routh Canonical Form," *IEEE Trans. Automatic Control*, Vol. AC-12, pp. 463–464, Aug. 1967.

7. V. KRISHNAMURTHI, "Correlation between Routh's Stability Criterion and Relative Stability of Linear Systems," *IEEE Trans. Automatic Control*, Vol. AC-17, pp. 144–145, Feb. 1972.

8. V. KRISHNAMURTHI, "Gain Margin of Conditionally Stable Systems from Routh's Stability Criterion," *IEEE Trans. Automatic Control*, Vol. AC-17, pp. 551–552, Aug. 1972.

9. K. J. KHATWANI, "On Routh-Hurwitz Criterion," *IEEE Trans. Automatic Control*, Vol. AC-26, p. 583, April 1981.

10. S. K. PILLAI, "The ϵ Method of the Routh-Hurwitz Criterion," *IEEE Trans. Automatic Control*, Vol. AC-26, p. 584, April 1981.

PROBLEMS

6.1. A pair of complex-conjugate poles in the s-plane is required to meet the various specifications below. For each specification, sketch the region in the s-plane in which the poles may be located.
   (a) $\zeta \geq 0.5$, $\omega_n \geq 5$ rad/sec (positive damping)
   (b) $0 \leq \zeta \leq 0.707$, $\omega \leq 5$ rad/sec (positive damping)

(c) $\zeta \le 0.5, 1 \le \omega_n \le 4$ rad/sec (positive damping)

(d) $0.5 \le \zeta \le 0.707, \omega_n \le 10$ rad/sec (positive and negative damping)

6.2. Determine the step, ramp, and parabolic error constants for the following control systems with unity feedback. The open-loop transfer functions are given as

(a) $G(s) = \dfrac{500}{(1 + 0.1s)(1 + 5s)}$

(b) $G(s) = \dfrac{K}{s(1 + 0.1s)(1 + 0.5s)}$

(c) $G(s) = \dfrac{10K}{s(s^2 + 4s + 200)}$

(d) $G(s) = \dfrac{K(1 + 2s)(1 + 4s)}{s^2(s^2 + 2s + 10)}$

6.3. For the systems in Problem 6.2, determine the steady-state error for a unit step input, a unit ramp input, and an acceleration input $t^2/2$.

6.4. The open-loop transfer function of a control system with unity feedback is

$$G(s) = \frac{1000}{s(1 + 0.1s)}$$

Evaluate the error series for the system. Determine the steady-state error of the system when the following inputs are applied:

(a) $r(t) = t^2 u_s(t)/2$

(b) $r(t) = (1 + 2t + t^2)u_s(t)$

Show that the steady-state error obtained from the error series is equal to the inverse Laplace transform of $E(s)$ with the terms generated by the poles of $E(s)/R(s)$ discarded.

6.5. In Problem 6.4, if a sinusoidal input $r(t) = \sin \omega t$ is applied to the system at $t = 0$, determine the steady-state error of the system by use of the error series for $\omega = 10$ rad/sec. What are the limitations in the error series when $r(t)$ is sinusoidal?

6.6. A machine-tool contouring control system is to cut the piecewise-linear contour shown in Fig. 6P-6 with a two-axis control system. Sketch the reference input of each axis of the two-axis system as a function of time.

6.7. A machine-tool contouring control system is to cut a perfect circular arc with a two-axis control system. Determine the reference inputs of the two systems that will accomplish this.

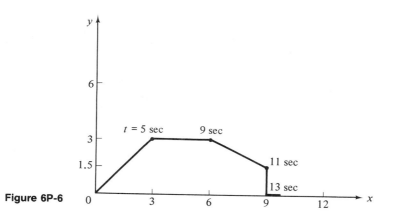

Figure 6P-6

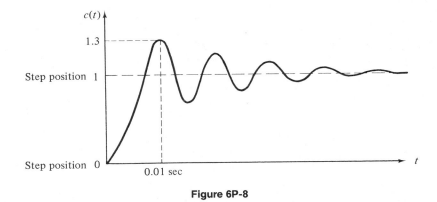

Figure 6P-8

6.8. A step motor gives a single response shown in Fig. 6P-8 after a pulse excitation is applied. Find a linear second-order transfer function to model the motor for this operation.

6.9. The attitude control of the missile shown in Fig. 4P-11 is accomplished by thrust vectoring. The transfer function between the thrust angle and the angle of attack can be represented by

$$G_p(s) = \frac{\Theta(s)}{\delta(s)} = \frac{K}{s^2 - a}$$

(refer to Problem 4.11) where K and a are positive constants. The attitude-control system is represented by the block diagram in Fig. 6P-9.

(a) In Fig. 6P-9, consider that only the attitude sensor loop is in operation ($K_t = 0$). Determine the performance of the overall system with respect to the relative values of K, K_s, and a.

(b) Consider that both loops are in operation. Determine the minimum values of K_t and K_s in terms of K and a so that the missile will not tumble.

(c) It is desired that the closed-loop system should have a damping ratio of ζ and a natural undamped frequency of ω_n. Find the values of K_t and K_s in terms of ζ, ω_n, a, and K.

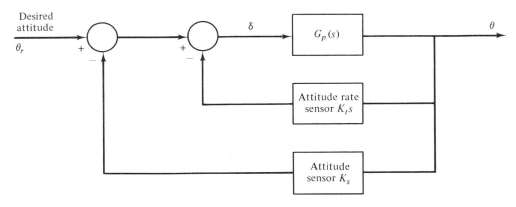

Figure 6P-9

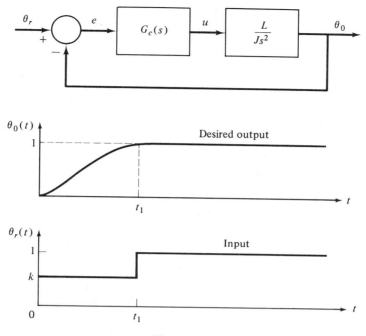

Figure 6P-11

6.10. A controlled process is represented by the following state equations:

$$\dot{x}_1 = x_1 - 5x_2$$
$$\dot{x}_2 = 8x_1 + u$$

The control is obtained from state feedback such that

$$u = -g_1 x_1 - g_2 x_2$$

where g_1 and g_2 are real constants.
(a) Find the locus in the g_1-versus-g_2 plane on which the overall system has a natural undamped frequency of $\sqrt{2}$ rad/sec.
(b) Find the locus in the g_1-versus-g_2 plane on which the overall system has a damping ratio of 70.7 percent.
(c) Find the values of g_1 and g_2 such that $\zeta = 0.707$ and $\omega_n = \sqrt{2}$ rad/sec.
6.11. The block diagram of a missile attitude-control system is shown in Fig. 6P-11; control is represented by $u(t)$, and the dynamics of the missile are represented by

$$\frac{\Theta_0(s)}{U(s)} = \frac{L}{Js^2}$$

The attitude controller is represented by $G_c(s)$, and $\Theta_0(s)$ is the actual heading or output.

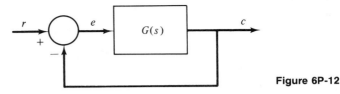

Figure 6P-12

(a) With $G_c(s) = 1$, determine the response of the system, $\theta_0(t)$, when the input $\theta_r(t)$ is a unit step function. Assume zero initial conditions. Discuss the effects of L and J on this response.

(b) Let $G_c(s) = (1 + T_d s)$, $L = 1$, and $J = 500$. Determine the value of T_d so that the system is critically damped.

(c) It is desired to obtain an output response with no overshoot (see Fig. 6P-11); the response time may not be minimum. Let $G_c(s) = 1$, and the system is controlled through the input $\theta_r(t)$, which is chosen to be of the form shown. Determine the values of k and t_1 so that the desired output is obtained. Use the same values of L and J as given above.

6.12. The block diagram of a feedback control system is shown in Fig. 6P-12. It is desired that:

(a) The steady-state error due to a unit step function input is zero.

(b) The characteristic equation of the overall system is

$$s^3 + 4s^2 + 6s + 10 = 0$$

Find the third-order open-loop transfer function $G(s)$ so that the foregoing two requirements are satisfied simultaneously.

6.13. For the feedback control system shown in Fig. 6P-12, it is required that:

(a) The steady-state error due to a unit-ramp-function input be equal to 1.5.

(b) The dominant roots of the characteristic equation of the third-order system are at $-1 + j1$ and $-1 - j1$. Find the third-order open-loop transfer function $G(s)$ so that the foregoing two conditions are satisfied.

6.14. For the feedback control system shown in Fig. 6P-12, it is required that the dominant roots of the characteristic equation of the third-order system be at $-\zeta\omega_n + j\omega_n\sqrt{1 - \zeta^2}$ and $-\zeta\omega_n - j\omega_n\sqrt{1 - \zeta^2}$. The steady-state error due to a step input must be zero. Find the minimum steady-state error of the system due to a unit ramp function input in terms of ζ and ω_n.

6.15. The block diagram of a feedback control system is shown in Fig. 6P-15. Find the characteristic equation of the system. Construct a parameter plane of K_P versus K_D (K_P on the vertical axis and K_D on the horizontal axis) and show the following regions in the plane.

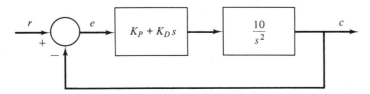

Figure 6P-15

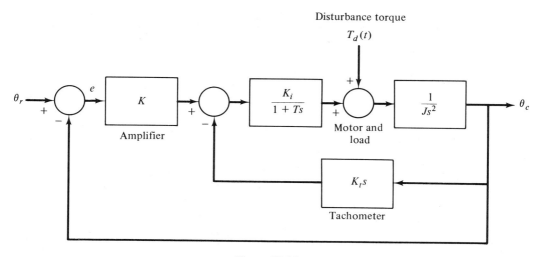

Figure 6P-16

(a) Unstable region.
(b) Stable region.
(c) Trajectory on which the damping ratio is unity (critical damping).
(d) Region in which the system is overdamped (damping ratio > 1).
(e) Region in which the system is underdamped (damping ratio < 1).
(f) Trajectory on which the parabolic error constant K_a is 40 sec^{-2}.
(g) Trajectory on which the natural undamped frequency ω_n is 40 rad/sec.
(h) Trajectory on which the system is either uncontrollable or unobservable.

6.16. The block diagram of a feedback control system is shown in Fig. 6P-16. The fixed parameters of the system are

$$T = 0.1 \qquad J = 0.01 \qquad K_i = 10$$

It is assumed that all the units of these parameters are consistent so that no conversions are necessary.

(a) Determine how the values of K and K_t affect the steady-state error [$e(t)$ is the error] when $\theta_r(t) = tu_s(t)$. Set $T_d = 0$.
(b) Determine how the values of K and K_t affect the steady-state value of $\theta_c(t)$ when the disturbance torque T_d is a unit step function, $T_d(t) = u_s(t)$. In this case set $\theta_r(t) = 0$.
(c) Set $K_t = 0.01$ and with the system parameters given above, find the minimum steady-state value of $\theta_c(t)$ that you can actually get by varying K, when the disturbance torque T_d is a unit step function. Give the value of this minimum steady-state value of θ_c and the corresponding value of K. Assume that $\theta_r = 0$ for this part. From the transient response standpoint, would you operate the system with this value of K? Explain.
(d) Assume that it is desired to operate the system with K as selected in part (c). Find the value of K_t so that the complex roots of the characteristic equation will have a real part of -2.5. Find these three roots.

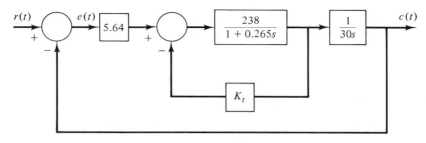

Figure 6P-17

6.17. The block diagram of a motor control system with tachometer feedback is shown in Fig. 6P-17.
(a) Find the natural frequency of the system.
(b) Find the value of the tachometer constant K_t so that the damping is critical.

6.18. By means of the Routh–Hurwitz criterion, determine the stability of the systems that have the following characteristic equations. In each case, determine the number of roots of the equation that are in the right-half s-plane.
(a) $s^3 + 20s^2 + 9s + 400 = 0$
(b) $s^3 + 20s^2 + 2s + 200 = 0$
(c) $3s^4 + 10s^3 + 5s^2 + s + 2 = 0$
(d) $s^4 + 2s^3 + 6s^2 + 8s + 8 = 0$
(e) $s^6 + 2s^5 + 8s^4 + 17s^3 + 20s^2 + 16s + 16 = 0$

6.19. The characteristic equations for certain feedback control systems are given below. In each case, determine the values of K that correspond to a stable system.
(a) $s^4 + 22s^3 + 10s^2 + s + K = 0$
(b) $s^4 + 20Ks^3 + 5s^2 + (10 + K)s + 15 = 0$
(c) $s^3 + (K + 0.5)s^2 + 5Ks + 50 = 0$

6.20. The conventional Routh–Hurwitz criterion gives only the location of the roots of a polynomial with respect to the right half and the left half of the s-plane. The open-loop transfer function of a unity feedback control system is given as

$$G(s) = \frac{K}{s(1 + Ts)}$$

It is desired that all the roots of the system's characteristic equation lie in the region to the left of the line, $s = -a$. This will assure not only that a stable system is obtained, but also that the system has a minimum amount of damping. Extend the Routh–Hurwitz criterion to this case, and determine the values of K and T required so that there are no roots to the right of the line $s = -a$.

6.21. The loop transfer function of a feedback control system is given by

$$G(s)H(s) = \frac{K(s + 5)}{s(1 + Ts)(1 + 2s)}$$

The parameters K and T may be represented in a plane with K as the horizontal axis and T as the vertical axis. Determine the region in which the closed-loop system is stable.

6.22. The open-loop transfer function of a unity feedback control system is given by

$$G(s) = \frac{K(s + 5)(s + 40)}{s^3(s + 200)(s + 1000)}$$

Discuss the stability of the closed-loop system as a function of K. Determine the values of K that will cause sustained oscillations in the closed-loop system. What are the frequencies of oscillations?

6.23. A controlled process is represented by the following state equations:

$$\dot{x}_1 = x_1 - 3x_2$$
$$\dot{x}_2 = 8x_1 + u$$

The control is obtained from state feedback such that

$$u = -g_1 x_1 - g_2 x_2$$

where g_1 and g_2 are real constants. Determine the region in the g_2 versus g_1 plane in which the overall system is stable.

6.24. Given a linear time-invariant system that is described by the following state equations:

$$\dot{\mathbf{x}}(t) = \mathbf{A}\mathbf{x}(t) + \mathbf{B}u(t)$$

where

$$\mathbf{A} = \begin{bmatrix} 0 & 1 & 0 \\ 0 & 0 & 1 \\ 0 & -3 & -2 \end{bmatrix} \quad \mathbf{B} = \begin{bmatrix} 0 \\ 0 \\ 1 \end{bmatrix}$$

The closed-loop system is implemented by state feedback, so that

$$u(t) = -\mathbf{G}\mathbf{x}(t)$$

where $\mathbf{G}$ is the feedback matrix, $\mathbf{G} = [g_1 \quad g_2 \quad g_3]$ with g_1, g_2, and g_3 equal to real constants. Determine the constraints on the elements of $\mathbf{G}$ so that the overall system is asymptotically stable.

6.25. Given the system $\dot{\mathbf{x}} = \mathbf{A}\mathbf{x} + \mathbf{B}u$, where

$$\mathbf{A} = \begin{bmatrix} 1 & 0 & 0 \\ 0 & -2 & 0 \\ 0 & 0 & 3 \end{bmatrix} \quad \mathbf{B} = \begin{bmatrix} 2 \\ 0 \\ 1 \end{bmatrix}$$

Consider that state feedback may be implemented. Is the system stabilizable by state feedback?

6.26. The block diagram of a control system is shown in Fig. 6P-26, where $r(t)$ is the reference input and $n(t)$ is disturbance input.
   (a) Find the steady-state error $\lim_{t \to \infty} e(t)$ when $n(t) = 0$ and $r(t) =$ unit ramp function. Assume that the values of a and K are chosen such that the closed-loop system is stable.
   (b) Find the steady-state output $\lim_{t \to \infty} c(t)$ when $r(t) = 0$, and $n(t) =$ unit step function. Assume stable system.

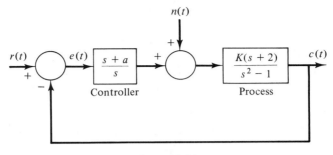

Figure 6P-26

(c) Find the requirements on a and K so that the system is stable. Express the stable region in the K-versus-a plane. (Use K as vertical and a as horizontal axis.)

6.27. The schematic diagram shown in Fig. 6P-27 represents a control system whose purpose is to hold the level of the liquid in the tank at a fixed level. The liquid level is controlled by a float whose position is connected to the wiper arm of a potentiometer error detector. The error signal between the reference level and the actual level of the liquid is fed into the controller whose transfer function is $G_c(s)$. The system parameters are given as follows:

DC motor:

Armature resistance	$R = 10$ ohms
Armature inductance	$L = $ negligible
Torque constant	$K_i = 10$ oz-in./A
Back emf constant	$K_b = 0.075$ V/rad/sec
Rotor inertia	$J_m = 0.005$ oz-in.-sec^2
Load and motor friction	negligible
Load inertia	$J_L = 7.5$ oz-in.-sec^2
Gear ratio	$n = n_1/n_2 = 1/100$

Motor equations:

$$e_a = Ri + e_b \qquad\qquad e_b = K_b\omega_m$$

$$\omega_m = \frac{d\theta_m}{dt} \qquad\qquad T_m = K_i i = J\dot{\omega}_m$$

$$\theta_c = \frac{n_1\theta_m}{n_2} = n\theta_m \qquad\qquad J = J_m + n^2 J_L$$

Dynamics of tank: There are N inlets to the tank from the reservoir. All the inlet valves have the same characteristics, and are controlled simultaneously by θ_c. The

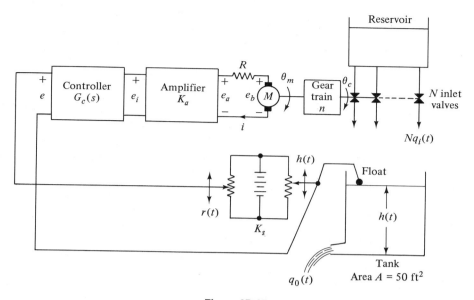

Figure 6P-27

equations that govern the volume of flow are

$$q_I(t) = K_I N \theta_c(t) \qquad K_I = 10 \text{ ft}^3/\text{sec-rad}$$

$$q_O(t) = K_O h(t) \qquad K_O = 50 \text{ ft}^3/\text{sec}$$

$$h(t) = \frac{\text{volume of tank}}{\text{area of tank}} = \frac{1}{A} \int [q_I(t) - q_O(t)] \, dt$$

Amplifier gain:

$$K_a = 50$$

Error detector:

$$K_s = 1 \text{ V/ft} \qquad e(t) = K_s[r(t) - h(t)]$$

(a) Draw a block diagram of the overall system, showing the functional relationship between the transfer functions.

(b) Let the transfer function of the controller, $G_c(s)$, be unity. Find the open-loop transfer function $G(s) = H(s)/E(s)$ and the closed-loop transfer function $M(s) = H(s)/R(s)$. Find the characteristic equation of the closed-loop system. Note that all the system parameters are specified except for N, the number of inlets.

(c) Apply the Routh–Hurwitz criterion to the characteristic equation and determine what is the maximum number of inlets (integer) so that the system is still asymptotically stable; $G_c(s) = 1$.

(d) Since one of the poles of the open-loop transfer function is far to the left (relatively) on the negative real axis in the s-plane (very small time constant), it is suggested that this pole can be neglected as far as the transient response is concerned.

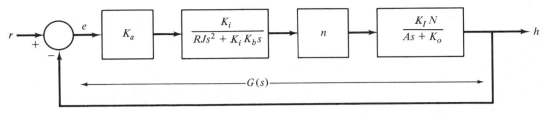

Figure 6P-28

Approximate the third-order open-loop transfer function $G(s)$ by a second-order transfer function by properly neglecting the pole with the "smallest" time constant. Once we have a second-order system, we may use the equations in the text to predict the maximum overshoot and peak time t_{max}. Compute the maximum overshoot and peak time for the second-order approximating system you obtained, with $N = 2$ and then $N = 8$.

(e) Obtain the unit step response of the second-order approximating system with $N = 2$ and then with $N = 8$. Compare the maximum overshoot and peak time with the results obtained in part (d).

(f) Obtain the unit step responses of the third-order closed-loop system (restoring the third pole) with $N = 2$ and then with $N = $. Assume zero initial conditions. Compare the maximum overshoots and peak times with those of the second-order system. Also, compare the unit step responses. What can you say about the second-order approximation? How do the approximating results depend on the values of N?

6.28. The block diagram shown in Fig. 6P-28 represents the tank-level control system described in Problem 6.27. The following data are given:

$$K_a = 50, \quad K_I = 50, \quad K_b = 0.075, \quad J = 0.006, \quad R = 10, \quad K_i = 10, \quad n = \tfrac{1}{100}$$

The values of A, N, and K_O will be assigned in the following.

(a) Let the tank area A be a variable parameter (in addition to N). For $K_O = 50$, find the ranges of A and N for the closed-loop system to be stable. Since this is a two-dimensional problem, represent the results by sketching the stable region in the N-versus-A plane. (Use N as a vertical axis and A as horizontal axis.) This is called the "parameter plane" and is a favorite tool of some control system engineers. Also sketch the region by considering that realistically N is an integer. It is best to apply the Routh–Hurwitz criterion for this part. If we have a tank that is infinitely large in its cross-sectional area, what is the maximum value of N for stability?

(b) It is more logical to investigate the relationships between n and K_O, which represent inflow and outflow. Find the stable region in the N-versus-K_O parameter plane (for K_O up to 100). Let $A = 50$.

6.29. For the two-phase permanent-magnet step motor described in Example 4-11, it is desirable to study the dynamic characteristics of the system near its stable equilibrium state. Let the stable equilibrium state be described by $\theta_m^* = 0$, $\omega_m^* = 0$, $v_a = V$, and $v_b = 0$. This corresponds to the condition in which phase a is energized, phase b is off, and the motor is at the stable position of $\theta_m^* = 0$. The eigenvalues of $\mathbf{A}^*$ of the linearized system will reveal the damping characteristics of the motor.

(a) Evaluate the matrix $\mathbf{A}^*$ as defined in Eq. (4-177) at the specified equilibrium condition. Find the characteristic equation of $\mathbf{A}^*$.

(b) For $K_m = 16$ oz-in./amp, $J = 0.0095$ oz-in.-sec^2, $N_r = 50$, $K_b = 0.00226$ V/rad/sec, $B = 0$, and $L = 0.005$ H, find the characteristic equation of $\mathbf{A}^*$ with the winding resistance R as a variable parameter. Plot the roots of the characteristic equation for $R = 0$ to $R = 100$. Comment on the effect of R on the damping of the system.

6.30. For the double inverted pendulum described in Problem 5.44,

(a) Find the state feedback matrix $\mathbf{G}$ such that

$$u = -\mathbf{Gx} \qquad \mathbf{G} = [\, g_1 \quad g_2 \quad g_3 \quad g_4 \quad g_5 \quad g_6 \,]$$

and the eigenvalues of $\mathbf{A} - \mathbf{BG}$ are at $-10, -20, -2 + j2, -2 - j2, -10 + j10$, and $-10 - j10$.

(b) Find the time responses of θ_1, θ_2, and y with the initial state vector

$$\mathbf{x}(0) = [1 \quad 0 \quad 1 \quad 0 \quad 1 \quad 0]'$$

chapter seven

Root Locus Technique

7.1 INTRODUCTION

In the design and analysis of control systems we often need to investigate a system's performance when one or more of its parameters vary over a given range. Our purpose may be to select the appropriate value of a system parameter, such as gain, or we may want to study parameter variations due to aging of the system components or environmental changes.

We demonstrated in the last chapter the importance of the characteristic equation in studying the dynamic performance of linear control systems. An important problem in linear control systems theory is the investigation of the trajectories of the roots of the characteristic equation—or, simply, the *root loci*—when a certain system parameter varies. In fact, several examples in Chapter 6 illustrated the usefulness of the root loci in the study of linear control systems.

The present chapter introduces the basic properties of the root loci and shows how to construct these loci by following some simple rules. For complex systems we can always rely on root-finding programs and a digital computer to find roots of an algebraic equation. For the root locus diagram, however, which often includes hundreds of data points, such a solution may consume a great deal of computer time if one does not have a clear understanding of the expected solutions, so that the problem can be set up properly.

The root locus technique is not confined to the study of control systems. The equation under study does not necessarily have to be the characteristic equation of a linear system. In general, the technique can be applied to study the behavior of roots of any high-order algebraic equations.

The general root locus problem can be formulated by referring to the following equation of the complex variable s:

$$F(s) = P(s) + KQ(s) = 0 \qquad (7\text{-}1)$$

where $P(s)$ is an nth order polynomial of s,

$$P(s) = s^n + a_1 s^{n-1} + \cdots + a_{n-1}s + a_n \qquad (7\text{-}2)$$

and $Q(s)$ is an mth-order polynomial of s; n and m are positive integers.

$$Q(s) = s^m + b_1 s^{m-1} + \cdots + b_{m-1}s + b_m \qquad (7\text{-}3)$$

K is a real constant which can vary from $-\infty$ to $+\infty$.

The coefficients $a_1, a_2, \ldots, a_n, b_1, b_2, \ldots, b_m$ are considered to be fixed. These coefficients can be real or complex, although our main interest here is in real coefficients.

Root loci of multiple-variable parameters can be treated by varying one parameter at a time. The resultant trajectories are called the *root contours*, and the subject is treated in Section 7.6.

By replacing s with z in Eqs. (7-1) through (7-3), we can construct the root loci of the characteristic equation of a linear digital control system in a similar fashion.

Although the loci of the roots of Eq. (7-1) when K varies between $-\infty$ and ∞ are generally referred to as the *root loci* in control systems literature, for the simplicity of reference, we define the following categories:

1. *Root Loci (RL)*: the portion of the root loci when K varies from 0 to ∞; i.e., K is positive.

2. *Complementary Root Loci (CRL)*: the portion of the root loci when K varies from $-\infty$ to 0; i.e., K is negative.

3. *Root Contours (RC)*: loci of roots when more than one parameter varies.

The *complete root loci* refers to the combination of the root loci and the complementary root loci—that is, when K varies from $-\infty$ to ∞.

7.2 BASIC PROPERTIES OF THE ROOT LOCI

Since our main interest is in the study of control systems, we refer to the closed-loop transfer function, which is expressed as[1]

$$\frac{C(s)}{R(s)} = \frac{G(s)}{1 + G(s)H(s)} \qquad (7\text{-}4)$$

The characteristic equation of the closed-loop system is obtained by setting the denominator polynomial of $C(s)/R(s)$ to zero, which is the same as setting the numerator of $1 + G(s)H(s)$ to zero. Thus, the roots of the characteristic equation

---

[1]Although we are using a single-loop configuration, the analysis can be applied to any closed-loop system by considering Eq. (7-1) as the characteristic equation.

must satisfy

$$1 + G(s)H(s) = 0 \qquad (7\text{-}5)$$

Suppose that $G(s)H(s)$ contains a variable parameter K such that the rational function can be written as

$$G(s)H(s) = \frac{KQ(s)}{P(s)} \qquad (7\text{-}6)$$

where $P(s)$ and $Q(s)$ are polynomials as defined in Eqs. (7-2) and (7-3), respectively. Thus, Eq. (7-5) becomes

$$1 + \frac{KQ(s)}{P(s)} = \frac{P(s) + KQ(s)}{P(s)} = 0 \qquad (7\text{-}7)$$

It is easy to see that the numerator polynomial of Eq. (7-7) is identical to Eq. (7-1). Thus, by considering that the loop transfer function $G(s)H(s)$ of a closed-loop system can be written in the form of Eq. (7-6), we have fitted the root loci of a control system into the form of the general root locus problem.

When the variable parameter K does not appear as a multiplying factor of $G(s)H(s)$, we can always condition the functions in the form of Eq. (7-1). Consider that the loop transfer function of a control system is of the form:

$$G(s)H(s) = \frac{s^2 + (3 + 2K)s + 5}{s(s + 1)(s + 2)} \qquad (7\text{-}8)$$

The characteristic equation of the closed-loop system is

$$s(s + 1)(s + 2) + s^2 + (3 + 2K)s + 5 = 0 \qquad (7\text{-}9)$$

Dividing both sides of the last equation by the terms that do not contain K, we get

$$1 + \frac{2Ks}{s(s + 1)(s + 2) + s^2 + 3s + 5} \qquad (7\text{-}10)$$

which is of the general form of Eq. (7-7), with

$$Q(s) = 2s \qquad (7\text{-}11)$$

and

$$P(s) = s^3 + 4s^2 + 5s + 5 \qquad (7\text{-}12)$$

The conclusion is that, given any loop transfer function $G(s)H(s)$ with one variable parameter K imbedded in it, we can always find the characteristic equation by equating the numerator polynomial of $1 + G(s)H(s)$ to zero, as in the step from Eq. (7-8) to Eq. (7-9). To isolate the variable parameter K, we first factor the terms in the characteristic equation with K, and then without K. This forms the $P(s)$ and the $Q(s)$ polynomials, as in Eq. (7-1). Finally, we divide both sides of the characteristic equation by $P(s)$, which contains the terms without K. This last step is essential in conditioning the characteristic equation so that the properties of the root loci of Eq. (7-1) can be obtained. As it turns out, all the properties of the root loci of Eq. (7-1) are derived from the characteristics of the function $Q(s)/P(s)$.

Since $G(s)H(s) = KQ(s)/P(s)$, this is another example in which the characteristics of the closed-loop system, in this case represented by the roots of the characteristic equation, are determined from the knowledge of the loop transfer function $G(s)H(s)$.

Now we are ready to investigate the conditions under which Eq. (7-5) or Eq. (7-7) is satisfied.

Let us express $G(s)H(s)$ as

$$G(s)H(s) = KG_1(s)H_1(s) \tag{7-13}$$

where $G_1(s)H_1(s)$ is equal to $Q(s)/P(s)$, and it does not contain the variable parameter K. Then, Eq. (7-5) is written

$$G_1(s)H_1(s) = -\frac{1}{K} \tag{7-14}$$

To satisfy the last equation, the following conditions must be met simultaneously:

Condition of magnitude:

$$|G_1(s)H_1(s)| = \frac{1}{|K|} \qquad -\infty < K < \infty \tag{7-15}$$

Condition on angles:

$$\underline{/G_1(s)H_1(s)} = (2k + 1)\pi \qquad (K \geq 0) \tag{7-16}$$

$$= \text{odd multiples of } \pi \text{ radians}$$

$$\underline{/G_1(s)H_1(s)} = 2k\pi \qquad (K \leq 0) \tag{7-17}$$

$$= \text{even multiples of } \pi \text{ radians}$$

where $k = 0, \pm 1, \pm 2, \dots$ (any integer).

In practice, the conditions stated in Eqs. (7-15) through (7-17) play different roles in the construction of the complete root loci. The condition on angles in Eq. (7-16) or Eq. (7-17) is used to determine the shape of the root loci in the s-plane. Once the root loci are drawn, the values of K on the loci are determined by using the condition on magnitude in Eq. (7-15).

The construction of the root loci is basically a graphical problem, although some of the rules of construction are arrived at analytically. The start of the graphical construction of the root loci is based on knowledge of the poles and zeros of the function $G(s)H(s)$. In other words, $G(s)H(s)$ must first be written as[2]

$$G(s)H(s) = KG_1(s)H_1(s) = \frac{K(s + z_1)(s + z_2)\dots(s + z_m)}{(s + p_1)(s + p_2)\dots(s + p_n)} \qquad (7\text{-}18)$$

where the zeros and the poles of $G(s)H(s)$ are real or in complex-conjugate pairs.

Using Eq. (7-18), the conditions stated in Eqs. (7-15), (7-16), and (7-17) become

$$|G_1(s)H_1(s)| = \frac{\displaystyle\prod_{i=1}^{m}|s + z_i|}{\displaystyle\prod_{j=1}^{n}|s + p_j|} = \frac{1}{|K|} \qquad -\infty < K < \infty \qquad (7\text{-}19)$$

and

$$\angle G_1(s)H_1(s) = \sum_{i=1}^{m}\angle s + z_i - \sum_{j=1}^{n}\angle s + p_j$$

$$= (2k + 1)\pi \qquad 0 \le K < \infty \qquad (7\text{-}20)$$

$$\angle G_1(s)H_1(s) = \sum_{i=1}^{m}\angle s + z_i - \sum_{j=1}^{n}\angle s + p_j$$

$$= 2k\pi \qquad -\infty < K \le 0 \qquad (7\text{-}21)$$

for $k = 0, \pm 1, \pm 2, \dots$.

It was mentioned earlier that Eqs. (7-20) and (7-21) may be used for the construction of the complete root loci in the s-plane. In other words, Eq. (7-20) implies that for any positive value of K, a point (e.g., s_1) in the s-plane is a point on the RL if the difference between the sums of the angles of the vectors drawn from the zeros and the poles of $G(s)H(s)$ to s_1 is an odd multiple of 180°. Similarly, for negative values of K, Eq. (7-21) shows that any point on the CRL must satisfy the condition that the difference between the sums of the angles of the vectors drawn from the zeros and the poles to the point is an even multiple of 180 degrees, or 0 degrees.

To illustrate the use of Eqs. (7-19) to (7-21) for the construction of the root loci, let us consider

$$G(s)H(s) = \frac{K(s + z_1)}{s(s + p_2)(s + p_3)} \qquad (7\text{-}22)$$

The locations of the poles and the zero of $G(s)H(s)$ are arbitrarily assumed, as

---

[2] We shall first consider that $G(s)H(s)$ is a rational function of s. For systems with time delays, $G(s)H(s)$ will contain exponential terms such as e^{-Ts}. In Section 7.7 the root loci problem is extended to systems with $G(s)H(s)$ containing e^{-Ts} as a multiplying factor.

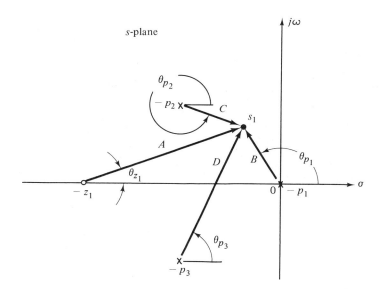

Figure 7-1 Pole–zero configuration of $G(s)H(s) = [K(s + z_1)] / [s(s + p_2)(s + p_3)]$.

shown in Fig. 7-1. Next, we select an arbitrary point, s_1, in the s-plane and draw vectors directing from the poles and the zero of $G(s)H(s)$ to the point s_1. If s_1 is indeed a point on the RL ($0 \leq K < \infty$), it must satisfy the following two conditions simultaneously: from Eq. (7-19),

$$\frac{|s_1 + z_1|}{|s_1| \, |s_1 + p_2| \, |s_1 + p_3|} = \frac{1}{|K|} \tag{7-23}$$

and from Eq. (7-20),

$$\underline{/s_1 + z_1} - \left(\underline{/s_1} + \underline{/s_1 + p_2} + \underline{/s_1 + p_3} \right) = (2k + 1)\pi \qquad k = 0, \pm 1, \pm 2, \ldots \tag{7-24}$$

Similarly, if s_1 is to be a point on the CRL ($-\infty < K \leq 0$), it must satisfy Eq. (7-21); that is,

$$\underline{/s + z_1} - \left(\underline{/s_1} + \underline{/s_1 + p_2} + \underline{/s_1 + p_3} \right) = 2k\pi \tag{7-25}$$

for $k = 0, \pm 1, \pm 2, \ldots$.

As shown in Fig. 7-1, the angles θ_{z_1}, θ_{p_1}, θ_{p_2}, and θ_{p_3} are the angles of the vectors measured with the positive real axis as zero reference. Equations (7-24) and (7-25) become

$$\theta_{z_1} - (\theta_{p_1} + \theta_{p_2} + \theta_{p_3}) = (2k + 1)\pi \qquad 0 \leq K < \infty \tag{7-26}$$

and

$$\theta_{z_1} - (\theta_{p_1} + \theta_{p_2} + \theta_{p_3}) = 2k\pi \qquad -\infty < K \le 0 \qquad (7\text{-}27)$$

respectively.

If s_1 is found to satisfy either Eq. (7-26) or Eq. (7-27), Eq. (7-23) is used to determine the value of K at the point. Rewriting Eq. (7-23), we have

$$|K| = \frac{|s_1| \, |s_1 + p_2| \, |s_1 + p_3|}{|s_1 + z_1|} \qquad (7\text{-}28)$$

where the factor $|s_1 + z_1|$ is the length of the vector drawn from the zero z_1 to the point s_1. If, as in Fig. 7-1, the vector lengths are represented by A, B, C, and D, Eq. (7-28) becomes

$$|K| = \frac{BCD}{A} \qquad (7\text{-}29)$$

The sign of K, of course, depends on whether s_1 is on the RL or the CRL. Consequently, given the pole–zero configuration of $G(s)H(s)$, the construction of the complete root locus diagram involves the following two steps:

1. A search for all the s_1 points in the s-plane that satisfy Eqs. (7-20) and (7-21).
2. The determination of the values of K at points on the root loci and the complementary root loci by use of Eq. (7-19).

From the basic principles of the root locus diagram discussed thus far, it may seem that the search for all the s_1 points in the s-plane that satisfy Eqs. (7-21) and (7-22) is a very tedious task. This would be true if we were to find all the points on the root loci by applying the conditions in Eqs. (7-21) and (7-22) manually. Years ago, when Evans[3] first invented the root locus technique, he devised a special tool, called the *Spirule*, which can be used to assist in adding and subtracting angles of vectors conveniently, according to Eq. (7-21) or Eq. (7-22). However, the Spirule is an effective tool only if the user knows already the general proximity of the roots in the s-plane. Indeed, when we set out to find a point s_1 on the root loci in the s-plane manually, we must first have some general knowledge of the approximate location of the point; then we can select a trial point and test it in Eqs. (7-21) and (7-22).

With the availability of digital computers and efficient root-finding subroutines, the Spirule has long become obsolete. However, even with the high-speed computer, the analyst should still have an in-depth understanding of the properties of the root loci in order to sketch easily the loci of simple and moderately complex systems and to interpret the computer results.

[3] Refer to reference 1 at the end of this chapter.

7.3 CONSTRUCTION OF THE COMPLETE ROOT LOCI

The following rules of construction are developed from the relation between the poles and zeros of $G(s)H(s)$ and the zeros of $1 + G(s)H(s)$. These rules should be regarded only as an aid to the construction of the RL and the CRL, as they do not give the exact plots.

$K = 0$ Points

Theorem 7-1. *The $K = 0$ points on the complete root loci are at the poles of $G(s)H(s)$.*

Proof. From Eq. (7-19),

$$|G_1(s)H_1(s)| = \frac{\prod_{i=1}^{m}|s + z_i|}{\sum_{j=1}^{n}|s + p_j|} = \frac{1}{|K|} \tag{7-30}$$

As K approaches zero, the magnitude of $|G_1(s)H_1(s)|$ approaches infinity, and, correspondingly, s approaches the poles of $G_1(s)H_1(s)$ or of $G(s)H(s)$; that is, s approaches $-p_j(j = 1, 2, \ldots, n)$. It is apparent that this property applies to both the RL and the CRL, since the sign of K has no bearing in Eq. (7-30). ■

Example 7-1

Consider the following equation:

$$s(s + 2)(s + 3) + K(s + 1) = 0 \tag{7-31}$$

When $K = 0$, the three roots of the equation are at $s = 0$, $s = -2$, and $s = -3$. These three points are also the poles of the function $G(s)H(s)$ if we divide both sides of Eq. (7-31) by the

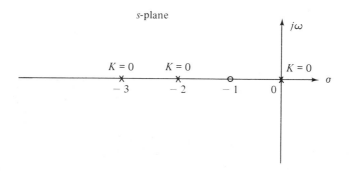

Figure 7-2 Points at which $K = 0$ on the complete root loci of $s(s + 2)(s + 3) + K(s + 1) = 0$.

terms that do not contain K, and establish the relationship

$$1 + G(s)H(s) = 1 + \frac{K(s+1)}{s(s+2)(s+3)} = 0 \qquad (7\text{-}32)$$

Thus

$$G(s)H(s) = \frac{K(s+1)}{s(s+2)(s+3)} \qquad (7\text{-}33)$$

The three $K = 0$ points on the complete root loci are as shown in Fig. 7-2.

■

$K = \pm\infty$ Points

Theorem 7-2. *The $K = \pm\infty$ points on the complete root loci are at the zeros of $G(s)H(s)$.*

Proof. Referring again to Eq. (7-30), as K approaches $\pm\infty$, the equation approaches zero. This corresponds to s approaching the zeros of $G(s)H(s)$; or s approaching $-z_i (i = 1, 2, \ldots, m)$.

■

Example 7-2

Consider again the equation

$$s(s+2)(s+3) + K(s+1) = 0 \qquad (7\text{-}34)$$

It is apparent that when K is very large, the equation can be approximated by

$$K(s+1) = 0 \qquad (7\text{-}35)$$

which has the root $s = -1$. Notice that this is also the zero of $G(s)H(s)$ in Eq. (7-33). Therefore, Fig. 7-3 shows the point $s = -1$ at which $K = \pm\infty$. However, $G(s)H(s)$ in this

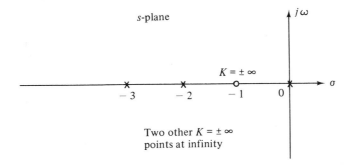

Figure 7-3 Points at which $K = \pm\infty$ on the complete root loci of $s(s+2)(s+3) + K(s+1) = 0$.

case also has two other zeros located at infinity, because for a rational function, the total number of poles and zeros must be equal if the poles and zeros at infinity are included. Therefore, for the equation in Eq. (7-31), the $K = \pm\infty$ points are at $s = -1$, ∞, and ∞.[4]

∎

Number of Branches on the Complete Root Loci

A branch of the complete root loci is the locus of one root when K takes on values between $-\infty$ and ∞. Since the number of branches of the complete root loci must equal the number of roots of the equation, the following theorem results:

Theorem 7-3. *The number of branches of the root loci of Eq. (7-1) is equal to the greater of n and m.*[5]

∎

Example 7-3

The number of branches on the complete root loci of

$$s(s + 2)(s + 3) + K(s + 1) = 0 \tag{7-36}$$

is three, since $n = 3$ and $m = 1$. Or, since the equation is of the third order in s, it must have three roots, and therefore three root loci.

∎

Symmetry of the Complete Root Loci

Theorem 7-4. *The complete root loci are symmetrical with respect to the real axis of the s-plane. In general, the complete root loci are symmetrical with respect to the axes of symmetry of the poles and zeros of $G(s)H(s)$.*

Proof. The proof of the first statement is self-evident, since, for real coefficients in Eq. (7-1), the roots must be real or in complex-conjugate pairs.

The reasoning behind the second statement on symmetry is also simple, since if the poles and zeros of $G(s)H(s)$ are symmetrical to an axis other than the real axis in the s-plane, we can regard this axis of symmetry as if it were the real axis of a new complex plane obtained through a linear transformation.

∎

Example 7-4

Let us consider the equation

$$s(s + 1)(s + 2) + K = 0 \tag{7-37}$$

---

[4] It is useful to consider that infinity in the s-plane is a point concept. We can visualize that the finite s-plane is only a small surface on a sphere with an infinite radius. Then, infinity in the s-plane is a point on the other side of the sphere which we face.

[5] Although in control systems the number of poles of $G(s)H(s)$ must be at least equal to or greater than that of the zeros, from a strict analytical standpoint there are no restrictions on the relative magnitudes of n and m.

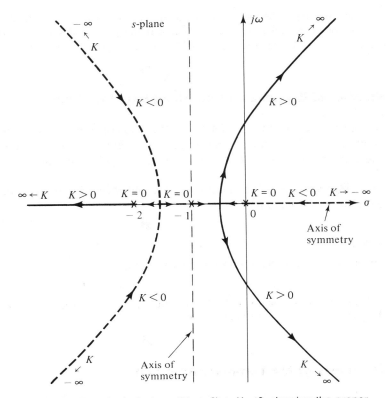

Figure 7-4 Root loci of $s(s + 1)(s + 2) + K = 0$, showing the properties of symmetry.

Dividing both sides of the equation by the terms that do not contain K leads to

$$G(s)H(s) = \frac{K}{s(s + 1)(s + 2)} \tag{7-38}$$

The complete root loci of Eq. (7-37) are sketched as shown in Fig. 7-4. Notice that since the poles of $G(s)H(s)$ are symmetrical with respect to the $s = -1$ axis (in addition to being always symmetrical with respect to the real axis), the complete root loci are symmetrical to the $s = -1$ axis and the real axis.

Example 7-5

When the pole-zero configuration of $G(s)H(s)$ is symmetrical with respect to a point in the s-plane, the complete root loci will also be symmetrical to that point. This is illustrated by the root locus plot of

$$s(s + 2)(s + 1 + j)(s + 1 - j) + K = 0 \tag{7-39}$$

as shown in Fig. 7-5.

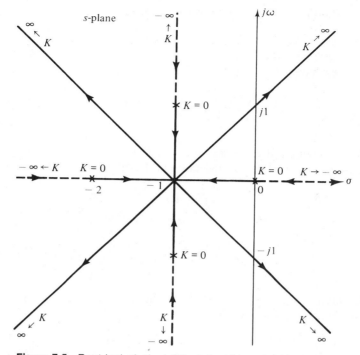

Figure 7-5 Root loci of $s(s + 2)(s + 1 + j)(s + 1 - j) + K = 0$, showing the properties of symmetry.

Asymptotes of the Complete Root Loci (Behavior of Root Loci at $s = \infty$)

The properties of the complete root loci near infinity in the s-plane are important, since when $n \neq m$, $2|n - m|$ of the loci will approach infinity in the s-plane.

Theorem 7-5. *For large values of s, the RL ($K \geq 0$) are asymptotic to straight lines or asymptotes with angles given by*

$$\theta_k = \frac{(2k + 1)\pi}{n - m} \tag{7-40}$$

where $k = 0, 1, 2, \ldots, |n - m| - 1$ and n and m are the poles and zeros of $G(s)H(s)$, respectively.[6]

For the CRL, $K \leq 0$, the angles of the asymptotes are

$$\theta_k = \frac{2k\pi}{n - m} \tag{7-41}$$

where $k = 0, 1, 2, \ldots, |n - m| - 1$.

---

[6]According to the defining equations of the root loci, $k = 0, \pm 1, \pm 2, \ldots$. However, since there are only $|n - m|$ asymptotes for each type of root loci, we need to assign only $|n - m|$ values to k.

Proof.   Let us divide both sides of Eq. (7-1) by

$$Q(s) = s^m + b_1 s^{m-1} + \cdots + b_{m-1}s + b_m$$

We then have

$$\frac{s^n + a_1 s^{n-1} + \cdots + a_{n-1}s + a_n}{s^m + b_1 s^{m-1} + \cdots + b_{m-1}s + b_m} + K = 0 \tag{7-42}$$

Carrying out the fraction of the left side of Eq. (7-42) by the process of long division, and for large s neglecting all but the first two terms, we have

$$s^{n-m} + (a_1 - b_1)s^{n-m-1} \cong -K \tag{7-43}$$

or

$$s\left(1 + \frac{a_1 - b_1}{s}\right)^{1/(n-m)} \cong (-K)^{1/(n-m)} \tag{7-44}$$

The factor $[1 + (a_1 - b_1)/s]^{1/(n-m)}$ in Eq. (7-44) is expanded by binomial expansion, and Eq. (7-44) becomes

$$s\left[1 + \frac{a_1 - b_1}{(n-m)s} + \cdots\right] \cong (-K)^{1/(n-m)} \tag{7-45}$$

Again, if only the first two terms in the last series are retained, we get

$$s + \frac{a_1 - b_1}{n - m} \cong (-K)^{1/(n-m)} \tag{7-46}$$

Now let $s = \sigma + j\omega$, and, using DeMoivre's algebraic theorem, Eq. (7-46) is written

$$\sigma + j\omega + \frac{a_1 - b_1}{n - m} \cong K^{1/(n-m)}\left[\cos\frac{(2k+1)\pi}{n-m} + j\sin\frac{(2k+1)\pi}{n-m}\right] \tag{7-47}$$

for $0 \le K < \infty$, and

$$\sigma + j\omega + \frac{a_1 - b_1}{n - m} \cong |K^{1/(n-m)}|\left[\cos\frac{2k\pi}{n-m} + j\sin\frac{2k\pi}{n-m}\right] \tag{7-48}$$

for $-\infty < K \le 0$, and $k = 0, \pm 1, \pm 2, \ldots$.

Equating the real and imaginary parts of both sides of Eq. (7-48), we have, for $0 \le K < \infty$,

$$\sigma + \frac{a_1 - b_1}{n - m} \cong K^{1/(n-m)}\cos\frac{(2k+1)\pi}{n-m} \tag{7-49}$$

and

$$\omega \cong K^{1/(n-m)}\sin\frac{(2k + 1)\pi}{n - m} \qquad (7\text{-}50)$$

Solving for $K^{1/(n-m)}$ from Eq. (7-50), we have

$$K^{1/(n-m)} \cong \frac{\omega}{\sin\dfrac{2k + 1}{n - m}\pi} \cong \frac{\sigma + \dfrac{a_1 - b_1}{n - m}}{\cos\dfrac{2k + 1}{n - m}\pi} \qquad (7\text{-}51)$$

or

$$\omega \cong \tan\frac{(2k + 1)\pi}{n - m}\left(\sigma + \frac{a_1 - b_1}{n - m}\right) \qquad (7\text{-}52)$$

Equation (7-52) represents a straight line in the s-plane, and the equation is of the form

$$\omega \cong M(\sigma - \sigma_1) \qquad (7\text{-}53)$$

where M represents the slope of the straight line or the asymptote, and σ_1 is the intersect with the σ axis.

From Eqs. (7-52) and (7-53) we have

$$M = \tan\frac{2k + 1}{n - m}\pi \qquad (7\text{-}54)$$

$k = 0, 1, 2, \ldots, |n - m| - 1$, and

$$\sigma_1 = -\frac{a_1 - b_1}{n - m} \qquad (7\text{-}55)$$

Note that these properties of the asymptotes are for the RL ($0 \le K < \infty$) only.

Similarly, from Eq. (7-48) we can show that for the CRL ($-\infty < K \le 0$),

$$M = \tan\frac{2k\pi}{n - m} \qquad (7\text{-}56)$$

$k = 0, 1, 2, \ldots, |n - m| - 1$, and the same expression as in Eq. (7-55) is obtained for σ_1. Therefore, the angular relations for the asymptotes, given by Eqs. (7-40) and (7-41), have been proven. This proof also provided a by-product, which is the intersect of the asymptotes with the real axis of the s-plane, and therefore resulted in the following theorem.

Intersection of the Asymptotes (Centroid)

Theorem 7-6. (a) *The intersection of the $2|n - m|$ asymptotes of the complete root loci lies on the real axis of the s-plane.*

(*b*) *The intersection of the asymptotes is given by*

$$\sigma_1 = \frac{b_1 - a_1}{n - m} \tag{7-57}$$

where a_1, b_1, n, *and* m *are defined in Eqs.* (7-2) *and* (7-3).

Proof. The proof of (*a*) is straightforward, since it has been established that the complete root loci are symmetrical to the real axis.

The proof of (*b*) is a consequence of Eq. (7-55). Furthermore, if we define a function $G(s)H(s)$ as in Eq. (7-6), Eq. (7-57) may be written as

$$
\begin{aligned}
\sigma_1 &= \frac{b_1 - a_1}{n - m} \\
&= \frac{\sum \text{ finite poles of } G(s)H(s) - \sum \text{ finite zeros of } G(s)H(s)}{\text{number of finite poles of } G(s)H(s)} \\
&\quad - \text{number of finite zeros of } G(s)H(s)
\end{aligned} \tag{7-58}
$$

since from the law of algebra,

$$
\begin{aligned}
-a_1 &= \text{sum of the roots of } s^n + a_1 s^{n-1} + \ldots + a_{n-1}s + a_n = 0 \\
&= \text{sum of the finite poles of } G(s)H(s)
\end{aligned} \tag{7-59}
$$

$$
\begin{aligned}
-b_1 &= \text{sum of the roots of } s^m + b_1 s^{m-1} + \ldots + b_{m-1}s + b_m = 0 \\
&= \text{sum of the finite zeros of } G(s)H(s)
\end{aligned} \tag{7-60}
$$

Since the poles and zeros are either real or complex-conjugate pairs, the imaginary parts always cancel each other. Thus in Eq. (7-58) the terms in the summations may be replaced by the real parts of the poles and zeros of $G(s)H(s)$, respectively.

It should be noted that Eq. (7-58) is valid for the RL as well as the CRL.

■

Example 7-6

Consider the equation

$$s(s + 4)(s^2 + 2s + 2) + K(s + 1) = 0 \tag{7-61}$$

This equation corresponds to the characteristic equation of a feedback control system with the loop transfer function

$$G(s)H(s) = \frac{K(s + 1)}{s(s + 4)(s^2 + 2s + 2)} \tag{7-62}$$

The pole-zero configuration of $G(s)H(s)$ is shown in Fig. 7-6. From the six theorems on the

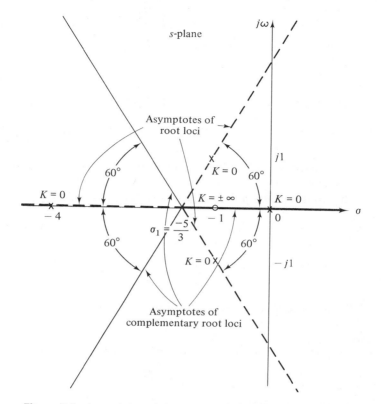

Figure 7-6 Asymptotes of the complete root loci of $s(s + 4)(s^2 + 2s + 2) + K(s + 1) = 0$.

construction of the complete root loci described so far, the following information concerning the RL and the CRL of Eq. (7-61) is obtained:

1. $K = 0$: The $K = 0$ points on the complete root loci are at the poles of $G(s)H(s)$: $s = 0$, $s = -4$, $s = -1 + j1$, and $s = -1 - j1$.
2. $K = \pm\infty$: The $K = \pm\infty$ points on the complete root loci are at the zeros of $G(s)H(s)$: $s = -1$, $s = \infty$, $s = \infty$, and $s = \infty$.
3. Since Eq. (7-61) is of the fourth order, there are four complete root loci.
4. The complete root loci are symmetrical to the real axis.
5. For large values of s, the RL are asymptotic to straight lines with angles measured from the real axis: Root loci ($K \geq 0$), Eq. (7-40),

$$k = 0 \qquad \theta_0 = \frac{180°}{3} = 60°$$

$$k = 1 \qquad \theta_1 = \frac{540°}{3} = 180°$$

$$k = 2 \qquad \theta_2 = \frac{900°}{3} = 300°$$

The angles of the asymptotes of the complementary root loci are given by Eq. (7-41):

$$k = 0 \qquad \theta_0 = \frac{0°}{3} = 0°$$

$$k = 1 \qquad \theta_1 = \frac{360°}{3} = 120°$$

$$k = 2 \qquad \theta_2 = \frac{720°}{3} = 240°$$

The asymptotes of the complementary root loci may be obtained by extending the asymptotes of the root loci.

6. The six asymptotes of the complete root loci intersect at

$$\sigma_1 = \frac{\sum \text{ finite poles of } G(s)H(s) - \sum \text{ finite zeros of } G(s)H(s)}{n - m}$$
$$= \frac{(0 - 4 - 1 + j1 - 1 - j1) - (-1)}{4 - 1} = -\frac{5}{3} \qquad (7\text{-}63)$$

The asymptotes are sketched as shown in Fig. 7-6.

Example 7-7
The asymptotes of the complete root loci for several different equations are shown in Fig. 7-7.

Root Loci on the Real Axis

Theorem 7-7. (a) *Root loci: On a given section of the real axis, RL ($K \geq 0$) are found in the section only if the total number of real poles and zeros of $G(s)H(s)$ to the right of the section is odd.*

(b) *Complementary root loci: On a given section of the real axis, CRL ($K \leq 0$) are found in the section only if the total number of real poles and zeros of $G(s)H(s)$ to the right of the section is even. Alternatively, we can state that CRL will be found in sections on the real axis not occupied by the root loci.*

In all cases the complex poles and zeros of $G(s)H(s)$ do not affect the existence properties of the root loci on the real axis.

Proof. The proof of the theorem is based on the following observations:

1. At any point (e.g., s_1) on the real axis, the angles of the vectors drawn from the complex-conjugate poles and zeros of $G(s)H(s)$ add up to be zero. Therefore, the only contribution to the angular relations in Eqs. (7-20) and (7-21) is from the real poles and zeros of $G(s)H(s)$.

2. Only the real poles and zeros of $G(s)H(s)$ that lie to the right of the point s_1 may contribute to Eqs. (7-20) and (7-21), since real poles and zeros that lie to the left of the point contribute zero degrees.

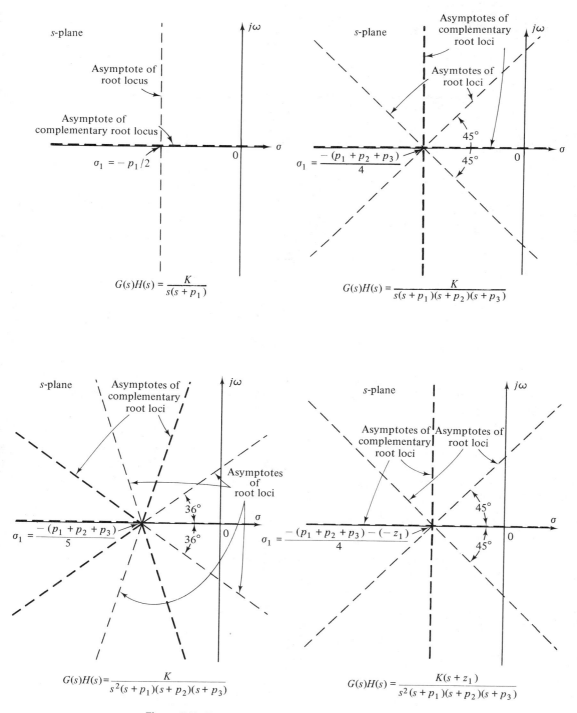

Figure 7-7 Examples of the asymptotes of the root loci.

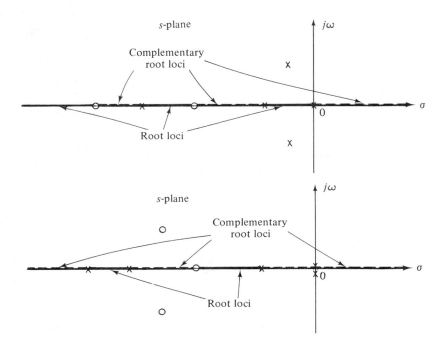

Figure 7-8 Properties of root loci on the real axis.

3. Each real pole of $G(s)H(s)$ to the right of the point s_1 contributes -180 degrees and each zero to the right of the point contributes 180 degrees to Eqs. (7-20) and (7-21).

The last observation shows that for s_1 to be a point on the RL, there must be an odd number of poles and zeros of $G(s)H(s)$ to the right of s_1, and for s_1 to be a point on the CRL, the total number of poles and zeros of $G(s)H(s)$ to the right of s_1 must be even. The following example illustrates the properties of the complete root loci on the real axis of the s-plane.

Example 7-8

The complete root loci on the real axis in the s-plane are shown in Fig. 7-8 for two different pole-zero configurations of $G(s)H(s)$. Notice that the entire real axis is occupied by either the RL or the CRL.

Angles of Departure (from Poles) and the Angles of Arrival (at Zeros) of the Complete Root Loci

The angle of departure (arrival) of the complete root locus at a pole (zero) of $G(s)H(s)$ denotes the behavior of the root loci near that pole (zero). For the RL $(K \geq 0)$ these angles can be determined by use of Eq. (7-20). For instance, in the

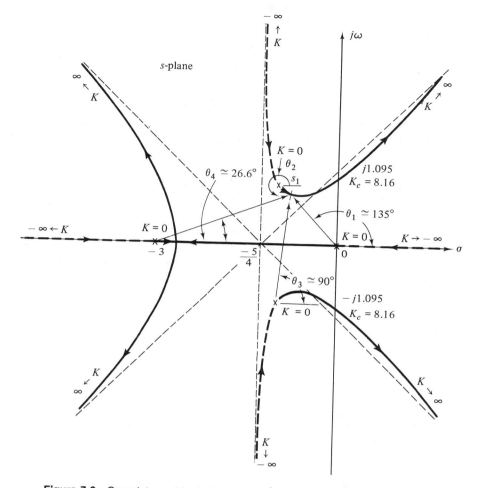

Figure 7-9 Complete root loci of $s(s + 3)(s^2 + 2s + 2) + K = 0$ to illustrate the angles of departure or arrival.

pole–zero configuration of $G(s)H(s)$ given in Fig. 7-9, it is desired to determine the angle at which the root locus leaves the pole at $-1 + j1$. Notice that the unknown angle θ_2 is measured with respect to the real axis. Let us assume that s_1 is a point on the root locus leaving the pole at $-1 + j1$ and is very near the pole. Then s_1 must satisfy Eq. (7-20). Thus

$$\underline{/G(s_1)H(s_1)} = -(\theta_1 + \theta_2 + \theta_3 + \theta_4) = (2k + 1)180° \qquad (7-64)$$

Since s_1 is very close to the pole at $-1 + j1$, the angles of the vectors drawn from the other three poles are determined from Fig. 7-9, and Eq. (7-64) becomes

$$-(135° + \theta_2 + 90° + 26.6°) = (2k + 1)180° \qquad (7-65)$$

We can simply set k equal to zero, since the same result is obtained for all other values. Therefore,

$$\theta_2 = -431.6°$$

which is the same as -71.6 degrees.

When the angle of the root locus at a pole or zero of $G(s)H(s)$ is determined, the angle of the CRL at the same point differs from this angle by 180 degrees, since Eq. (7-21) must now be used.

Intersection of the Root Loci with the Imaginary Axis

The points where the complete root loci intersect the imaginary axis of the s-plane, and the corresponding values of K, may be determined by means of the Routh–Hurwitz criterion. For complex situations with multiple intersections on the imaginary axis, the critical values of K and ∞ can be more easily determined by finding the phase-crossover points on the Bode plot, as described in Chapter 9.

■

Example 7-9

The complete root loci of the equation

$$s(s + 3)(s^2 + 2s + 2) + K = 0 \qquad (7\text{-}66)$$

are drawn in Fig. 7-9. The RL intersect the $j\omega$ axis at two conjugate points. Applying the Routh–Hurwitz criterion to Eq. (7-66), we have, by solving the auxiliary equation, $K_c = 8.16$ and $\omega_c = \pm 1.095$ rad/sec.

■

Breakaway Points (Saddle Points) on the Complete Root Loci

Breakaway points or saddle points on the root loci of an equation correspond to multiple-order roots of the equation. Figure 7-10(a) illustrates a case in which two branches of the root loci meet at the breakaway point on the real axis and then depart from the axis in opposite directions. In this case the breakaway point represents a double root of the equation to which the root loci belong. Figure 7-10(b) shows another common situation where a breakaway point may occur.

In general a breakaway point may involve more than two root loci. Figure 7-11 illustrates a situation where the breakaway point represents a fourth-order root.

A root locus diagram can, of course, have more than one breakaway point. Moreover, the breakaway points need not always be on the real axis. However, because of the conjugate symmetry of the root loci, the breakaway points must either be real or in complex-conjugate pairs.

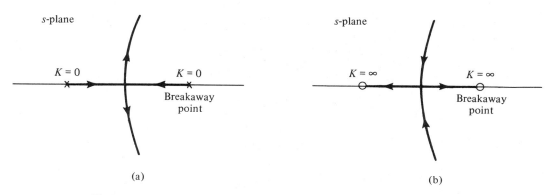

Figure 7-10   Examples of breakaway points on the real axis in the *s*-plane.

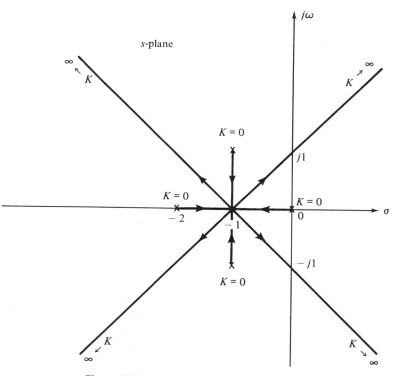

Figure 7-11   Fourth-order breakaway point.

Because of the symmetry of the root loci, it is easy to see that the root loci in Fig. 7-10(a) and (b) break away at 180° apart, whereas in Fig. 7-11 the four root loci depart with angles 90° apart. In general, if *n* root loci ($-\infty < K < \infty$) approach or leave a breakaway point, they must be $180/n$ degrees apart.

   Several graphical and analytical methods are available for the determination of the location of the breakaway points. A method that seems to be the most general is presented below.

Theorem 7-8. *The breakaway points on the complete root loci of* $1 +$ $KG_1(s)H_1(s) = 0$ *must satisfy*

$$\frac{dG_1(s)H_1(s)}{ds} = 0 \tag{7-67}$$

Proof.   From Eq. (7-1),

$$Q(s) + KP(s) = 0 \tag{7-68}$$

Equation (7-6) is repeated as follows:

$$G(s)H(s) = \frac{KP(s)}{Q(s)} \tag{7-69}$$

If we consider that K is varied by an increment ΔK, Eq. (7-68) becomes

$$Q(s) + (K + \Delta K)P(s) = 0 \tag{7-70}$$

Dividing both sides of Eq. (7-70) by $Q(s) + KP(s)$, we have

$$1 + \frac{\Delta KP(s)}{Q(s) + KP(s)} = 0 \tag{7-71}$$

which can be written

$$1 + \Delta KF(s) = 0 \tag{7-72}$$

where

$$F(s) = \frac{P(s)}{Q(s) + KP(s)} \tag{7-73}$$

Since the denominator of $F(s)$ is the same as the left-hand side of Eq. (7-68), at points very close to an nth-order root $s = s_i$ of Eq. (7-68), which corresponds to a breakaway point of n loci, $F(s)$ can be approximated by

$$F(s) \cong \frac{A_i}{(s - s_i)^n} = \frac{A_i}{(\Delta s)^n} \tag{7-74}$$

where A_i is a constant.
    Substituting Eq. (7-74) into Eq. (7-72) gives

$$1 + \frac{\Delta KA_i}{(\Delta s)^n} = 0 \tag{7-75}$$

from which we obtain

$$\frac{\Delta K}{\Delta s} = -\frac{(\Delta s)^{n-1}}{A_i} \qquad (7\text{-}76)$$

Taking the limit on both sides of the last equation as ΔK approaches zero, we have

$$\lim_{\Delta K \to 0} \frac{\Delta K}{\Delta s} = \frac{dK}{ds} = 0 \qquad (7\text{-}77)$$

We have shown that at a breakaway point on the root loci, dK/ds is zero.[7]
    Now, since the roots of Eq. (7-68) must also satisfy

$$1 + KG_1(s)H_1(s) = 0 \qquad (7\text{-}78)$$

or

$$K = -\frac{1}{G_1(s)H_1(s)} \qquad (7\text{-}79)$$

it is simple to see that $dK/ds = 0$ is equivalent to

$$\frac{dG_1(s)H_1(s)}{ds} = 0 \qquad (7\text{-}80)$$

It is important to point out that the condition for the breakaway point given by Eq. (7-80) is necessary but not sufficient. In other words, all breakaway points must satisfy Eq. (7-80), but not all solutions of Eq. (7-80) are breakaway points. To be a breakaway point, the solution of Eq. (7-80) must also satisfy Eq. (7-68); or, Eq. (7-80) must be a factor of Eq. (7-68) for some real K.
    In general, the following conclusions can be made with regard to the solutions of Eq. (7-80):

1. All real solutions of Eq. (7-80) are breakaway points on the root loci $(-\infty < K < \infty)$, since the entire real axis of the s-plane is occupied by the complete root loci.

2. The complex-conjugate solutions of Eq. (7-80) are breakaway points only if they also satisfy Eq. (7-68). This uncertainty does not cause difficulty in the effective use of Eq. (7-80), since the other properties of the root loci are usually sufficient to provide information on the general proximity of the breakaway points. This information may also be helpful in solving for the roots of a high-order equation, as a result of Eq. (7-67), by trial and error.

---

[7]The quantity $(ds/s)/(dK/K)$ is defined as the root sensitivity [44–46] of an equation with respect to incremental variation of the parameter K. In this case it is proved that at the breakaway points of the root loci, the roots have infinite sensitivity.

The following examples are devised to illustrate the application of Eq. (7-67) for the determination of breakaway points on the root loci.

∎

Example 7-10

Consider the problem of sketching the root loci of the second-order equation

$$s(s + 2) + K(s + 4) = 0 \qquad (7-81)$$

Based on some of the theorems on root loci, the root loci of Eq. (7-81) are easily sketched, as shown in Fig. 7-12. It can be proven that the complex part of the loci is described by a circle. The two breakaway points are all on the real axis, one between 0 and -2 and the other between -4 and $-\infty$.

When we divide both sides of Eq. (7-81) by $s(s + 2)$, the term that does not contain K, we get,

$$G_1(s)H_1(s) = \frac{(s + 4)}{s(s + 2)} \qquad (7-82)$$

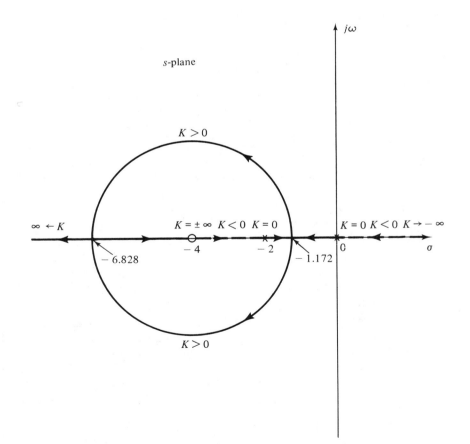

Figure 7-12   Root loci of $s(s + 2) + K(s + 4) = 0$.

Therefore, from Eq. (7-67), the breakaway points must satisfy

$$\frac{dG_1(s)H_1(s)}{ds} = \frac{s(s+2) - 2(s+1)(s+4)}{s^2(s+2)^2} = 0 \tag{7-83}$$

or

$$s^2 + 8s + 8 = 0 \tag{7-84}$$

Solving Eq. (7-84), we find that the two breakaway points of the root loci are at $s = -1.172$ and $s = -6.828$. Note also that the breakaway points happen to occur all on the RL ($K > 0$).

Example 7-11

Consider the equation

$$s^2 + 2s + 2 + K(s+2) = 0 \tag{7-85}$$

The equivalent $G(s)H(s)$ is obtained by dividing both sides of Eq. (7-85) by $s^2 + 2s + 2$,

$$G(s)H(s) = \frac{K(s+2)}{s^2 + 2s + 2} \tag{7-86}$$

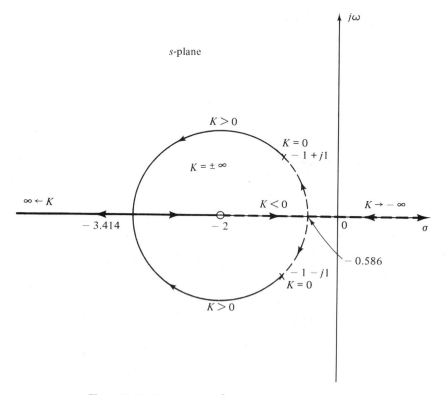

Figure 7-13 Root loci of $s^2 + 2s + 2 + K(s+2) = 0$.

Based on the poles and zeros of $G(s)H(s)$, the complete root loci of Eq. (7-85) are sketched as shown in Fig. 7-13. The diagram shows that both the RL and the CRL possess a breakaway point. These breakaway points are determined from

$$\frac{dG_1(s)H_1(s)}{ds} = \frac{d}{ds}\left[\frac{(s+2)}{s^2+2s+2}\right]$$

$$= \frac{s^2+2s+2-2(s+1)(s+2)}{(s^2+2s+2)^2} = 0 \qquad (7\text{-}87)$$

or

$$s^2+4s+2=0 \qquad (7\text{-}88)$$

Upon solving Eq. (7-88), the breakaway points are found to be at $s = -0.586$ and $s = -3.414$. Notice that in this case $s = -3.414$ is a breakaway point on the RL, whereas $s = -0.586$ is a breakaway point on the CRL.

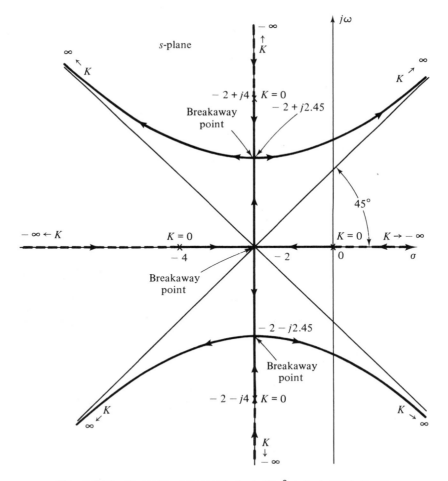

Figure 7-14   Complete root loci of $s(s+4)(s^2+4s+20)+K=0$.

Example 7-12

Figure 7-14 shows the complete root loci of the equation

$$s(s + 4)(s^2 + 4s + 20) + K = 0 \tag{7-89}$$

Dividing both sides of Eq. (7-89) by the terms that do not contain K, we have

$$1 + G(s)H(s) = 1 + \frac{K}{s(s + 4)(s^2 + 4s + 20)} = 0 \tag{7-90}$$

Since the poles of $G(s)H(s)$ are symmetrical about the axes $\sigma = -2$ and $\omega = 0$ in the s-plane, the complete root loci of the equation are also symmetrical with respect to these two axes.

Taking the derivative of $G_1(s)H_1(s)$ with respect to s, we get

$$\frac{dG_1(s)H_1(s)}{ds} = -\frac{4s^3 + 24s^2 + 72s + 80}{[s(s + 4)(s^2 + 4s + 20)]^2} = 0 \tag{7-91}$$

or

$$s^3 + 6s^2 + 18s + 20 = 0 \tag{7-92}$$

Because of the symmetry of the poles of $G(s)H(s)$, one of the breakaway points is easily determined to be at $s = -2$. The other two breakaway points are found by solving Eq. (7-92) using this information; they are $s = -2 + j2.45$ and $s = -2 - j2.45$.

Example 7-13

In this example we shall show that the solutions of Eq. (7-67) do not necessarily represent breakaway points on the root loci. The complete root loci of the equation

$$s(s^2 + 2s + 2) + K = 0 \tag{7-93}$$

are shown in Fig. 7-15; neither the RL nor the CRL have any breakaway point in this case. However, writing Eq. (7-93) as

$$1 + KG_1(s)H_1(s) = 1 + \frac{K}{s(s^2 + 2s + 2)} = 0 \tag{7-94}$$

and applying Eq. (7-67), we have

$$\frac{dG_1(s)H_1(s)}{ds} = \frac{d}{ds}\frac{1}{s(s^2 + 2s + 2)} = 0 \tag{7-95}$$

which gives

$$3s^2 + 4s + 2 = 0 \tag{7-96}$$

The roots of Eq. (7-96) are $s = -0.667 + j0.471$ and $s = -0.667 - j0.471$. These two roots

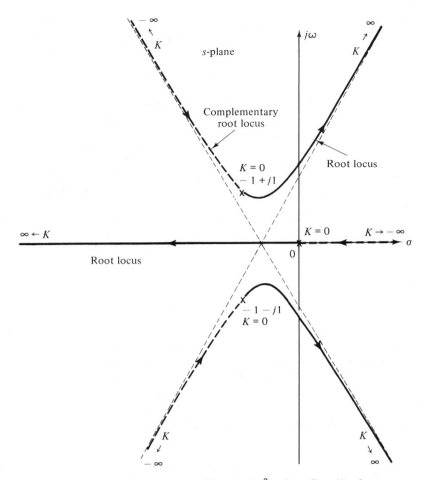

Figure 7-15 Complete root loci of $s(s^2 + 2s + 2) + K = 0$.

do not represent breakaway points on the root loci, since they do not satisfy Eq. (7-93) for any real values of K.

∎

Calculation of K on the Root Loci

Once the root loci have been constructed, the values of K at any point s_1 on the loci can be determined by use of the defining equation of Eq. (7-19); that is,

$$|K| = \frac{1}{|G_1(s_1)H_1(s_1)|} \qquad (-\infty < K < \infty) \qquad (7\text{-}97)$$

or

$$|K| = \frac{\prod\limits_{j=1}^{n} |s_1 + p_j|}{\prod\limits_{i=1}^{m} |s_1 + z_i|} \qquad (-\infty < K < \infty) \qquad (7\text{-}98)$$

or

$$|K| = \frac{\text{product of lengths of vectors drawn from the poles of } G_1(s)H_1(s) \text{ to } s_1}{\text{product of lengths of vectors drawn from the zeros of } G_1(s)H_1(s) \text{ to } s_1}$$
$$(7\text{-}99)$$

Equations (7-98) and (7-99) can be evaluated either graphically or analytically. Usually, if the root loci are already drawn accurately, the graphical method is more convenient. For example, the root loci of the equation

$$s^2 + 2s + 2 + K(s + 2) = 0 \qquad (7\text{-}100)$$

are shown in Fig. 7-16. The value of K at the point s_1 is given by

$$K = \frac{A \cdot B}{C} \qquad (7\text{-}101)$$

where A and B are the lengths of the vectors drawn from the poles of $G(s)H(s) = K(s + 2)/(s^2 + 2s + 2)$ to the point s_1 and C is the length of the vector drawn from the zero of $G(s)H(s)$ to s_1. In the illustrated case s_1 is on the RL, so K is positive. If s_1 is a point on the CRL, K should have a negative sign.

The value of K at the point where the root loci intersect the imaginary axis can also be found by the method just described. However, the Routh–Hurwitz criterion usually represents a more direct method of computing this critical value of K.

The 11 rules on the construction of root locus diagrams described above should be regarded only as important properties of the root loci. Remember earlier it was pointed out that the usefulness of most of these rules of construction depends on first writing Eq. (7-1) in the form

$$(s + p_1)(s + p_2) \cdots (s + p_n) + K(s + z_1)(s + z_2) \cdots (s + z_m) = 0 \quad (7\text{-}102)$$

Then, except for extremely complex cases, these rules are usually adequate for the analyst to make a reasonably accurate sketch of the root loci just short of plotting

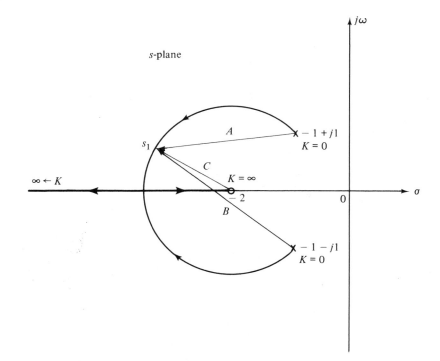

Figure 7-16 Graphical method of finding the values of K on the root loci.

them point by point. In complicated situations, one has to rely on a computer as a more practical means of constructing the root loci.

The following example illustrates the rules of construction for the root loci.

■

Example 7-14

Consider the equation

$$s(s + 5)(s + 6)(s^2 + 2s + 2) + K(s + 3) = 0 \qquad (7\text{-}103)$$

The complete root loci $(-\infty < K < \infty)$ of the system are to be constructed. Using the rules of construction, the following properties of the root loci are determined:

1. The $K = 0$ points on the complete root loci are at $s = 0$, -5, -6, $-1 + j1$, and $-1 - j1$. Notice that these points are the poles of $G(s)H(s)$, where

$$G(s)H(s) = \frac{K(s + 3)}{s(s + 5)(s + 6)(s^2 + 2s + 2)} \qquad (7\text{-}104)$$

2. The $K = \pm\infty$ points on the complete root loci are at $s = -3, \infty, \infty, \infty, \infty$, which are the zeros of $G(s)H(s)$.

3. There are five separate branches on the complete root loci.
4. The complete root loci are symmetrical with respect to the real axis of the s-plane.
5. The angles of the asymptotes of the RL at infinity are given by [Eq. (7-40)]

$$\theta_k = \frac{(2k+1)\pi}{n-m} = \frac{(2k+1)\pi}{5-1} \qquad 0 \le K < \infty \qquad (7\text{-}105)$$

for $k = 0, 1, 2, 3$. Thus the four root loci that approach infinity in the s-plane as K approaches $+\infty$ should approach asymptotes with angles of 45, -45, 135, and -135 degrees, respectively. The angles of the asymptotes of the CRL at infinity are given by [Eq. (7-41)]

$$\theta_k = \frac{2k\pi}{n-m} = \frac{2k\pi}{5-1} \qquad -\infty < K \le 0 \qquad (7\text{-}106)$$

Therefore, as K approaches $-\infty$, four CRL should approach infinity along asymptotes with angles of 0, 90, 180, and 270 degrees.

6. The intersection of the asymptotes is given by [Eq. (7-57)]:

$$\sigma_1 = \frac{b_1 - a_1}{n-m} = \frac{\sum \text{poles of } G(s)H(s) - \sum \text{zeros of } G(s)H(s)}{n-m}$$

$$= \frac{(0 - 5 - 6 - 1 + j1 - 1 - j1) - (-3)}{4} = -2.5 \qquad (7\text{-}107)$$

The results from these six step are illustrated in Fig. 7-17.

In general, the rules on the asymptotes do not indicate on which side of the asymptote the root locus will lie. Therefore, the asymptotes indicate nothing more than the behavior of the root loci as $s \to \infty$. In fact, a root locus can cross an asymptote. The segments of the RL and the CRL shown in Fig. 7-17 can be accurately made only if information in addition to the asymptotes is obtained.

7. Complete root loci on the real axis: There are RL $(0 \le K < \infty)$ on the real axis between $s = 0$ and $s = -3$, $s = -5$ and $s = -6$. There are CRL $(-\infty < K \le 0)$ on the remaining portion of the real axis, that is, between $s = -3$ and $s = -5$, and $s = -6$ and $s = -\infty$ (see Fig. 7-18).

8. Angles of departure: The angle of departure; θ, of the RL leaving the pole at $-1 + j1$ is determined using Eq. (7-20). If s_1 is a point on the RL leaving $-1 + j1$, and s_1 is very close to $-1 + j1$ as shown in Fig. 7-19, Eq. (7-20) gives

$$\underline{/s_1 + 3} - \left(\underline{/s_1} + \underline{/s_1 + 1 + j1} + \underline{/s_1 + 5} + \underline{/s_1 + 6} \right.$$

$$\left. + \underline{/s_1 + 1 - j1} \right) = (2k+1)180° \quad (7\text{-}108)$$

or

$$26.6° - (135° + 90° + 14° + 11.4° + \theta) \cong (2k+1)180° \qquad (7\text{-}109)$$

for $k = 0, \pm 1, \pm 2, \ldots$. Therefore,

$$\theta \cong -43.8° \qquad (7\text{-}110)$$

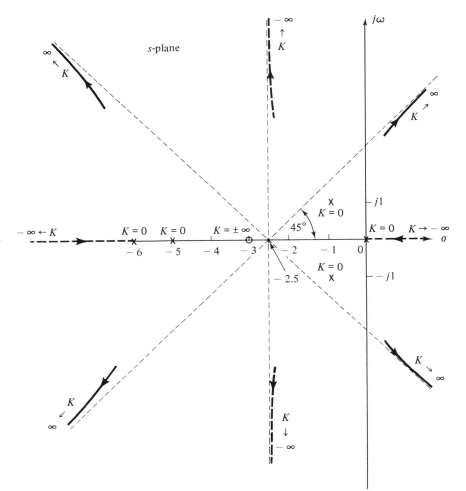

Figure 7-17 Preliminary calculations of the root loci of $s(s + 5)(s + 6)(s^2 + 2s + 2) + K(s + 3) = 0$.

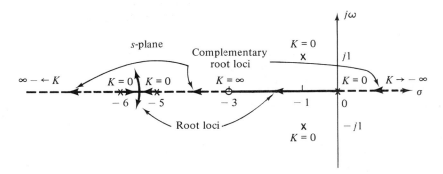

Figure 7-18 Complete root loci on the real axis of $s(s + 5)(s + 6)(s^2 + 2s + 2) + K(s + 3) = 0$.

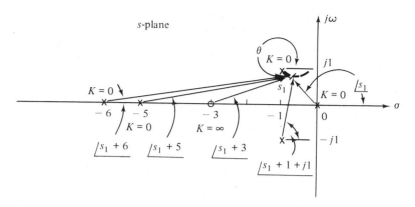

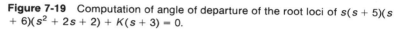

Figure 7-19 Computation of angle of departure of the root loci of $s(s + 5)(s + 6)(s^2 + 2s + 2) + K(s + 3) = 0$.

Similarly, Eq. (7-21) is used to determine the angle of arrival of the complementary root locus arriving at the point $-1 + j1$. If this angle is designated as θ', it is easy to see that θ' differs from θ by 180 degrees; that is,

$$\theta' = 180° - 43.8° = 136.2° \qquad (7\text{-}111)$$

9. The intersection of the root loci of the imaginary axis is determined by the Routh–Hurwitz criterion. Equation (7-103) is rewritten

$$s^5 + 13s^4 + 54s^3 + 82s^2 + (60 + K)s + 3K = 0 \qquad (7\text{-}112)$$

the Routh tabulation is

s^5	1	54	$60 + K$
s^4	13	82	$3K$
s^3	47.7	$0.769K$	0
s^2	$65.6 - 0.212K$	$3K$	0
s^1	$\dfrac{3940 - 105K - 0.163K^2}{65.6 - 0.212K}$	0	0
s^0	$3K$	0	

For Eq. (7-112) to have no roots in the right half of the s-plane, the quantities in the first column of the Routh tabulation should be of the same sign. Therefore, the following inequalities must be satisfied:

$$65.6 - 0.212K > 0 \qquad \text{or} \quad K < 309 \qquad (7\text{-}113)$$

$$3940 - 105K - 0.163K^2 > 0 \qquad \text{or} \quad K < 35 \qquad (7\text{-}114)$$

$$K > 0 \qquad (7\text{-}115)$$

Hence all the roots of Eq. (7-112) will stay in the left half of the s-plane if K lies between 0 and 35, which means that the root loci of Eq. (7-112) cross the imaginary axis when $K = 35$ and $K = 0$. The coordinate at the crossover point on the imaginary axis that corresponds to $K = 35$ is determined from the auxiliary equation

$$A(s) = (65.6 - 0.212K)s^2 + 3K = 0 \qquad (7\text{-}116)$$

Substituting $K = 35$ into Eq. (7-116), we have

$$58.2s^2 + 105 = 0 \qquad (7\text{-}117)$$

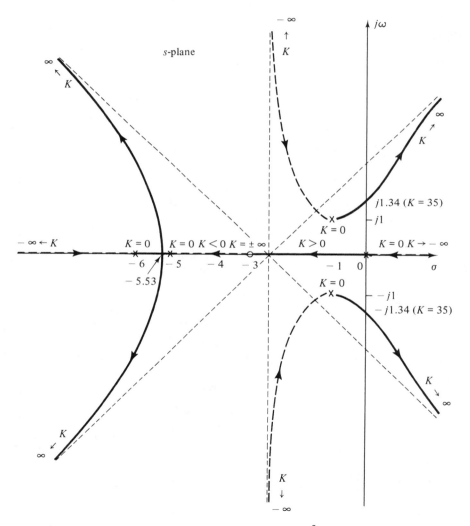

Figure 7-20 Complete root loci of $s(s + 5)(s + 6)(s^2 + 2s + 2) + K(s + 3)$
= 0.

Table 7-1   Rules of Construction of Root Loci

1. $K = 0$ points	The $K = 0$ points on the complete root loci are at the poles of $G(s)H(s)$. (The poles include those at infinity.)
2. $K = \pm \infty$ points	The $K = \pm \infty$ points on the complete root loci are at the zeros of $G(s)H(s)$. (The zeros include those at infinity.)
3. Number of separate root loci	The total number of root loci is equal to the order of the equation $F(s) = 0$.
4. Symmetry of root loci	The complete root loci of systems with rational transfer functions with constant coefficients are symmetrical with respect to the axes of symmetry of the poles and zeros of $G(s)H(s)$.
5. Asymptotes of root loci as $s \to \infty$	For large values of s, the RL ($K > 0$) are asymptotic to straight lines with angles given by $$\theta_k = \frac{(2k + 1)\pi}{n - m}$$ and for the CRL ($K < 0$) $$\theta_k = \frac{2k\pi}{n - m}$$ where $k = 0, 1, 2, \ldots, \|n - m\| - 1$.
6. Intersection of the asymptotes (centroids)	(a) The intersection of the asymptotes lies only on the real axis in the s-plane. (b) The point of intersection of the asymptotes on the real axis is given by (for all values of K) $$\sigma_1 = \frac{\sum \text{ real parts of poles of } G(s)H(s) - \sum \text{ real parts of zeros of } G(s)H(s)}{n - m}$$
7. Root loci on the real axis	On a given section on the real axis in the s-plane, RL is found for $K \geq 0$ in the section only if the total number of real poles and real zeros of $G(s)H(s)$ to the right of the section is *odd*. If the total number of real poles and zeros to the right of a given section is *even*, CRL ($K \leq 0$) are found in the section.
8. Angles of departure and arrival	The angle of departure of the root locus ($K \geq 0$) from a pole or the angle of arrival at a zero of $G(s)H(s)$ can be determined by assuming a point s_1 that is on the root locus associated with the pole, or zero, and which is very close to the pole, or zero,

Table 7-1   (continued)

and applying the equation

$$\underline{/G(s_1)H(s_1)} = \sum_{i=1}^{m} \underline{/s_1 + z_i} - \sum_{j=1}^{n} \underline{/s_1 + p_j}$$

$$= (2k + 1)\pi \qquad k = 0, \pm 1, \pm 2, \ldots$$

The angle of departure or arrival of a complementary root locus is determined from

$$\underline{/G(s_1)H(s_1)} = \sum_{i=1}^{m} \underline{/s_1 + z_i} - \sum_{j=1}^{n} \underline{/s_1 + p_j}$$

$$= 2k\pi \qquad k = 0, \pm 1, \pm 2, \ldots$$

9. Intersection of the root loci with the imaginary axis

The values of ω and K at the crossing points of the root loci on the imaginary axis of the s-plane may be obtained by use of the Routh–Hurwitz criterion. The Bode plot of $G(s)H(s)$ may also be used.

10. Breakaway points (saddle points)

The breakaway points on the complete root loci are determined by finding the roots of $dK/ds = 0$, or $dG(s)H(s)/ds = 0$. These are necessary conditions only.

11. Calculation of the values of K on the root loci

The absolute value of K at any point s_1 on the complete root loci is determined from the equation

$$|K| = \frac{1}{|G(s_1)H(s_1)|}$$

$$= \frac{\text{product of lengths of vectors drawn from the poles of } G(s)H(s) \text{ to } s_1}{\text{product of lengths of vectors drawn from zeros of } G(s)H(s) \text{ to } s_1}$$

which yields

$$s = \pm j1.34$$

10. Breakaway points: Based on the information obtained from the preceding nine steps, a trial sketch of the root loci indicates that there can be only one breakaway point on the entire root loci, and the point lies between the two poles of $G(s)H(s)$ at $s = -5$ and -6.

Applying $dK/ds = 0$ to Eq. (7-112) gives

$$s^5 + 13.5s^4 + 66s^3 + 142s^2 + 123s + 45 = 0 \qquad (7-118)$$

Since there is only one breakaway point, there is no need to solve this fifth-order equation. It is simpler to solve for the breakaway point by trial and error, since we know that the desired root is between -5 and -6. After a few trial-and-error calculations, the root of the last equation that corresponds to the breakaway point is found to be $s = -5.53$.

The other four roots of Eq. (7-118) are found by a digital computer program, $s = -0.656 \pm j0.468$ and $s = -3.33 \pm j1.204$. We can show that these solutions are not points on the RL or the CRL and are not breakaway points.

From the information obtained in these 10 steps, the complete root locus diagram is sketched as shown in Fig. 7-20.

◼

In this section we have described 11 important properties of the root loci. These properties have been regarded as rules when they are used in aiding the construction of the root locus diagram. Of course, there are other minor properties of the root loci which are not mentioned here. However, in general, it is found that these 11 rules are adequate in helping to obtain a reasonably accurate sketch of the complete root loci just short of actually plotting them.

For each reference, the 11 rules of construction are tabulated in Table 7-1.

7.4 SOME IMPORTANT ASPECTS OF THE CONSTRUCTION OF THE ROOT LOCI

One of the important aspects of the root locus techniques is that for most control systems with moderate complexity, the analyst or designer may conduct a quick study of the system in the s-plane by making a sketch of the root loci using some or all of the rules of construction. In general, it is not necessary to make an exact plot of the root loci. Therefore, time may be saved by skipping some of the rules, and the sketching of the root locus diagram becomes an art that depends to some extent on the experience of the analyst.

In this section we shall present some of the important properties of the root loci which may be helpful in the construction of the root locus diagram.

Effect of Adding Poles and Zeros to *G(s)H(s)*

In Chapter 6 the effects of the derivative and integral control were illustrated by means of the root locus diagram. From the fundamental viewpoint we may investigate the effects to the root loci when poles and zeros are added to $G(s)H(s)$.

Addition of Poles. In general we may state that adding a pole to the function $G(s)H(s)$ in the left half of the s-plane has the effect of pushing the original root loci toward the right-half plane. Although it is difficult to make a precise statement and provide the necessary proof, we can illustrate the point by several examples. Let us consider the function

$$G(s)H(s) = \frac{K}{s(s + a)} \qquad a > 0 \qquad (7\text{-}119)$$

The zeros of $1 + G(s)H(s)$ are presented by the root locus diagram of Fig. 7-21(a). These root loci are constructed based on the poles of $G(s)H(s)$ at $s = 0$ and

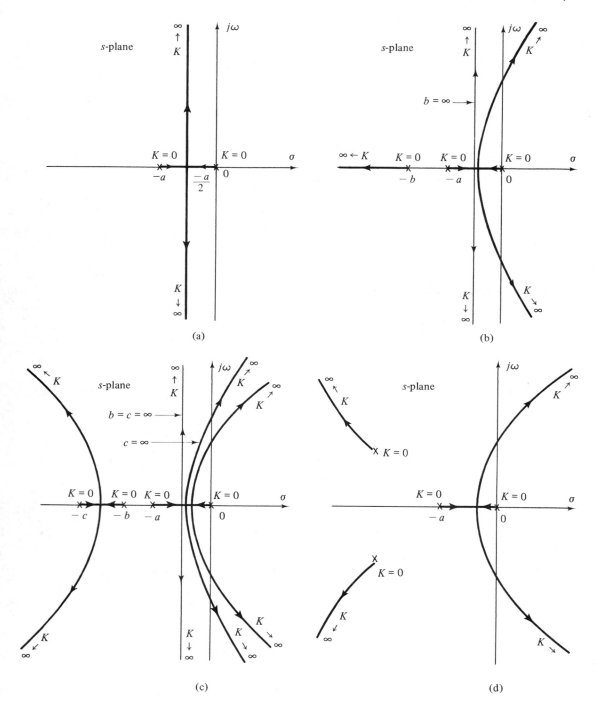

Figure 7-21   Root locus diagrams that show the effects of adding poles to $G(s)H(s)$.

$s = -a$. Now let us introduce a pole at $s = -b$ so that

$$G(s)H(s) = \frac{K}{s(s+a)(s+b)} \qquad b > a \qquad (7\text{-}120)$$

Figure 7-21(b) shows that the additional pole causes the complex part of the root loci to bend toward the right half of the s-plane. The angles of the asymptotes are changed from ± 90 degrees to ± 60 degrees. The breakaway point is also moved to the right. For instance, if $a = 1$ and $b = 2$, the breakaway point is moved from -0.5 to -0.422 on the real axis. If $G(s)H(s)$ represents the loop transfer function of a feedback control system, the system with the root loci in Fig. 7-21(b) may become unstable if the value of K exceeds the critical value, whereas the system represented by the root loci of Fig. 7-21(a) is always stable. Figure 7-21(c) shows the root loci when another pole is added to $G(s)H(s)$ at $s = -c$. The system is now of the fourth order, and the two complex root loci are moved farther to the right. The angles of the asymptotes of these two loci are $\pm 45°$. For a feedback control system, the stability condition of the system becomes even more restricted. Figure 7-21(d) illustrates that the addition of a pair of complex-conjugate poles to the original two-pole configuration will result in a similar effect. Therefore, we may draw a general conclusion that the addition of poles to the function $G(s)H(s)$ has the effect of moving the root loci toward the right half of the s-plane.

Addition of Zeros. Adding zeros to the function $G(s)H(s)$ has the effect of moving the root loci toward the left half of the s-plane. For instance, Fig. 7-22(a) shows the root locus diagram when a zero at $s = -b$ is added to the function $G(s)H(s)$ of Eq. (7-119) with $b > a$; the resultant root loci are bent toward the left and form a circle. Therefore, if $G(s)H(s)$ represents the loop transfer function of a feedback control system, the relative stability of the system is improved by the addition of the zero. Figure 7-22(b) illustrates that a similar effect will result if a pair of complex-conjugate zeros is added to the function of Eq. (7-119). Figure 7-22(c) shows the root locus diagram when a zero at $s = -c$ is added to the transfer function of Eq. (7-120).

Effects of Movements of Poles and Zeros

It was mentioned earlier that the construction of the root locus diagram depends greatly on the understanding of the principle of the technique rather than just the rigid rules of construction. In this section we show that in all cases the study of the effects of the movement of the poles and zeros of $G(s)H(s)$ on the root loci is an important and useful subject. Again, the best way to illustrate the subject is to use a few examples.

■

Example 7-15

Consider the equation

$$s^2(s+a) + K(s+b) = 0 \qquad (7\text{-}121)$$

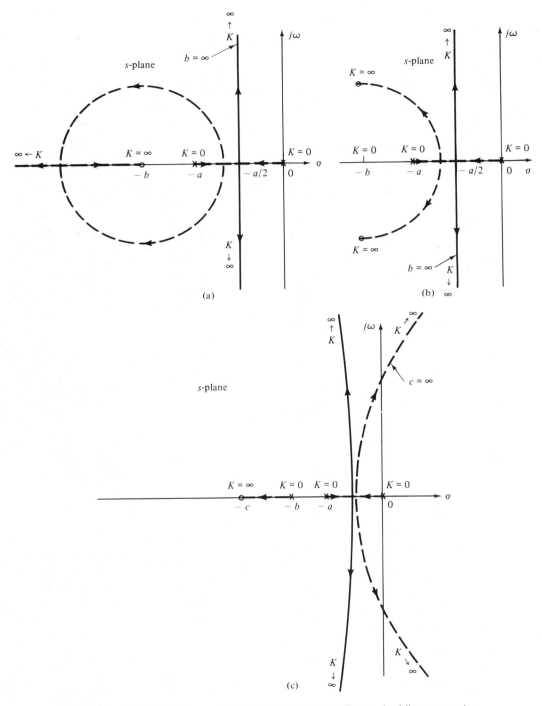

Figure 7-22 Root locus diagrams that show the effects of adding a zero to $G(s)H(s)$.

which is easily converted to the form of $1 + G(s)H(s) = 0$, with

$$G(s)H(s) = \frac{K(s+b)}{s^2(s+a)} \tag{7-122}$$

Let us set $b = 1$ and investigate the root loci of Eq. (7-121) for several values of a.

Figure 7-23(a) illustrates the root loci of Eq. (7-121) with $a = 10$ and $b = 1$. The two breakaway points are found at $s = -2.5$ and -4.0. It can be shown that for arbitrary a the nonzero breakaway points are given by

$$s = -\frac{a+3}{4} \pm \frac{1}{4}\sqrt{a^2 - 10a + 9} \tag{7-123}$$

When $a = 9$, Eq. (7-123) indicates that the breakaway points converge to one point at $s = -3$, and the root locus diagram becomes that of Fig. 7-23(b). It is interesting to note that a change of the pole from -10 to -9 equals a considerable change to the root loci. For values of a less than 9, the values of s as given by Eq. (7-123) no longer satisfy the equation in Eq. (7-121), which means that there are no finite, nonzero, breakaway points. Figure 7-23(c) illustrates this case with $a = 8$. As the pole at $s = -a$ is moved farther to the right, the complex portion of the root loci is pushed farther toward the right-half plane. When $a = b$, the pole at $s = -a$ and the zero at $-b$ cancel each other, and the root loci degenerate into a second-order one and lie on the imaginary axis. These two cases are shown in Fig. 7-23(d) and (e), respectively.

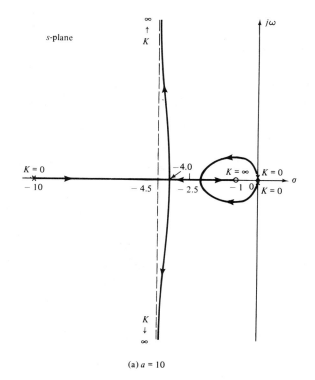

Figure 7-23 Root locus diagrams that show the effects of moving a pole of $G(s)H(s)$. $G(s)H(s) = [K(s+1)]/[s^2(s + a)]$. (a) $a = 10$. (b) $a = 9$. (c) $a = 8$. (d) $a = 3$. (e) $a = 1$.

(a) $a = 10$

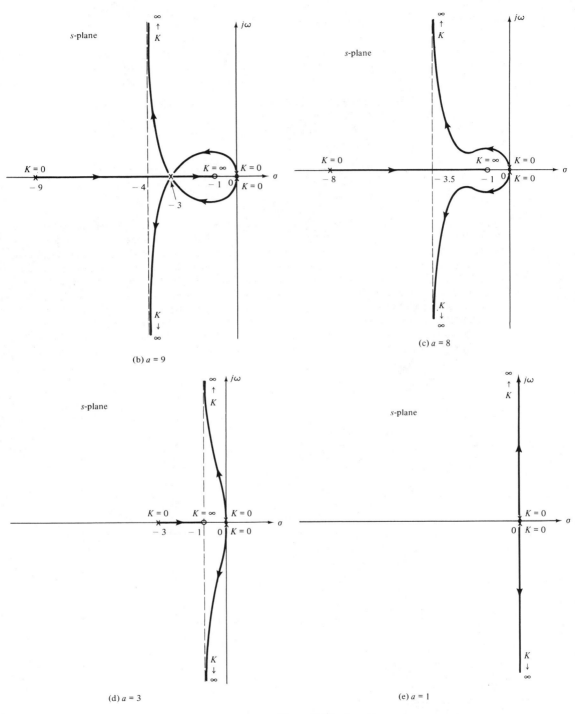

(b) $a = 9$

(c) $a = 8$

(d) $a = 3$

(e) $a = 1$

Figure 7-23 (cont.)

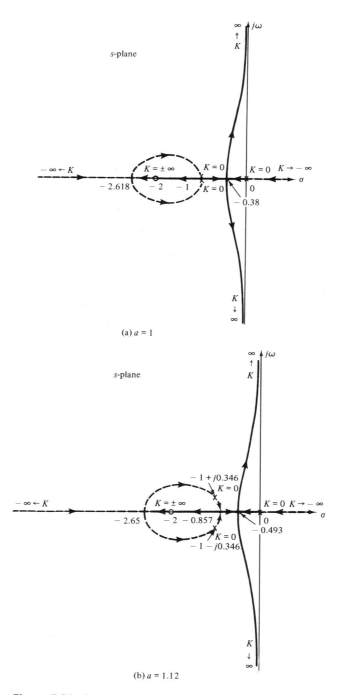

(a) $a = 1$

(b) $a = 1.12$

Figure 7-24 Root locus diagrams that show the effects of moving a pole of $G(s)H(s)$. $G(s)H(s) = K(s + 2) / s(s^2 + 2s + a)$. (a) $a = 1$. (b) $a = 1.12$. (c) $a = 1.185$. (d) $a = 3$.

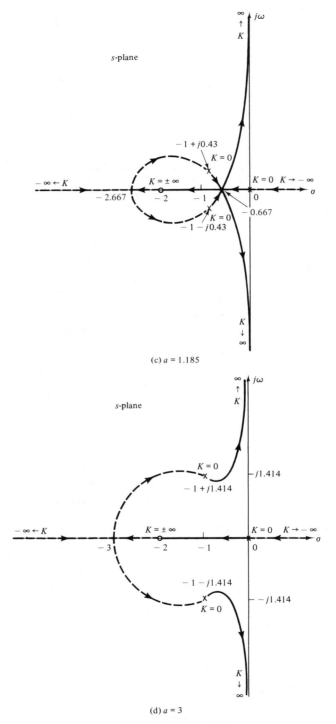

(c) $a = 1.185$

(d) $a = 3$

Figure 7-24   (cont.)

Example 7-16

Consider the equation

$$s(s^2 + 2s + a) + K(s + 2) = 0 \qquad (7\text{-}124)$$

which is converted to the form of $1 + G(s)H(s) = 0$, with

$$G(s)H(s) = \frac{K(s + 2)}{s(s^2 + 2s + a)} \qquad (7\text{-}125)$$

The objective is to study the complete root loci $(-\infty < K < \infty)$ for various values of $a(> 0)$. As a start, let $a = 1$ so that the poles of $G(s)H(s)$ are at $s = 0$, -1, and -1. The complete root loci for this case are sketched in Fig. 7-24(a). By setting $dG(s)H(s)/ds$ to zero, the breakaway points are found at $s = -0.38$, -1, and -2.618.

As the value of a is increased from unity, the two double poles of $G(s)H(s)$ at $s = -1$ will move vertically up and down. The sketch of the root loci is governed mainly by the knowledge of the breakaway points. We can show that $dG(s)H(s)/ds$ leads to

$$s^3 + 4s^2 + 4s + a = 0 \qquad (7\text{-}126)$$

As the value of a increases, the breakaway points at $s = -0.38$ and $s = -2.618$ move to the left, whereas the breakaway point at $s = -1$ moves toward the right. Figure 7-24(b) shows the complete root loci with $a = 1.12$; that is,

$$G(s)H(s) = \frac{K(s + 2)}{s(s^2 + 2s + 1.12)} \qquad (7\text{-}127)$$

Since the real parts of the poles and zeros of $G(s)H(s)$ are not affected by the value of a, the intersect of the asymptotes is always at the origin of the s-plane. The breakaway points when $a = 1.12$ are at $s = -0.493$, -0.857, and -2.65. These are obtained by solving Eq. (7-126).

By solving for a double-root condition in Eq. (7-126) when $a = 1.185$, it can be shown that the two breakaway points that lie between $s = 0$ and $s = -1$ converge to a point. The root loci for this situation are sketched as shown in Fig. 7-24(c).

When a is greater than 1.185, Eq. (7-126) yields one real root and two complex-conjugate roots. Although complex breakaway points do occur quite often in root loci, we can easily show in the present case that these complex roots do not satisfy the original equation of Eq. (7-124) for any real K. Thus, the root loci have only one breakaway point, as shown in Fig. 7-24(d) for $a = 3$. The transition between the cases in Fig. 7-24(c) and (d) should be apparent.

■

7.5 ROOT CONTOUR — MULTIPLE-PARAMETER VARIATION

The root locus technique discussed thus far is restricted to only one variable parameter in K. However, in many control systems problems, the effects of varying several parameters must be studied. For example, when designing a controller that is represented by a transfer function with poles and zeros, it is necessary to investigate the effects on the performance of the overall system when these poles and zeros take on various values.

In Section 7.4 the root locus diagrams of equations with two variable parameters are studied by assigning different values to one of the parameters. In this section the multiparameter problem is investigated through a more systematic method of embedding. When more than one parameter varies continuously from $-\infty$ to ∞, the loci of the root are referred to as the *root contours*. It will be shown that the same conditions and rules of the root loci are still applicable to the construction of the root contours.

The principle of the root contours can be illustrated by considering the equation

$$Q(s) + K_1 P_1(s) + K_2 P_2(s) = 0 \qquad (7\text{-}128)$$

where K_1 and K_2 are the variable parameters and $Q(s)$, $P_1(s)$, and $P_2(s)$ are polynomials of s. The first step involves the setting of one of the parameters equal to zero. Let us set K_2 equal to zero. Then Eq. (7-128) becomes

$$Q(s) + K_1 P_1(s) = 0 \qquad (7\text{-}129)$$

The root loci of this equation may be obtained by dividing both sides of the equation by $Q(s)$. Thus

$$1 + \frac{K_1 P_1(s)}{Q(s)} = 0 \qquad (7\text{-}130)$$

or

$$1 + G_1(s)H_1(s) = 0 \qquad (7\text{-}131)$$

The construction of the root loci depends on the pole-zero configuration of

$$G_1(s)H_1(s) = \frac{K_1 P_1(s)}{Q(s)} \qquad (7\text{-}132)$$

Next, we restore the value of K_2, and divide both sides of Eq. (7-128) by the terms that do not contain K_2. We have

$$1 + \frac{K_2 P_2(s)}{Q(s) + K_1 P_1(s)} = 0 \qquad (7\text{-}133)$$

which is of the form:

$$1 + G_2(s)H_2(s) = 0 \qquad (7\text{-}134)$$

Now the root contours of Eq. (7-128) are constructed based upon the poles and zeros of

$$G_2(s)H_2(s) = \frac{K_2 P_2(s)}{Q(s) + K_1 P_1(s)} \qquad (7\text{-}135)$$

However, one important feature is that the poles of $G_2(s)H_2(s)$ are identical to the roots of Eq. (7-129) or of Eq. (7-131). Thus the root contours of the original equation must all start ($K_2 = 0$) at the points that lie on the root loci of Eq. (7-131). This is the reason why one root contour problem is considered to be embedded in another. The same procedure may be extended to more than two variable parameters.

■

Example 7-17

Consider the equation

$$s^3 + K_2 s^2 + K_1 s + K_1 = 0 \qquad (7\text{-}136)$$

where K_1 and K_2 are the variable parameters and with values that lie between 0 and ∞.
As a first step we let $K_2 = 0$; Eq. (7-136) becomes

$$s^3 + K_1 s + K_1 = 0 \qquad (7\text{-}137)$$

which is converted to

$$1 + \frac{K_1(s + 1)}{s^3} = 0 \qquad (7\text{-}138)$$

The root loci of Eq. (7-137) are drawn from the poles and zeros of

$$G_1(s)H_1(s) = \frac{K_1(s + 1)}{s^3} \qquad (7\text{-}139)$$

as shown in Fig. 7-25(a).
Next, we let K_2 vary between zero and infinity while holding K_1 at a constant nonzero value. Dividing both sides of Eq. (7-136) by the terms that do not contain K_2, we have

$$1 + \frac{K_2 s^2}{s^3 + K_1 s + K_1} = 0 \qquad (7\text{-}140)$$

Thus the root contours of Eq. (7-136) when K_2 varies may be drawn from the pole-zero configuration of

$$G_2(s)H_2(s) = \frac{K_2 s^2}{s^3 + K_1 s + K_1} \qquad (7\text{-}141)$$

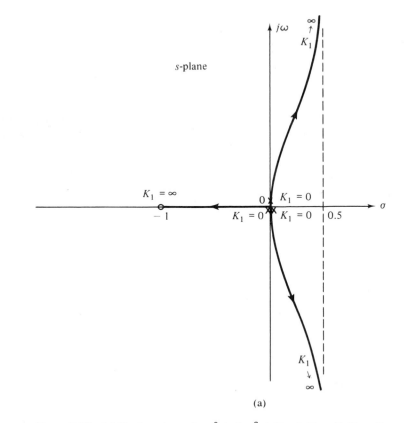

(a)

Figure 7-25   (a) Root contours for $s^3 + K_2 s^2 + K_1 s + K_1 = 0$, $K_2 = 0$.

The zeros of $G_2(s)H_2(s)$ are at $s = 0,0$; but the poles are the zeros of $1 + G_1(s)H_1(s)$ which have been found on the contours of Fig. 7-25(a). Thus for fixed K_1 the root contours when K_2 varies must all emanate from the root contours of Fig. 7-25(a).

Example 7-18

Consider the loop transfer function

$$G(s)H(s) = \frac{K}{s(1 + Ts)(s^2 + 2s + 2)} \qquad (7\text{-}142)$$

of a closed-loop control system. It is desired to construct the root contours of the characteristic equation with K and T as variable parameters.

The characteristic equation of the system is written

$$s(1 + Ts)(s^2 + 2s + 2) + K = 0 \qquad (7\text{-}143)$$

First, we shall set T equal to zero. The characteristic equation becomes

$$s(s^2 + 2s + 2) + K = 0 \qquad (7\text{-}144)$$

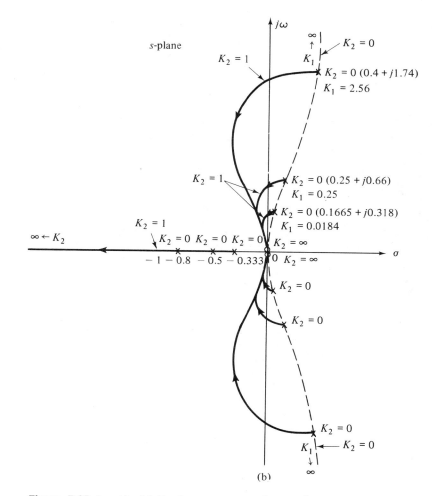

Figure 7-25 (cont.) (b) Root contours for $s^3 + K_2 s^2 + K_1 s + K_1 = 0$. K_2 varies, K_1 = constant.

The root contours of this equation when K varies are drawn based on the poles and zeros of

$$G_1(s) H_1(s) = \frac{K}{s(s^2 + 2s + 2)} \tag{7-145}$$

as shown in Fig. 7-26(a).

For the root contours with T varying and K held constant, we write Eq. (7-143) as

$$1 + G_2(s) H_2(s) = 1 + \frac{Ts^2(s^2 + 2s + 2)}{s(s^2 + 2s + 2) + K} = 0 \tag{7-146}$$

Therefore, the root contours when T varies are constructed from the pole-zero configuration of $G_2(s) H_2(s)$. When $T = 0$, the points on the root contours are at the poles of

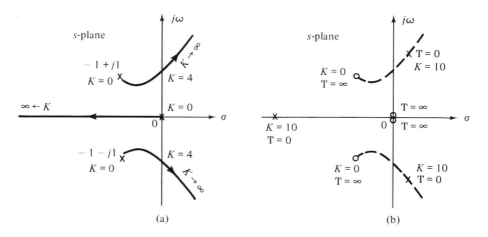

Figure 7-26 (a) Root loci for $s(s^2 + 2s + 2) + K = 0$ (b) Pole-zero configuration of $G_2(s)H_2(s) = [Ts^2(s^2 + 2s + 2)] / s(s^2 + 2s + 2) + K]$.

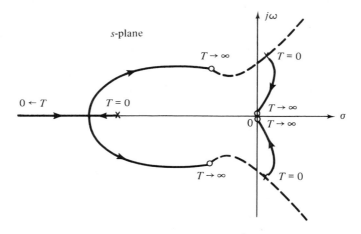

Figure 7-27 Root contours for $s(1 + sT)(s^2 + 2s + 2) + K = 0$; $K > 4$.

$G_2(s)H_2(s)$, which are the points on the root loci of Eq. (7-144), as shown in Fig. 7-26(b) for $K = 10$. The $T = \infty$ points on the root contours are at the zeros of $G_2(s)H_2(s)$, and these are at $s = 0, 0, -1 + j1$, and $-1 - j1$. The root contours for the system are sketched in Figs. 7-27, 7-28, and 7-29 for three different values of K; when $K = 0.5$ and $T = 0.5$, the characteristic equation has a quadruple root at $s = -1$.

Example 7-19
As an example illustrating the effect of the variation of a zero of $G(s)H(s)$, consider

$$G(s)H(s) = \frac{K(1 + Ts)}{s(s + 1)(s + 2)} \qquad (7\text{-}147)$$

The characteristic equation of the system is

$$s(s + 1)(s + 2) + K(1 + Ts) = 0 \qquad (7\text{-}148)$$

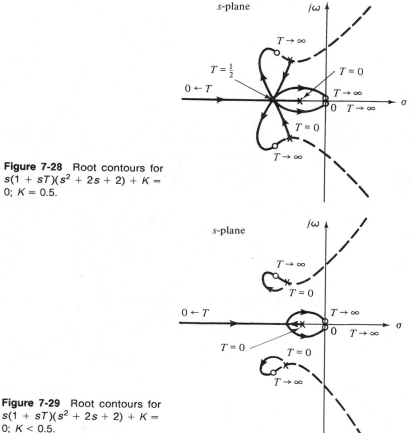

Figure 7-28 Root contours for $s(1 + sT)(s^2 + 2s + 2) + K = 0$; $K = 0.5$.

Figure 7-29 Root contours for $s(1 + sT)(s^2 + 2s + 2) + K = 0$; $K < 0.5$.

Let us first consider the effect of varying the parameter K. Setting $T = 0$ in Eq. (7-148) yields

$$s(s + 1)(s + 2) + K = 0 \qquad (7\text{-}149)$$

which leads to

$$1 + \frac{K}{s(s + 1)(s + 2)} = 0 \qquad (7\text{-}150)$$

The root loci of Eq. (7-149) are sketched in Fig. 7-30, based on the pole-zero configuration of

$$G_1(s)H_1(s) = \frac{K}{s(s + 1)(s + 2)} \qquad (7\text{-}151)$$

When T varies between zero and infinity, we write Eq. (7-148) as

$$1 + G_2(s)H_2(s) = 1 + \frac{TKs}{s(s + 1)(s + 2) + K} = 0 \qquad (7\text{-}152)$$

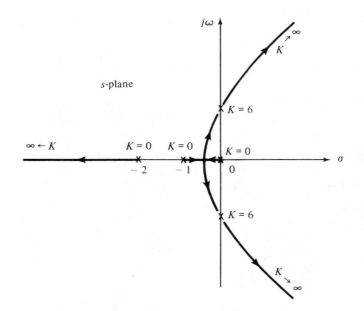

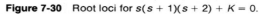

Figure 7-30   Root loci for $s(s + 1)(s + 2) + K = 0$.

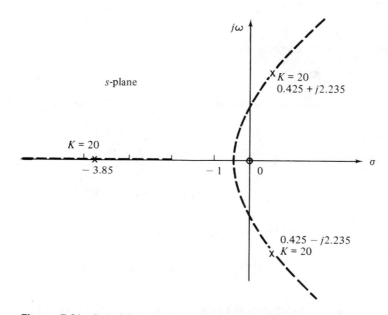

Figure 7-31 Pole-zero configuration of $G_2(s)H_2(s) = TKs / [s(s + 1)(s + 2) + K]$, $K = 20$.

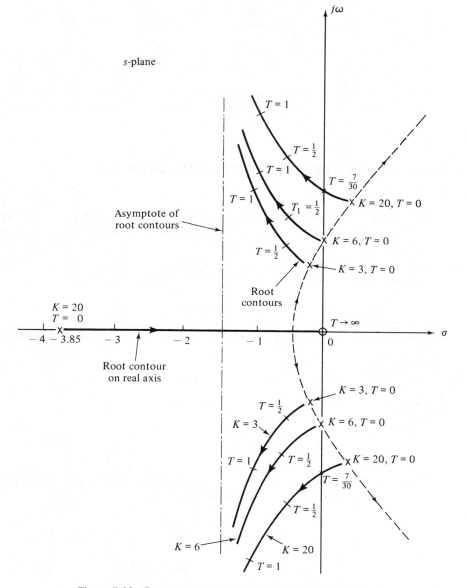

Figure 7-32   Root contours of $s(s + 1)(s + 2) + K + KTs = 0$.

The points that correspond to $T = 0$ on the root contours are at the roots of $s(s + 1)(s + 2) + K = 0$, whose loci are sketched as shown in Fig. 7-30. If we choose $K = 20$, the pole-zero configuration of $G_2(s)H_2(s)$ is shown in Fig. 7-31. The root contours of Eq. (7-148) for $0 \leq T < \infty$ are sketched in Fig. 7-32 for three values of K. The intersection of the asymptotes of the root contours is obtained from Eq. (7-58); that is,

$$\sigma_1 = \frac{-3.85 + 0.425 + 0.425}{3 - 1} = -1.5 \qquad (7\text{-}153)$$

Therefore, the intersection of the asymptotes is always at $s = -1.5$ because the sum of the poles of $G_2(s)H_2(s)$ is always equal to -3, regardless of the value of K, and the sum of the zeros of $G_2(s)H_2(s)$ is zero.

The root contours in Fig. 7-32 show that adding a zero to the open-loop transfer function generally improves the relative stability of the closed-loop system by moving the characteristic equation roots toward the left in the s-plane. As shown in Fig. 7-32, for $K = 20$, the system is stabilized for all values of T greater than 0.2333. However, the largest damping ratio that the system can have by increasing T is approximately 30 percent.

■

7.6 ROOT LOCI OF SYSTEMS WITH PURE TIME DELAY [50]

In Chapter 4 we investigated the modeling of systems with pure time delays and pointed out that the time delay between signals at various points of a system can be represented by the exponential term e^{-Ts} as its transfer function, where T is the delay in seconds. Therefore, we shall assume that the characteristic equation of a typical closed-loop system with pure time delay may be written

$$Q(s) + KP(s)e^{-Ts} = 0 \qquad (7\text{-}154)$$

where $Q(s)$ and $P(s)$ are polynomials of s. An alternative condition of Eq. (7-154) is

$$1 + KG_1(s)H_1(s)e^{-Ts} = 0 \qquad (7\text{-}155)$$

where

$$G_1(s)H_1(s) = \frac{P(s)}{Q(s)} \qquad (7\text{-}156)$$

Thus, similar to the development in Section 7.2, in order to satisfy Eq. (7-156), the following conditions must be met simultaneously:

$$e^{-T\sigma}|G_1(s)H_1(s)| = \frac{1}{|K|} \qquad\qquad -\infty < K < \infty \qquad (7\text{-}157)$$

$$\underline{/G_1(s)H_1(s)} = (2k+1)\pi + \omega T \qquad\qquad K \geq 0 \qquad (7\text{-}158)$$

$$\underline{/G_1(s)H_1(s)} = 2k\pi + \omega T \qquad\qquad K \leq 0 \qquad (7\text{-}159)$$

where $s = \sigma + j\omega$ and $k = 0, \pm 1, \pm 2, \ldots$. Note that the condition for any point $s = s_1$ in the s-plane to be a point on the complete root loci is given in Eqs. (7-158) and (7-159), which differ from the conditions of Eqs. (7-16) and (7-17) by the term ωT. When $T = 0$, Eqs. (7-158) and (7-159) revert to Eqs. (7-16) and (7-17). Since ω is a variable in the s-plane, the angular conditions of Eqs. (7-158) and (7-159) are no longer constant in the s-plane but depend upon the point at which a root of Eq.

(7-154) may lie. Viewing the problem from another standpoint, it is recognized that if $T = 0$, given a value of K, there are only n points in the s-plane that will satisfy either Eq. (7-158) or Eq. (7-159) for all possible values of k, where n is the highest order of $P(s)$ and $Q(s)$. However, for $T \neq 0$, the angular conditions in Eqs. (7-158) and (7-159) depend on ω, which varies along the vertical axis in the s-plane. Thus, for a given K, there may be more than n points which satisfy the angular conditions in the s-plane, as k takes on all possible integral values. In fact, there are an infinite number of these points, since Eq. (7-154), which is transcendental, is known to have an infinite number of roots.

The difficulty with the construction of the root loci of Eq. (7-154) is that many of the rules of construction developed originally for systems without time delay are no longer valid for the present case. It is of interest to investigate how some of the rules of construction given in Section 7.3 may be modified to apply to the time-delay case.

K = 0 Points

Theorem 7-9. *The $K = 0$ points on the complete root loci of Eq. (7-155) are at the poles of $G_1(s)H_1(s)$ and $\sigma = -\infty$.*

Proof. Equation (7-157) is repeated,

$$e^{-T\sigma}|G_1(s)H_1(s)| = \frac{1}{|K|} \tag{7-160}$$

Thus, if K equals zero, s approaches the poles of $G_1(s)H_1(s)$, or σ, which is the real part of s, approaches $-\infty$.

K = ±∞ Points

Theorem 7-10. *The $K = \pm\infty$ points on the complete root loci of Eq. (7-155) are at the zeros of $G_1(s)H_1(s)$ and $\sigma = \infty$.*

Proof. Referring again to Eq. (7-160), the proof of this theorem becomes evident.

Number of Branches on the Complete Root Loci

The number of branches on the root loci of Eq. (7-154) is infinite, since the equation has an infinite number of roots.

Symmetry of the Complete Root Loci

The complete root loci are symmetrical with respect to the real axis of the s-plane. This is explained by expanding e^{-Ts} into an infinite series; then Eq. (7-154) again becomes a polynomial with a real coefficient but with infinite order.

Table 7-2

K	$n - m$	$K = 0$ Asymptotes	$K = \pm\infty$ Asymptotes
≥ 0	Odd	$N =$ even integers $= 0, \pm 2, \pm 4, \ldots$	$N =$ odd integers $= \pm 1, \pm 3, \pm 5, \ldots$
	Even	$N =$ odd integers $= \pm 1, \pm 3, \pm 5, \ldots$	$N =$ odd integers $= \pm 1, \pm 3, \pm 5, \ldots$
≤ 0	Odd	$N =$ odd integers $= \pm 1, \pm 3, \pm 5, \ldots$	$N =$ even integers $= 0, \pm 2, \pm 4, \ldots$
	Even	$N =$ even integers $= 0, \pm 2, \pm 4, \ldots$	$N =$ even integers $= 0, \pm 2, \pm 4, \ldots$

Asymptotes of the Complete Root Loci

Theorem 7-11. *The asymptotes of the root loci of Eq. (5-154) are infinite in number and all are parallel to the real axis of the s-plane. The intersects of the asymptotes with the imaginary axis are given by*

$$\omega = \frac{N\pi}{T} \tag{7-161}$$

where N is tabulated in Table 7-2 for the various conditions indicated.

$$n = \text{number of finite poles of } G_1(s)H_1(s)$$
$$m = \text{number of finite zeros of } G_1(s)H_1(s)$$

Proof. Since as $s \to \infty$ on the root loci, K either approaches zero or $\pm\infty$, Theorems 7-9 and 7-10 show that the asymptotes are at $\sigma = \infty$ ($K = \pm\infty$) and $\sigma = -\infty$ ($K = 0$). The intersections of the asymptotes with the $j\omega$ axis and the conditions given in Table 7-2 are arrived at by use of Eqs. (7-158) and (7-159).

Root Loci on the Real Axis

The property of the root loci of Eq. (7-154) on the real axis is the same as stated in Theorem 7-7, because on the real axis, $\omega = 0$, the angular conditions of Eqs. (7-158) and (7-159) revert to those of Eqs. (7-16) and (7-17), respectively.

Angles of Departure and Arrival

Angles of departure and arrival are determined by use of Eqs. (7-158) and (7-159).

Intersection of the Root Loci with the Imaginary Axis

Since Eq. (7-154) is not an algebraic equation of s, the intersection of its loci with the imaginary axis *cannot* be determined by use of the Routh–Hurwitz criterion. The determination of all the points of intersection of the root loci with the $j\omega$ axis is a difficult task, since the root loci have an infinite number of branches. However, we shall show in the following section that only the intersections nearest the real axis are of interest for stability studies.

Breakaway Points

Theorem 7-12. *The breakaway points on the complete root loci of Eq. (7-154) must satisfy*

$$\frac{dG_1(s)H_1(s)e^{-Ts}}{ds} = 0 \tag{7-162}$$

Proof. The proof of this theorem is similar to that of Theorem 7-8.

Determination of the Values of *K* on the Root Loci

The value of K at any point $s = s_1$ on the root loci is determined from Eq. (7-157); that is,

$$|K| = \frac{e^{T\sigma_1}}{|G_1(s_1)H_1(s_1)|} \tag{7-163}$$

where σ_1 is the real part of s_1.

■

Example 7-2

Consider the equation

$$s + Ke^{-Ts} = 0 \tag{7-164}$$

It is desired to construct the complete root loci of this equation for a fixed value of T. Dividing both sides of Eq. (7-164) by s, we get

$$1 + \frac{Ke^{-Ts}}{s} = 0 \tag{7-165}$$

which is of the form of Eq. (7-155) with

$$G_1(s)H_1(s) = \frac{1}{s} \tag{7-166}$$

The following properties of the root loci of Eq. (7-164) are determined:

1. The $K = 0$ points: From Theorem 7-9, $K = 0$ at $s = 0$ and at $\sigma = -\infty$. Using Theorem 7-11 and Table 7-2, we have
   $K \geq 0$: K approaches zero as σ approaches $-\infty$ at $\omega = 0, \pm 2\pi/T, \pm 4\pi/T, \ldots$
   $K \leq 0$: K approaches zero as σ approaches $-\infty$ at $\omega = \pm \pi/T, \pm 3\pi/T, \pm 5\pi/T, \ldots$

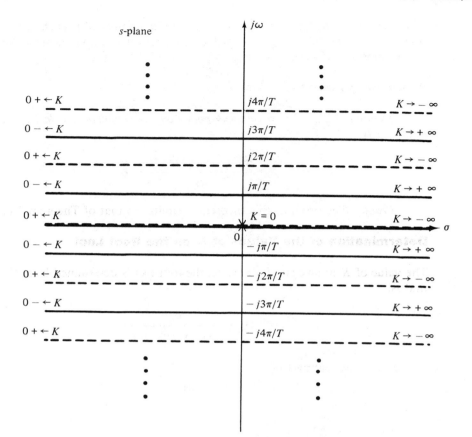

Figure 7-33 Asymptotes of the complete root loci of the equation $s + Ke^{-Ts}$ = 0.

2. The $K = \pm\infty$ points: From Theorem 7-10, $K = \pm\infty$ at $\sigma = \infty$. Using Theorem 7-11 and Table 7-2, we have

   $K \geq 0$: K approaches $+\infty$ as σ approaches $+\infty$ at $\omega = \pm\pi/T, \pm3\pi/T, \ldots$

   $K \leq 0$: K approaches $-\infty$ as σ approaches $+\infty$ at $\omega = 0, \pm2\pi/T, \pm4\pi/T, \ldots$

   The $K = 0$, $K = \pm\infty$ points, and the asymptotes of the root loci are shown in Fig. 7-33. The notation of 0^{+} is used to indicate the asymptotes of the RL, and 0^{-} is for the CRL.

3. The RL $(K \geq 0)$ occupy the negative real axis, and the CRL $(K \leq 0)$ occupy the positive real axis.

4. The intersections of the root loci with the $j\omega$ axis are relatively easy to determine for this simple problem.

   Since $G_1(s)H_1(s)$ has only a simple pole at $s = 0$, for any point s_1 on the positive $j\omega$ axis,

$$\angle G_1(s_1)H_1(s_1) = -\frac{\pi}{2} \qquad (7\text{-}167)$$

and for any point s_1 on the negative $j\omega$ axis,

$$\angle G_1(s_1)H_1(s_1) = \frac{\pi}{2} \qquad (7\text{-}168)$$

Thus, for $K \geq 0$, Eq. (7-158) gives the condition of RL on the $j\omega$ axis ($\omega > 0$),

$$-\frac{\pi}{2} = (2k + 1)\pi + \omega T \qquad (7\text{-}169)$$

$k = 0, \pm 1, \pm 2, \ldots$. The values of ω that correspond to the points at which the root loci cross the positive $j\omega$ axis are

$$\omega = \frac{\pi}{2T}, \frac{5\pi}{2T}, \frac{9\pi}{2T}, \ldots \qquad (7\text{-}170)$$

For $K \geq 0$, and $\omega < 0$,

$$\frac{\pi}{2} = (2k + 1)\pi + \omega T \qquad (7\text{-}171)$$

and for $k = 0, \pm 1, \pm 2, \ldots$, the crossover points are found to be at

$$\omega = -\frac{\pi}{2T}, -\frac{5\pi}{2T}, -\frac{9\pi}{2T}, \ldots \qquad (7\text{-}172)$$

Similarly, for $K \leq 0$, the conditions for the complementary root loci to cross the $j\omega$ axis are

$$-\frac{\pi}{2} = 2k\pi + \omega T \qquad \omega > 0 \qquad (7\text{-}173)$$

$$\frac{\pi}{2} = 2k\pi + \omega T \qquad \omega < 0 \qquad (7\text{-}174)$$

The crossover points are found by substituting $k = 0, \pm 1, \pm 2, \ldots$ into the last two equations. We have

$$\omega = \pm\frac{3\pi}{2T}, \pm\frac{7\pi}{2T}, \pm\frac{11\pi}{2T}, \ldots \qquad (7\text{-}175)$$

5. Breakaway points: The breakaway points on the complete root loci are determined by the use of Eq. (7-162). Thus,

$$\frac{dG_1(s)H_1(s)e^{-Ts}}{ds} = \frac{d}{ds}\left(\frac{e^{-Ts}}{s}\right) = 0 \qquad (7\text{-}176)$$

or

$$\frac{-Te^{-Ts}s - e^{-Ts}}{s^2} = 0 \qquad (7\text{-}177)$$

from which we have

$$e^{-Ts}(Ts + 1) = 0 \qquad (7\text{-}178)$$

Therefore, the finite breakaway point is at $s = -1/T$.

6. The values of K at the crossover point on the $j\omega$ axis are found by using Eq. (7-154). Since $\sigma = 0$ on the $j\omega$ axis, we have

$$|K| = \frac{1}{|G_1(j\omega_c)H_1(j\omega_c)|} = |\omega_c| \qquad (7\text{-}179)$$

where ω_c is a crossover point.

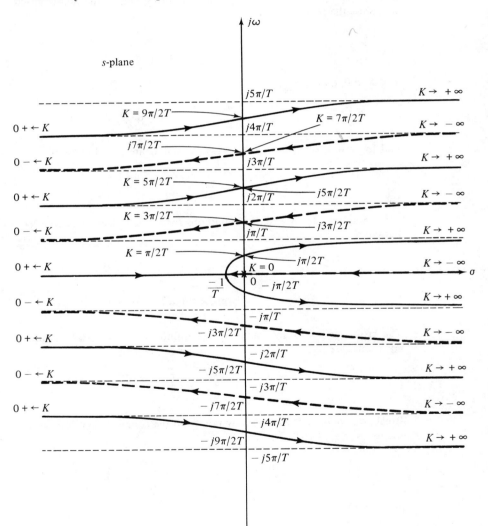

Figure 7-34   Complete root loci for $s + Ke^{-Ts}0$.

Based on the properties accumulated above, the complete root loci of Eq. (7-157) are sketched as shown in Fig. 7-34.

Although the equation of Eq. (7-154) has an infinite number of roots, and therefore the root loci have an infinite number of branches, from the system analysis standpoint, only the branches that lie between $-\pi/T \leq \omega \leq \pi/T$ are of interest. We shall refer to these as the *primary branches*. One reason is that the critical value of K at the crossover point on this portion of the root loci is equal to $\pi/2T$, whereas the critical value of K at the next branch at $\omega = \pm 5\pi/2T$ is $5\pi/2T$, which is much greater. Therefore, $K = \pi/2T$ is the critical value for stability. Another reason for labeling the primary branches as the dominant loci is that for any value of K less than the critical value of $\pi/2T$, the corresponding roots on the other branches are all far to the left in the s-plane. Therefore, the transient response of the system, which has Eq. (7-164) as its characteristic equation, is predominantly controlled by the roots on the primary branches.

Example 7-21

As a slightly more complex problem in the construction of the root loci of a system with pure time delay, let us consider the control system shown in Fig. 7-35. The loop transfer function of the system is

$$G(s)H(s) = \frac{Ke^{-Ts}}{s(s+1)}$$

The characteristic equation is

$$s^2 + s + Ke^{-Ts} = 0$$

In order to construct the complete root loci, the following properties are assembled by using the rules given in Theorems 7-9 through 7-12.

1. The $K = 0$ points: From Theorem 7-9, $K = 0$ at $s = 0$, $s = -1$, and $\sigma = -\infty$. Using Theorem 7-11 and Table 7-2, we have
   $K \geq 0$: K approaches zero as σ approaches $-\infty$ at $\omega = \pm \pi/T, \pm 3/\pi T, \pm 5\pi/T, \ldots$
   $K \leq 0$: K approaches zero as σ approaches $-\infty$ at $\omega = 0, \pm 2\pi/T, \pm 4\pi/T, \ldots$
2. The $K = \pm\infty$ points: From Theorem 7-10, $K = \pm\infty$ at $\sigma = \infty$. Using Theorem 7-11 and Table 7-2, we have
   $K \geq 0$: K approaches $+\infty$ as σ approaches $+\infty$ at $\omega = \pm \pi/T, \pm 3\pi/T, \ldots$
   $K \leq 0$: K approaches $-\infty$ as σ approaches $+\infty$ at $\omega = 0, \pm 2\pi/T, \pm 4\pi/T, \ldots$
   Notice that the $K = 0$ asymptotes depend upon $n - m$, which is even in this case, but the $K = \pm\infty$ asymptotes depend only on the sign of K and not on $n - m$.
3. The RL ($K \geq 0$) occupy the region between $s = 0$ and $s = -1$ on the real axis. The rest of the real axis is occupied by the CRL ($K \leq 0$).

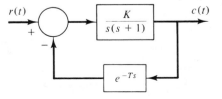

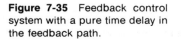

Figure 7-35 Feedback control system with a pure time delay in the feedback path.

4. Breakaway points: The breakaway points of the complete root loci are found from Eq. (7-162); that is,

$$\frac{d}{ds}\left[\frac{e^{-Ts}}{s(s+1)}\right] = 0$$

from which we have the two breakaway points at

$$s = \frac{1}{2T}\left[-(T+2) \pm \sqrt{T^2 + 4}\right]$$

For $T = 1$ sec the two breakaway points are at

$$s = -0.382 \qquad s = -2.618$$

where it is easily verified that one belongs to the RL, and the other is on the CRL.

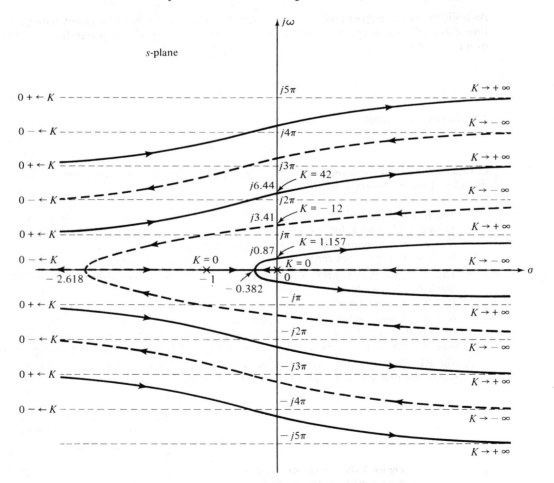

Figure 7-36 Complete root loci for $s^2 + s + Ke^{-Ts} = 0$, $T = 1$.

The complete root loci of the system are sketched as shown in Fig. 7-36 for $T = 1$ sec. Notice that from the system analysis standpoint, only the portion of the root loci that lies between $\omega = \pi$ and $\omega = -\pi$ is of significance. The closed-loop system of Fig. 7-35 is stable for $0 \le K < 1.157$. Therefore, the other RL branches, including the CRL, are perhaps only of academic interest.

■

Approximation of Systems with Pure Time Delay

Although the discussions given in the preceding section point to a systematic way of constructing the root loci of a closed-loop system with pure time delay, in general, for complex systems, the problem can still be quite difficult. We shall investigate ways of approximating the time delay term, e^{-Ts}, by a polynomial or a rational function of s. One method is to approximate e^{-Ts}, as follows:

$$e^{-Ts} \cong \frac{1}{[1 + (Ts/n)]^n} \tag{7-180}$$

Since e^{-Ts} has an infinite number of poles, the approximation is perfect when n becomes infinite. Figure 7-37 illustrates the effect of the approximation when the input to the pure time delay is a unit step function.

If Eq. (7-180) is used as the approximation for the root locus problem, only the primary branches of the root loci will be realized. However, this will be adequate for the great majority of practical problems, since only the primary branches will contain the dominant eigenvalues of the system.

Let us approximate the exponential term of Eq. (7-164) by the right side of Eq. (7-180). Figure 7-38 illustrates the dominant root loci for $n = 2, 3,$ and 4; $T = 1$, together with the primary branch of the exact root loci. The approximating root loci approach the exact ones as n becomes large.

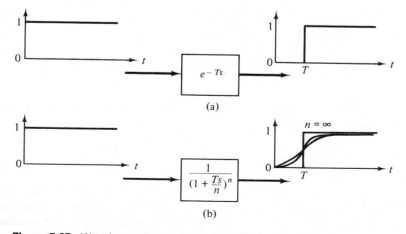

Figure 7-37 Waveforms that illustrate the effect of approximating a pure time delay by finite number of poles.

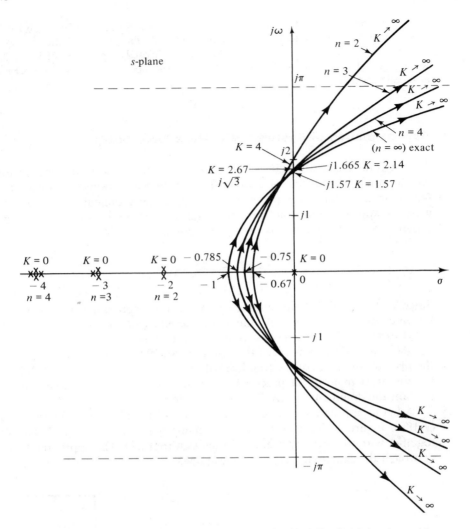

Figure 7-38 Approximation of the root loci of $s + Ke^{-s} = 0$ by those of $(1 + s/n)n + K = 0$.

Another way of approximating the pure time delay transfer relation is to use a power series; that is,

$$e^{-Ts} = 1 - Ts + \frac{T^2 s^2}{2!} - \frac{T^3 s^3}{3!} + \cdots \qquad (7\text{-}181)$$

The difference between this approximation and that of Eq. (7-180) is that in the former, the accuracy improves as the order n becomes larger, whereas in the present case, the validity of the approximation depends on the smallness of T. It is apparent that Eq. (7-181) can be conveniently applied if only a few terms of the series are used.

7.7 ROOT LOCI OF DISCRETE-DATA CONTROL SYSTEMS [47, 48]

The root locus technique for continuous-data systems can be readily applied to discrete-data systems without requiring any modifications. The characteristic equation roots of the discrete-data system having the block diagram of Fig. 7-39 must satisfy

$$1 + GH^*(s) = 0 \tag{7-182}$$

if the roots are defined in the s-plane, or

$$1 + GH(z) = 0 \tag{7-183}$$

if the z-plane is referred to.
  Since

$$GH^*(s) = \frac{1}{T} \sum_{n=-\infty}^{\infty} G(s + jn\omega_s) H(s + jn\omega_s) \tag{7-184}$$

which is an infinite series, the poles and zeros of $GH^*(s)$ in the s-plane will be infinite in number. This evidently makes the construction of the root loci of Eq. (7-182) more difficult. However, as an illustrative example of the difference between the characteristics of the root loci of continuous-data and discrete-data systems, let us consider that for the system of Fig. 7-39.

$$G(s)H(s) = \frac{K}{s(s + 1)} \tag{7-185}$$

Using Eq. (7-184), we have

$$GH^*(s) = \frac{1}{T} \sum_{n=-\infty}^{\infty} \frac{K}{(s + jn\omega_s)(s + jn\omega_s + 1)} \tag{7-186}$$

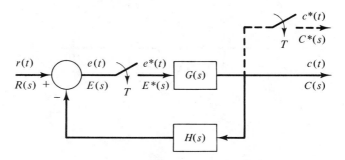

Figure 7-39 Discrete-data control system.

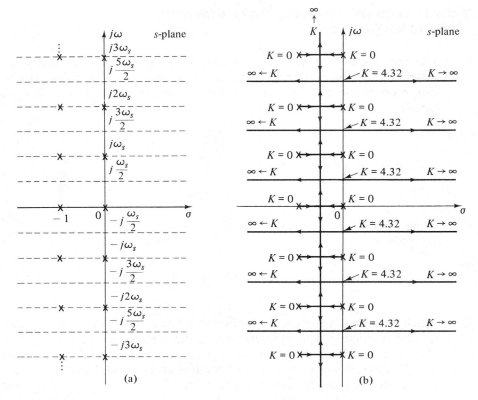

Figure 7-40 Pole configuration of $GH^*(s)$ and the root locus diagram in the s-plane for the discrete-data system in Fig. 7-39 with $G(s)H(s) = K / [s(s + 1)]$, $T = 1$ sec.

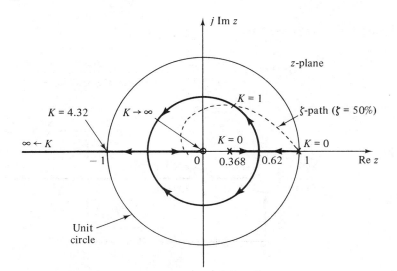

Figure 7-41 Root locus diagram of discrete-data control system without zero-order hold. $G(s)H(s) = K / [s(s + 1)]$, $T = 1$ sec.

which has poles at $s = -jn\omega_s$ and $s = -1 - jn\omega_s$, where n takes on all integers between $-\infty$ and ∞. The pole configuration of $GH*(s)$ is shown in Fig. 7-40(a). Using the rules of construction outlined earlier, the root loci of $1 + GH*(s) = 0$ for positive K are drawn as shown in Fig. 7-40(b) for $T = 1$. The root loci contain an infinite number of branches, and these clearly indicate that the closed-loop system is unstable for all values of K greater than 4.32. In contrast, it is well known that the same system without sampling is stable for all positive values of K.

The root locus problem for discrete-data systems is simplified if the root loci are constructed in the z-plane using Eq. (7-183). Since Eq. (7-183) is, in general, a polynomial in z with constant coefficients, the number of root loci is finite, and the same rules of construction for continuous-data systems are directly applicable.

As an illustrative example of the construction of root loci for discrete-data systems in the z-plane, let us consider the system of Fig. 7-39 with $T = 1$ sec, and $G(s)H(s)$ as given by Eq. (7-185). Taking the z-transform of Eq. (7-185) we have

$$GH(z) = \frac{0.632\,Kz}{(z-1)(z-0.368)} \tag{7-187}$$

which has a zero at $z = 0$ and poles at $z = 1$ and $z = 0.368$. The root loci for the closed-loop system are constructed based on the pole-zero configuration of Eq. (7-187) and are shown in Fig. 7-41. Notice that when the value of K exceeds 4.32, one of the roots of the characteristic equation moves outside the unit circle in the z-plane, and the system becomes unstable.

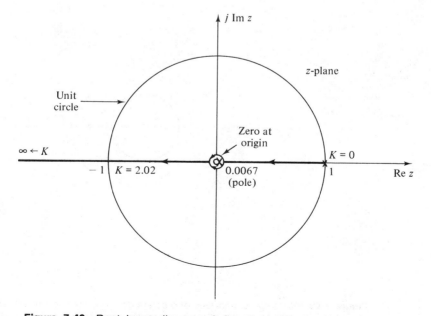

Figure 7-42 Root locus diagram of discrete-data control system without zero-order hold. $G(s)H(s) = K/[s(s+1)]$, $T = 5$ sec.

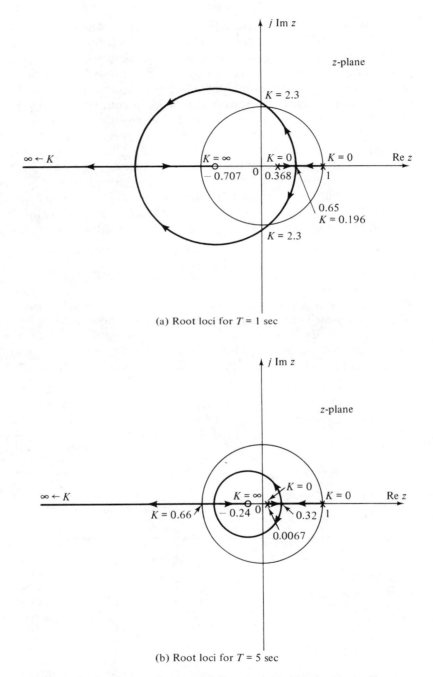

(a) Root loci for $T = 1$ sec

(b) Root loci for $T = 5$ sec

Figure 7-43 Root locus diagrams of discrete-data control system with sample-and-hold. $G(s)H(s) = K/[s(s + 1)]$. (a) Root loci for $T = 1$ sec. (b) Root loci for $T = 5$ sec.

For the same system, if the sampling period is changed to $T = 5$ sec, the z-transform of $G(s)H(s)$ becomes

$$GH(z) = \frac{0.993Kz}{(z - 1)(z - 0.0067)} \qquad (7\text{-}188)$$

The root loci for this case are drawn as shown in Fig. 7-42. It should be noted that although the complex part of the root loci for $T = 5$ sec takes the form of a smaller circle than that when $T = 1$ sec, the system is actually less stable. The marginal value of K for stability for $T = 5$ sec is 2.02 as compared to the marginal K of 4.32 for $T = 1$ sec.

The constant-damping-ratio path may be superimposed on the root loci to determine the required value of K for a specified damping ratio. In Fig. 7-41 the constant-damping-ratio path for $\zeta = 0.5$ is drawn, and the intersection with the root loci gives the desired value of $K = 1$. Thus for all values of K less than 1, the damping ratio of the system will be greater than 50 per cent.

As another example, let us consider that a zero-order hold is inserted between the sampler and the controlled process $G(s)$ in the system of Fig. 7-39. For the loop transfer function of Eq. (7-185), the z-transform with the zero-order hold is

$$G_{h0}GH(z) = \frac{K[(T - 1 + e^{-T})z - Te^{-T} + 1 - e^{-T}]}{(z - 1)(z - e^{-T})} \qquad (7\text{-}189)$$

The root loci of the system with sample-and-hold for $T = 1$ sec and $T = 5$ sec are shown in Fig. 7-43(a) and (b), respectively. In this case the marginal value of stability for K is 2.3 for $T = 1$ sec and 0.66 for $T = 5$ sec. In this case the zero-order hold reduces the stability margin of the discrete-data system.

7.8 ROOT SENSITIVITY — ROBUSTNESS OF SYSTEM

In the design of control systems, not only it is important to arrive at a system with the appropriate performance characteristics, but often the system must be insensitive to parameter variations. For instance, a system may perform satisfactorily at a certain forward gain K, but if it is very sensitive to the variation of K, it may get into the undesirable performance region or become unstable if K varies by a small amount. In formal control system terminology, a system that is insensitive to parameter variations is called a *robust* system.

One way of expressing the robustness of a system is to investigate the variation of the characteristic equation roots with respect to the change of a certain system parameter. In Sec. 7.3 we defined the *root sensitivity* as the sensitivity of the characteristic equation roots with respect to the variation of some parameter such as the gain K. Specifically, the root sensitivity of s with respect to K is defined as

$$S_K^s = \frac{\partial s/s}{\partial K/K} = \frac{\partial s}{\partial K} \cdot \frac{K}{s} \qquad (7\text{-}190)$$

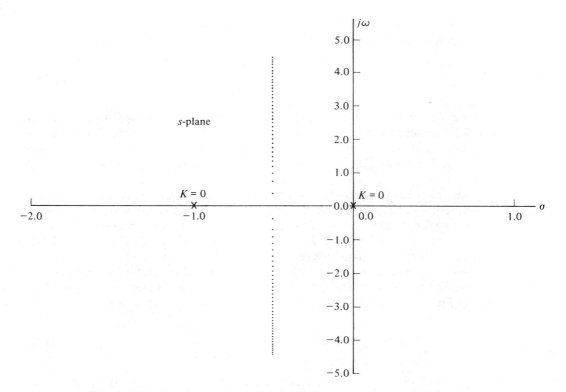

Figure 7-44 Root loci of $s(s + 1) + K = 0$ showing the root sensitivity with respect to K.

In terms of the root loci of the characteristic equation, we can regard root sensitivity as the unit incremental change of the roots along the root loci per unit incremental change in K.

As an illustration, Fig. 7-44 shows the root loci of

$$s(s + 1) + K = 0 \qquad\qquad (7\text{-}191)$$

with K incremented uniformly over 50 values from 0 to 20. The dots represent the distinct roots on the loci. Thus, we can see that as K increases uniformly, the root sensitivity becomes smaller. Similarly, Fig. 7-45 illustrates the root loci of

$$s(s + 1)(s + 2) + K = 0 \qquad\qquad (7\text{-}192)$$

as the value of K is incremented uniformly from 0 to 12 over 200 increments. Again, the dots show that as K increases, the root sensitivity decreases.

An analytical expression of the root sensitivity S_K^s is difficult to obtain for the general case. For the second-order system of Eq. (7-191), we can write

$$\frac{\partial K}{\partial s} = -2s - 1 \qquad\qquad (7\text{-}193)$$

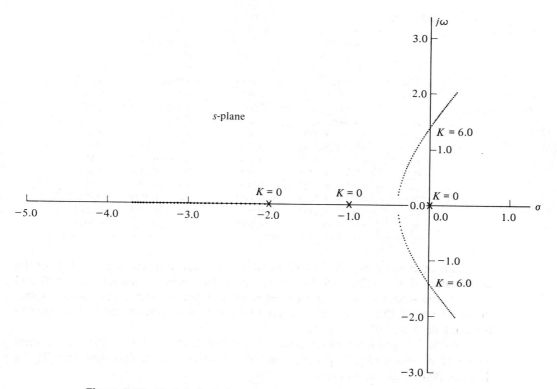

Figure 7-45 Root loci of $s(s + 1)(s + 2) + K = 0$ showing the root sensitivity with respect to K.

Since $K = -s(s + 1)$, the root sensitivity becomes

$$S_K^s = \frac{\partial s}{\partial K} \frac{K}{s} = \frac{s + 1}{2s + 1} \tag{7-194}$$

where, in general, $s = \sigma + j\omega$, and s must take on the values of the roots of Eq. (7-191). For the roots on the real axis, $\omega = 0$; thus, Eq. (7-194) leads to

$$|S_K^s|_{\omega=0} = \left| \frac{\sigma + 1}{2\sigma + 1} \right| \tag{7-195}$$

and it was shown in Sec. 7.3 that at the breakaway point, $K = 0.25$, $\sigma = -0.5$, the root sensitivity is infinite.

When the two roots are complex, $\sigma = -0.5$ for all values of ω; Eq. (7-194) gives

$$|S_K^s|_{\sigma=-0.5} = \left(\frac{0.25 + \omega^2}{4\omega^2} \right)^{1/2} \tag{7-196}$$

Table 7-3

K	Root No. 1	$|S_K^s(1)|$	Root No. 2	$|S_K^s(2)|$
0	0	1	-1	0
0.04	-0.0417	1.0454	-0.9583	0.454
0.16	-0.200	1.3333	-0.800	0.333
0.24	-0.400	3.000	-0.600	2.000
0.25	-0.500	∞	-0.500	∞
0.28	$-0.5 + j0.1732$	1.527	$-0.5 - j0.1732$	1.527
0.36	$-0.5 + j0.3317$	0.9045	$-0.5 - j0.3317$	0.9045
0.40	$-0.5 + j0.3873$	0.8165	$-0.5 - j0.3873$	0.8165
0.80	$-0.5 + j0.7416$	0.603	$-0.5 - j0.7416$	0.603
1.2	$-0.5 + j0.9747$	0.562	$-0.5 - j0.9747$	0.562
2.0	$-0.5 + j1.3229$	0.5345	$-0.5 - j1.3229$	0.5345
4.0	$-0.5 + j1.9365$	0.5164	$-0.5 - j1.9365$	0.5164
∞	$-0.5 + j\infty$	0.5000	$-0.5 - j\infty$	0.5000

It is interesting to note that although the two real roots reach $\sigma = -0.5$ for the same value of $K = 0.25$, and each root travels the same distance from $s = 0$ and $s = -1$, the sensitivities of the two real roots are not the same. The reason is that the two roots of $P(s) = s(s + 1)$ in Eq. (7-191) are not symmetrical with respect to the imaginary axis of the s-plane.

Table 7-3 gives the magnitudes of the two roots of Eq. (7-191) for several values of K, where $|S_K^s(1)|$ denotes the root sensitivity of the first root and $|S_K^s(2)|$ denotes that of the second root.

REFERENCES

General Subjects

1. W. R. EVANS, "Graphical Analysis of Control Systems," *Trans. AIEE*, Vol. 67, pp. 547–551, 1948.

2. W. R. EVANS, "Control System Synthesis by Root Locus Method," *Trans. AIEE*, Vol. 69, pp. 66–69, 1950.

3. W. R. EVANS, *Control System Dynamics*, McGraw-Hill Book Company, New York, 1954.

4. Y. CHU and V. C. M. YEH, "Study of the Cubic Characteristic Equation by the Root Locus Method," *Trans. ASME*, Vol. 76, pp. 343–348, Apr. 1954.

5. J. G. TRUXAL, *Automatic Feedback Systems Synthesis*, McGraw-Hill Book Company, New York, 1955.

6. K. S. NARENDRA, "Inverse Root Locus, Reversed Root Locus, or Complementary Root Locus," *IRE Trans. Automatic Control*, pp. 359–360, Sept. 1961.

7. P. MOSNER, J. VULLO, and K. BLEY, "A Possible Source of Error in Root-Locus Analysis of Nonminimum Phase Systems," *IRE Trans. Automatic Control*, Vol. AC-7, pp. 87–88, Apr. 1962.

8. B. J. Matkowsky and A. H. Zemanian, "A Contribution to Root-Locus Techniques for the Analysis of Transient Responses," *IRE Trans. Automatic Control*, Vol. AC-7, pp. 69–73, Apr. 1962.

9. J. D. S. Muhlenberg, "Quartic Root Locus Types," *IEEE Trans. Automatic Control*, Vol. AC-12, pp. 228–229, Apr. 1967.

10. H. M. Power, "Root Loci Having a Total of Four Poles and Zeros," *IEEE Trans. Automatic Control*, Vol. AC-16, pp. 484–486, Oct. 1971.

Construction and Properties of the Root Locus

11. C. C. MacDuffe, *Theory of Equations*, John Wiley & Sons, Inc., New York, pp. 29–104, 1954.

12. G. A. Bendrikov and K. F. Teodorchik, "The Laws of Root Migration for Third and Fourth Power Linear Algebraic Equations as the Free Terms Change Continuously," *Automation and Remote Control*, Vol. 16, No. 3, 1955.

13. J. E. Gibson, "Build a Dynamic Root-Locus Plotter," *Control Eng.*, Feb. 1956.

14. C. S. Lorens and R. C. Titsworth, "Properties of Root Locus Asymptotes," *IRE Trans. Automatic Control*, AC-5, pp. 71–72, Jan. 1960.

15. K. F. Teodorchik and G. A. Bendrikov, "The Methods of Plotting Root Paths of Linear Systems and for Qualitative Determination of Path Type," *Proc. IFAC Cong.*, Vol. 1, pp. 8–12, 1960.

16. C. A. Stapleton, "On Root Locus Breakaway Points," *IRE Trans. Automatic Control*, Vol. AC-7, pp. 88–89, Apr. 1962.

17. M. J. Remec, "Saddle-Points of a Complete Root Locus and an Algorithm for Their Easy Location in the Complex Frequency Plane," *Proc. Natl. Electronics Conf.*, Vol. 21, pp. 605–608, 1965.

18. C. F. Chen, "A New Rule for Finding Breaking Points of Root Loci Involving Complex Roots," *IEEE Trans. Automatic Control*, Vol. AC-10, pp. 373–374, July 1965.

19. V. Krishnan, "Semi-analytic Approach to Root Locus," *IEEE Trans. Automatic Control*, Vol. AC-11, pp. 102–108, Jan. 1966.

20. R. H. LaBounty and C. H. Houpis, "Root Locus Analysis of a High-Gain Linear System with Variable Coefficients; Application of Horowitz's Method," *IEEE Trans. Automatic Control*, Vol. AC-11, pp. 255–263, Apr. 1966.

21. R. J. Fitzgerald, "Finding Root-Locus Breakaway Points with the Spirule," *IEEE Trans. Automatic Control*, Vol. AC-11, pp. 317–318, Apr. 1966.

22. J. D. S. Muhlenberg, "Synthetic Division for Gain Calibration on the Complex Root Locus," *IEEE Trans. Automatic Control*, Vol. AC-11, pp. 628–629, July 1966.

23. J. Feinstein and A. Fregosi, "Some Invariances of the Root Locus," *IEEE Trans. Automatic Control*, Vol. AC-14, pp. 102–103, Feb. 1969.

24. A. Fregosi and J. Feinstein, "Some Exclusive Properties of the Negative Root Locus," *IEEE Trans. Automatic Control*, Vol. AC-14, pp. 304–305, June 1969.

25. H. M. Power, "Application of Bilinear Transformation to Root Locus Plotting," *IEEE Trans. Automatic Control*, Vol. AC-15, pp. 693–694, Dec. 1970.

Analytical Representation of Root Loci

26. G. A. BENDRIKOV and K. F. TEODORCHIK, "The Analytic Theory of Constructing Root Loci," *Automation and Remote Control*, pp. 340–344, Mar. 1959.

27. K. STEIGLITZ, "An Analytical Approach to Root Loci," *IRE Trans. Automatic Control*, Vol. AC-6, pp. 326–332, Sept. 1961.

28. C. WOJCIK, "Analytical Representation of Root Locus," *Trans. ASME, J. Basic Engineering*, Ser. D, Vol. 86, Mar. 1964.

29. C. S. CHANG, "An Analytical Method for Obtaining the Root Locus with Positive and Negative Gain," *IEEE Trans. Automatic Control*, Vol. AC-10, pp. 92–94, Jan. 1965.

30. B. P. BHATTACHARYYA, "Root Locus Equations of the Fourth Degree," *Internat. J. Control*, Vol. 1, No. 6, pp. 533–556, 1965.

31. J. D. S. MUHLENBERG, "Some Expressions for Root Locus Gain Calibration," *IEEE Trans. Automatic Control*, Vol. AC-12, pp. 796–797, Dec. 1967.

Computer-Aided Plotting of Root Loci

32. D. J. DODA, "The Digital Computer Makes Root Locus Easy," *Control. Eng.*, May 1958.

33. Z. KLAGSBRUNN and Y. WALLACH, "On Computer Implementation of Analytic Root-Locus Plotting," *IEEE Trans. Automatic Control*, Vol. AC-13, pp. 744–745, Dec. 1968.

34. R. H. ASH and G. R. ASH, "Numerical Computation of Root Loci Using the Newton-Raphson Technique," *IEEE Trans. Automatic Control*, Vol. AC-13, pp. 576–582, Oct. 1968.

Root Locus Diagram for Design

35. Y. CHU, "Synthesis of Feedback Control Systems by Phase Angle Loci," *Trans. AIEE*, Vol. 71, Part II, 1952.

36. F. M. REZA, "Some Mathematical Properties of Root Loci for Control Systems Design," *Trans. AIEE Commun. Electronics*, Vol. 75, Part I, pp. 103–108, Mar. 1956.

37. J. A. ASELTINE, "Feedback System Synthesis by the Inverse Root-Locus Method," *IRE Natl. Convention Record*, Part 2, pp. 13–17, 1956.

38. G. A. BENDRIKOV and K. F. TEODORCHIK, "The Theory of Derivative Control in Third Order Linear Systems," *Automation and Remote Control*, pp. 516–517, May 1957.

39. R. J. HRUBY, "Design of Optimum Beam Flexural Damping in a Missile by Application of Root-Locus Techniques," *IRE Trans. Automatic Control*, Vol. AC-5, pp. 237–246, Aug. 1960.

40. A. KATAYAMA, "Semi-graphical Design of Servomechanisms Using Inverse Root-Locus," *J. IEE*, Vol. 80, pp. 1140–1149, Aug. 1960.

41. D. A. CALAHAN, "A Note on Root-Locus Synthesis," *IRE Trans. Automatic Control*, Vol. AC-7, p. 84, Jan. 1962.

42. A. KATAYAMA, "Adaptive Control System Design Using Root-Locus," *IRE Trans. Automatic Control*, Vol. AC-7, pp. 81–83, Jan. 1962.

43. J. R. MITCHELL and W. L. MCDANIEL, JR., "A Generalized Root Locus Following Technique," *IEEE Trans. Automatic Control*, Vol. AC-15, pp. 483–485, Aug. 1970.

Root Sensitivity

44. J. G. TRUXAL and I. M. HOROWITZ, "Sensitivity Considerations in Active Network Synthesis," *Proc. Second Midwest Symposium on Circuit Theory*, East Lansing, Mich., Dec. 1956.

45. R. Y. HUANG, "The Sensitivity of the Poles of Linear Closed-Loop Systems," *Trans. AIEE Appl. Ind.*, Vol. 77, Part 2, pp. 182–187, Sept. 1958.

46. H. UR, "Root Locus Properties and Sensitivity Relations in Control Systems," *IRE Trans. Automatic Control*, Vol. AC-5, pp. 57–65, Jan. 1960.

Root Locus for Discrete-Data Systems

47. M. MORI, "Root Locus Method of Pulse Transfer Function for Sampled-Data Control Systems," *IRE Trans. Automatic Control*, Vol. AC-3, pp. 13–20, Nov. 1963.

48. B. C. KUO, *Digital Control Systems*, Holt, Rinehart and Winston, New York, 1980.

Root Locus for Nonlinear Systems

49. M. J. ABZUG, "A Root-Locus Method for the Analysis of Nonlinear Servomechanisms," *IRE Trans. Automatic Control*, Vol. AC-4, No. 3, pp. 38–44, Dec. 1959.

Root Locus for Systems with Time Delays

50. Y. CHU, "Feedback Control System with Dead-Time Lag or Distributed Lag by Root-Locus Method," *Trans. AIEE*, Vol. 70, Part II, p. 291, 1951.

PROBLEMS

7.1. Sketch the root locus diagram for each of the following feedback control systems. In each case, determine everything about the root loci for $-\infty < K < \infty$. Indicate the starting points, the ending points, and the direction of increasing values of K. The poles and zeros of $G(s)H(s)$ of the systems are given as follows:

(a) Poles at 0, -3, and -4; zero at -5.
(b) Poles at 0, 0, -4, and -4; zero at -5.
(c) Poles at $-1 + j1$ and $-1 - j1$; zero at -2.
(d) Poles at 0, $-1 + j1$, $-1 - j1$; zero at -2.
(e) Poles at 0, -3, $-1 + j1$, and $-1 - j1$; no finite zeros.
(f) Poles at 0, 0, -12, and -12; zeros at $-6 + j5$, $-6 - j5$.
(g) Poles at 0, 0, -12 and -12; zeros at -6, -6.
(h) Poles at 0, $j1$, $-j1$; zeros at -1, -1.
(i) Poles at 0, 0, -12, and -12; zeros at -4, -8.
(j) Poles at $-1, 1$; zeros at $0, 0$.
(k) Poles at 0, 0, 0, and 1; zeros at -1, -2, -3.
(l) Poles at $j1$, $-j1$, $j2$, $-j2$; zeros at -1 and 1.

7.2. The open-loop transfer function of a unity-feedback control system is given by

$$G(s) = \frac{K(s + 2)}{s(s^2 + 2s + 2)(s + 5)(s + 6)}$$

   (a) Sketch the root locus diagram as a function of K ($-\infty < K < \infty$).
   (b) Determine the value of K that makes the relative damping ratio of the closed-loop complex characteristic equation roots equal to 0.4.

7.3. A unity feedback control system has an open-loop transfer function

$$G(s) = \frac{K(1 + 0.2s)(1 + 0.025s)}{s^3(1 + 0.001s)(1 + 0.004s)}$$

Sketch the complete root locus diagram for $-\infty < K < \infty$. Indicate the crossing points of the loci on the $j\omega$ axis, and the corresponding values of K at these points.

7.4. A unity feedback control system has an open-loop transfer function

$$G(s) = \frac{K}{s(1 + 0.02s)(1 + 0.05s)}$$

   (a) Sketch the root locus diagram of the system for $0 \le K < \infty$.
   (b) Determine the marginal value of K for stability.
   (c) Determine the value of K when the system has two equal real characteristic roots.

7.5. The open-loop transfer function of a feedback control system with unity feedback is

$$G(s) = \frac{K}{(s + 10)^n}$$

Sketch the root loci of the characteristic equation of the closed-loop system for $-\infty < K < \infty$, with
   (a) $n = 3$
   (b) $n = 4$.
Show all important information on the root loci.

7.6. The characteristic equation of the liquid-level control system described in Problem 6.28 is written

$$0.06s(s + 12.5)(As + K_0) + 250N = 0$$

   (a) For $A = K_0 = 50$, sketch the root locus diagram of the characteristic equation as N varies from 0 to ∞.
   (b) For $N = 20$ and $K_0 = 50$, sketch the root locus diagram of the characteristic equation as A varies from $-\infty$ to ∞.
   (c) For $A = 50$ and $N = 20$, sketch the root loci when K_0 varies from $-\infty$ to ∞.

7.7. The block diagram of a feedback control system is shown in Fig. 7P-7.
   (a) Sketch the root loci for $K \ge 0$ when the switch is open. Determine the stability of the system as a function of K. Show this information on the root loci.
   (b) Close the switch S so that the minor feedback loop is in effect. Set $K = 1$ and show by a root locus plot how the system is stabilized when K_t varies.

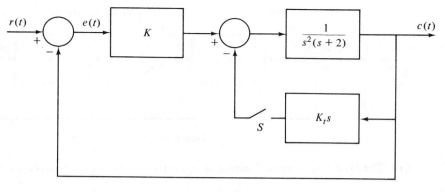

Figure 7P-7

7.8. The transfer functions of a single-loop feedback control system are given as

$$G(s) = \frac{K}{s^2(s + 2)(s + 3)} \quad \text{and} \quad H(s) = 1$$

(a) Sketch the loci of the zeros of $1 + G(s)H(s)$ for $0 \le K < \infty$. Indicate the crossing points of the loci on the $j\omega$ axis and the corresponding values of K at these points.

(b) The transfer function of $H(s)$ is now changed to $H(s) = 1 + 2s$. Show the effects on the root loci due to this change in $H(s)$.

7.9. The characteristic equation of a feedback control system is

$$s^3 + 2s^2 + (K + 1)s + 3K = 0$$

Sketch the root loci of this equation for $-\infty < K < \infty$.

7.10. A unity-feedback control system has the open-loop transfer function

$$G(s) = \frac{K(s^2 - 1)(s + 2)}{s(s^2 + 2s + 2)}$$

Sketch the root loci and the complementary root loci of the characteristic equation. Label all important points and information on the loci.

7.11. For the control system shown in Fig. 7P-11, sketch the root loci of the characteristic equation for $-\infty < K < \infty$. Label all important information on the loci.

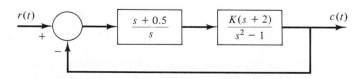

Figure 7P-11

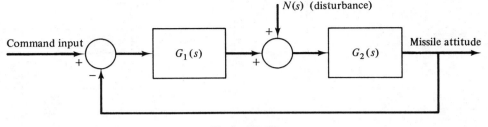

Figure 7P-13

7.12. The open-loop transfer function of a control system with unity feedback is written

$$G(s) = \frac{1000(1 + K_D s)}{s(1 + 0.1s)(1 + 0.002s)}$$

Sketch the root locus diagram of the characteristic equation of the closed-loop system for $0 \le K_D < \infty$. Determine the value (or values) of K_D so that the damping ratio of the complex roots of the characteristic equation is 0.707.

7.13. The block diagram of an attitude autopilot in a missile flight control system is shown in Fig. 7P-13. The transfer functions of the airframe dynamics are:

$$G_1(s) = \frac{K}{(s + 5)(s - 5)} \qquad G_2(s) = \frac{s + 1}{s}$$

Sketch the root loci of the characteristic equation of the closed-loop system for $-\infty < K < \infty$. Include all the important information on the root loci.

7.14. Given the equation

$$s^3 + as^2 + Ks + K = 0$$

It is desired to investigate the root loci of this equation for $-\infty < K < \infty$ and for several values of a.
(a) Sketch the root loci ($-\infty < K < \infty$) for $a = 10$.
(b) Repeat part (a) for $a = 3$.
(c) Determine the value of a so that there is only one nonzero breakaway point on the entire root loci for $-\infty < K < \infty$. Sketch the loci.
In sketching the root loci you should apply all known rules whenever they are applicable.

7.15. The open-loop transfer function of a control system with unity feedback is

$$G(s) = \frac{K(s + a)}{s^2(s + 2)}$$

Determine the values of a so that the root locus diagram will have zero, one, and two breakaway points, respectively, not counting the one at $s = 0$. Sketch the root loci for $-\infty < K < \infty$ for all three cases.

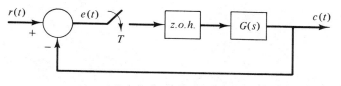

Figure 7P-17

7.16. Condition the following equation so that it is in the form of $1 + G(s)H(s) = 0$. Use the root locus method to show that the real root of the equation is between -1 and -2. Find the roots using this information and the trial-and-error method.

$$s^3 + 2s^2 + 3s + 3 = 0$$

7.17. For the sampled-data control system shown in Fig. 7P-17.

$$G(s) = \frac{K}{s(1 + 0.2s)}$$

(a) Sketch the root loci for the system $(0 < K < \infty)$ without the zero-order hold, for $T = 0.1$ sec and $T = 1$ sec. Determine the marginal value of K for stability in each case.

(b) Repeat part (a) when the system has a zero-order hold.

7.18. The following polynomial in z represents the characteristic equation of a certain discrete-data control system. Sketch the root loci $(-\infty < K < \infty)$ for the system. Determine the marginal value of K for stability.

$$z^3 + Kz^2 + 1.5Kz - (K + 1) = 0$$

7.19. Sketch the root loci $(0 \leq K < 0)$ in the z-plane for the discrete-data control system shown in Fig. 7P-19.

7.20. Given the characteristic equation of a closed-loop control system,

$$(s + 2)(s + 3) + K = 0$$

(a) Find the root sensitivity $S_K^s = K \cdot \partial s/s \cdot \partial K$.

(b) Find S_K^s for roots on the real axis. Find the points on the real axis where $S_K^s = 0$ and $S_K^s = \infty$.

7.21. Repeat Problem 7.20 with $s(s + 1)(s + 2) + K = 0$.

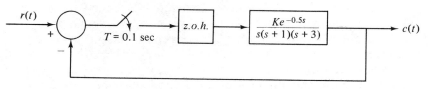

Figure 7P-19

chapter eight

Time-Domain Design
of Control Systems

8.1 INTRODUCTION

In this chapter some of the time-domain design methods of control systems are discussed. The time-domain design refers to the utilization of the time-domain properties of the system to be designed. The material presented in the preceding chapters on controllability, stability, and root locus are all used here for the purpose of design.

As discussed in Chapter 6, the time-domain characteristics of a control system are represented by the transient and the steady-state responses of the system when certain test signals are applied. Depending on the objectives of the design, these test signals are usually in the form of a step function or a ramp function. For a step function input, the percent peak overshoot, rise time, and settling time are often used to measure the performance of the system. Qualitatively, the damping ratio and the natural undamped frequency may be used to indicate the relative stability of the system. These quantities are defined strictly only for a second-order system. For high-order systems, these parameters are meaningful only if the corresponding pair of eigenvalues or the poles of the closed-loop transfer function dominates the dynamic response of the system. Therefore, for design in the time domain, the design criteria often include the peak overshoot as a design parameter.

In general, the dynamics of a linear controlled process can be represented by the block diagram shown in Fig. 8-1. The design objective is to have the controlled variables, represented by the output vector, $c(t)$, behave in certain desirable ways. The problem essentially involves the determination of the control signal $u(t)$ over the prescribed time interval so that the design objectives are all satisfied.

Most of the established design methods in control systems rely on the so-called *fixed-configuration design* in that the designer at the outset decides the basic composition of the overall system, and the place where the controller is to be

464

Figure 8-1 Controlled process.

positioned relative to the components of the controlled process. The problem then involves the design of the elements of the controller.

Figure 8-2 shows several commonly used system configurations with controller compensation. The most commonly used system configuration is shown in Fig. 8-2(a). In this case the controller is placed in series with the process, and the configuration is referred to as *series* or *cascade compensation*. In Fig. 8-2(b) the controller is placed in the minor feedback path, and the scheme is called *feedback compensation*. Figure 8-2(c) shows a system that generates the control signal by feeding back the state variables through constant gains; the scheme is called *state feedback*. The problem with state feedback control is that for high-order systems the large number of state variables involved would require a large number of transducers to sense for feedback. Thus, the actual implementation of the state-feedback control scheme may be costly. Even for low-order systems, often not all the state variables are directly accessible, and an *observer* may be necessary for practical implementation of estimating the state variables from the output variables.

The compensation schemes shown in Figs. 8-2(a), (b), and (c) all have one degree of freedom in that there is only one controller in each system, even though the controller may have more than one parameter. The disadvantage with a one-degree-of-freedom controller is that the performance criteria that can be realized are limited. For example, if a system is designed to achieve a certain amount of relative stability, it may have poor sensitivity to parameter variations. Or, if the roots of the characteristic equation are selected to provide a certain amount of relative damping, the peak overshoot of the step response may still be excessive, owing to the zero in the closed-loop transfer function, as demonstrated in Sec. 7.9.

Figures 8-2(d), (e), and (f) show compensation schemes that have two degrees of freedom. The configuration shown in Fig. 8-2(d) is called the series-feedback compensation. Figures 8-2(e) and (f) show the so-called *feedforward compensation*. In Fig. 8-2(e) the controller G_{c1} is placed in series with the closed-loop system, which has a controller G_{c2} in the forward path. In Fig. 8-2(f) the feedforward controller G_{c1} is in parallel with the forward path of the system. The key to the feedforward compensation is that the controller G_{c1} is not in the loop of the system, so that it does not affect the roots of the characteristic equation. By proper design of the controller G_{c1} the zeros of the closed-loop transfer function can be either canceled or placed properly to achieve desired system performance over a wider range of specifications.

Although the systems illustrated in Fig. 8-2 all have continuous-time control, the same configurations can be applied to digital control, in which case the controllers are all digital.

Because of the lack of straightforward or unique relationships between the time-domain specifications and the system equations or transfer functions of systems with order higher than the second, a general design procedure in the time

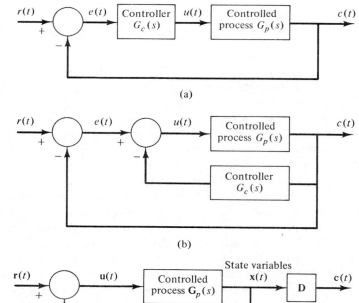

(a)

(b)

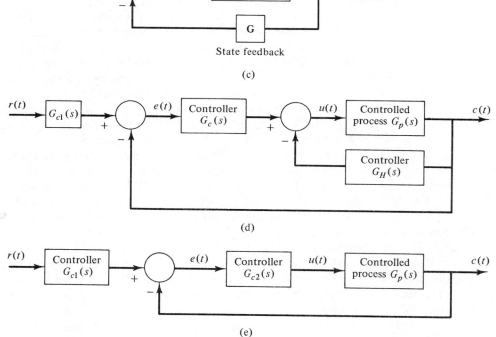

(c)

(d)

(e)

Figure 8-2 Various controller configurations in control system compensation. (a) Series or cascade comparison. (b) Feedback compensation. (c) State-feedback control. (d) Series-feedback compensation (two degrees of freedom). (e) Forward compensation with series compensation (two degrees of freedom). (f) Feedforward compensation (two degrees of freedom).

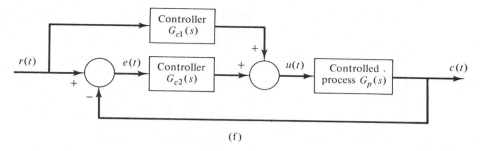

(f)

Figure 8.2 (cont.)

domain is very difficult to establish. J. G. Truxal [1] in his 1950 book *Control Systems Synthesis* did detail a time-domain synthesis method for linear control systems. However, modern practicing engineers still rely to a great extent on the known relations between the time-domain specifications and performance characteristics, and they generally use established controller configurations to solve day-to-day design problems.

8.2 TIME-DOMAIN DESIGN OF THE PID CONTROLLER

In this section we introduce the effects of basic controller operations other than the simple multiplication by a constant, such as in the case of an amplifier. In all the control systems examples we have discussed thus far, the controller has been typically simply an amplifier with a constant gain K. This type of control action is formally known as the "proportional control," since the actuating signal at the output of the controller is simply related to the input of the controller by a proportional constant. From a mathematical standpoint, a linear continuous-data controller should also be able to take a time derivative or a time integral of the input signal, in addition to the proportional and other simple algebraic operations, such as addition and subtraction. Therefore, we can simply describe a linear continuous-data controller to be a device that contains such components as adders (plus or minus), amplifiers, attenuators, differentiators, and integrators. The designer's task is to determine which of these components should be used, what their parameter values are, and how they should be connected. For example, one of the best-known controllers used in practice is the PID controller, where PID stands for proportional, integral, and derivative. The transfer function of a PID controller can be written

$$G_c(s) = K_P + K_D s + \frac{K_I}{s} \tag{8-1}$$

and the design problem is to determine the values of the constants K_P, K_D, and K_I so that the performance of the system is as prescribed.

We investigate next the effects of derivative control and integral control.

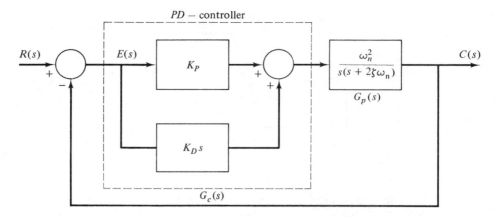

Figure 8-3 Feedback control system with proportional-derivative control.

Effects of Derivative Control on the Time Response of Feedback Control Systems

Figure 8-3 shows the block diagram of a feedback control system that has a second-order process with transfer function $G_p(s)$, and the controller with proportional-derivative control (PD controller). The transfer function of the PD controller is

$$G_c(s) = K_P + K_D s \qquad (8\text{-}2)$$

The open-loop transfer function of the overall system is

$$G(s) = G_c(s)G_p(s) = \frac{C(s)}{E(s)} = \frac{\omega_n^2(K_P + K_D s)}{s(s + 2\zeta\omega_n)} \qquad (8\text{-}3)$$

Clearly, Eq. (8-3) shows that the derivative control is equivalent to the addition of a simple zero at $s = -K_P/K_D$ to the open-loop transfer function.

The effect of the derivative control on the transient response of a feedback control system can be investigated by referring to the time responses shown in Fig. 8-4. Let us assume that the unit step response of a system with only proportional control is as shown in Fig. 8-4(a). The corresponding error signal $e(t)$ and its time derivative $de(t)/dt$ are as shown in Fig. 8-4(b) and (c), respectively. Notice that the response shown in Fig. 8-4(a) has a relatively high peak overshoot and is rather oscillatory. This may be objectionable for many practical control purposes. For a system that is driven by a motor of some kind, this large overshoot and subsequent oscillations are due to the excessive amount of torque developed by the motor and the lack of damping in the time interval $0 < t < t_1$, during which the error signal $e(t)$ is positive. For the time interval $t_1 < t < t_3$, $e(t)$ is negative, and the corresponding motor torque is negative. This negative torque tends to slow down the output acceleration and eventually causes the direction of the output member to

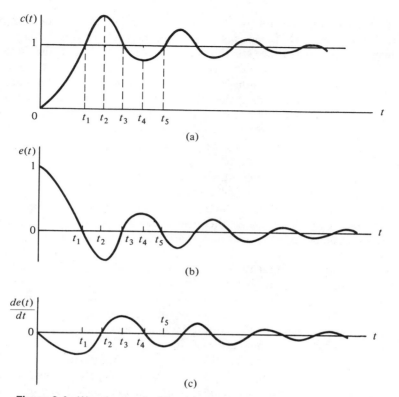

Figure 8-4 Waveforms of $c(t)$, $e(t)$ and $de(t)/dt$, showing the effect of derivative control. (a) Step response. (b) Error signal. (c) Time rate of change of error signal.

reverse and undershoot during $t_3 < t < t_5$. During the time interval $t_3 < t < t_5$ the motor torque is again positive, thus tending to reduce the undershoot in the response caused by the negative torque in the previous interval. Since the system is assumed to be stable, the error amplitude is reduced with each oscillation, and the output eventually is settled to its final desired value.

Considering the explanation given above, we can say that the contributing factors to a high overshoot are as follows: (1) The positive correcting torque in the interval $0 < t < t_1$ is too large, and (2) the retarding torque in the time interval $t_1 < t < t_2$ is inadequate. Therefore, in order to reduce the overshoot in the step response, a logical approach is to decrease the amount of positive correcting torque and to increase the retarding torque. Similarly, in the time interval $t_2 < t < t_4$, the negative corrective torque should be reduced, and the retarding torque, which is now in the positive direction, should be increased in order to improve the undershoot.

The derivative control as represented by the system of Fig. 8-3 gives precisely the compensation effect described in the last paragraph. Let us consider that the proportional type of control system whose signals are described in Fig. 8-4 is now modified so that the torque developed by the motor is proportional to the signal $e(t) + K_D de(t)/dt$. In other words, in addition to the error signal, a signal that is proportional to the time rate of change of error is applied to the motor. As shown in Fig. 8-4(c), for $0 < t < t_1$, the time derivative of $e(t)$ is negative; this will reduce the original torque developed due to $e(t)$ alone. For $t_1 < t < t_2$, both $e(t)$ and $de(t)/dt$ are negative, which means that the negative retarding torque developed will be greater than that of the proportional case. Therefore, all these effects will result in a smaller overshoot. It is easy to see that $e(t)$ and $de(t)/dt$ have opposite signs in the time interval $t_2 < t < t_3$; therefore, the negative torque that originally contributes to the undershoot is reduced also.

Since $de(t)/dt$ represents the slope of $e(t)$, the derivative control is essentially an anticipatory type of control. Normally, in a linear system, if the slope of $e(t)$ or $c(t)$ due to a step input is large, a high overshoot will subsequently occur. The derivative control measures the instantaneous slope of $e(t)$, predicts the large overshoot ahead of time, and makes a proper correcting effort before the overshoot actually occurs.

It is apparent that the derivative control will affect the steady-state error of a system only if the steady-state error varies with time. If the steady-state error of a system is constant with respect to time, the time derivative of this error is zero, and the derivative control has no effect on the steady-state error. But if the steady-state error increases with time, a torque is again developed in proportion to $de(t)/dt$, which will reduce the magnitude of the error.

As an illustrative example, let us consider that the microprocessor controller of the printwheel control system shown in Fig. 6-22 is programmed to implement a PD control. For the present we shall assume that the control operation is described by the analog transfer function of Eq. (8-2). We should keep in mind, of course, that this is not strictly true; Eq. (8-2) can be approached by a digital program only if the sampling period of the microprocessor is very small. Later in this chapter we investigate how a PID controller is implemented digitally.

With the inductance L_a set to zero, the open-loop transfer function of the system is obtained by setting τ_a to zero in Eq. (6-122) and replacing K by the right-hand side of Eq. (8-2). We have

$$G(s) = \frac{\Theta_c(s)}{\Theta_e(s)} = \frac{K_s K_i (K_P + K_D s)}{R_a B s (1 + \tau s) + K_b K_i s}$$

$$= \frac{400(K_P + K_D s)}{s(s + 48.5)} \tag{8-4}$$

Figure 6-24 illustrates the unit step responses of the closed-loop system with only the proportional control, with $K = K_P = 1$, 2.94, and 100. Figure 8-5 illustrates the unit step response of the system with the PD controller when $K_P = 2.94$ and $K_D = 0.0502$. For comparison, the response of the system with only the proportional control and $K_P = 2.94$ is also shown in the figure. Notice that in the case of the relatively low value of K_P, the effect of the derivative control is the increase of the damping, and the step response is slowed down. The value of

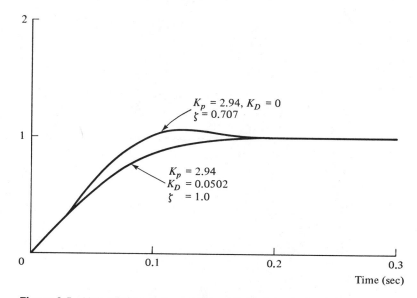

Figure 8-5 Unit step responses of the printwheel control system in Fig. 6-22 with PD control, $K_P = 2.94$.

$K_D = 0.0502$ is chosen so that the damping ratio of the system is at critical ($\zeta = 1.0$). This is done by writing the characteristic equation of the closed-loop system,

$$s^2 + (48.5 + 400K_D)s + 400K_P = 0 \qquad (8\text{-}5)$$

For $K_P = 2.94$, $\omega_n = \sqrt{400K_P} = \sqrt{1176}$ rad/sec. For $\zeta = 1.0$, the value of K_D is found from

$$2\zeta\omega_n = 48.5 + 400K_D = 68.59 \qquad (8\text{-}6)$$

Thus,

$$K_D = 0.0502$$

In practice, there would be little justification to apply derivative control to the system when $K_P = 2.94$, since the resulting slower rise time may not be desirable. Figure 8-6 shows the unit step responses when the proportional constant K_P is 100. Without derivative control, the step response has an overshoot of 68 percent. When $K_D = 0.0502$, we see that there is only a slight improvement on the transient response; the peak overshoot is reduced to approximately 40 percent. The corresponding damping ratio as calculated from Eq. (8-6) is improved from 0.12125 to 0.1715. For critical damping, we set K_D to 0.8788, and the unit step response is shown in Fig. 8-6. In this case, the response exhibits no overshoot and the rise time is quite short, as compared with the responses in Fig. 8-5. Thus, we see that the PD controller when designed properly could yield a system with fast rise time and little or no overshoot.

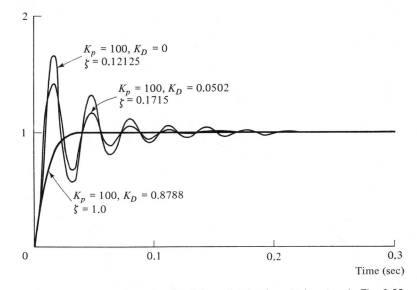

Figure 8-6 Unit step responses of the printwheel control system in Fig. 6-22 with PD control, $K_P = 100$.

Analytically, the design of the PD control can be carried out by the root contour method introduced in the last chapter. This is done by referring to the characteristic equation of Eq. (8-5). Since the equation has two parameters in K_D and K_P, we can first let $K_D = 0$, and the equation becomes

$$s^2 + 48.5s + 400K_P = 0$$

The root loci of the last equation as K_P varies are shown in Fig. 8-7. In a design problem, the value of K_P is usually determined from the specification given on the steady-state error. For this type 1 system, the steady-state error due to a step-function input is zero; thus only the ramp error constant K_v is a meaningful specification. For $K_P = 100$, for example,

$$K_v = \lim_{s \to 0} sG(s) = \lim_{s \to 0} \frac{400(K_P + K_D s)}{s + 48.5} = 824.74 \qquad (8\text{-}7)$$

which means that if a unit-ramp function input is applied to the closed-loop system, the steady-state error will be equal to $1/K_v = 0.00121$.

When $K_D \neq 0$, the characteristic equation of Eq. (8-5) is conditioned as follows:

$$1 + G_{eq}(s) = 1 + \frac{400K_D s}{s^2 + 48.5s + 400K_P} = 0 \qquad (8\text{-}8)$$

The root contours of Eq. (8-5) with $K_P = $ constant and as K_D varies, are constructed based on the pole-zero configuration of $G_{eq}(s)$, and two are shown in Fig. 8-8 for $K_P = 2.94$ and 100. We see that when $K_P = 2.94$ and $K_D = 0$, the character-

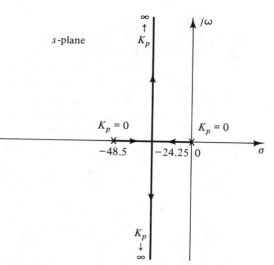

Figure 8-7 Root loci of Eq. (8-6).

istic equation roots are at $-24.25 + j24.25$, $-24.25 - j24.25$, and the damping ratio of the closed-loop system is 0.707. When the value of K_D is increased, the two roots move toward the real axis along a circular arc. When $K_D = 0.0502$, the roots are real and equal, -34.29, and the damping is a critical. When K_D is increased beyond 0.0502, the two roots become real and unequal, and the system is over-damped.

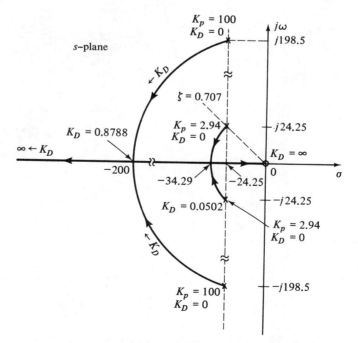

Figure 8-8 Root contours of Eq. (8-5) when $K_P = 2.94$ and 100; K_D varies.

When $K_P = 100$ and $K_D = 0$, the two roots are at $-24.25 + j198.5$ and $-24.25 - j198.5$. As K_D increases, the roots again move toward the real axis. When K_D is equal to 0.8788, the two roots become equal at -200. Thus, the root contours of Fig. 8-8 clearly show the damping effect of the PD controller.

Another analytic way of studying the effects of the parameters K_P and K_D is to evaluate the performance characteristics in the parameter plane of K_P and K_D. From the characteristic equation of Eq. (8-5), we have

$$\omega_n = 20\sqrt{K_P} \tag{8-9}$$

$$\zeta = \frac{1.21 + 10K_D}{\sqrt{K_P}} \tag{8-10}$$

Applying the stability criterion to Eq. (8-5), we find that for stability,

$$K_P > 0$$

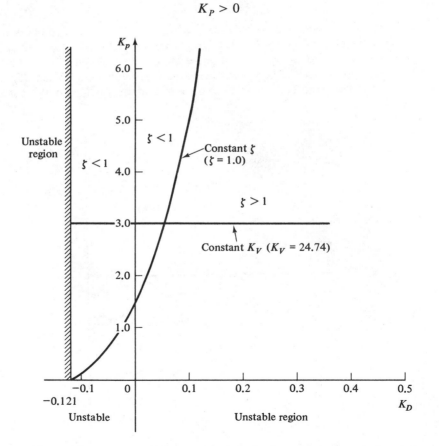

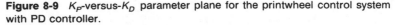

Figure 8-9 K_P-versus-K_D parameter plane for the printwheel control system with PD controller.

and

$$K_D > -0.121 \qquad (8\text{-}11)$$

The boundary of stability in the K_P-versus-K_D parameter plane is shown in Fig. 8-9. By setting $\zeta = 1$ in Eq. (8-10), the trajectory that corresponds to critical damping is also shown in Fig. 8-9. From Eq. (8-7), the ramp error constant K_v is written

$$K_v = \frac{400 K_P}{48.5} = 8.25 K_P \qquad (8\text{-}12)$$

Thus, the constant-K_v trajectory in the K_P-versus-K_D parameter plane are simply horizontal lines. Without doubt, Fig. 8-9 gives a clear picture as to how K_P and K_D should be chosen as related to the performance specifications.

From the practical implementation standpoint, the PD controller cannot be physically realized by passive R, L, C, elements alone, since the transfer function has one zero but no poles. However, for one who is familiar with analog circuit design, it would be a simple task to construct a PD controller using op amps, resistors, and capacitors. The practical difficulty with the analog-circuit implementation of the PD controller is that the differentiator in the PD controller is a high-pass filter, and the design usually has to contend with noise problems.

Effects of Integral Control on the Time Response of Feedback Control Systems

The integral part of the PID controller produces a signal that is proportional to the time integral of the input of the controller. Figure 8-10 shows the block diagram of a feedback control system that has a second-order process with transfer function $G_p(s)$, and a controller with proportional-integral control (PI controller). The

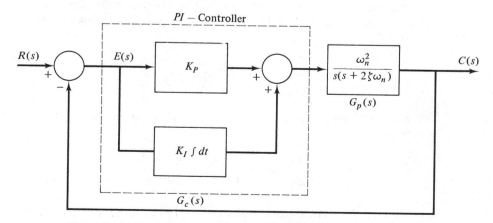

Figure 8-10 Feedback control system with proportional-integral control.

transfer function of the PI controller is

$$G_c(s) = K_P + \frac{K_I}{s} \tag{8-13}$$

The open-loop transfer function of the overall system is

$$G(s) = G_c(s)G_p(s) = \frac{\omega_n^2(K_P s + K_I)}{s^2(s + 2\zeta\omega_n)} \tag{8-14}$$

In this case the PI controller is equivalent to adding a zero at $s = -K_I/K_P$ and a pole at $s = 0$ to the open-loop transfer function. One obvious effect of the integral control is that it increases the order of the system by one. More important is that it increases the type of the system by one. Therefore, the steady-state error of the original system without integral control is improved by one order; that is, if the steady-state error to a given input is constant, the integral control reduces it to zero (provided that the final system is stable, of course). In the case of Eq. (8-14), the closed-loop system in Fig. 8-10 will now have a zero steady-state error when the reference input is a ramp function. However, because the system is now of the third order, it *may* be less stable than the original second-order system or even become unstable if the parameters K_P and K_I are not properly chosen.

In the case of a system with PD control, the value of K_P is important because for a type 1 system it governs the ramp-error constant, and thus affects the magnitude of the nonzero steady-state error when the input is a ramp. Thus, if K_P is too large, the stability of the system may be adversely affected, and the steady-state error is inversely proportional to K_P.

When a type 1 system is converted to a type 2 system by a PI controller, the proportional constant K_P no longer affects the steady-state error, and the latter is always zero for a ramp function input. The problem then is to choose the proper combination of K_P and K_I so that the transient response is satisfactory.

For the printwheel control system, let the controller be described by the transfer function of Eq. (8-13). The open-loop transfer function of the overall system is

$$G(s) = \frac{400(K_P s + K_I)}{s^2(s + 48.5)} \tag{8-15}$$

Let $K_P = 100$ and $K_I = 10$. The closed-loop transfer function of the system becomes

$$\frac{\Theta_c(s)}{\Theta_r(s)} = \frac{40,000(s + 0.1)}{s^3 + 48.5s^2 + 40,000s + 4000} \tag{8-16}$$

The roots of the characteristic equation are found to be $s = -0.10001$, $-24.2 + j198.5$, and $-24.2 - j198.5$. We see that the real root is very close to the zero of the

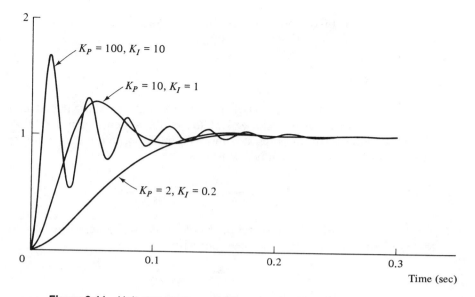

Figure 8-11 Unit step responses of the printwheel control system in Fig. 6-22 with PI controller.

closed-loop transfer function, and thus for all practical purposes, the closed-loop transfer function can be approximated by

$$\frac{\Theta_c(s)}{\Theta_r(s)} \cong \frac{40{,}000}{s^2 + 48.5s + 40{,}000} \tag{8-17}$$

Figure 8-11 shows the unit step response of the system, and as expected, it is very close to the response shown in Fig. 8-6 when $K_P = 100$ and $K_D = 0$. To improve on the transient response, we can reduce the value of K_P, since it no longer affects the steady-state error. Figure 8-11 shows the unit step response when $K_P = 10$ and $K_I = 1.0$. The peak overshoot is now approximately 27 percent. By decreasing the value of K_P further, it is possible to achieve a unit step response with little or no overshoot, such as that shown in Fig. 8-11 with $K_P = 2$ and $K_I = 0.2$. Notice that in these various controller configurations we have reduced the value of K_I somewhat in proportion to the reduction of K_P. This is not absolutely necessary, but was done so that the real pole of the transfer function will be close to its zero, and thus the transient response can better be predicted by the two complex poles.

The effect of K_I on the stability of the system can be investigated by applying the Routh's test to the characteristic equation

$$s^3 + 48.5s^2 + 400K_Ps + 400K_I = 0 \tag{8-18}$$

The result is that the closed-loop system is stable for $0 \leq K_I < 48.5K_P$.

The effects of the PI controller on the system performance can better be described by the root loci and root contours. With reference to the characteristic equation of Eq. (8-18), when $K_P = 100$, the root contours for $0 \leq K_I < \infty$ are constructed from the pole-zero configuration of

$$G_{eq}(s) = \frac{400K_I}{s(s^2 + 48.5s + 40{,}000)} \qquad (8\text{-}19)$$

and are shown in Fig. 8-12(a). These root contours clearly show the adverse effect of the integral control on the relative stability of the system when the value of K_I is very large. For design purposes, it is more illuminating to use the open-loop transfer function of Eq. (8-15), which is rewritten as

$$G(s) = \frac{400K_P(s + K_I/K_P)}{s^2(s + 48.5)} \qquad (8\text{-}20)$$

Figure 8-12(b) shows the root loci of the closed-loop characteristic equation of Eq. (8-18) when K_P varies from 0 to ∞, but with

$$\frac{K_I}{K_P} \ll 48.5 \qquad (8\text{-}21)$$

The key to successful design of the PI controller lies in the satisfaction of the inequality of Eq. (8-21). As shown by the root loci in Fig. 8-12(b), when the value of K_I/K_P is relatively very small, the transient response of the type 2 system can be made to track that of the original second-order system, since the shape of the root loci in Fig. 8-12(b) at points far away from the pole and zero of the PI controller is very similar to that of the original system. This explains how the values of K_P and K_D are selected for the designs that produced the responses in Fig. 8-11.

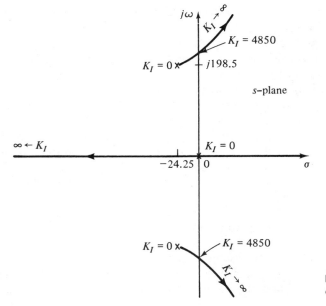

Figure 8-12(a) Root contours of Eq. (8-18) with $K_P = 100$.

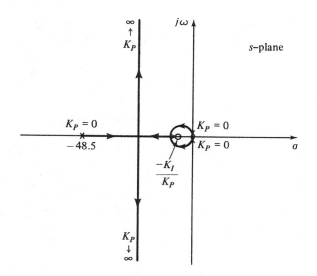

Figure 8-12(b) Root loci of Eq. (8-18) with K_I / K_P equal to a small number.

While the PI controller improves the steady-state error by one order, simultaneously allowing a transient response with little or no overshoot, the rise time of the response can be quite slow, as shown in Fig. 8-11. This is not surprising, since the PI controller is essentially a low-pass filter which attenuates high-frequency signals. This leads to the motivation of using a PID controller so that the best properties of each of the controllers can be utilized.

Figure 8-13 shows the unit step response of the printwheel control system with the controller described by

$$G_c(s) = K_P + K_D s + \frac{K_I}{s} \tag{8-22}$$

where $K_P = 100$, $K_D = 0.8788$, and $K_I = 10$. In this case, the system is of type 2, which assures that the steady-state error is zero to a ramp function input, and the transient response is quite respectable, with a fast rise time and a peak overshoot of just 7 percent.

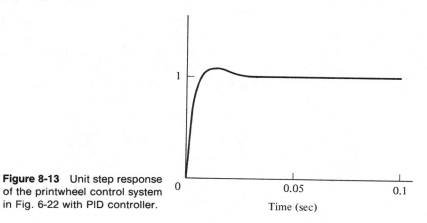

Figure 8-13 Unit step response of the printwheel control system in Fig. 6-22 with PID controller.

8.3 TIME-DOMAIN DESIGN OF THE PHASE-LEAD AND PHASE-LAG CONTROLLERS

The PID controller represents the simplest form of controllers that utilize the derivative and integration operations in the compensation of control systems. In general, we can regard the design of controllers for control systems as a filter design problem; then we can come up with a great deal of varieties. Since root loci are used often for designs, it would be advantageous to describe the controller by the poles and zeros of its transfer function. For the PD controller in Eq. (8-2), it has a zero at $s = -K_P/K_D$. For the PI controller, the transfer function of Eq. (8-13) has a pole at $s = 0$ and a zero at $s = -K_I/K_P$. The PID controller in Eq. (8-22) has a pole at $s = 0$ and two zeros from the function $K_D s^2 + K_P s + K_I$.

From the filtering standpoint, we have pointed out that the PD controller is a high-pass filter, whereas the PI controller is a low-pass filter. The PID controller is band-pass or band-attenuate, depending on the controller parameters. The high-pass filter is often referred to as a phase-lead controller since positive phase is introduced to the system over some appropriate frequency range. The low-pass filter is also known as a phase-lag controller, since the corresponding phase introduced is negative. These ideas related to filtering and phase shifts are explored further in Chapter 10, where the subject of frequency-domain design is treated.

There are obvious advantages of using only passive network elements in a controller. The transfer function of a simple controller that can be realized by passive resistive-capacitive network elements is

$$G_c(s) = \frac{s + z_1}{s + p_1} \tag{8-23}$$

In Eq. (8-23) the controller is high-pass or phase-lead if $p_1 > z_1$; low-pass or phase-lag if $p_1 < z_1$.

We outline design procedures of using the controller of Eq. (8-23) in the following sections.

Phase-Lead Controller

A network realization of the phase-lead controller of Eq. (8-23) ($p_1 > z_1$) is shown in Fig. 8-14. Although the network may be simplified further, and still be representing a low-pass filter by eliminating R_1, the dc signals would be completely blocked in the steady state, and thus the resulting controller would not be acceptable for control systems.

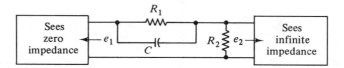

Figure 8-14 Passive phase-lead network.

The transfer function of the network is derived as follows by assuming that the source impedance which the lead network sees is zero, and the output load impedance is infinite. This assumption is necessary in the derivation of the transfer function of any four-terminal network.

$$\frac{E_2(s)}{E_1(s)} = \frac{R_2 + R_1 R_2 Cs}{R_1 + R_2 + R_1 R_2 Cs} \qquad (8\text{-}24)$$

or

$$\frac{E_2(s)}{E_1(s)} = \frac{R_2}{R_1 + R_2} \frac{1 + R_1 Cs}{1 + \dfrac{R_1 R_2}{R_1 + R_2} Cs} \qquad (8\text{-}25)$$

Let

$$a = \frac{R_1 + R_2}{R_2} \qquad a > 1 \qquad (8\text{-}26)$$

and

$$T = \frac{R_1 R_2}{R_1 + R_2} C \qquad (8\text{-}27)$$

Then Eq. (8-25) becomes

$$\frac{E_2(s)}{E_1(s)} = \frac{s + 1/aT}{s + 1/T} \qquad a > 1 \qquad (8\text{-}28)$$

As seen from Eq. (8-28), the transfer function of the phase-lead network has a real zero at $s = -1/aT$ and a real pole at $s = -1/T$. These are represented in the s-plane as shown in Fig. 8-15. By varying the values of a and T, the pole and zero may be located at any point on the negative real axis in the s-plane. Since $a > 1$, the

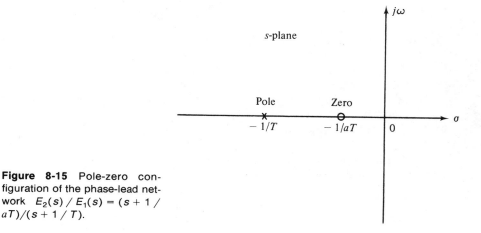

Figure 8-15 Pole-zero configuration of the phase-lead network $E_2(s) / E_1(s) = (s + 1 / aT)/(s + 1 / T)$.

zero is always located to the right of the pole, and the distance between them is determined by the constant a.

The zero being to the right of the pole is the reason that the phase-lead controller can improve the relative and absolute stability of a closed-loop control system. We shall use the following example to illustrate the design of phase-lead controller for control systems.

■

Example 8-1

The block diagram shown in Fig. 8-16 describes the components of a sun-seeker control system. The system may be mounted on a space vehicle so that it will track the sun with high accuracy. The variable θ_r represents the reference angle of the solar ray, and θ_0 denotes the vehicle axis. The objective for the sun-seeker control system is to maintain the error between θ_r and θ_0, α, near zero. The parameters of the system are given as

$$R_F = 10,000$$
$$K_b = 0.0125 \text{ V/rad/sec}$$
$$K_i = 0.0125 \text{ N-m/A}$$
$$R_a = 6.25$$
$$J = 10^{-6} \text{ kg} = \text{m}^2$$
$$K_s = 0.1 \text{ A/rad}$$
$$K = \text{to be determined}$$
$$B = 0$$
$$n = 800$$

The open-loop transfer function of the uncompensated system is

$$\frac{\Theta_0(s)}{\alpha(s)} = \frac{K_s R_F K K_i / n}{R_a J s^2 + K_i K_b s} \tag{8-29}$$

Substituting the numerical values of the system parameters, Eq. (8-29) gives

$$\frac{\Theta_0(s)}{\alpha(s)} = \frac{2500 K}{s(s + 25)} \tag{8-30}$$

The specifications of the system are given as follows:

1. The steady-state value of $\alpha(t)$ due to a unit ramp function input for $\theta_r(t)$ should be less than or equal to 0.01 rad per rad/sec of the final steady-state output velocity. In other words, the steady-state error due to a ramp input should be less than or equal to 1 percent.
2. The peak overshoot should be less than 10 percent.

The loop gain of the system is determined from the steady-state error requirement. Applying the final-value theorem to $\alpha(t)$, we have

$$\lim_{t \to \infty} \alpha(t) = \lim_{s \to 0} s\alpha(s) = \lim_{s \to 0} \frac{s\Theta_r(s)}{1 + [\Theta_0(s)/\alpha(s)]} \tag{8-31}$$

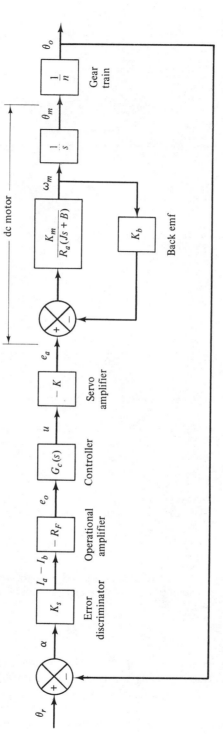

Figure 8-16 Block diagram of a sun-seeker control system.

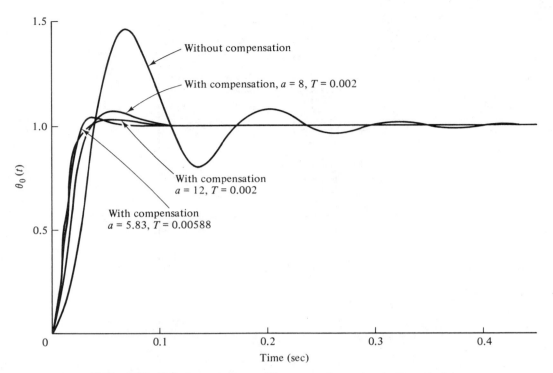

Figure 8-17 Unit step responses of the sun-seeker system in Example 8-1.

For the unit ramp input, $\Theta_r(s) = 1/s^2$; using Eq. (8-30), Eq. (8-31) gives

$$\lim_{t \to \infty} \alpha(t) = \frac{0.01}{K} \tag{8-32}$$

Thus, for the steady-state error to be less than or equal to 0.01, K must be greater than or equal to 1. For $K = 1$, the worst case, the characteristic equation of the uncompensated system is

$$s^2 + 25s + 2500 = 0 \tag{8-33}$$

Thus the damping ratio of the uncompensated system is merely 25 percent, which corresponds to a peak overshoot of over 44.4 percent. Figure 8-17 shows the unit step response of the system with $K = 1$.

■

The block diagram of Fig. 8-16 already indicates that a series controller is applied. Let us consider using the phase-lead controller of Eq. (8-28), although in this case a PD controller would also be able to satisfy the performance specifications given above.

Let us use the root contour method to aid the selection of the controller parameters a and T.

The open-loop transfer function of the system with the phase-lead controller is written

$$G(s) = \frac{2500(1 + aTs)}{s(s + 25)(1 + Ts)} \tag{8-34}$$

We have set the loop gain of the system to compensate for the attenuation $1/a$ of the phase-lead controller, so that the ramp-error constant K_v is maintained at 100.

To begin with the root contour design, we first set a to zero in Eq. (8-34). Then, the characteristic equation of the compensated system becomes

$$s(s + 25)(1 + Ts) + 2500 = 0 \tag{8-35}$$

Since T is the variable parameter, we divide both sides of Eq. (8-35) by the terms that do not contain T. We have

$$1 + \frac{Ts^2(s + 25)}{s^2 + 25s + 2500} = 0 \tag{8-36}$$

This equation is of the form of $1 + G_1(s) = 0$, where $G_1(s)$ is an equivalent transfer function that can be used to study the root loci of Eq. (8-35). The root contour of Eq. (8-35) is drawn as shown in Fig. 8-18 starting with the poles and zeros of $G_1(s)$. Of significance is that the poles of $G_1(s)$ are the eigenvalues of the system when $a = 0$ and $T = 0$. As can be seen from the figure, the factor $1 + Ts$ in the denominator of Eq. (8-34) alone would not improve the system performance at all. In fact, the eigenvalues of the system are moved toward the right half of the s-plane, and the system becomes unstable when T is greater than 0.0133. To achieve the full effect of the phase-lead compensation, we must restore the value of a in Eq. (8-34). The characteristic equation of the compensated system now becomes

$$s(s + 25)(1 + Ts) + 2500(1 + aTs) = 0 \tag{8-37}$$

Now we must consider the effect of varying a while keeping T constant. This is accomplished by dividing both sides of Eq. (8-37) by the terms that do not contain a. We have

$$1 + \frac{2500aTs}{s(s + 25)(1 + Ts) + 2500} = 0 \tag{8-38}$$

This equation is again of the form of $1 + G_2(s) = 0$, and for a given T, the root loci of Eq. (8-38) can be obtained based upon the poles and zeros of

$$G_2(s) = \frac{2500aTs}{s(s + 25)(1 + Ts) + 2500} \tag{8-39}$$

Notice also that the denominator of $G_2(s)$ is identical to the left side of Eq. (8-35),

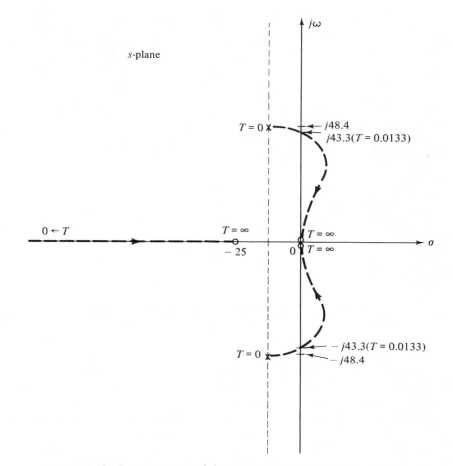

Figure 8-18 Root contours of the sun-seeker system of Example 8-1 with $a = 0$, and T varies from 0 to ∞.

which means that the poles of $G_2(s)$ must lie on the root contours of Fig. 8-18, for a given T. In other words, the root contours of Eq. (8-37) as a varies must start from the points on the trajectories of Fig. 8-18. These root contours end at $s = 0, \infty, \infty$, which are the zeros of $G_2(s)$. The complete root contours of the system with phase-lead compensation are now sketched in Fig. 8-19.

From the root contours of Fig. 8-19, we can see that for effective phase-lead compensation, the value of T should be small. For large values of T, the natural frequency of the system increases very rapidly as a increases, while very little improvement is made on the damping of the system. We must remember that these remarks are made with respect to the phase-lead design that corresponds to values of a greater than unity.

As seen from the root contours of Fig. 8-19, there may be many sets of values of a and T that will satisfy the transient response requirement, but in any case the value of T must be very small. Referring to Eq. (8-34), for very small T, the term

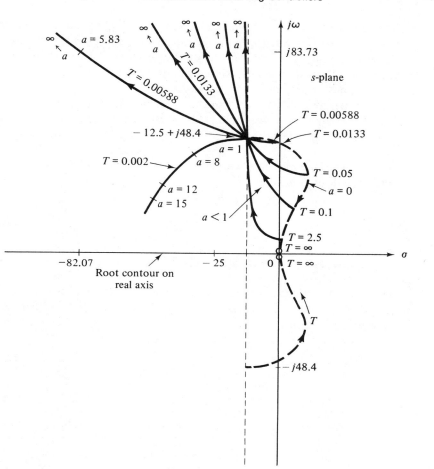

Figure 8-19 Root contours of the sun-seeker system with a phase-lead controller, $G_c(s) = (1 + aTs)/(1 + Ts)$, Example 8-1.

$(1 + Ts)$ in the denominator can be neglected. As a very rough approximation the characteristic equation of Eq. (8-37) becomes

$$s^2 + 25(1 + 100aT)s + 2500 = 0 \qquad (8\text{-}40)$$

Let us select the damping ratio for the approximating second-order system to be 0.707. Then, Eq. (8-40) gives

$$\zeta = 0.707 = 25aT + 0.25 \qquad (8\text{-}41)$$

from which we have

$$aT = 0.0183$$

This gives us an idea on what aT should be. For instance, if we choose T to be 0.00588, then $a = 3.1$; if $T = 0.002$, the corresponding a is 9.14. Keep in mind that the actual pole at $-1/T$ which was neglected will have an adverse effect on the transient response, so that it would be safer to select a larger value of a than those calculated above.

In the following we present three sets of results for different combinations of a and T. The first two cases yield peak overshoots in the compensated system that are less than the specified 10 percent. The corresponding unit step responses are shown in Fig. 8-17.

If we select the case with $a = 5.83$ and $T = 0.00588$, the transfer function of the phase-lead controller is

$$G_c(s) = \frac{s + 1/aT}{s + 1/T} = \frac{s + 29.17}{s + 170} \tag{8-42}$$

It should be noted that certain types of systems cannot be compensated satisfactorily by a phase-lead controller. Usually, when a system is unstable or has very low relative stability, the phase-lead controller may not be very effective. The sun-seeker system considered in this example happens to be one for which the phase-lead controller works. The following example will illustrate a typical situation under which phase-lead compensation is ineffective.

T	a	Peak Overshoot (%)	Characteristic Equation Roots
0.00588	5.83	7.3	$-30.93, \quad -82.07 \pm j83.73$
0.002	12	6.6	$-433.6, \quad -45.68 \pm j28.21$
0.002	8	13.8	$-460.3, \quad -32.35 \pm j40.85$

∎

Example 8-2

Let the process of a control system with unity feedback be described by

$$G_p(s) = \frac{K}{s(1 + 0.1s)(1 + 0.2s)} \tag{8-43}$$

where $K = 100$. We can show that the closed-loop system without any controller is unstable. The marginal value of K for stability is 15. Figure 8-20 shows the root loci of the system. When $K = 100$, the two roots of the characteristic equation are at $3.8 + j14.4$ and $3.8 - j14.4$.

Let the series controller be represented by

$$G_c(s) = \frac{1 + aTs}{1 + Ts} \qquad a > 1 \tag{8-44}$$

Then the open-loop transfer function becomes

$$G_c(s)G_p(s) = \frac{5000(1 + aTs)}{s(s + 5)(s + 10)(1 + Ts)} \tag{8-45}$$

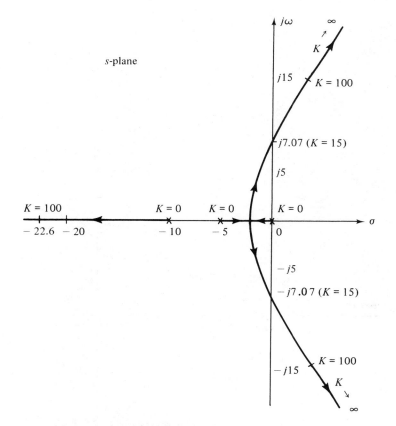

Figure 8-20 Root loci of $s(s + 5)(s + 10) + 50K = 0$.

First, we set $a = 0$ while we vary T from zero to infinity. The characteristic equation becomes

$$s(s + 5)(s + 10)(1 + Ts) + 5000 = 0 \qquad (8\text{-}46)$$

The root contours of Eq. (8-46) are constructed from the pole-zero configuration of

$$G_1(s) = \frac{Ts^2(s + 5)(s + 10)}{s(s + 5)(s + 10) + 5000} \qquad (8\text{-}47)$$

Thus, in Fig. 8-21 the poles of $G_1(s)$ are labeled as $T = 0$ points, and the zeros of $G_1(s)$ are points at which $T = \infty$.

Next, we restore the value of a in Eq. (8-45), and the root contours of Eq. (8-46) become the trajectories on which $a = 0$. In other words, the characteristic equation of the overall system is written

$$s(s + 5)(s + 10)(1 + Ts) + 5000(1 + aTs) = 0 \qquad (8\text{-}48)$$

When a is considered the variable parameter, we divide both sides of Eq. (8-48) by the terms that do not contain a; we have

$$1 + \frac{5000\,aTs}{s(s + 5)(s + 10)(1 + Ts) + 5000} = 0 \qquad (8\text{-}49)$$

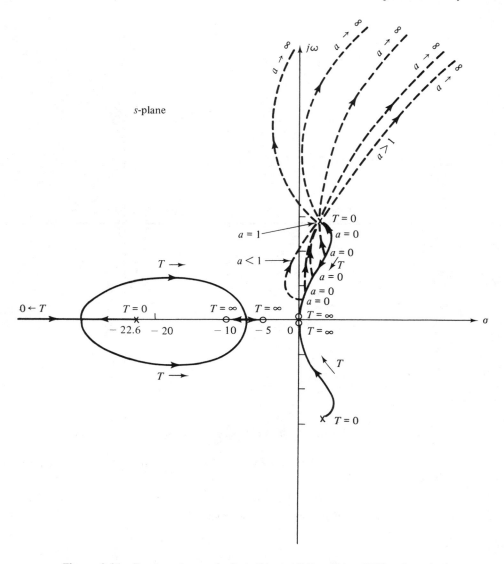

Figure 8-21 Root contours of $s(s + 5)(s + 10)(1 + Ts) + 5000 = 0$, and of $s(s + 5)(s + 10)(1 + Ts) + 5000(1 + aTs) = 0$.

Thus, as we have stated, the root contours with a varying, start ($a = 0$) at the poles of

$$G_2(s) = \frac{5000\,aTs}{s(s + 5)(s + 10)(1 + Ts) + 5000} \qquad (8\text{-}50)$$

The dominant part of the root loci of Eq. (8-48) is sketched in Fig. 8-21. Notice that, since for phase-lead compensation the value of a is limited to greater than 1, the root contours that correspond to this range are mostly in the right-half plane. It is apparent from this root contour plot that the ineffectiveness of phase-lead compensation, in this case, may be attributed to the eigenvalues of the uncompensated system being in the right-half plane.

■

Phase-Lag Controller

In contrast to using a high-pass filter or phase-lead controller for improving the performance of control systems, we may use a low-pass filter or phase-lag controller. The PI controller discussed in Section 8.2 is one of the simplest forms of the phase-lag controller. In the following transfer function, if we set $a < 1$, we have a phase-lag controller.

$$\frac{E_2(s)}{E_1(s)} = \frac{1 + aTs}{1 + Ts} \qquad a < 1 \tag{8-51}$$

Notice that Eq. (8-51) is identical to Eq. (8-44), except that a is less than 1 in the present case. Figure 8-22 shows a simple RC network that realizes the transfer function of Eq. (8-51). If we assume that the input impedance of the network is zero and the output impedance that the network sees is infinite, the transfer function of the network is written

$$\frac{E_2(s)}{E_1(s)} = \frac{1 + R_2Cs}{1 + (R_1 + R_2)Cs} \tag{8-52}$$

Comparing Eqs. (8-51) and (8-52), we have

$$aT = R_2C \tag{8-53}$$

and

$$a = \frac{R_2}{R_1 + R_2} \qquad a < 1 \tag{8-54}$$

It is worth pointing out that the phase-lead network of Fig. 8-14, which has the transfer function of Eq. (8-28), has a zero-frequency attenuation of $1/a$ ($a > 1$), whereas in the present case the zero-frequency gain of the transfer function of Eq. (8-51) is unity, but at infinite frequency there is an attenuation of a ($a < 1$). In the case of the phase-lead compensation we always assume that the attenuation of $1/a$

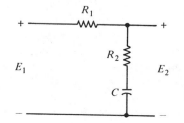

Figure 8-22 *RC* phase-lag network.

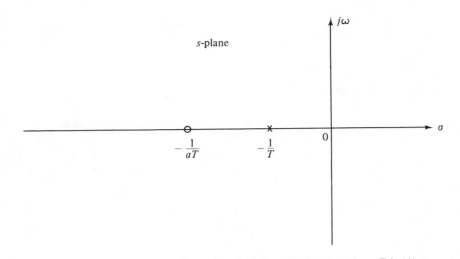

Figure 8-23 Pole-zero configuration of the transfer function $(1 + aTs)/(1 + Ts)$, $a < 1$, of a phase-lag network.

is taken care of by amplifiers located in the forward path of the system, so that this attenuation need never be of concern during design.

The transfer function of the phase-lag controller in Eq. (8-51) has a real zero at $s = -1/aT$ and a real pole at $s = -1/T$. As shown in Fig. 8-23, since a is less than unity, the pole is always located to the right of the zero, and the distance between them is determined by a.

The design philosophy of phase-lag compensation based on the time-domain specifications is best illustrated by the root locus diagram of the system considered in Example 8-1, which is now shown in Fig. 8-24. Figure 8-24(a) shows the root loci of the uncompensated system, based on the open-loop transfer function of Eq. (8-30). As shown in the figure, to achieve a damping ratio of 0.707, K has to be set at 0.125. However, to satisfy the steady-state error requirement, K must be greater or equal to 1; this yields a system with low relative stability. The philosophy of the phase-lag control is that for effective control, the pole and zero of the controller transfer function should be placed very close together, and then the combination should be located relatively close to the origin of the s-plane. Figure 8-24(b) shows such an arrangement, with $1/T = 0.4$ and $1/aT = 2.0$. For the root loci of the compensated system, as K increases from 0, the two roots that leave from $s = 0$ and $s = -0.4$ move toward each other on the real axis. After reaching the breakaway point, they separate, forming a small loop around the zero at $s = -2$, and then meet again at the breakaway point to the left of the -2 point. In the meantime the root that leaves the open-loop pole at $s - 25$ travels to the right as K increases; it meets one of the two roots from the other two loci at the third breakaway point at $s = -11.25$, and then the two roots become complex conjugate as K approaches infinity. It is important to note that since the pole and zero of the phase-lag controller are placed close together and are very near the origin, they have very little effect on the *shape* of the original root loci, especially at points far away from the origin of the s-plane. Thus in Fig. 8-24(b), other than the small loop near $s = 0$, the rest of the root loci are very similar to that of Fig. 8-24(a). However, the values of K that correspond to similar points on the two root loci are different. It is simple to show that for the values of a and T chosen, the values of K on the root loci of the compensated system in Fig. 8-24(b) at points relatively far away from the controller pole-zero combination are five times greater than those values of K at similar points on the uncompensated root loci in Fig. 8-24(a). For instance, at the root $s = -11.6 + j18$ on the compensated root loci of Fig. 8-24(b), the value of K is 1; the comparable point on the uncompensated root loci in Fig. 8-24(a) is at $s = -12.5 + j18.4$, and the value of K at that point is 0.2. A more formal way of demonstrating the relationship is to refer to the open-loop transfer functions of the original system and the compensated system. The value of K at any point s_1 on the root loci of the uncompensated system is written by applying the root locus condition of Eq. (7-91) to Eq. (8-30),

$$|K| = \frac{|s_1| \, |s_1 + 25|}{2500} \tag{8-55}$$

Assuming that the comparable point s_1 on the compensated root loci is far from the pole-zero combination of the phase-lag controller, the value of K at s_1 is given by

$$|K| = \frac{|s_1| \, |s_1 + 0.4| \, |s_1 + 25|}{500|s_1 + 2|} \tag{8-56}$$

or

$$|K| \cong \frac{|s_1| \, |s_1 + 25|}{500} \tag{8-57}$$

since the distance from s_1 to -0.4 will be approximately the same as that from s_1 to -2. This argument also points to the fact that the exact location of the pole and the zero of the phase-lag network is not significant as long as they are close to the origin, and that the distance between the pole and the zero is a fixed desired quantity. In the case of the last example, the ratio between $1/aT$ and $1/T$ is 5.

Based on the discussions given above, we may outline a root locus design procedure for the phase-lag design of control systems as follows. Since the design will be carried out in the s-plane, the specifications on the transient response or the relative stability should be given in terms of the damping ratio of the dominant roots, and other quantities, such as rise time, bandwidth, and maximum overshoot, which can be correlated with the location of the eigenvalues.

1. Sketch the root loci of the characteristic equation of the uncompensated system.

2. Determine on these root loci where the desired eigenvalues should be located to achieve the desired relative stability of the system. Find the value of K that corresponds to these eigenvalues.

3. Compare the value of K required for steady-state performance and the K found in the last step. The ratio of these two Ks is a ($a < 1$), which is the desired ratio between the pole and the zero of the phase-lag controller.

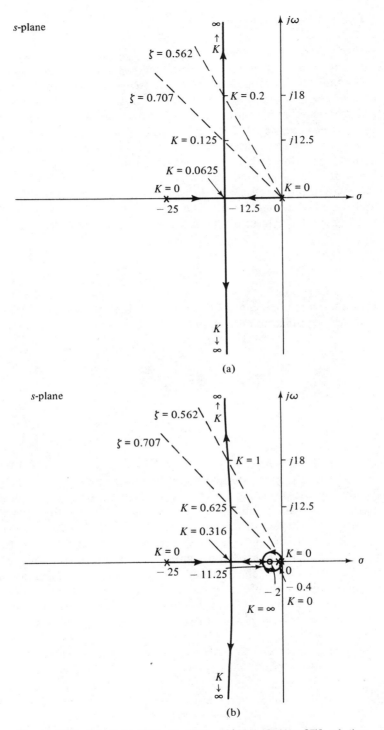

Figure 8-24 (a) Root loci of $G(s) = 2500K / [s(s + 25)]$ of the sun-seeker system. (b) Root loci of $G(s) = 500(s + 2) / [s(s + 25)(s + 0.4)]$ of the sun-seeker system.

4. The exact value of T is not critical as long as it is relatively large. We may choose the value of $1/aT$ to be many orders of magnitudes smaller than the smallest pole of the process transfer function.

   Let us design the system in Example 8-1 by means of the phase-lag controller. For the specification on the peak overshoot we estimate that the relative damping ratio of the system should be approximately 0.707. Also, the steady-state performance specification requires that K should be equal to 1 or greater.

   We first construct the root loci of the uncompensated system based on the pole-zero configuration of Eq. (8-30), as shown in Fig. 8-24(a). These root loci show that a damping ratio of 0.707 is attained by setting K of the original system to 0.125. Based on the development given above on the principle of design of the phase-lag controller via the root loci, we set the value of a to the ratio of the two values of K, that is,

$$a = \frac{K \text{ to realize the desired damping}}{K \text{ to realize the steady-state performance}} = \frac{0.125}{1} = \frac{1}{8} \qquad (8\text{-}58)$$

   We can also arrive at this relationship for a with the following derivations. The desired relative damping is achieved by setting K to 0.125 in the original system. The open-loop transfer function of Eq. (8-30) becomes

$$\frac{\Theta_0(s)}{\alpha(s)} = \frac{312.5}{s(s+25)} \qquad (8\text{-}59)$$

The open-loop transfer function of the compensated system is

$$\frac{\Theta_0(s)}{\alpha(s)} = \frac{250K(1+aTs)}{s(s+25)(1+Ts)}$$

$$= \frac{250aK(s+1/aT)}{s(s+25)(s+1/T)} \qquad (8\text{-}60)$$

If the values of aT and T are chosen to be large, Eq. (8-60) is approximately

$$\frac{\Theta_0(s)}{\alpha(s)} = \frac{2500aK}{s(s+25)} \qquad (8\text{-}61)$$

from the transient-response standpoint. [We cannot apply this approximation to the steady-state response, since as s approaches zero, Eq. (8-60) reverts to Eq. (8-30).] Since K is necessarily equal to unity, to have the right side of Eqs. (8-59) and (8-61) equal to each other, $a = \frac{1}{8}$, as is already concluded in Eq. (8-58). Theoretically, the value of T can be arbitrarily large. However, if the value of T is too large, we may encounter difficulties in realizing the phase-lag controller by physical components.

The following table summarizes the results when the value of a is set at $\frac{1}{8}$, and $T = 5$, 10, and 20.

a	T	Peak Overshoot (%)	Characteristic Equation Roots	
0.125	5	15	-1.818,	$-11.69 \pm j11.76$
0.125	10	10	-0.849,	$-12.13 \pm j12.14$
0.125	20	7.2	-0.412,	$-12.32 \pm j12.32$

When $T = 5$, the complex roots of the characteristic equation correspond to a relative damping ratio of approximately 0.707, but the real root at -1.818 still has some influence on the transient response, and the peak overshoot is 15 percent. When the value of T is increased, the relative damping is closer to 0.707, and the real root is moved toward the origin of the s-plane, so that the peak overshoot is reduced to 10 percent. When $T = 20$, the real root is moved farther to the right, and the peak overshoot is reduced to 7.2 percent. It is interesting to realize that when T becomes very large, the two complex roots should eventually become $-12.5 \pm j12.5$, which is the same as for the roots of the characteristic equation of the original second-order system with $K = 0.125$; and the overshoot should approach 4.32 percent, which corresponds to that of a second-order system with $\zeta = 0.707$.

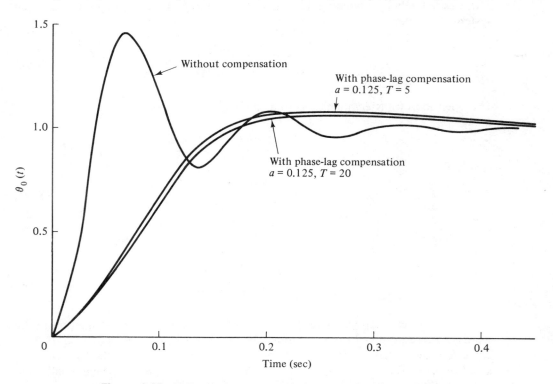

Figure 8-25 Unit step responses of the sun-seeker control system with phase-lag control.

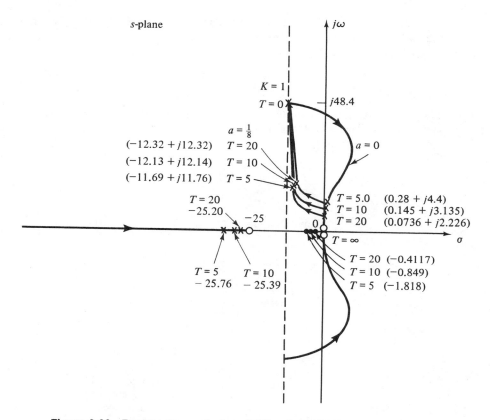

Figure 8-26 Root contours of $s(s + 25)(1 + Ts) + 2500K(1 + aTs) = 0$; $K = 1$.

If we choose $a = 0.125$ and $T = 20$, the transfer function of the phase-lag controller becomes

$$G_c(s) = \frac{1 + 2.5s}{1 + 20s} \qquad (8\text{-}62)$$

Figure 8-25 illustrates the unit step responses of the uncompensated system and the compensated system with $a = 0.125$, $T = 5$ and 20.

As an alternative, the root locus design of the phase-lag control can also be carried out by means of the root contour method. The root contour design conducted earlier by Eqs. (8-34) through (8-39) for the phase-lead controller and Figs. 8-18 and 8-19 is still valid for the phase-lag control, except that in the present case $a < 1$. Thus in Fig. 8-19 only the portions of the root contours that correspond to $a < 1$ are applicable for the phase-lag compensation. These root contours show that for effective phase-lag control, the value of T should be relatively large. In Fig. 8-26 we illustrate further that the complex poles of the closed-loop transfer function are rather insensitive to the value of T when the latter is relatively large.

The time responses of Fig. 8-25 point out a major disadvantage of the phase-lag control. Since the phase-lag controller is essentially a low-pass filter, the rise time of the compensated system is usually increased. However, we shall show by the following example that the phase-lag controller is more versatile in its applications than the phase-lead controller.

Example 8-3

Consider the system given in Example 8-2 for which the phase-lead control was proven to be ineffective. The open-loop transfer function of the original system and the performance specifications are given as follows:

$$G_p(s) = \frac{K}{s(1 + 0.1s)(1 + 0.2s)} \tag{8-63}$$

$$K_v = 100 \text{ sec}^{-1}$$

relative damping ratio = 0.707

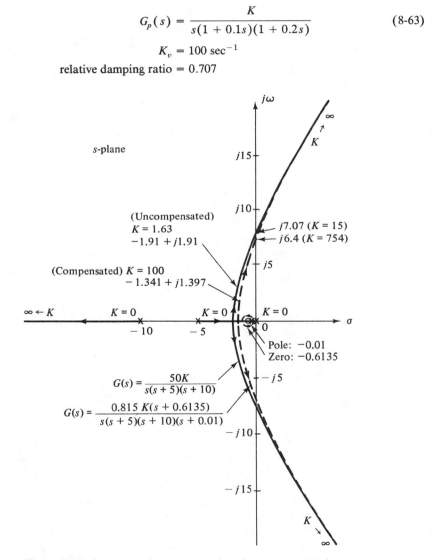

Figure 8-27 Root loci of compensated and uncompensated systems in Example 8-3.

The root loci of the uncompensated system are drawn as shown in Fig. 8-27. For $K_v = 100$, K has to be equal to 100, which corresponds to an unstable system. When $K = 1.63$, the uncompensated characteristic equation roots are at -11.118, $-1.91 + j1.91$, and $-1.91 - j1.91$, which corresponds to a relative damping ratio of 0.707. Thus, using the design method described earlier, we set

$$a = \frac{1.63}{100} = 0.0163 \qquad (8\text{-}64)$$

Let us set T to be arbitrarily large at 100. Then, the phase-lag controller is described by the transfer function

$$G_c(s) = \frac{1 + 1.635s}{1 + 100s} \qquad (8\text{-}65)$$

and the open-loop transfer function of the compensated system is

$$G(s) = G_c(s)G_p(s) = \frac{0.815K(s + 0.6135)}{s(s + 5)(s + 10)(s + 0.01)} \qquad (8\text{-}66)$$

where $K = 100$.

Figure 8-27 illustrates the root loci of the compensated system. In this case, when $K = 100$, the roots of the compensated characteristic equation are at -11.13, -1.198, $-1.341 + j1.397$, and $-1.341 - j1.397$. The relative damping ratio of the complex roots is

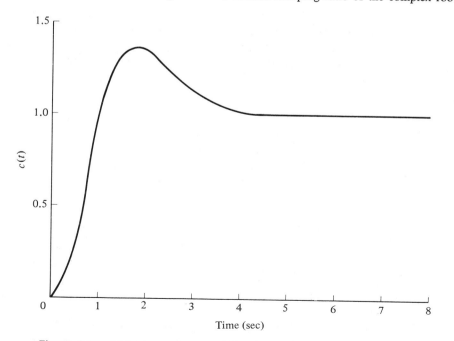

Figure 8-28 Unit step response of the system with phase-lag control in Example 8-3.

slightly less than 0.7. This can be improved by selecting a smaller value for a or a larger T. The unit step response of the system with the phase-lag controller is shown in Fig. 8-28. The peak overshoot of the response is approximately 36 percent.

■

Lag – Lead Controller

We have learned from preceding sections that the series phase-lead controller generally improves the rise time and the damping but increases the natural frequency of the closed-loop system. On the other hand, phase-lag control when applied properly improves the overshoot and relative stability, but usually results in a longer rise time. Therefore, each of these control schemes has its advantages, disadvantages, and limitations, since there are many systems that cannot be satisfactorily improved by either scheme. It is natural, therefore, whenever necessary, to consider using a combination of the lead and the lag controllers, so that the advantages of both schemes are utilized.

The transfer function of a simple lag-lead (or lead-lag) controller can be written

$$G_c(s) = \left(\frac{1 + aT_1 s}{1 + T_1 s} \right) \left(\frac{1 + bT_2 s}{1 + T_2 s} \right) \qquad (8\text{-}67)$$

$$| \leftarrow \text{lead} \rightarrow | \leftarrow \text{lag} \rightarrow |$$

where $a > 1$ and $b < 1$, and the attenuation factor $1/a$ of the phase-lead controller is not included in the equation if we assume that adequate loop gain is available in the system to compensate for this loss.

As an illustrative example, we can cascade a phase-lead controller to the phase-lag controller used in Example 8-3. The transfer function of the controller is now

$$G_c(s) = \left(\frac{1 + aT_1 s}{1 + T_1 s} \right) \left(\frac{1 + 1.635 s}{1 + 100 s} \right) \qquad (8\text{-}68)$$

Let us choose a to be 10 and T_1 to be 0.01. The open-loop transfer function of the compensated system is

$$G(s) = \frac{815(s + 0.6135)(s + 10)}{s(s + 5)(s + 10)(s + 100)(s + 0.01)} \qquad (8\text{-}69)$$

The characteristic equation roots are

$$-100.1 \qquad -10.0, \qquad -2.398, \qquad -1.264 + j0.698 \quad \text{and} \quad -1.264 - j0.698$$

Figure 8-29 shows the unit step response of the system with the lag-lead controller. It is apparent that the system now has an improved rise time and overshoot over the system with the phase-lag controller.

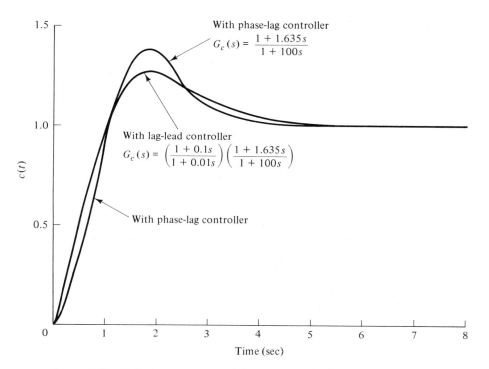

Figure 8-29 Unit step responses of the system with phase-lag control and lag-lead controller in Example 8-3.

Usually it is not necessary to cascade the lead and the lag networks of Figs. 8-14 and 8-22 for the realization of Eq. (8-67) if a and b need not be specified independently. A network that has lag-lead characteristics, but with fewer number of elements, is shown in Fig. 8-30. The transfer function of the network is

$$G_c(s) = \frac{E_2(s)}{E_1(s)} = \frac{(1 + R_1C_1s)(1 + R_2C_2s)}{1 + (R_1C_1 + R_1C_2 + R_2C_2)s + R_1R_2C_1C_2s^2} \quad (8\text{-}70)$$

Comparing Eq. (8-67) with Eq. (8-70), we have

$$aT_1 = R_1C_1 \quad (8\text{-}71)$$
$$bT_2 = R_2C_2 \quad (8\text{-}72)$$
$$T_1T_2 = R_1R_2C_1C_2 \quad (8\text{-}73)$$

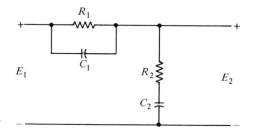

Figure 8-30 Lag-lead network.

From Eqs. (8-71) and (8-72), we have

$$abT_1T_2 = R_1R_2C_1C_2 \qquad (8\text{-}74)$$

Thus

$$ab = 1 \qquad (8\text{-}75)$$

which means that a and b cannot be specified independently.

8.4 POLE–ZERO CANCELLATION CONTROL

Many controlled processes have transfer functions that contain one or more pairs of complex-conjugate poles. These processes are more difficult to control, especially if the complex poles are very close to the imaginary axis in the s-plane. One immediate thought is to use a controller that has a transfer function with zeros so selected as to cancel the undesired complex poles of the controlled process, and the poles of the controller are placed at the desired locations in the s-plane. For instance, if the transfer function of a process is

$$G_p(s) = \frac{K}{s(s^2 + s + 10)} \qquad (8\text{-}76)$$

in which the complex-conjugate poles may cause stability problems in the closed-loop system, especially if the value of K is large, the suggested series controller may be of the form

$$G_c(s) = \frac{s^2 + s + 10}{s^2 + 2\zeta\omega_n s + \omega_n^2} \qquad (8\text{-}77)$$

The constants ζ and ω_n are determined according to the performance specifications of the closed-loop system.

There are practical difficulties with the pole–zero cancellation design scheme which should not be left unmentioned. The problem is that in practice rarely is exact cancellation of poles and zeros of the process possible by zeros and poles of the controller. One reality is that the transfer function of the process, $G_p(s)$, is usually determined through testing and physical modeling; linearization and approximations are unavoidable. Thus, the "true" poles (and zeros) of the transfer function of the process may not be accurately represented. In fact, the true order of the system may even be higher than that represented by the transfer function used for modeling purposes. Another difficulty is that the dynamic properties of the process may vary, even very slowly, due to aging of the system components or changes in the operating environment, so that the poles and zeros of the transfer function may move during the operation of the system. For these and other reasons, even if we could precisely design the poles and zeros of the transfer function of the controller, exact pole–zero cancellation is almost never possible in practice. We show in the following that in most cases, exact cancellation is not really necessary to effectively improve the performance of a control system using the pole–zero cancellation control scheme.

Let us assume that a process is represented by

$$G_p(s) = \frac{K}{s(s + p_1)(s + \bar{p}_1)} \tag{8-78}$$

where p_1 and $\bar{p}_1$ are the two poles that are to be canceled; p_1 and $\bar{p}_1$ can be real or in complex-conjugate pairs. Let the transfer function of the series controller be

$$G_c(s) = \frac{(s + p_1 + \varepsilon)(s + \bar{p}_1 + \bar{\varepsilon})}{s^2 + 2\zeta\omega_n s + \omega_n^2} \tag{8-79}$$

where ε is a complex number whose magnitude is very small, and $\bar{\varepsilon}$ is its complex conjugate. The open-loop transfer function of the compensated system is

$$G(s) = G_c(s)G_p(s) = \frac{K(s + p_1 + \varepsilon)(s + \bar{p}_1 + \bar{\varepsilon})}{s(s + p_1)(s + \bar{p}_1)(s^2 + 2\zeta\omega_n s + \omega_n^2)} \tag{8-80}$$

Because of the inexact cancellation we cannot discard the terms $(s + p_1)(s + \bar{p}_1)$ in the denominator of Eq. (8-80). The closed-loop transfer function is written

$$\frac{C(s)}{R(s)} = \frac{G(s)}{1 + G(s)}$$

$$= \frac{K(s + p_1 + \varepsilon)(s + \bar{p}_1 + \bar{\varepsilon})}{s(s + p_1)(s + \bar{p}_1)(s^2 + 2\zeta\omega_n s + \omega_n^2) + K(s + p_1 + \varepsilon)(s + \bar{p}_1 + \bar{\varepsilon})} \tag{8-81}$$

The root locus diagram in Fig. 8-31 explains the effect of inexact pole–zero cancellation. Notice that the two closed-loop poles as a result of inexact cancellation lie between the pairs of poles and zeros at $s = -p_1, -\bar{p}_1$ and $-p_1 - \varepsilon, -\bar{p}_1 - \bar{\varepsilon}$, respectively. Thus, these closed-loop poles are very close to the open-loop poles and zeros that are meant to be canceled. Equation (8-81) can be approximately written as

$$\frac{C(s)}{R(s)} = \frac{K(s + p_1 + \varepsilon)(s + \bar{p}_1 + \bar{\varepsilon})}{(s + p_1 + \delta)(s + \bar{p}_1 + \bar{\delta})(s^3 + 2\zeta\omega_n s^2 + \omega_n^2 s + K)} \tag{8-82}$$

where δ and $\bar{\delta}$ are a pair of very small complex-conjugate numbers that depend on ε, $\bar{\varepsilon}$, and all the other parameters.

The partial-fraction expansion of Eq. (8-82) is written

$$\frac{C(s)}{R(s)} = \frac{K_1}{s + p_1 + \delta} + \frac{K_2}{s + \bar{p}_1 + \bar{\delta}} + \text{terms due to the remaining poles} \tag{8-83}$$

We can show that K_1 is proportional to $\varepsilon - \delta$, which is a very small number.

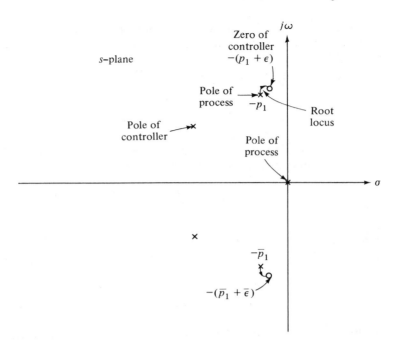

Figure 8-31 Pole–zero configuration of inexact pole–zero cancellation.

Similarly, K_2 is a very small number. This exercise simply shows that although we cannot cancel the poles at $-p_1$ and $-\bar{p}_1$ precisely, the resulting transient terms due to the inexact cancellation will have insignificant amplitudes, so that unless the controller zeros earmarked for cancellation are too far off the target, the effect can be neglected for all practical purposes. Another way of looking at this problem is that the zeros of $G(s)$ are retained as the zeros of the closed-loop transfer function $C(s)/R(s)$, so that from Eq. (8-82) we see that the two pairs of poles and zeros are close enough to be canceled from the transient response standpoint.

There are cases, however, that could cause difficulty due to the inexact cancellation of poles and zeros. If the undesirable poles of the open-loop process are very close to or right on the imaginary axis of the s-plane, inexact cancellation may result in an unstable system. Figure 8-32(a) illustrates a situation in which the relative positions of the pole and zeros intended for cancellation result in a stable system, whereas in Fig. 8-32(b), the inexact cancellation is unacceptable. Although the relative distance between the poles and zeros intended for cancellation is small, which results in terms in the time response that have very small amplitudes, these responses will grow as time increases and the system will be unstable in the long run.

Transfer functions with complex poles and zeros can be realized by various types of electrical networks. The bridged-T networks shown in Fig. 8-33 have the advantage of containing only RC elements. In the following discussion, the network shown in Fig. 8-33(a) is referred to as the bridged-T type 1, and that of Fig. 8-33(b) is referred to as type 2.

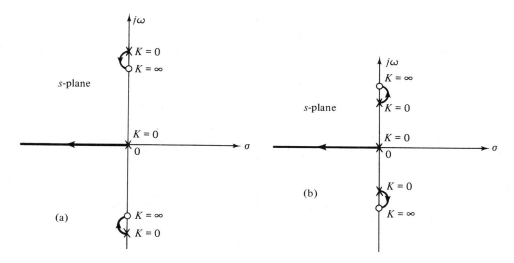

Figure 8-32 Root locus diagrams showing the effects of inexact pole–zero cancellations.

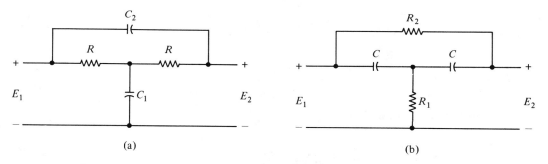

Figure 8-33 Two basic types of bridged-T network. (a) Type 1 network. (b) Type 2 network.

With the assumption of zero input source impedance and infinite output impedance, the transfer function of the bridged-T type 1 network is given by

$$\frac{E_2(s)}{E_1(s)} = \frac{1 + 2RC_2s + R^2C_1C_2s^2}{1 + R(C_1 + 2C_2)s + R^2C_1C_2s^2} \tag{8-84}$$

and that of the bridged-T type 2 network is

$$\frac{E_2(s)}{E_1(s)} = \frac{1 + 2R_1Cs + C^2R_1R_2s^2}{1 + C(R_2 + 2R_1)s + C^2R_1R_2s^2} \tag{8-85}$$

When these two equations are compared, it is apparent that the two networks have similar transfer characteristics. In fact, if R, C_1, and C_2 in Eq. (8-84) are replaced by C, R_2, and R_1, respectively, Eq. (8-84) becomes the transfer function of the type 2 network given by Eq. (8-85).

It is useful to study the behavior of the poles and zeros of the transfer functions of the bridged-T networks of Eqs. (8-84) and (8-85) when the network parameters are varied. Owing to the similarity of the two networks, only type 1 will be studied here.

Equation (8-84) is written

$$\frac{E_2(s)}{E_1(s)} = \frac{s^2 + \dfrac{2}{RC_1}s + \dfrac{1}{R^2C_1C_2}}{s^2 + \dfrac{C_1 + 2C_2}{RC_1C_2}s + \dfrac{1}{R^2C_1C_2}} \tag{8-86}$$

If both the numerator and the denominator polynomials of Eq. (8-86) are written in the standard form

$$s^2 + 2\zeta\omega_n s + \omega_n^2 = 0 \tag{8-87}$$

we have, for the numerator,

$$\omega_{nz} = \pm\frac{1}{R\sqrt{C_1C_2}} \tag{8-88}$$

$$\zeta_z = \sqrt{\frac{C_2}{C_1}} \tag{8-89}$$

and for the denominator,

$$\omega_{np} = \pm\frac{1}{R\sqrt{C_1C_2}} = \omega_{nz} \tag{8-90}$$

$$\zeta_p = \frac{C_1 + 2C_2}{2\sqrt{C_1C_2}} = \frac{1 + 2C_2/C_1}{2\sqrt{C_2/C_1}} = \frac{1 + 2\zeta_z^2}{2\zeta_z} \tag{8-91}$$

The loci of the poles and zeros of $E_2(s)/E_1(s)$ of Eq. (8-86) when C_1, C_2, and R vary individually are sketched in Fig. 8-34. When R varies, the numerator and the denominator of Eq. (8-86) contain R in the form of R^2, and the root locus method cannot be applied directly to this nonlinear problem. Fortunately, in this case, the equations that are of the form of Eq. (8-87) are of the second order and can be solved easily. Therefore, the pole and zero loci of Fig. 8-34 show that the two zeros of the bridged-T network type 1 can be either real or complex; for complex zeros, C_2 must be less than C_1. The poles of the transfer function always lie on the negative real axis.

The ζ and ω_n parameters of the denominator and the numerator of the transfer function of the type 2 network may be obtained by replacing R, C_1, and C_2

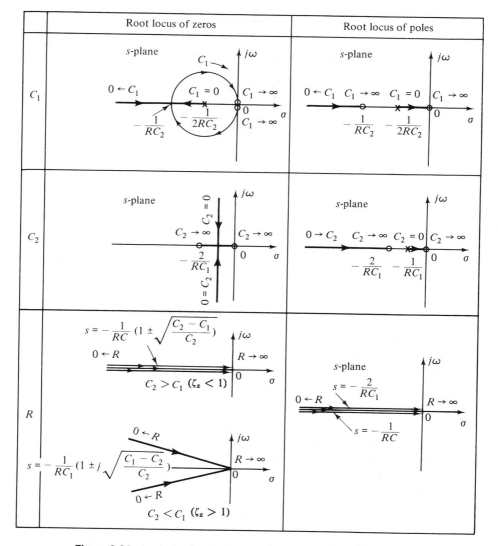

Figure 8-34 Loci of poles and zeros of the bridged-T type 1 network.

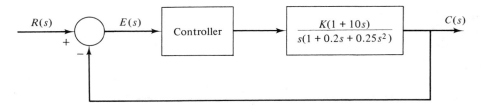

Figure 8-35 Feedback control system for Example 8-4.

in Eqs. (8-90) through (8-91) by C, R_2, and R_1, respectively. Thus,

$$\omega_{nz} = \pm \frac{1}{C\sqrt{R_1 R_2}} \qquad (8\text{-}92)$$

$$\zeta_z = \sqrt{\frac{R_1}{R_2}} \qquad (8\text{-}93)$$

$$\omega_{np} = \omega_{nz} \qquad (8\text{-}94)$$

$$\zeta_p = \frac{R_2 + 2R_1}{2\sqrt{R_1 R_2}} = \frac{1 + 2\zeta_z^2}{2\zeta_z} \qquad (8\text{-}95)$$

The root loci shown in Fig. 8-34 can still be used for the type 2 bridged-T network if the corresponding symbols are altered.

■

Example 8-4

The control system shown in Fig. 8-35 is selected to demonstrate the use of the bridged-T network as a controller and the principle of improving system performance by pole–zero cancellation.

The controlled process has the transfer function

$$G_p(s) = \frac{K(1 + 10s)}{s(1 + 0.2s + 0.25s^2)} \qquad (8\text{-}96)$$

The root locus diagram of the uncompensated closed-loop system is shown in Fig. 8-36. It is shown that although the closed-loop system is always stable, the damping of the system is very low, and the step response of the system will be quite oscillatory for any positive K. Figure 8-38 illustrates the unit step response of the system when $K = 1$. Notice that the zero at $s = -0.1$ of the open-loop transfer function causes the closed-loop system to have an eigenvalue at $s = -0.091$, which corresponds to a time constant of 11 sec. Thus, this small eigenvalue causes the step response of Fig. 8-38 to oscillate about a level that is below unity, and it would take a long time for the response to reach the desired steady state.

Let us select the type 1 bridged-T network as series compensation to improve the relative stability of the system. The complex-conjugate zeros of the network should be so placed that they will cancel the undesirable poles of the controlled process. Therefore, the transfer function of the bridged-T network should be

$$G_c(s) = \frac{s^2 + 0.8s + 4}{s^2 + 2\zeta_p \omega_{np} s + \omega_{np}^2} \qquad (8\text{-}97)$$

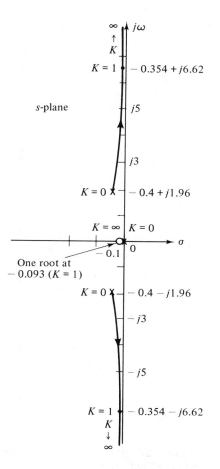

Figure 8-36 Root loci of the feedback control system in Example 8-4 with $G_p(s) = [K(1 + 10s)] / [s(1 + 0.2s + 0.25s^2)]$.

Using Eqs. (8-88) and (8-89), we have

$$\omega_{nz} = 2 = \frac{1}{R\sqrt{C_1 C_2}} \tag{8-98}$$

$$\zeta_z = \sqrt{\frac{C_2}{C_1}} = 0.2 \tag{8-99}$$

From Eqs. (8-90) and (8-91),

$$\omega_{np} = \omega_{nz} = 2 \tag{8-100}$$

$$\zeta_p = \frac{1 + 2\zeta_z^2}{2\zeta_z} = 2.7 \tag{8-101}$$

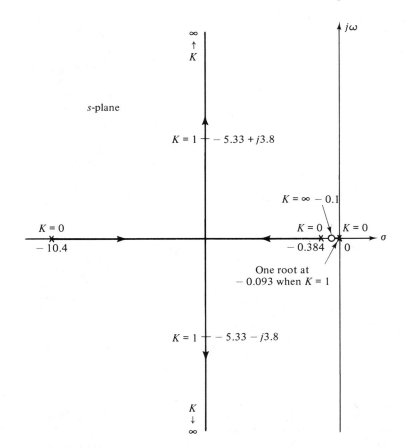

Figure 8-37 Root loci of feedback control system in Example 8-4 with bridged-T controller.

The transfer function of the type 1 bridged-T network is

$$G_c(s) = \frac{s^2 + 0.8s + 4}{(s + 0.384)(s + 10.42)} \qquad (8\text{-}102)$$

The open-loop transfer function of the compensated system is

$$G(s) = G_c(s)G_p(s)\frac{40K(s + 0.1)}{s(s + 0.384)(s + 10.42)} \qquad (8\text{-}103)$$

The root loci of the compensated system are sketched as shown in Fig. 8-37. The root loci show that for K greater than 0.64, two of the eigenvalues of the system are complex. However, the dynamic response of the system is still dominated by the eigenvalue that lies near the origin of the s-plane. Figure 8-38 illustrates the unit step response of the compensated system when $K = 1$. In this case the complex eigenvalues of the compensated system cause the response to oscillate only slightly, but the eigenvalue at $s = -0.093$ causes

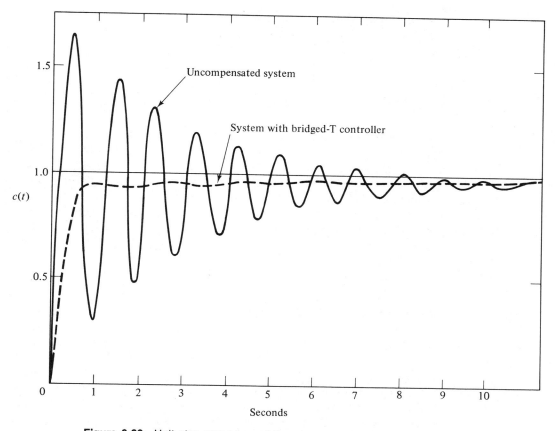

Figure 8-38 Unit step responses of the compensated and uncompensated systems in Example 8-4.

the system response to take a very long time in reaching its steady state of unity. It is apparent that the relative stability of the system is greatly improved by the bridged-T controller through pole–zero cancellation.

◼

Let us consider that the design of the bridged-T compensation resulted in an inexact pole–zero cancellation, so that the transfer function of the controller is

$$G_c(s) = \frac{s^2 + 0.76s + 3.8}{(s + 0.366)(s + 10.4)} \qquad (8\text{-}104)$$

Since the complex poles of the controlled process are not canceled by the zeros of $G_c(s)$, the open-loop transfer function of the system becomes

$$G(s) = \frac{40K(s + 0.1)(s^2 + 0.76s + 3.8)}{s(s + 0.366)(s + 10.4)(s^2 + 0.8s + 4)} \qquad (8\text{-}105)$$

For all practical purposes, the complex poles and zeros of $G(s)$ of Eq. (8-105) can still be considered as close enough for cancellation. For $K = 1$, the closed-loop transfer function is

$$\frac{C(s)}{R(s)} = \frac{40(s + 0.1)(s^2 + 0.76s + 3.8)}{(s + 0.093)(s^2 + 0.81s + 3.805)(s^2 + 10.66s + 42.88)} \tag{8-106}$$

Since the zeros of the open-loop transfer function are retained as zeros of the closed-loop transfer function, Eq. (8-106) verifies that near cancellation of poles and zeros in the open-loop transfer function will result in the same situation for the closed-loop transfer function. Therefore, we may conclude that from the transient response standpoint, the effectiveness of the cancellation compensation is not diminished even if the pole–zero cancellation is not exact. Alternatively, we can show that if the partial-fraction expansion of Eq. (8-106) is carried out, the coefficients that correspond to the roots of $s^2 + 0.81s + 3.805 = 0$ will be very small, so that the contribution to the time response from these roots will be negligible.

8.5 FORWARD AND FEEDFORWARD COMPENSATION

As described in Sec. 8.1, the forward and feedforward compensation schemes shown in Figs. 8-2(e) and (f), respectively, provide two degrees of freedom to the design of system controllers.

The two-degree-of-freedom compensation offers additional flexibility when a multiple number of design criteria have to be satisfied simultaneously.

Referring to Fig. 8-2(e), the closed-loop transfer function is

$$\frac{C(s)}{R(s)} = \frac{G_{c1}(s)G_{c2}(s)G_p(s)}{1 + G_{c2}(s)G_p(s)} \tag{8-107}$$

and the error transfer function is

$$\frac{E(s)}{R(s)} = \frac{1}{1 + G_{c2}(s)G_p(s)} \tag{8-108}$$

Thus, the controller $G_{c2}(s)$ can be designed so that the error transfer function will have certain desirable characteristics, while the controller $G_{c1}(s)$ can be selected to satisfy performance requirements with reference to the input–output relationship. Another way of describing the flexibility of a two-degree-of-freedom design is that the controller $G_{c2}(s)$ is usually selected to provide a certain degree of system stability, but since the zeros of $G_{c2}(s)$ always become the zeros of the closed-loop transfer function, unless some of the zeros are canceled by the poles of the process transfer function $G_p(s)$, these zeros may cause a large overshoot in the system output even when the relative damping as determined by the characteristic equation

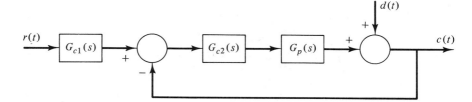

Figure 8-39 A control system with forward compensation and disturbance input.

is good. In this case the transfer function $G_{c1}(s)$ may be used for the control or cancellation of the undesirable zeros of the closed-loop transfer function, while keeping the characteristic equation intact. Of course, we can also introduce zeros in $G_{c1}(s)$ to cancel some of the undesirable poles of the closed-loop transfer function that are the result of the compensation by $G_{c2}(s)$.

It should be kept in mind that while the forward and feedforward compensation may sound powerful, in that they can be used to add or delete poles or zeros directly to the closed-loop transfer function, there is a fundamental question involving the basic characteristics of feedback. If the forward or feedforward controller is so powerful, then why do we need feedback at all? Since $G_{c1}(s)$ in the systems of Figs. 8-2(e) and (f) is outside the feedback loop, the system is susceptible to parameter variations in $G_{c1}(s)$. Therefore, in reality, these types of compensation cannot be satisfactorily applied to all situations.

Another situation under which the forward or feedforward controller may be desirable is shown in Fig. 8-39. The disturbance-output transfer function is written

$$\frac{C(s)}{D(s)} = \frac{1}{1 + G_{c2}(s)G_p(s)} \tag{8-109}$$

In this case the controller $G_{c2}(s)$ is designed so that the output will be insensitive to the disturbance input $d(t)$ while maintaining stability. The controller $G_{c1}(s)$ is selected so that the output follows the input $r(t)$ accurately.

The feedforward compensation scheme shown in Fig. 8-2(f) serves the same purpose as the forward compensation, and the difference between the two configurations depends on system and hardware considerations.

As an illustration of the design of the forward and feedforward compensation, consider the system studied in Example 8-3. The unit-step response of the system with the phase-lag controller of Eq. (8-65) has a peak overshoot of 36 percent. The closed-loop transfer function of the phase-lag compensated system is

$$\frac{C(s)}{R(s)} = \frac{81.5(s + 0.6135)}{s^4 + 15.01s^3 + 50.15s^2 + 82s + 50} \tag{8-110}$$

We can show that the overshoot of the step response of the system is due to the zero at $s = -0.6135$. Figure 8-40 shows the unit-step response if the closed-loop transfer

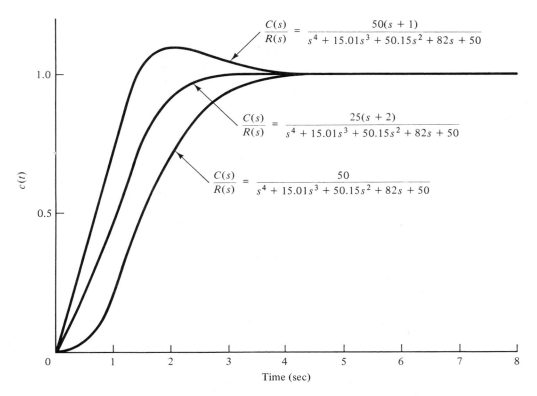

Figure 8-40 Time responses of the system with the closed-loop transfer function in Eq. (8-115).

function is

$$\frac{C(s)}{R(s)} = \frac{50}{s^4 + 15.01s^3 + 50.15s^2 + 82s + 50} \qquad (8\text{-}111)$$

Clearly, without the zero at $s = -0.6135$, the step response no longer has any overshoot, although the rise time is slower.

To realize the closed-loop transfer function of Eq. (8-110), the transfer function of the forward-compensation controller is

$$G_{c1}(s) = \frac{0.6135}{s + 0.6135} \qquad (8\text{-}112)$$

If instead the feedforward configuration is chosen, we have

$$\frac{[G_{c1}(s) + G_{c2}(s)]G_p(s)}{1 + G_{c2}(s)G_p(s)} = \frac{0.6135}{s + 0.6135} \cdot \frac{G_{c2}(s)G_p(s)}{1 + G_{c2}(s)G_p(s)} \qquad (8\text{-}113)$$

Solving for $G_{c1}(s)$ from the last equation yields

$$G_{c1}(s) = \frac{-0.01635}{s + 0.01} \qquad (8\text{-}114)$$

The rise time of the step response can be improved by properly positioning the zero of $C(s)/R(s)$, so that the closed-loop transfer function is of the form:

$$\frac{C(s)}{R(s)} = \frac{50(1 + T_z s)}{s^4 + 15.01s^3 + 50.15s^2 + 82s + 50} \qquad (8\text{-}115)$$

with the term $(1 + T_z s)$ contributed by $G_{c1}(s)$. The corresponding forward compensation $G_{c1}(s)$ has the form:

$$G_{11}(s) = \frac{0.6135(1 + T_z s)}{s + 0.6135} \qquad (8\text{-}116)$$

For the feedforward controller, from Eq. (8-113),

$$G_{cf}(s) = \frac{0.01635(0.6135T_z - 1)s}{s + 0.01} \qquad (8\text{-}117)$$

Figure 8-40 illustrates unit-step responses of the system with the closed-loop transfer function in Eq. (8-115). The forward controller or the feedforward controller cancels the zero of $C(s)/R(s)$ at $s = -0.6135$ and adds a zero at $s = -1/T_z$. As the value of T_z decreases, the peak overshoot also decreases.

8.6 MINOR-LOOP FEEDBACK CONTROL

All of the control schemes discussed in preceding sections have utilized series or cascade controllers. Although series controllers are the most commonly used because of their simplicity in implementation, depending on the nature of the system there may sometimes be certain advantages in, or reasons for, placing the controller in a minor feedback loop, as shown in Fig. 8-2(d). For example, a tachometer may be coupled directly to a dc motor not only for the purpose of speed indication, but more often for improving the stability of the closed-loop system by feeding back the output signal of the tachometer. In principle, the PID controller or the phase-lead and phase-lag controllers discussed earlier can all, with varying degree of effectiveness, be applied to the minor-loop-feedback controller configuration. Some details of these control schemes are discussed in the following paragraphs.

Rate-Feedback or Tachometer-Feedback Control

The principle of using the derivative of the actuating signal to improve the damping of a closed-loop system can be applied to the output signal to achieve a similar effect. In other words, the derivative of the output signal is fed back and added

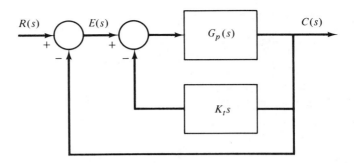

Figure 8-41 Control system with rate feedback control.

algebraically to the actuating signal of the system. In practice, if the output variable is mechanical displacement, a tachometer may be used which converts the mechanical displacement into an electrical signal that is proportional to the derivative of the displacement. Figure 8-41 shows the block diagram of a control system with a secondary path that feeds back the derivative of the output. The transfer function of the tachometer is denoted by $K_t s$, where K_t is the tachometer constant, usually expressed in volts per unit velocity. The derivative signal of the output can also be generated electronically, even if the output is a mechanical displacement. In Chapter 4 a method of generating the velocity signal from data obtained from an incremental encoder was described.

The effects of rate or tachometer feedback can be illustrated simply by means of an example. Consider that the process of the system shown in Fig. 8-41 has the transfer function

$$G_p(s) = \frac{\omega_n^2}{s(s + 2\zeta\omega_n)} \tag{8-118}$$

The closed-loop transfer function of the system is written

$$\frac{C(s)}{R(s)} = \frac{\omega_n^2}{s^2 + (2\zeta\omega_n + K_t\omega_n^2)s + \omega_n^2} \tag{8-119}$$

and the characteristic equation is

$$s^2 + (2\zeta\omega_n + K_t\omega_n^2)s + \omega_n^2 = 0 \tag{8-120}$$

From Eq. (8-119) it is apparent that the effect of the tachometer feedback is the increase of the damping of the closed-loop system, since K_t appears in the same term as the damping ratio ζ. In this respect the rate-feedback control has exactly the same effect as the PD control. However, the closed-loop transfer function of the closed-loop system with PD control in Fig. 8-3 is

$$\frac{C(s)}{R(s)} = \frac{\omega_n^2(K_P + K_D s)}{s^2 + (2\zeta\omega_n + K_D\omega_n^2)s + \omega_n^2 K_P} \tag{8-121}$$

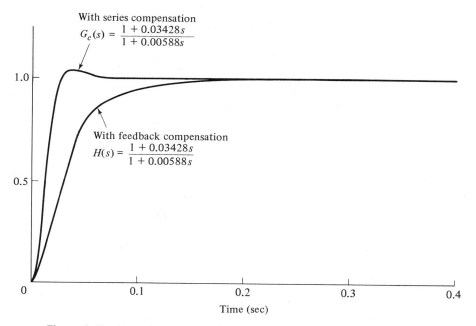

Figure 8-42 Unit step responses of the sun-seeker system with phase-lead controller in the forward path and the feedback path.

Comparing the two transfer functions in Eqs. (8-119) and (8-121), we see that the two characteristic equations are identical if $K_P = 1$ and $K_D = K_t$. However, Equation (8-121) has a zero at $s = -K_P/K_D$ whereas Eq. (8-119) does not. Thus, even if $K_D = K_t$ and $K_P = 1$ and all the other constants in the two transfer functions are the same, the responses of the two systems will not be identical. In fact, the response of the system with rate feedback and is represented by the transfer function of Eq. (8-119) is uniquely defined by the characteristic equation, whereas the response of the system with the PD controller, as described by the transfer function of Eq. (8-121), also depends on the zero at $s = -K_P/K_D$. Figure 8-42 illustrates the effects of a zero of the transfer function of a closed-loop system. This shows the danger of concentrating totally on the roots of the characteristic equation when carrying out the design of control systems.

With reference to the steady-state analysis, the open-loop transfer function of the system with rate feedback is

$$\frac{C(s)}{E(s)} = G(s) = \frac{\omega_n^2}{s(s + 2\zeta\omega_n + K_t\omega_n^2)} \tag{8-122}$$

Since the system is still of type 1, the basic characteristics of the steady-state error are not altered; that is, when the input is a step function, the steady-state error is zero, and so on. However, for a unit ramp function input, the steady-state error of the system in Fig. 8-41 is $(2\zeta + K_t\omega_n)/\omega_n$, whereas that of the system with the PD control in Fig. 8-3 is $2\zeta/\omega_n$.

Phase-Lead and Phase-Lag Feedback Controls

The phase-lead and phase-lag controllers discussed in Section 8.3, as well as more complex controller dynamics, can be applied in the feedback loop. In general, since there may be several controller parameters to be determined, no unique methods of design are available. Let us use some examples to illustrate the key effects of incorporating these controllers in the feedback configuration.

Consider that for the sun-seeker system in Example 8-1, instead of placing the controller in the forward path, we adopt the minor-loop feedback control as shown in Fig. 8-2(b). The closed-loop transfer function of the compensated system is written

$$\frac{\Theta_0(s)}{\Theta_r(s)} = \frac{G_p(s)}{1 + G_p(s)H(s)} \tag{8-123}$$

where

$$G_p(s) = \frac{2500}{s(s + 25)} \tag{8-124}$$

and

$$H(s) = \frac{1 + aTs}{1 + Ts} \tag{8-125}$$

which could represent a phase-lead or a phase-lag controller, depending on the value of a. Substituting Eqs. (8-124) and (8-125) into Eq. (8-123), we get

$$\frac{\Theta_0(s)}{\Theta_r(s)} = \frac{2500(1 + Ts)}{s(s + 25)(1 + Ts) + 2500(1 + aTs)} \tag{8-126}$$

With the controller in the forward path, the closed-loop transfer function is

$$\frac{\Theta_0(s)}{\Theta_r(s)} = \frac{G_c(s)G_p(s)}{1 + G_c(s)G_p(s)} \tag{8-127}$$
$$= \frac{2500(1 + aTs)}{s(s + 25)(1 + Ts) + 2500(1 + aTs)}$$

We see that the denominator polynomials of the closed-loop transfer functions of the two control schemes are identical, but the numerators are different. This means that if we use the characteristic equation roots or the root loci as a design guide, the two equations will have different responses for the same input.

To further investigate the meaning of this finding, let us first consider that the controller is as given in Eq. (8-42) (phase-lead); thus,

$$H(s) = \frac{s + 29.17}{s + 170} \tag{8-128}$$

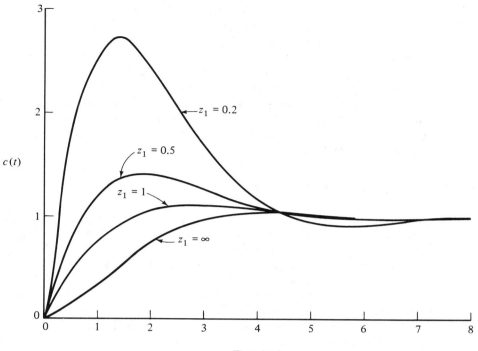

Figure 8-43   Unit step responses of the system with the transfer function

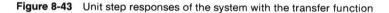

$$\frac{C(s)}{R(s)} = \frac{1 + s/z_1}{s^2 + 1.414s + 1}$$

Substituting Eq. (8-128) into Eq. (8-126) and simplifying, we get

$$\frac{\Theta_0(s)}{\Theta_r(s)} = \frac{2500s + 425{,}170}{s^3 + 195s^2 + 18{,}826.7s + 425{,}170} \tag{8-129}$$

The roots of the characteristic equation are $s = -30.93$, $-82.07 + j83.73$, and $-82.07 - j83.73$, which are identical to those of the system with the same phase-lead controller in the forward path. The unit step response of the system with the feedback controller is shown in Fig. 8-42. When compared with the response of the system with the series controller, we see that the step response of the system with the same feedback controller has a slower rise time, and no overshoot. This is due to the fact that Eq. (8-129) has a zero at $s = -1/T = -170$, whereas for the series controller, the zero is at $s = -1/aT = -29.17$, which is closer to the origin.

As shown in Sec. 6.7, as a zero of the closed-loop transfer function is moved closer to the origin, the peak overshoot may increase dramatically. Figure 8-43 illustrates the unit step response of a second-order system as the zero of the closed-loop transfer function varies. Notice that the peak overshoot can be several hundred percent, and theoretically can reach infinity as the zero approaches $s = 0$. In view of this effect, it would be simple to reason that placing a phase-lag controller in the minor feedback loop would not produce any useful compensation, since in this case the zero at $s = -1/aT$ is intentionally placed very close to the origin, it will again increase the overshoot substantially even though the characteristic equation roots are chosen properly. In fact, we can show that if the phase-lag controller designed in Eq. (8-62) for the sun-seeker system is incorporated in a minor-loop feedback configuration, the characteristic equation roots would still be at -0.412, $-12.32 + j12.32$, and $-12.32 - j12.32$, but the peak overshoot would be in excess of 600 percent!

8.7 STATE-FEEDBACK CONTROL

A majority of the design techniques in modern control theory is based on the state-feedback configuration. That is, instead of using controllers with fixed configurations in the forward or feedback path, control is achieved by feeding back the state variables through constant gains. The block diagram of a system with state-feedback control is shown in Fig. 8-2(c).

We can show that the PID control and the rate-feedback control discussed earlier are all special cases of the state-feedback control scheme. In the case of rate-feedback control, let us consider the second-order process described in Eq. (8-118). The process is decomposed by direct decomposition to be represented by the state diagram of Fig. 8-44(a). If the states x_1 and x_2 are physically accessible, these variables may be fed back through constant gains g_1 and g_2, respectively, to form the control, as shown in Fig. 8-44(b). The closed-loop transfer function of the system is written

$$\frac{C(s)}{R(s)} = \frac{\omega_n^2}{s^2 + (2\zeta\omega_n + g_2)s + g_1} \tag{8-130}$$

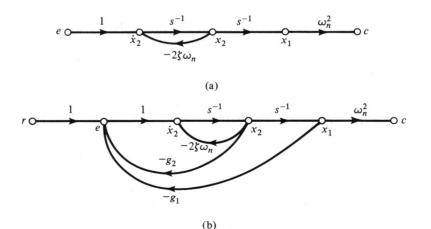

(a)

(b)

Figure 8-44 Control of a second-order system by state feedback.

Comparing this transfer function with that of the system with rate feedback, Eq. (8-130), we notice that the two transfer functions are identical if $g_1 = \omega_n^2$ and $g_2 = K_t\omega_n^2$. In fact, if the system is to have zero steady-state error, g_1 should equal ω_n^2. The value of g_2 is selected to satisfy the damping requirements.

For the system with PD control shown in Fig. 8-3, the closed-loop transfer function is

$$\frac{C(s)}{R(s)} = \frac{\omega_n^2(K_P + K_Ds)}{s^2 + (2\zeta\omega_n + K_D\omega_n^2)s + \omega_n^2 K_P} \tag{8-131}$$

Thus the characteristic equations of the systems described by Eqs. (8-130) and (8-131) would be identical if $g_2 = K_D\omega_n^2$ and $g_1 = \omega_n^2 K_P$. The numerators of the two transfer functions are different, however.

If the reference input $r(t)$ is zero, the class of systems is commonly described as *regulators*. Under this condition the control objective is to drive any arbitrary initial conditions of the system to zero in some prescribed manner, such as as quickly as possible. Then the regulator system with the PD controller is the same as the state-feedback control.

Since the PI control increases the order of the system by one, it cannot be made equivalent to state feedback through constant gains. We show in Section 8.9 that if we combine state feedback with integral control, we can again realize PI control in the sense of state-feedback control.

8.8 POLE-PLACEMENT DESIGN THROUGH STATE FEEDBACK

When root loci are utilized for the design of control systems, the general problem may be described as that of pole placement; the pole here refers to that of the closed-loop transfer function, which is also the same as the roots of the characteristic equation. Knowing the relation between the closed-loop poles and the system performance, we can effectively design the system by specifying the location of these poles. The design methods discussed in the preceding sections are all characterized by the property that the poles are selected based on what can be achieved with the fixed controller configuration and the physical range of the controller parameters. A natural question would be: Under what condition can the poles be placed arbitrarily? This is an entirely new design philosophy and freedom which apparently can be achieved only under certain conditions.

When we have a process of the second order or higher, the PD, PI, the first-order phase-lead, or the phase-lag controller would not be able to control independently all of the three or more poles of the system, since there are only two free parameters in each of these controllers.

To investigate the condition required for arbitrary pole placement in an nth-order system, let us consider that a linear process is described by the following state equation:

$$\dot{\mathbf{x}}(t) = \mathbf{A}\mathbf{x}(t) + \mathbf{B}u(t) \tag{8-132}$$

where $\mathbf{x}(t)$ is the $n \times 1$ state vector, and $u(t)$ is the scalar control input. The state-feedback control is given by

$$u(t) = -\mathbf{G}\mathbf{x}(t) + r(t) \tag{8-133}$$

where $\mathbf{G}$ is the $1 \times n$ feedback matrix with constant-gain elements. Combining Eqs. (8-132) and (8-133), the closed-loop system is represented by the state equation

$$\dot{\mathbf{x}}(t) = (\mathbf{A} - \mathbf{B}\mathbf{G})\mathbf{x}(t) + \mathbf{B}r \tag{8-134}$$

It will be shown in the following that if the pair $[\mathbf{A}, \mathbf{B}]$ is completely controllable, then a matrix $\mathbf{G}$ exists that can give an arbitrary set of eigenvalues of $(\mathbf{A} - \mathbf{B}\mathbf{G})$; that is, the n roots of the characteristic equation

$$|\lambda\mathbf{I} - \mathbf{A} + \mathbf{B}\mathbf{G}| = 0 \tag{8-135}$$

can be arbitrarily placed.

It has been shown in Chapter 5 that if a system is state controllable, it can always be represented in the phase-variable canonical form; that is, in Eq. (8-132),

$$\mathbf{A} = \begin{bmatrix} 0 & 1 & 0 & \cdots & 0 \\ 0 & 0 & 1 & \cdots & 0 \\ \vdots & \vdots & \vdots & \cdots & \vdots \\ 0 & 0 & 0 & \cdots & 1 \\ -a_1 & -a_2 & -a_3 & \cdots & -a_n \end{bmatrix} \qquad \mathbf{B} = \begin{bmatrix} 0 \\ 0 \\ 0 \\ \vdots \\ 1 \end{bmatrix}$$

The reverse is also true in that if the system is represented in the phase-variable canonical form, it is always state controllable. To show this, we form the following matrices:

$$\mathbf{A}\mathbf{B} = \begin{bmatrix} 0 \\ 0 \\ 0 \\ \vdots \\ 1 \\ -a_n \end{bmatrix} \qquad \mathbf{A}^2\mathbf{B} = \begin{bmatrix} 0 \\ 0 \\ 0 \\ \vdots \\ 1 \\ -a_n \\ a_n^2 - a_{n-1} \end{bmatrix} \qquad \mathbf{A}^3\mathbf{B} = \begin{bmatrix} 0 \\ 0 \\ 0 \\ \vdots \\ 1 \\ -a_n \\ a_n^2 - a_{n-1} \\ -a_n^3 + a_{n-1}a_n - a_{n-2} \end{bmatrix}$$

Continuing with the matrix product through $\mathbf{A}^{n-1}\mathbf{B}$, it will become apparent that regardless of what the values of $a_1, a_2, \ldots, a_n$, are, the determinant of $\mathbf{S} = [\mathbf{B} \quad \mathbf{A}\mathbf{B} \quad \mathbf{A}^2\mathbf{B} \cdots \mathbf{A}^{n-1}\mathbf{B}]$ will always be equal to -1, since $\mathbf{S}$ is a triangular matrix with 1s on the main diagonal. Therefore, we have proved that if the system is represented by the phase-variable canonical form, it is always state controllable.

The feedback matrix $\mathbf{G}$ can be written

$$\mathbf{G} = [g_1 \quad g_2 \quad \cdots \quad g_n] \tag{8-136}$$

Then

$$\mathbf{A} - \mathbf{BG} = \begin{bmatrix} 0 & 1 & 0 & \cdots & 0 \\ 0 & 0 & 1 & \cdots & 0 \\ 0 & 0 & 0 & \cdots & 0 \\ \cdots & \cdots & \cdots & \cdots & \cdots \\ 0 & 0 & 0 & & 1 \\ -a_1 - g_1 & -a_2 - g_2 & \cdots & & -a_n - g_n \end{bmatrix} \tag{8-137}$$

The eigenvalues of $\mathbf{A} - \mathbf{BG}$ are then found from the characteristic equation

$$|\lambda \mathbf{I} - (\mathbf{A} - \mathbf{BG})| = \lambda^n + (a_n + g_n)\lambda^{n-1} + (a_{n-1} + g_{n-1})\lambda^{n-2} + \cdots + (a_1 + g_1)$$
$$= 0 \tag{8-138}$$

Clearly, the eigenvalues can be arbitrarily chosen by the choice of $g_1, g_2, \ldots, g_n$.

■

Example 8-5

Consider that the transfer function of a linear process is

$$G(s) = \frac{C(s)}{E(s)} = \frac{20}{s^2(s+1)} \tag{8-139}$$

Figure 8-45 (a) shows the state diagram of $G(s)$, and Fig. 8-45(b) shows the state diagram with feedback from all three states. The closed-loop transfer function of the system is

$$\frac{C(s)}{R(s)} = \frac{20}{s^3 + (g_3 + 1)s^2 + g_2 s + g_1} \tag{8-140}$$

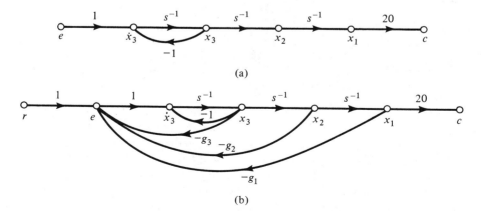

(a)

(b)

Figure 8-45 Control of a third-order system by state feedback.

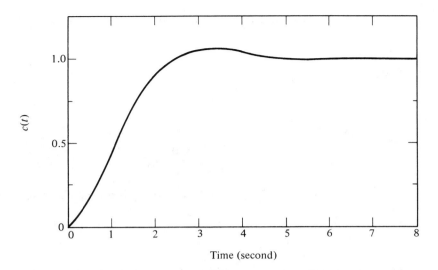

Figure 8-46   Unit step response of the control system in Example 8-5.

Let us assume that we desire to have zero steady-state error with the input being a unit step function, and in addition, two of the closed-loop poles must be at $s = -1 + j$ and $s = -1 - j$. The steady-state requirement fixes the value of g_1 at 20, and only g_2 and g_3 need to be determined from the eigenvalue location. Since the transfer function does not have common poles and zeros, the process is completely state controllable.

The characteristic equation of the system is

$$s^3 + (g_3 + 1)s^2 + g_2 s + 20 = (s + 1 - j)(s + 1 + j)(s + a) \qquad (8\text{-}141)$$

Equating the coefficients of the corresponding terms in Eq. (8-141), we get

$$g_2 = 22 \qquad \text{and} \qquad g_3 = 11$$

and the third pole is at $s = -10$.

Since the complex closed-loop poles have a damping ratio of 0.707, and the third pole is quite far to the left of these poles, the system acts like a second-order system. Figure 8-46 shows that the unit step response has an overshoot of 4 percent.

■

In general, it is not necessary to transform a system first into the phase-variable canonical form for the purpose of the state feedback. General methods for the determination of the feedback matrix **G** are available as long as the process is completely controllable. The following development describes a method that determines **G** based on the matrices **A** and **B**, and the characteristic equations of the open-loop system and the closed-loop system with a single input. Let us define the following functions:

$$\Delta_o(s) = |s\mathbf{I} - \mathbf{A}| = \text{characteristic equation of } \mathbf{A} \text{ (open-loop system)} \qquad (8\text{-}142)$$

$$\Delta_c(s) = |s\mathbf{I} - \mathbf{A} + \mathbf{BG}| = \text{characteristic equation of } \mathbf{A} - \mathbf{BG} \text{ (closed-loop system)}$$
$$(8\text{-}143)$$

$$\Delta(s) = 1 + \mathbf{G}(s\mathbf{I} - \mathbf{A})^{-1}\mathbf{B} \qquad (8\text{-}144)$$

where $\mathbf{G}$ is $1 \times n$ and $\mathbf{B}$ is $n \times 1$.

First, we show that

$$\Delta(s) = \frac{\Delta_c(s)}{\Delta_o(s)} \qquad (8\text{-}145)$$

This relation is proved by writing

$$s\mathbf{I} - \mathbf{A} + \mathbf{BG} = (s\mathbf{I} - \mathbf{A})\left[\mathbf{I} + (s\mathbf{I} - \mathbf{A})^{-1}\mathbf{BG}\right] \qquad (8\text{-}146)$$

Taking the determinant on both sides of the last equation, we get

$$\Delta_c(s) = |s\mathbf{I} - \mathbf{A} + \mathbf{BG}| = \Delta_o(s)\left|\mathbf{I} + (s\mathbf{I} - \mathbf{A})^{-1}\mathbf{BG}\right| \qquad (8\text{-}147)$$

Since

$$\left|\mathbf{I} + (s\mathbf{I} - \mathbf{A})^{-1}\mathbf{BG}\right| = \left|\mathbf{I} + \mathbf{BG}(s\mathbf{I} - \mathbf{A})^{-1}\right| = \left|\mathbf{I} + \mathbf{G}(s\mathbf{I} - \mathbf{A})^{-1}\mathbf{B}\right|$$
$$(8\text{-}148)$$
$$= 1 + \mathbf{G}(s\mathbf{I} - \mathbf{A})^{-1}\mathbf{B} = \Delta(s)$$

where the identity matrices are of different dimensions, Eq. (8-147) becomes

$$\Delta_c(s) = \Delta_o(s)\Delta(s) \qquad (8\text{-}149)$$

Now we write Eq. (8-144) as

$$\Delta(s) = 1 + \mathbf{G}\frac{\text{Adj}(s\mathbf{I} - \mathbf{A})\mathbf{B}}{\Delta_o(s)} \qquad (8\text{-}150)$$

where $\text{Adj}(s\mathbf{I} - \mathbf{A})$ denotes the adjoint matrix of $s\mathbf{I} - \mathbf{A}$. Let

$$\mathbf{k}(s) = [\text{Adj}(s\mathbf{I} - \mathbf{A})]\mathbf{B} \qquad n \times 1 \qquad (8\text{-}151)$$

Then, Eq. (8-150) becomes

$$\Delta(s) = \frac{\Delta_o(s) + \mathbf{Gk}(s)}{\Delta_o(s)} \qquad (8\text{-}152)$$

and using Eq. (8-149), we get

$$\mathbf{G}\mathbf{k}(s) = \Delta_c(s) - \Delta_o(s) \tag{8-153}$$

Thus, knowing $\mathbf{k}(s)$, $\Delta_c(s)$ and $\Delta_o(s)$, the feedback gain matrix $\mathbf{G}$ can be solved from Eq. (8-153) if the process is completely controllable.

■

Example 8-6

Consider that a linear process is described by the state equation

$$\dot{\mathbf{x}}(t) = \mathbf{A}\mathbf{x}(t) + \mathbf{B}u(t) \tag{8-154}$$

where

$$\mathbf{A} = \begin{bmatrix} 1 & 0 & 0 \\ -1 & 0 & 2 \\ 0 & -1 & 1 \end{bmatrix} \qquad \mathbf{B} = \begin{bmatrix} 1 \\ 0 \\ 0 \end{bmatrix}$$

From Eq. (8-151) we have

$$\mathbf{k}(s) = \begin{bmatrix} s^2 - s + 2 \\ -(s - 1) \\ 1 \end{bmatrix} \tag{8-155}$$

The open-loop characteristic equation is

$$\Delta_o(s) = |s\mathbf{I} - \mathbf{A}| = s^3 - 2s^2 + 3s - 2 \tag{8-156}$$

Let the desired closed-loop eigenvalues be $s = -2$, -1, and -1. Then the closed-loop characteristic equation is

$$\Delta_c(s) = s^3 + 4s^2 + 5s + 2 \tag{8-157}$$

Substituting Eqs. (8-155), (8-156), and (8-157) into Eq. (8-153), the elements of $\mathbf{G}$ are found to be $g_1 = 6$, $g_2 = -8$, and $g_3 = 0$.

■

8.9 STATE FEEDBACK WITH INTEGRAL CONTROL [4]

The state feedback structured in the preceding two sections has a serious deficiency in that it does not improve the type of the system. As a result, the state-feedback control with constant-gain feedback is generally useful only for regulator systems for which there are no inputs. In general, there is a large class of control systems that has inputs which must be followed, and often there are undesirable noise or disturbances that the system must suppress or eliminate. One remedy for this problem is to introduce integral control together with the state feedback with constant gains.

Let us consider the following design problem.

Given a linear system that is represented by the following dynamic equations:

$$\dot{\mathbf{x}}(t) = \mathbf{A}\mathbf{x}(t) + \mathbf{B}\mathbf{u}(t) + \mathbf{F}\mathbf{w} \qquad (8\text{-}158)$$

where $\mathbf{x}(t) = n \times 1$ state vector
$\mathbf{u}(t) = p \times 1$ control vector
$\mathbf{w} = m \times 1$ input and disturbance vector

and $\mathbf{A} = n \times n$, $\mathbf{B} = n \times p$, $\mathbf{F} = n \times m$, and $\mathbf{w}$ is defined as a constant vector whose elements are composed of input signals and disturbances. In practice, the magnitudes of the input signals are given, but the magnitudes of some or all of the constant disturbances may be unknown.

The objective of the design is to find feedback controls from the state variables such that the state $\mathbf{x}(t)$ is driven to any desired state (set point) as time approaches infinity. For instance, we may want the state variable $x_1(t)$ to be driven to a set point $r(t) = R$ as t approaches infinity, while the overall system is asymptotically stable.

For the problem described, it is convenient to define the output equation as

$$\mathbf{c}(t) = \mathbf{D}\mathbf{x}(t) + \mathbf{H}\mathbf{w} \qquad (8\text{-}159)$$

where $\mathbf{c}(t) = q \times 1$ output vector, $\mathbf{D} = q \times n$, and $\mathbf{H} = q \times m$ coefficient matrices. Then, to drive the state $x_1(t)$ to the set point $r(t)$ is equivalent to driving

$$\mathbf{c}(t) = w_1 - x_1(t) \qquad (8\text{-}160)$$

to zero as t approaches infinity, where

$$w_1 = r(t) = R = \text{constant} \qquad (8\text{-}161)$$

Thus, the design problem may be regarded as "output regulation."

As indicated earlier in the state-feedback design, the present "output regulation" problem cannot be achieved by simply using constant-gain state feedback, since the latter cannot improve the *type* of a system. Since a state-feedback control system with constant-gain feedback from all the states corresponds to a type-0 system, we must introduce integral control in order to have the output of the system track any input.

Let the control $\mathbf{u}(t)$ be given by

$$\mathbf{u}(t) = -\mathbf{G}_1\mathbf{x}(t) - \mathbf{G}_2\int\mathbf{c}(t)\,dt \qquad (8\text{-}162)$$

where $\mathbf{G}_1$ is a $p \times n$ feedback-gain matrix and $\mathbf{G}_2$ is a $p \times q$ feedback gain matrix, both with constant elements.

Equation (8-162) shows that the control is realized by feeding back the states $\mathbf{x}(t)$ through a constant-gain feedback matrix $\mathbf{G}_1$, and in addition, the integrals of the "outputs" $\mathbf{c}(t)$ are fed back through the constant-gain matrix $\mathbf{G}_2$. The block

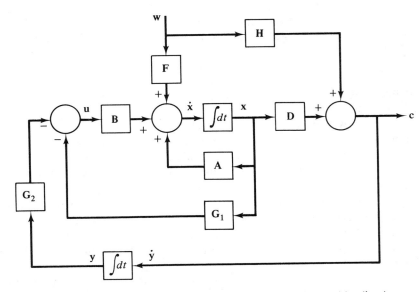

Figure 8-47 Control system with state feedback and integral feedback.

diagram in Fig. 8-47 illustrates the elements of the overall feedback control system. Naturally, when $\mathbf{G_2} = \mathbf{0}$, the system becomes a state regulator with state feedback.

Since the feedback system now has q additional integrators, the overall system is of the $(n + q)$th order. Let $\mathbf{y}(t)$ denote the $q \times 1$ state vector of the integral control, as shown in Fig. 8-47, the $(n + q)$ state equations are written in vector-matrix form directly from Fig. 8-47.

$$\begin{bmatrix} \dot{\mathbf{x}}(t) \\ \dot{\mathbf{y}}(t) \end{bmatrix} = \begin{bmatrix} \mathbf{A} & \mathbf{0} \\ \mathbf{D} & \mathbf{0} \end{bmatrix} \begin{bmatrix} \mathbf{x}(t) \\ \mathbf{y}(t) \end{bmatrix} - \begin{bmatrix} \mathbf{B} \\ \mathbf{0} \end{bmatrix} [\mathbf{G_1} \quad \mathbf{G_2}] \begin{bmatrix} \mathbf{x}(t) \\ \mathbf{y}(t) \end{bmatrix} + \begin{bmatrix} \mathbf{F} \\ \mathbf{H} \end{bmatrix} \mathbf{w} \qquad (8\text{-}163)$$

Or

$$\dot{\hat{\mathbf{x}}}(t) = \hat{\mathbf{A}}\hat{\mathbf{x}}(t) - \hat{\mathbf{B}}\mathbf{G}\hat{\mathbf{x}}(t) + \hat{\mathbf{F}}\mathbf{w} \qquad (8\text{-}164)$$

where

$$\hat{\mathbf{x}}(t) = \begin{bmatrix} \mathbf{x}(t) \\ \mathbf{y}(t) \end{bmatrix} \qquad (n + q) \times 1 \qquad (8\text{-}165)$$

$$\hat{\mathbf{A}} = \begin{bmatrix} \mathbf{A} & \mathbf{0} \\ \mathbf{D} & \mathbf{0} \end{bmatrix} \qquad (n + q) \times (n + q) \qquad (8\text{-}166)$$

$$\hat{\mathbf{B}} = \begin{bmatrix} \mathbf{B} \\ \mathbf{0} \end{bmatrix} \qquad (n + q) \times p \qquad (8\text{-}167)$$

$$\mathbf{G} = [\mathbf{G_1} \quad \mathbf{G_2}] \qquad p \times (n + q) \qquad (8\text{-}168)$$

$$\hat{\mathbf{F}} = \begin{bmatrix} \mathbf{F} \\ \mathbf{H} \end{bmatrix} \qquad (n + q) \times m \qquad (8\text{-}169)$$

Equation (8-164) is also written as

$$\dot{\hat{\mathbf{x}}}(t) = (\hat{\mathbf{A}} - \hat{\mathbf{B}}\mathbf{G})\hat{\mathbf{x}}(t) + \hat{\mathbf{F}}\mathbf{w} \tag{8-170}$$

which is of the form of state feedback from the state $\hat{\mathbf{x}}(t)$.

The following example illustrates the application of the design with state feedback and output integral control.

■

Example 8-7

Consider a dc motor control system that is described by the following equations:

$$\dot{\omega} = \frac{-B}{J}\omega + \frac{K_i}{J}i_a - \frac{1}{J}T_L \tag{8-171}$$

$$\dot{i}_a = \frac{-K_b}{L}\omega - \frac{R}{L}i_a + \frac{1}{L}e_a \tag{8-172}$$

where i_a = armature current, A
e_a = armature applied voltage, V
ω = motor velocity, rad/sec
B = viscous-friction coefficient of motor and load = 0
J = moment of inertia of motor and load = 0.02 N-m/rad/sec^2
K_i = motor torque constant = 1 N-m/A
K_b = motor back emf constant = 1 V/rad/sec
T_L = constant load torque (magnitude not known), N-m
L = armature inductance = 0.005 H
R = armature resistance = 1 Ω
The design problem is to find the control $u(t) = e_a(t)$ such that

(1) $$\lim_{t \to \infty} i_a(t) = 0 \quad \text{and} \quad \lim_{t \to \infty} \dot{\omega}(t) = 0 \tag{8-173}$$

and

(2) $$\lim_{t \to \infty} \omega(t) = \text{constant set point } r \tag{8-174}$$

Let the state variables be defined as $x_1(t) = \omega(t)$ and $x_2(t) = i_a(t)$. The vector **w** is defined as

$$\mathbf{w} = \begin{bmatrix} T_L \\ r \end{bmatrix} = \begin{bmatrix} w_1 \\ w_2 \end{bmatrix} \tag{8-175}$$

Let the output variable be defined as

$$c(t) = r - \omega(t) = w_2 - x_1(t) \tag{8-176}$$

Then the condition in Eq. (8-174) is equivalent to

$$\lim_{t \to \infty} c(t) = 0$$

which is the condition of output regulation.

Since the original system has two state variables in x_1 and x_2, and one output in $c(t)$, the control is of the form

$$u(t) = -g_1 x_1(t) - g_2 x_2(t) - g_3 \int c(t)\, dt$$

$$= -\mathbf{G}_1 \mathbf{x}(t) - \mathbf{G}_2 \int c(t)\, dt \tag{8-177}$$

where

$$\mathbf{G}_1 = [\, g_1 \quad g_2 \,] \qquad \mathbf{G}_2 = g_3 \tag{8-178}$$

g_1, g_2, and g_3 are real constants.

However, with the integral control, the closed-loop system is of the third order, and there are three poles to be specified. Let the closed-loop poles be placed at -300, $-10 + j10$, and $-10 - j10$.

We must now find the values of g_1, g_2, and g_3 so that the conditions in Eqs. (8-173), (8-174) and the pole-placement requirements are met simultaneously. Expressing the state equations in Eqs. (8-171) and (8-172) in the form of Eq. (8-158), we have

$$\dot{\mathbf{x}}(t) = \begin{bmatrix} \dot{\omega}(t) \\ \dot{i}_a(t) \end{bmatrix} = \begin{bmatrix} -\dfrac{B}{J} & \dfrac{K_i}{J} \\ -\dfrac{K_b}{L} & -\dfrac{R}{L} \end{bmatrix} \mathbf{x}(t) + \begin{bmatrix} 0 \\ \dfrac{1}{L} \end{bmatrix} u(t) + \begin{bmatrix} -\dfrac{1}{J} & 0 \\ 0 & 0 \end{bmatrix} \begin{bmatrix} w_1 \\ w_2 \end{bmatrix} \tag{8-179}$$

The output equation is

$$c(t) = [\, -1 \quad 0 \,]\mathbf{x}(t) + [\, 0 \quad 1 \,]\mathbf{w} \tag{8-180}$$

Substituting the values of the system parameters into Eq. (8-179), we have

$$\mathbf{A} = \begin{bmatrix} -\dfrac{B}{J} & \dfrac{K_i}{J} \\ -\dfrac{K_b}{L} & -\dfrac{R}{L} \end{bmatrix} = \begin{bmatrix} 0 & 50 \\ -200 & -200 \end{bmatrix} \tag{8-181}$$

$$\mathbf{B} = \begin{bmatrix} 0 \\ \dfrac{1}{L} \end{bmatrix} = \begin{bmatrix} 0 \\ 200 \end{bmatrix} \tag{8-182}$$

$$\mathbf{F} = \begin{bmatrix} -\dfrac{1}{J} & 0 \\ 0 & 0 \end{bmatrix} = \begin{bmatrix} -50 & 0 \\ 0 & 0 \end{bmatrix} \tag{8-183}$$

From Eq. (8-180),

$$\mathbf{D} = [\, -1 \quad 0 \,] \qquad \mathbf{H} = [\, 0 \quad 1 \,]$$

Define the transformation according to Eq. (8-163); then

$$\hat{\mathbf{A}} = \begin{bmatrix} \mathbf{A} & \mathbf{0} \\ \mathbf{D} & 0 \end{bmatrix} = \begin{bmatrix} 0 & 50 & 0 \\ -200 & -200 & 0 \\ -1 & 0 & 0 \end{bmatrix} \qquad (8\text{-}184)$$

$$\hat{\mathbf{B}} = \begin{bmatrix} \mathbf{B} \\ 0 \end{bmatrix} = \begin{bmatrix} 0 \\ 200 \\ 0 \end{bmatrix} \qquad (8\text{-}185)$$

We can show that both $[\mathbf{A}, \mathbf{B}]$ and $[\hat{\mathbf{A}}, \hat{\mathbf{B}}]$ are controllable.
Let the control of the transformed system be

$$u(t) = -\mathbf{G}_1 \mathbf{x}(t) - \mathbf{G}_2 \int c(t) \, dt$$

$$= -\mathbf{G} \begin{bmatrix} \mathbf{x}(t) \\ y(t) \end{bmatrix} \qquad (8\text{-}186)$$

where

$$\mathbf{G} = \begin{bmatrix} g_1 & g_2 & g_3 \end{bmatrix} \qquad (8\text{-}187)$$

and

$$\mathbf{G}_1 = \begin{bmatrix} g_1 & g_2 \end{bmatrix} \qquad (8\text{-}188)$$
$$\mathbf{G}_2 = g_3 \qquad (8\text{-}189)$$

The coefficient matrix of the closed-loop system becomes

$$\hat{\mathbf{A}} - \hat{\mathbf{B}}\mathbf{G} = \begin{bmatrix} 0 & 50 & 0 \\ -200 - 200 g_1 & -200 - 200 g_2 & -200 g_3 \\ -1 & 0 & 0 \end{bmatrix} \qquad (8\text{-}190)$$

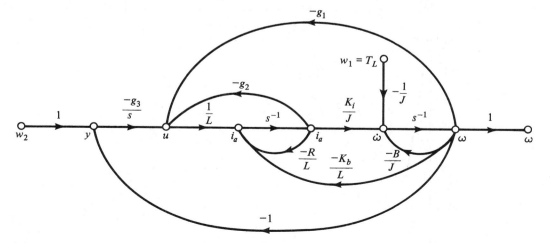

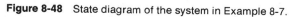

Figure 8-48 State diagram of the system in Example 8-7.

The characteristic equation is

$$\Delta_c(s) = |s\mathbf{I} - A + BG| = s^3 + 200(1 + g_2)s^2 + 10{,}000(1 + g_1)s - 10{,}000\,g_3 = 0 \tag{8-191}$$

For the three assigned poles, Eq. (8-191) must equal

$$\Delta_c(s) = s^3 + 320s^2 + 6200s + 60{,}000 = 0 \tag{8-192}$$

Thus, the three feedback gains are

$$g_1 = -0.38 \qquad g_2 = 0.6 \qquad g_3 = -6$$

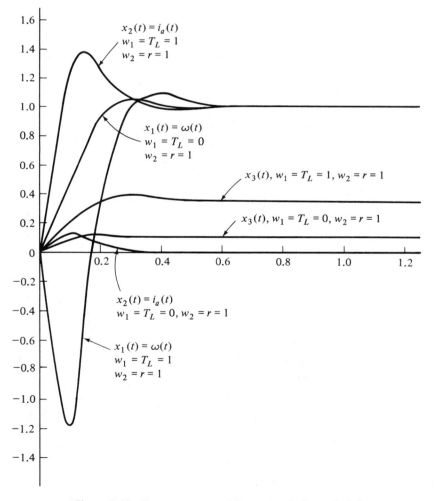

Figure 8-49   Time responses of the system in Example 8-7.

Figure 8-48 shows the state diagram of the overall designed system. Applying Mason's gain formula between the inputs w_1 and w_2 and the states $\Omega(s)$ and $I_a(s)$ on the state diagram, we have

$$
\begin{bmatrix} \Omega(s) \\ I_a(s) \end{bmatrix} = \frac{1}{\Delta_c(s)} \begin{bmatrix} -\frac{1}{J}\left(s^2 + \frac{R}{L}s + \frac{g_2}{L}s\right) & -\frac{g_3 K_i}{JL} \\ -\frac{1}{J}\left(-\frac{K_b}{L}s - \frac{g_1}{L}s + \frac{g_3}{L}\right) & -\frac{g_3}{L}\left(s + \frac{B}{J}\right) \end{bmatrix} \begin{bmatrix} \dfrac{w_1}{s} \\ \dfrac{w_2}{s} \end{bmatrix} \tag{8-193}
$$

Applying the final-value theorem to the last equation, the final values of the state vector are found to be

$$
\lim_{t \to \infty} \begin{bmatrix} \omega(t) \\ i_a(t) \end{bmatrix} = \lim_{s \to 0} s \begin{bmatrix} \Omega(s) \\ I_a(s) \end{bmatrix} = \begin{bmatrix} 0 & 1 \\ 1 & \dfrac{B}{K_i} \\ \dfrac{1}{K_i} & \end{bmatrix} \begin{bmatrix} w_1 \\ w_2 \end{bmatrix} \tag{8-194}
$$

Therefore, the motor velocity $\omega(t)$ will approach the constant reference set point $r = w_2$ as t approaches infinity, independent of the disturbance torque, $w_1 = T_L$. Figure 8-49 illustrates the responses of all the three state variables. The large overshoot in the response $\omega(t)$ is caused by the disturbance torque of 1 N-m. The figure also shows the responses when $T_L = 0$.

■

8.10 DIGITAL IMPLEMENTATION OF CONTROLLERS

Since digital control systems have many advantages over continuous-data systems, quite often, in practice, controllers that are designed in the analog domain are implemented digitally. Ideally, if the designer has intended to use digital control, the system should be designed so that when the design is complete the dynamics of the controller would already be described by a z-transfer function or difference equations. However, there are situations under which the controller of an existing system is analog, and the system already operates in a satisfactory fashion, but the availability and advantages of digital control suggest that the controller be implemented by digital elements. Thus, the problems discussed in this section are two-fold: first, we investigate how continuous-data controllers such as the PID, the phase-lag and phase-lead controllers, and others can be approximated by digital controllers; second, the problem of implementing digital controllers by digital processors is investigated.

Digital Implementation of the PID Controller

The block diagram of the analog PID controller is shown in Fig. 8-50. The proportional component of the PID controller, K_P, is implemented digitally by a gain element. Since a digital computer or processor always has a finite digital word length, the constant K_P cannot be realized with an infinite resolution. Most digital computers are based on the binary-number system, so that multiplying and dividing

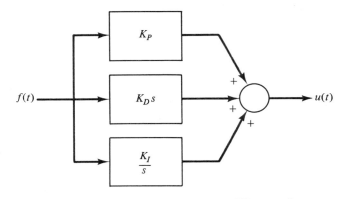

Figure 8-50 Block diagram of analog PID controller.

by 2 would be most convenient to execute simply by a left shift or a right shift of the binary word.

The time derivative of a function $f(t)$ at $t = kT$ can be approximated numerically by use of the values of $f(t)$ measured at $t = kT$ and $t = (k - 1)T$; that is,

$$\left.\frac{df(t)}{dt}\right|_{t=kT} \cong \frac{1}{T}[f(kT) - f(k - 1)T] \qquad (8\text{-}195)$$

To find the z-transform function of the derivative operation described numerically above, we take the z-transform on both sides of Eq. (8-195). We have

$$\delta\left[\left.\frac{df(t)}{dt}\right|_{t=kT}\right] = \frac{1}{T}(1 - z^{-1})F(z) = \frac{z - 1}{Tz}F(z) \qquad (8\text{-}196)$$

Thus, the z-transform function of the digital differentiator is written

$$G_D(z) = K_D\frac{z - 1}{Tz} \qquad (8\text{-}197)$$

where K_D is the proportional constant of the derivative controller. Replacing z by e^{Ts} in Eq. (8-197) we can show that as the sampling period T approaches zero, $G_D(z)$ approaches $K_D s$, which is the transfer function of the analog derivative controller.

For the integrator we normally have a number of choices of digital approximation, as we know that there are many numerical-integration algorithms. Rectangular integration is a simple method of approximating the area of a function numerically. As shown in Fig. 8-49(a), the integral of $f(t)$ can be approximated by the area under the rectangular pulses. It is simple to see that this approximation is equivalent to the sample-and-hold (zero-order) operation described in Chapter 3. Figure 8-51(b) shows the block diagram representation of rectangular integration. From

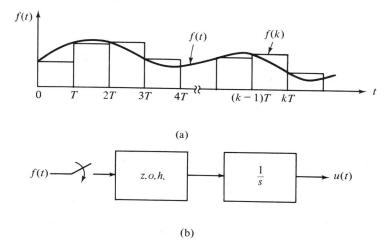

(a)

(b)

Figure 8-51 (a) Rectangular integration. (b) Equivalent rectangular integration with sample-and-hold.

Fig. 8-51(b), the z-transfer function of the digital integrator can be written

$$G_I(z) = K_I\delta\left[\frac{1 - e^{-Ts}}{s} \cdot \frac{1}{s}\right] = \frac{K_IT}{z - 1} \qquad (8\text{-}198)$$

Again, we can show that as T approaches zero, $G_I(z)$ approaches K_I/s, the transfer function of the analog integral controller.

In practice, there are other numerical integration rules, such as the trapezoidal integration, Simpson's rule, and so on. Some of these may be too complex to be implemented by a microcomputer, so that from a practical standpoint, the simplest algorithm often becomes the most preferable.

Using the methods presented above, the block diagram of a digital PID controller is shown in Fig. 8-52.

Once the transfer function of a digital controller is determined, the controller can be implemented by a digital processor or computer. The operator z^{-1} is

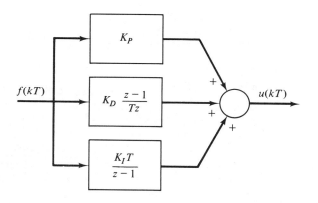

Figure 8-52 Block diagram of a digital PID controller.

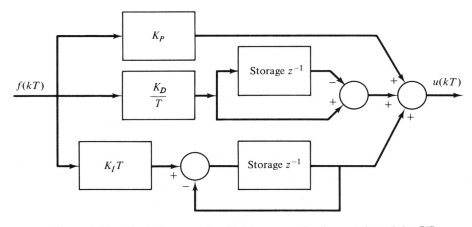

Figure 8-53 Block diagram of a digital program implementation of the PID controller.

interpreted as a time delay of T seconds, where T is the sampling period. In practice, this time delay is implemented by storing a variable in some convenient storage location in the computer and then taking it out after T seconds have elapsed. Once this relation is established, we can easily identify the digital program of any physically realizable transfer function. For the digital differentiator, the transfer function is written

$$G_D(z) = \frac{K_D}{T}(1 - z^{-1}) \qquad (8\text{-}199)$$

and for the digital integrator, it is

$$G_I(z) = \frac{K_I T z^{-1}}{1 - z^{-1}} \qquad (8\text{-}200)$$

Figure 8-53 shows a block diagram representation of the digital program of the PID controller in Fig. 8-51.

Digital Implementation of Lead and Lag Controllers

In principle, any continuous-data controller can be made into a digital controller simply by adding sample-and-hold units at the input and the output terminals. Figure 8-54 illustrates the scheme with $G_c(s)$ being the transfer function of the continuous-data controller and $G_c(z)$ the equivalent digital controller. The sampling period T should be sufficiently small so that the dynamic characteristics of the continuous-data controller are not lost through the digitization. The system configuration shown in Fig. 8-54 actually suggests that given the continuous-data controller $G_c(s)$, the equivalent digital control $G_c(z)$ can be obtained as shown. On

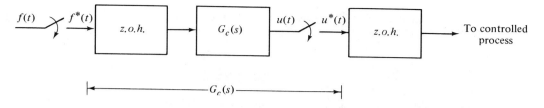

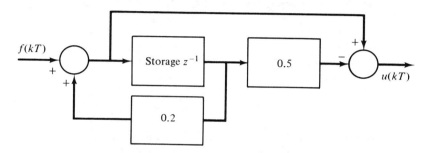

Figure 8-54 Realization of digital controller by an analog controller with sample-and-hold.

Figure 8-55 Digital program realization of Eq. (8-202).

the other hand, given the digital controller $G_c(z)$, we can realize it by using an analog controller $G_c(s)$ and sample-and-hold units, as shown in Fig. 8-54.

As an illustrative example, we consider that the continuous-data controller in Fig. 8-54 is represented by the transfer function

$$G_c(s) = \frac{s + 1}{s + 1.61} \tag{8-201}$$

The transfer function $G_c(z)$ is written

$$G_c(z) = (1 - z^{-1}) \mathcal{J}\left[\frac{s + 1}{s(s + 1.61)}\right] = \frac{z - 0.5}{z - 0.2} \tag{8-202}$$

Figure 8-55 shows the digital-program implementation of Eq. (8-202).

REFERENCES

1. J. G. TRUXAL, *Control Systems Synthesis*, McGraw-Hill Book Company, New York, 1950.

2. W. M. WONHAM, "On Pole Assignment in Multi-Input Controllable Linear Systems," *IEEE Trans. Automatic Control*, Vol. AC-12, pp. 660–665, Dec. 1967.

3. J. C. WILLEMS and S. K. MITTER, "Controllability, Observability, Pole Allocation, and State Reconstruction," *IEEE Trans. Automatic Control*, Vol. AC-16, pp. 582–95, Dec. 1971.

4. H. W. SMITH and E. J. DAVISON, "Design of Industrial Regulators," *Proc. IEE* (*London*), Vol. 119, pp. 1210–1216, Aug. 1972.

PROBLEMS

8.1. Find the transfer function of the controller $G_c(s)$ for the control system shown in Fig. 8P-1, so that the following specifications are satisfied:
   (a) The steady-state error due to a step function input is zero.
   (b) The characteristic equation of the closed-loop system is

$$s^3 + 4s^2 + 6s + 4 = 0$$

The transfer function of the controlled process is

$$G_p(s) = \frac{4}{s(s^2 + 4s + 6)}$$

8.2. For the system described in Problem 8-1, find the transfer function of the controller $G_c(s)$ so that the following specifications are satisfied:
   (a) The ramp error constant is 0.9.
   (b) The dominant roots of the third-order characteristic equation of the closed-loop system are at $-1 + j1$ and $-1 - j1$.

8.3. For the system described in Problem 8-1, the transfer function of the controller $G_c(s)$ is of the form

$$G_c(s) = \frac{K(1 + Ts)}{s}$$

   (a) Find the values of K and T so that the following specifications are satisfied:
      1. Ramp error constant $K_v = 4$.
      2. The magnitudes of the imaginary parts of the complex characteristic equation roots of the closed-loop system must not be greater than 4 rad/sec.
   (b) Sketch the root loci of the closed-loop characteristic equation with the value of K as determined above and $T \geq 0$. Determine the unit step response of the system with zero initial conditions.

8.4. A considerable amount of effort is being spent by automobile manufacturers to meet the exhaust emission performance standards of various governmental agencies. Modern automotive power plant systems consist of an internal combustion engine which has an internal cleanup device called a catalytic converter. Such a system requires control of such variables as engine air-fuel ratio (A/F), ignition spark timing, exhaust gas

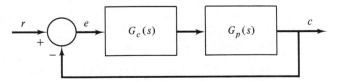

Figure 8P-1

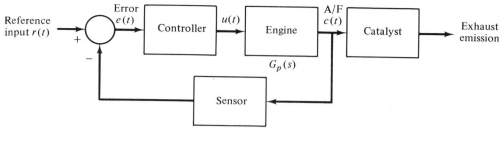

Figure 8P-4

recirculation, and injection air. The control system problem considered in this exercise deals with the control of the air-fuel ratio A/F. In general, depending on fuel composition and other factors, a typical stoichiometric A/F is 14.7 : 1, that is, 14.7 g of air to each gram of fuel. An A/F greater or less than stoichiometry will cause high hydrocarbons, carbon monoxide, and oxides of nitrogen in the tailpipe emission.

The control system whose block diagram is shown in Fig. 8P-4 is devised to control the air-fuel ratio so that a desired output variable is maintained for a given command signal. Refer to the block diagram in Fig. 8P-4, which shows that the sensor senses the composition of the exhaust-gas mixture entering the catalytic converter. The electronic controller detects the difference or the error between the command and the sensor signals and computes the control signal necessary to achieve the desired exhaust-gas composition. The output variable $c(t)$ denotes the effective air-fuel ratio A/F. The transfer function of the engine is given by

$$\frac{C(s)}{U(s)} = \frac{e^{-T_d s}}{1 + \tau s} = G_p(s)$$

where T_d is the time delay (0.2 sec). The time constant τ is 0.1 sec. The time-delay term $e^{-T_d s}$ can be represented by a power series,

$$e^{-T_d s} = \frac{1}{e^{T_d s}} = \frac{1}{1 + T_d s + T_d^2 s^2 / 2! + \cdots}$$

The power series in the denominator of this equation can then be truncated to approximate the time delay term. The transfer function of the sensor is considered to be unity.

(a) Approximate $e^{-T_d s}$ by just the first two terms of the power series; i.e.,

$$e^{-T_d s} \cong \frac{1}{1 + T_d s}$$

Find the transfer function $G_p(s) = C(s)/U(s)$. Decompose this transfer function and express the system between C and U by the following state equations:

$$\dot{\mathbf{x}} = \mathbf{A}\mathbf{x} + \mathbf{B}u$$

where $\mathbf{A}$ and $\mathbf{B}$ are in phase-variable canonical form. Find $\mathbf{A}$ and $\mathbf{B}$.

(b) Repeat part (a) by approximating $e^{-T_d s}$ by the first three terms of the power series.

(c) Assuming that there is no controller in the system; i.e., $u = e$, find the characteristic equation of the closed-loop system (overall system between r and c) and the closed-loop eigenvalues. Do this for parts (a) and (b).

(d) Define the state feedback control as

$$u = -\mathbf{G}\mathbf{x} + r$$

where $\mathbf{x}$ is the state vector you defined in parts (a) and (b); r is the reference input, and $\mathbf{G}$ is the feedback matrix and is of the form:

$$\mathbf{G} = [\, g_1 \quad g_2 \quad g_3 \quad \cdots \quad g_n \,]$$

For part (a) where you have a second-order system, $n = 2$, find the elements of $\mathbf{G}$ so that the eigenvalues of the closed-loop system are at -5 and -5. With this feedback control, find the steady-state value of $c(t)$ when the input $r(t)$ is a unit-step function.

(e) Repeat part (d) for the system in part (b). (Now you have a third-order system so that $\mathbf{G} = [\, g_1 \quad g_2 \quad g_3 \,]$). The desired closed-loop eigenvalues are at $-5, -5 + j5, -5 - j5$.

8.5. For the automobile emission control system described in Problem 8.4, let the time delay term be approximated by the first two terms of the power series, as in part (a). Let the controller be described by $G_c(s) = K = $ constant. Find the value of K so that the damping ratio of the closed-loop system is 0.707. With this value of K, what is the steady-state error of the system when the reference input $r(t)$ is a unit step function? Now let the controller be a PID controller with the transfer function

$$G_c(s) = K_P + K_D s + \frac{K_1}{s}$$

Find the values of K_P, K_D, and K_I so that the roots of the characteristic equation of the closed-loop system are at -50, $-5 + j5$, and $-5 - j5$. Can you approximate the system just designed by a second-order system? If so, what is the equivalent closed-loop transfer function of this second-order system? Find the peak overshoot, if any, and the peak time T_{max} of the second-order system. For the system with the PID controller, find the steady-state error when the input is a unit step function, and when the input is a unit ramp function.

8.6. The telescope for tracking stars and asteroids on the space shuttle may be modeled as a pure mass M. It is suspended by magnetic bearings so that there is no friction, and its attitude is controlled by magnetic actuators located at the base of the payload. The dynamic model for the control of the z-axis motion is shown in Fig. 8P-6(a). The control of the rotation motion is independent and is not considered here. The linear spring shown is used to model the wire cable attachment, which exerts a spring force on the mass. Since there are electrical components on the telescope, electric power must be brought to the telescope through the cable. The force produced by the magnetic actuators is denoted by $f(t)$. The force equation of motion is

$$f(t) - K_s z(t) = M\frac{d^2 z(t)}{dt^2}$$

where $K_s = 1$ N-m and $M = 100$ kg (all units are consistent).

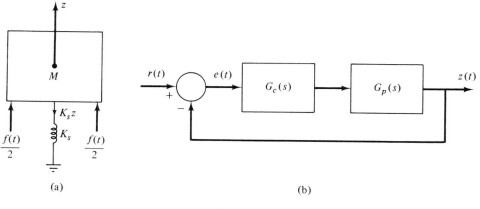

Figure 8P-6

(a) Show that the natural response of the system is oscillatory without damping. Find the natural undamped frequency of the system.

(b) In order to stabilize the system, one control engineer suggests the use of state feedback through constant feedback gains. Let $x_1 = z$ and $x_2 = dz/dt$, and

$$f(t) = r(t) - g_1 x_1 - g_2 x_2$$

where $r(t)$ is the reference input. Find the ranges of the values of g_1 and g_2 so that the closed-loop system will not have any characteristic equation roots in the right-half s-plane or on the imaginary axis on the s-plane. What is the steady-state error in $z(t)$ when the reference input $r(t)$ is a unit step function?

(c) It is suggested that instead of the state feedback, a PID controller of the form

$$G_c(s) = K_P + K_D s + \frac{K_I}{s}$$

be used as a cascade controller, as shown in Fig. 8P-6(b). Find the values of K_P, K_D, and K_I so that all of the following performance criteria are satisfied.

(1) Zero steady-state error to a step input.

(2) Steady-state error to a ramp input is equal to 1 percent of the magnitude of the ramp input.

(3) The quadratic roots of the closed-loop characteristic equation correspond to a relative damping ratio of 0.707 and a natural undamped frequency of 1 rad/sec.

8.7. An inventory control system is modeled by the following state equations:

$$\frac{dx_1}{dt} = -2x_2$$

$$\frac{dx_2}{dt} = -2u$$

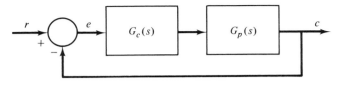

Figure 8P-7

where x_1 = level of inventory
      x_2 = rate of sales of product
      u = production rate
The output equation is $c(t) = x_1(t)$.

(a) Let 1 unit of time be 1 day. Given the initial level of inventory to be 10,000 units and the initial rate of sales of product to be 100 units per day, if the production rate is zero ($u = 0$) for all t, determine how many days it would take to deplete the entire inventory.

(b) If the production rate u is constant and is 10 units per day, with the same initial conditions as given in part (a), find x_1 and x_2 for all t greater than or equal to zero.

(c) It is desired to control the inventory by using a reference command input and closing the loop. One engineer suggests the system configuration shown in Fig. 8P-7, where r = reference input,

$$G_p(s) = \frac{C(s)}{U(s)}$$

where $C(s)$ and $U(s)$ are the Laplace transforms of $c(t)$ and $u(t)$, respectively. The controller is the PD controller,

$$G_c(s) = K_P + K_D s$$

Find the values of K_P and K_D so that the following two conditions are satisfied simultaneously.
(1) $e_{ss} = \lim_{t \to \infty} e(t) = 0$ when $r(t)$ = unit step function.
(2) The roots of the characteristic equation of the closed-loop system are at -2 and -2.

(d) Another engineer suggests the use of state-feedback control. Instead of using the PD-controller as in part (c), he suggests that the control u be of the form

$$u = -\mathbf{G}\mathbf{x} + r$$

where $\mathbf{G} = [g_1 \quad g_2]$; g_1 and g_2 are real constants. Find the values of g_1 and g_2 so that the two conditions stated in part (c) are satisfied.

8.8. Consider the liquid-level control system of Problem 6-27, Fig. 6P-27. It is found from Problem 6-27 that when $N = 8$, the closed-loop system is stable, but the unit step response has a high overshoot and is quite oscillatory. This problem deals with the design of controls to improve the performance of the system.

(a) For $N = 8$, apply state feedback control; let $G_c(s) = 1$ in Fig. 6P-27,

$$e_i(t) = r(t) - \mathbf{G}\mathbf{x}(t)$$

where

$$\mathbf{x}(t) = \begin{bmatrix} h(t) \\ \theta_m(t) \\ \dot{\theta}_m(t) \end{bmatrix} \qquad \mathbf{G} = [g_1 \quad g_2 \quad g_3]$$

Find $\mathbf{G}$ such that the steady-state value of $h(t)$ is unity when $r(t) = $ unit step function, and the two complex characteristic equation roots of the closed-loop system are at $-1 + j1$ and $-1 - j1$. The third root is unspecified but is constrained. Plot the time response of $h(t)$ when $r(t) = $ unit step function. The initial conditions are zero.

(b) Instead of using state feedback, the controller $G_c(s)$ is to be a PD controller,

$$G_c(s) = K_P + K_D s$$

Find the values of K_P and K_D such that the two complex eigenvalues of the closed-loop system are at $-1 + j1$ and $-1 - j1$, and the output $h(t)$ must follow a unit step input without steady-state error. Plot the time response of $h(t)$ when $r(t) = $ unit step function. The initial conditions are zero.

8.9. Figure 8P-9 shows the block diagram of the speed control system for an electrical power generating system. The speed governor valve controls the steam flow input to the turbine. The turbine drives the generator, which puts out electric power at a frequency proportional to the generator speed ω_g. The desired steady-state speed of the generator is 1200 rpm, at which the generated output voltage is 60 Hz. $J = 100$.

(a) Let the speed governor valve gain K be set at 10 rad/V. Determine the tachometer gain so that the complex eigenvalues of the closed-loop system correspond to a damping ratio of 0.707. Sketch the root loci as a function of K_T and indicate the location of the roots with the desired damping.

(b) Determine the desired reference input voltage, with the value of K_T set at the value determined in part (a), so that the generator speed is 1200 rpm ($T_L = 0$).

(c) In Fig. 8P-9, T_L denotes a load change. Determine the percent change in the steady-state speed due to a constant load change when K_T is as determined in part (a).

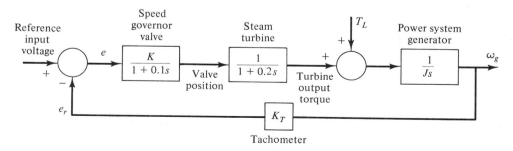

Figure 8P-9

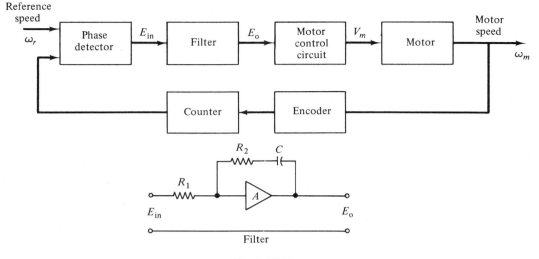

Figure 8P-10

(d) It is desired to keep the frequency variation due to load change within ± 0.1 percent. At the same time, the relative damping of the complex eigenvalues of the overall system must be approximately 0.707. Can both requirements be satisfied by only changing the values of K and K_T? If not, design a series controller in the forward path for this purpose. Sketch the root locus for the compensated system with $K = 10$, with K_T as the variable parameter.

8.10. The phase-lock loop technique is a popular method for *dc* motor speed control. A basic phase-lock loop motor speed control system is shown in Fig. 8P-10. The encoder produces digital pulses that represent motor speed. The pulse train from the encoder is compared with the reference frequency by a phase comparator or detector. The output of the phase detector is a voltage that is proportional to the phase difference between the reference speed and the actual motor speed. This error voltage upon filtering is used as the control signal for the motor. The system parameters and transfer functions are as follows:

Phase detector	$K_p = 0.0609$ V/rad
Motor control circuit gain	$K_a = 1$
Motor transfer function	$\dfrac{\omega_m(s)}{V_m(s)} = \dfrac{K_m}{s(1 + T_m s)}$ $\quad (K_m = 10,\ T_m = 0.04)$
Encoder gain	$K_e = 5.73$ pulses/rad
Filter transfer function	$\dfrac{E_0}{E_{in}} = \dfrac{R_2 Cs + 1}{R_1 Cs}$
	$R_1 = 1.745 \times 10^6\ \Omega,\ C_1 = 1\ \mu F$
Counter	$\dfrac{1}{N} = 1$

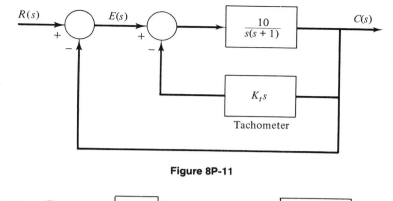

Figure 8P-11

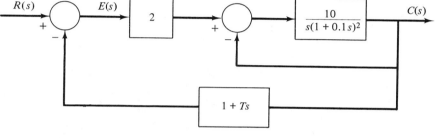

Figure 8P-12

Determine the value of R_2 so that the complex roots of the closed-loop system have a damping ratio of approximately 0.707. If the problem has more than one solution, use the one that corresponds to the smallest system bandwidth. Sketch the root locus diagram for the characteristic equation roots as R_2 varies.

8.11. Tachometer feedback is employed frequently in feedback control systems for the purpose of stabilization. Figure 8P-11 shows the block diagram of a system with tachometer feedback. Choose the tachometer gain constant K_t so that the relative damping ratio of the system is 70.7 percent.

8.12. The block diagram of a control system is shown in Fig. 8P-12. By means of the root contour method, show the effect of variation in the value of T on the location of the closed-loop poles of the system.

8.13. A computer-tape-drive system utilizing a permanent-magnet dc motor is shown in Fig. 8P-13(a). The system is modeled by the diagram shown in Fig. 8P-13(b). The constant K_L represents the spring constant of the elastic tape, and B_L denotes the viscous frictional coefficient between the tape and the capstans. The system parameters are as follows:

$$\frac{K_a}{R_a} = 36 \text{ oz-in.}/\text{V} \qquad J_m = 0.023 \text{ oz-in.-sec}^2$$

$$K_e = 6.92 \text{ oz-in.}/\text{rad}/\text{sec} \qquad K_L = 2857.6 \text{ oz-in.}/\text{rad}$$

$$B_L = 10 \text{ oz-in.-sec} \qquad J_L = 7.24 \text{ oz-in.-sec}^2$$

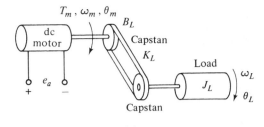

(a)

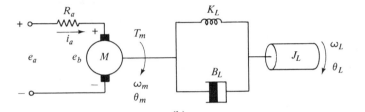

(b)

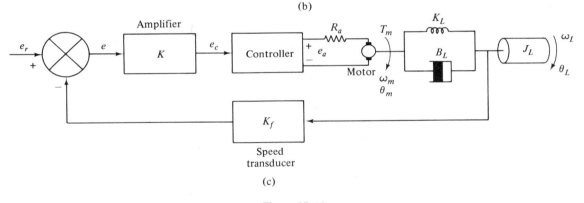

(c)

Figure 8P-13

where K_b = back-emf constant, $K_e = K_b K_a/R_a + B_m$, K_a = torque constant in oz-in./A, and B_m = motor viscous friction constant.

(a) Write the state equations of the system shown in Fig. 8P-13(b) with θ_L, ω_L, θ_m, and ω_m as the state variables in the prescribed order. Derive the transfer functions

$$\frac{\omega_m(s)}{E_a(s)} \quad \text{and} \quad \frac{\omega_L(s)}{E_a(s)}$$

(b) The objective of the system is to control the speed of the load, ω_L, accurately. Figure 8P-13(c) shows a closed-loop system in which the load speed is fed back through a speed transducer and compared with the reference input, with $K_f = 0.01$. Design a controller and select the amplifier gain so that the following specifications are satisfied: (1) no steady-state speed error when the input e_r is a step function; (2) the dominant roots of the characteristic equation correspond to a damping ratio of approximately 0.707; and (3) what should be the value of the input e_r if the

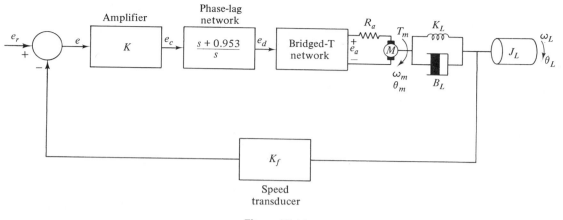

Figure 8P-14

steady-state speed is to be 100 rpm? Sketch the root loci of the designed system with K as the variable parameter.

8.14. The computer-tape-drive system described in Problem 8-14 has the following system parameters:

$$\frac{K_a}{R_a} = 36 \text{ oz-in./V} \qquad\qquad J_m = 0.023 \text{ oz-in.-sec}^2$$

$$K_e = \frac{K_b K_a}{R_a} + B_m = 6.92 \text{ oz-in./rad/sec} \qquad B_L = 0$$

$$K_L = 28{,}576 \text{ oz-in./rad} \qquad\qquad K_f = 0.01$$

$$J_L = 7.24 \text{ oz-in.-sec}^2$$

(a) Show that the closed-loop system without compensation has an oscillatory response in ω_L for any positive K.

(b) In order to reduce the speed oscillations, a bridged-T network is proposed to cancel the undesirable poles of the open-loop transfer function. Figure 8P-14 shows the block diagram of the overall system. The phase-lag controller preceding the bridged-T controller is for the purpose of ensuring zero steady-state speed error when a step input is applied. Determine the transfer function of the bridged-T network and the amplifier gain so that the dominant roots of the characteristic equation correspond to a damping ratio of approximately 70.7 percent.

8.15. The block diagram of a control system with state feedback is shown in Fig. 8P-15. The parameters g_1, g_2, and g_3 are real constants.
(a) Find the values of g_1, g_2, and g_3 so that:
   (1) The steady-state error $e_{ss} = \lim\limits_{t \to \infty} e(t)$ is zero when $r(t)$ is a unit step function.
   (2) The roots of the characteristic equation of the closed-loop system are at -1, -1, and -1.
(b) Instead of using state feedback, a cascade controller is implemented as shown in Fig. 8P-15(b). Find the transfer function for $G_c(s)$ in terms of the g_1, g_2, and g_3 you found in part (a) and the other system transfer functions.

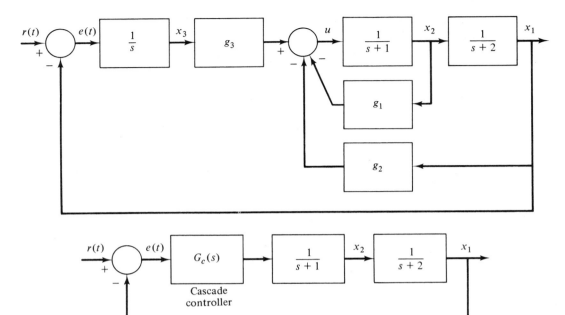

Figure 8P-15

8.16. The state equations of the linear perturbed idle-speed control system are given in part
(b) of Problem 5.25.
   (a) With $u = 0$, find the time response of ω when $T_D = 100$. What is the steady-state
   value of ω?
   (b) Apply state feedback,

$$u = -Gy = -[g_1 \quad g_2 \quad g_3 \quad g_4]y$$

   and find the elements of G so that the eigenvalues of the closed-loop system are at
   $-50, -50, -100, -100$. Repeat part (a).

8.17. The ball-suspension control system described in Problem 4-18 has the following
parameters:

$$R = 1, \quad L = 0.001, \quad M = 1, \quad g = 32.2, \quad K = 1, \quad E_{eq} = 10$$

The linearized state equations are expressed as

$$\delta \dot{x} = A^* \delta x + B^* \delta e$$

(a) Find the matrix G^* such that the state feedback

$$\delta e = -G^* \delta x$$

causes the closed-loop system $\delta \dot{x} = (A^* - B^*G^*) \delta x$ to have eigenvalues at -1,
$-1 + j1$, and $-1 - j1$.

(b) Find the time responses of $\delta i(t)$ and $\delta y(t)$, where $\delta i(t) = \delta x_1(t)$ and $\delta y(t) = \delta x_2(t)$ with the initial state

$$\delta x(0) = [0 \quad 0.1 \quad 0]$$

8.18. For the ball-suspension control system described in Problem 4.19,

(a) Find the matrix G^* so that the state feedback

$$\delta i = -G^* \, \delta x$$

places the eigenvalues of the linearized closed-loop system

$$\delta \dot{x} = (A^* - B^* G^*) \, \delta x$$

at -1, -1, -3, and -4. Plot the time responses of $\delta x_1(y_1)$ and $\delta x_3(y_2)$ with the initial condition

$$\delta x(0) = [0.1 \quad 0 \quad 0 \quad 0]$$

and then with

$$\delta x(0) = [0 \quad 0 \quad 0.2 \quad 0]$$

(b) Repeat part (a) by placing the eigenvalues at -10, -20, $-1 + j1$, and $-1 - j1$.

(c) Comment on the responses of the closed-loop systems with the two sets of initial states used in (b).

8.19. The temperature $x(t)$ in the electric furnace shown in Fig. 8P-19 is governed by the differential equation

$$\frac{dx(t)}{dt} = -x(t) + u(t) + w_2(t)$$

where $u(t)$ is the control and $w_2(t)$ is an unknown constant disturbance due to heat losses. It is desired that the equilibrium temperature follow a reference input $w_1 =$ constant. Design a control with state and dynamic feedback such that the roots of the characteristic equation of the closed-loop system are at $s = -1$, and

$$\lim_{t \to \infty} x(t) = w_1$$

Sketch $x(t)$ for $t \geq 0$ for $w_1 = 1$, $w_2 = -1$ and with zero initial conditions.

8.20. For the controlled process shown in Fig. 8P-20, design state and dynamic feedback control so that the state variable x_1 will follow a reference input $w_1 =$ constant as t approaches infinity. The noise signals w_2 and w_3 are unknown constants. The roots of the characteristic equation of the closed-loop system should all be at $s = -3$.

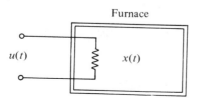

Furnace

$u(t)$

$x(t)$

Figure 8P-19

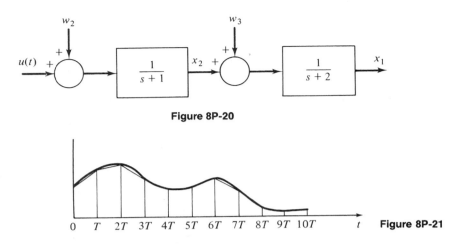

Figure 8P-20

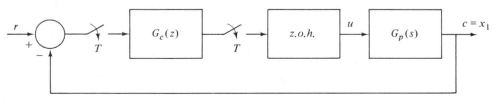

Figure 8P-21

8.21. The polygonal integration is featured by connecting the sampled values of a signal by straight lines, and then evaluating the areas under the polygonal waveforms, as shown in Fig. 8P-21. Show that the z-transfer function of the digital implementation of the polygonal integration is

$$G_I(z) = \frac{K_I T(z + 1)}{2(z - 1)}$$

where K_I is the constant of integral control. Show that

$$\lim_{T \to 0} G_I(z) = \frac{K_I}{s}$$

8.22. Consider the inventory control system described in Problem 8.7. The system is subject to digital control, and the block diagram of the overall system is shown in Fig. 8P-22. The controller is the digital PD controller with the transfer function

$$G_c(z) = K_P + \frac{K_D(z - 1)}{Tz}$$

(a) Find K_P and K_D in terms of the sampling period T so that two of the roots of the closed-loop characteristic equation are at $z = 0.5$ and 0.5. Find the other characteristic equation root.

(b) For $r(t) =$ unit step input, $T = 1$, find $c(kT)$ for $k = 0, 1, 2 \ldots$.

Figure 8P-22

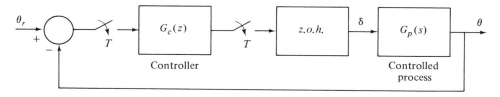

Figure 8P-23

8.23. Consider the missile attitude control system of Problem 4.11, part (b). The transfer function between the thrust angle δ and the angle of attack θ is

$$\frac{\Theta(s)}{\delta(s)} = \frac{T_s d_2}{J(s^2 - a)}$$

where $T_s d_2/J = 10$ and $a = 1$.

The block diagram of the digital control system is shown in Fig. 8P-23. The transfer function of the digital controller is

$$G_c(z) = \frac{z + z_1}{z + z_2}$$

The sampling period is 0.1 sec.

(a) Let $z_1 = 0.905$; find z_2 so that the characteristic equation of the closed-loop system has two equal real roots.

(b) For θ_r = unit step input, find $\theta(kT)$ for $k = 0, 1, 2, \ldots$.

chapter nine

Frequency-Domain Analysis of Control Systems

9.1 INTRODUCTION

It was pointed out earlier that in practice the performance of a control system is more realistically measured by its time-domain response characteristics. This is in contrast to the analysis and design of communication systems for which the frequency response is of more importance, since in this case most of the signals to be processed are either sinusoidal or can be represented by sinusoidal components. We have learned from Chapter 6 that analytically the time response of a control system is usually difficult to determine, especially in the case of high-order systems. In design problems the difficulties lie in the fact that there are no unified methods of arriving at a designed system given the time-domain specifications, such as peak overshoot, rise time, delay time, settling time, and so on. On the other hand, there are a wealth of graphical methods available in the frequency domain, all suitable for the analysis and design of linear control systems. It is important to realize that once the analysis and design are carried out in the frequency domain, the time-domain properties of the system can be interpreted based on the relationships that exist between the time-domain and the frequency-domain characteristics. Therefore, we may consider that the primary motivation of conducting control systems analysis and design in the frequency domain is because of convenience and the availability of the existing analytical tools.

The starting point of the frequency-domain analysis is the transfer function. First, we discuss the transfer function notation of control systems based on the state-variable representation and then on the classical approach.

In Section 3.3 the transfer function of a multivariable closed-loop system is derived. Referring to Fig. 3-6, the closed-loop transfer function matrix relation is written [Eq. (3-38)]

$$\mathbf{C}(s) = [\mathbf{I} + \mathbf{G}(s)\mathbf{H}(s)]^{-1}\mathbf{G}(s)\mathbf{R}(s) \tag{9-1}$$

where $\mathbf{C}(s)$ is a $q \times 1$ vector and $\mathbf{R}(s)$ is a $p \times 1$ vector. The closed-loop transfer function matrix is defined as

$$\mathbf{M}(s) = [\mathbf{I} + \mathbf{G}(s)\mathbf{H}(s)]^{-1}\mathbf{G}(s) \tag{9-2}$$

which is a $q \times p$ matrix.

Under the sinusoidal steady state, we set $s = j\omega$; then Eq. (9-2) becomes

$$\mathbf{M}(j\omega) = [\mathbf{I} + \mathbf{G}(j\omega)\mathbf{H}(j\omega)]^{-1}\mathbf{G}(j\omega) \tag{9-3}$$

The ijth element of $\mathbf{M}(j\omega)$ is defined as

$$M_{ij}(j\omega) = \frac{C_i(j\omega)}{R_j(j\omega)}\bigg|_{\text{all other inputs} = 0} \tag{9-4}$$

where i represents the row and j the column of $\mathbf{M}(j\omega)$.

State-Variable Representation

For a system that is represented by state equations,

$$\dot{\mathbf{x}}(t) = \mathbf{A}\mathbf{x}(t) + \mathbf{B}\mathbf{u}(t) \tag{9-5}$$
$$\mathbf{c}(t) = \mathbf{D}\mathbf{x}(t) + \mathbf{E}\mathbf{u}(t) \tag{9-6}$$

For the present case we assume that the feedback is described by

$$\mathbf{u}(t) = \mathbf{r}(t) - \mathbf{H}\mathbf{c}(t) \tag{9-7}$$

where $\quad \mathbf{x}(t) = n \times 1$ state vector
$\qquad \mathbf{u}(t) = p \times 1$ control vector
$\qquad \mathbf{c}(t) = q \times 1$ output vector
$\qquad \mathbf{r}(t) = p \times 1$ input vector

$\mathbf{A}, \mathbf{B}, \mathbf{D}$, and $\mathbf{E}$ are constant matrices of appropriate dimensions, and $\mathbf{H}$ is the $p \times q$ feedback matrix. The transfer function relation of the system is

$$\mathbf{C}(s) = \left[\mathbf{D}(s\mathbf{I} - \mathbf{A})^{-1}\mathbf{B} + \mathbf{E}\right]\mathbf{U}(s) \tag{9-8}$$

The open-loop transfer function matrix is defined as

$$\mathbf{G}(s) = \mathbf{D}(s\mathbf{I} - \mathbf{A})^{-1}\mathbf{B} + \mathbf{E} \tag{9-9}$$

The closed-loop transfer function relation is described by the equation

$$\mathbf{C}(s) = [\mathbf{I} + \mathbf{G}(s)\mathbf{H}]^{-1}\mathbf{G}(s)\mathbf{R}(s) \qquad (9\text{-}10)$$

Thus,

$$\mathbf{M}(s) = [\mathbf{I} + \mathbf{G}(s)\mathbf{H}]^{-1}\mathbf{G}(s) \qquad (9\text{-}11)$$

It should be noted that the matrix $\mathbf{H}$ in Eq. (9-11) has only constant elements.

In general, the elements of the transfer function matrices are rational functions of s. In Chapter 5 it is proved that if the system is completely controllable and observable, there will be no pole-zero cancellations in the transfer functions. Under this condition the poles of the transfer function will also be the eigenvalues of the system.

The analysis techniques in the frequency domain discussed in the following sections are conducted with the single-variable notation. Because linear systems satisfy the principle of superposition, these basic techniques can all be applied to multivariable systems.

For a single-loop feedback system, the closed-loop transfer function is written

$$M(s) = \frac{C(s)}{R(s)} = \frac{G(s)}{1 + G(s)H(s)} \qquad (9\text{-}12)$$

Under the sinusoidal steady state, we set $s = j\omega$; then Eq. (9-12) becomes

$$M(j\omega) = \frac{C(j\omega)}{R(j\omega)} = \frac{G(j\omega)}{1 + G(j\omega)H(j\omega)} \qquad (9\text{-}13)$$

The sinusoidal steady-state transfer relation $M(j\omega)$, which is a complex function of ω, may be expressed in terms of a real and an imaginary part; that is,

$$M(j\omega) = \text{Re}[M(j\omega)] + j\,\text{Im}[M(j\omega)] \qquad (9\text{-}14)$$

Or, $M(j\omega)$ can be expressed in terms of its magnitude and phase as

$$M(j\omega) = M(\omega)\underline{/\phi_m(\omega)} \qquad (9\text{-}15)$$

where

$$M(\omega) = \left| \frac{G(j\omega)}{1 + G(j\omega)H(j\omega)} \right| \qquad (9\text{-}16)$$

and

$$\phi_m(\omega) = \underline{/\dfrac{G(j\omega)}{1 + G(j\omega)H(j\omega)}}$$

$$= \underline{/G(j\omega)} - \underline{/1 + G(j\omega)H(j\omega)} \qquad (9\text{-}17)$$

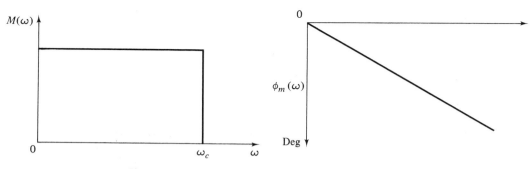

Figure 9-1 Gain-phase characteristics of an ideal low-pass filter.

Since the analysis is now in the frequency domain, some of the terminology used in communication systems may be applied to the present control system characterization. For instance, $M(\omega)$ of Eq. (9-16) may be regarded as the magnification of the feedback control system. The significance of $M(\omega)$ to a control system is similar to the gain or amplification of an electronic amplifier. In an audio amplifier, for instance, an ideal design criterion is that the amplifier must have a flat gain for all frequencies. Of course, realistically, the design criterion becomes that of having a flat gain in the audio frequency range. In control systems the ideal design criterion is similar. If it is desirable to keep the output $C(j\omega)$ identical to the input $R(j\omega)$ at all frequencies, $M(j\omega)$ must be unity for all frequencies. However, from Eq. (9-13) it is apparent that $M(j\omega)$ can be unity only when $G(j\omega)$ is infinite, while $H(j\omega)$ is finite and nonzero. An infinite magnitude for $G(j\omega)$ is, of course, impossible to achieve in practice, nor would it be desirable, since most control systems become unstable when its loop gain becomes very high. Furthermore, all control systems are subjected to noise. Thus, in addition to responding to the input signal, the system should be able to reject and suppress noise and unwanted signals. This means that the frequency response of a control system should have a cutoff characteristic in general, and sometimes even a band-pass characteristic.

The phase characteristics of the frequency response are also of importance. The ideal situation is that the phase must be a linear function of frequency within the frequency range of interest. Figure 9-1 shows the gain and phase characteristics of an ideal low-pass filter, which is impossible to realize physically. Typical gain and

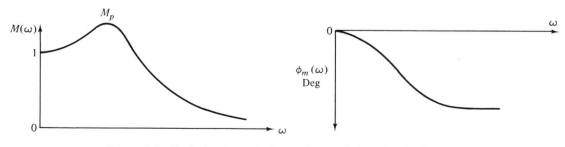

Figure 9-2 Typical gain and phase characteristics of a feedback control system.

phase characteristics of a feedback control system are shown in Fig. 9-2. The fact is that the great majority of control systems have the characteristics of a low-pass filter, so the gain decreases as the frequency increases.

9.2 NYQUIST STABILITY CRITERION

In Section 9.1 we introduced the idea that the frequency-domain analysis of a linear control system is based on the closed-loop transfer functions, such as those of Eq. (9-11) or Eq. (9-12). In the analysis problem we are again involved in determining the performance characteristics of the system, and the one of utmost importance is stability. Thus, we are interested in the location of the characteristic equation roots which are the poles of the closed-loop transfer function.

Thus far, two methods of determining stability by investigating the location of the roots of the characteristic equation have been indicated:

1. The roots of the characteristic equation are actually solved. These roots can actually be plotted as a function of some system parameter in the form of a root locus diagram. Thus, it is possible to study not only the absolute stability but also the relative stability of a closed-loop system via the location of the characteristic equation roots in the s-plane.

2. The relative positions of the characteristic equation roots with respect to the imaginary axis of the s-plane are determined by means of the Routh–Hurwitz criterion.

The basic difference between these two methods is that the root locus is generally more versatile, and is useful for analysis as well as design problems, whereas the Routh–Hurwitz criterion gives only an absolute answer on stability. Another important difference to realize is that the starting point of the root locus method is the loop transfer function of the closed-loop system, whereas for the Routh–Hurwitz test it is the characteristic equation.

The Nyquist criterion is a frequency-domain method which has the following features that make it desirable for the analysis as well as the design of control systems:

1. It provides the same amount of information on the absolute stability of a control system as does the Routh–Hurwitz criterion.

2. In addition to absolute system stability, the Nyquist criterion indicates the degree of stability of a stable system and gives an indication of how the system stability may be improved, if needed.

3. It gives information on the frequency-domain response of the system.

4. It can be used for a stability study of systems with time delay.

5. It can be modified for nonlinear systems.

We can formulate the Nyquist criterion by referring to the closed-loop transfer function matrix of Eq. (9-2), which may represent any linear time-invariant system.

The stability of the system can be studied by investigating the poles of the closed-loop transfer function matrix $\mathbf{M}(s)$. For asymptotic stability, all the poles of the transfer function matrix $\mathbf{M}(s)$ must be located in the left-half s-plane.

From Eq. (9-2) we see that the poles of $\mathbf{M}(s)$ are from the roots of the equation

$$|\mathbf{I} + \mathbf{G}(s)\mathbf{H}(s)| = 0 \tag{9-18}$$

as well as from the poles of $\mathbf{G}(s)$ that are not canceled by the poles of $\mathbf{G}(s)\mathbf{H}(s)$. The latter is possible if some of the poles of $\mathbf{G}(s)$ are canceled by the zeros of $\mathbf{H}(s)$. We can state that for asymptotic stability, all the roots of the characteristic equation must lie in the left-half s-plane. If all the poles of $\mathbf{G}(s)$ are in the left-half s-plane, then we can regard Eq. (9-18) as the characteristic equation for stability studies. In general, we must investigate all the poles of $\mathbf{M}(s)$, or we may define the characteristic equation from the Δ of Mason's gain formula, Eq. (3-79),

$$\Delta(s) = 1 + F(s) = 0 \tag{9-19}$$

where $F(s)$ is generally a rational function of s and is dependent only on the loops of the system. For a system with a single loop, Eq. (9-19) is known to be of the form

$$1 + G(s)H(s) = 0 \tag{9-20}$$

In short, the Nyquist criterion is a graphical method of determining the stability of a closed-loop system by investigating the properties of the frequency-domain plots of $F(s)$ or of $G(s)H(s)$.

Before embarking on the fundamentals of the Nyquist criterion, it is essential to summarize the pole-zero relationships with respect to the system functions.

1. Identification of poles and zeros:

   loop transfer function zeros = zeros of $F(s)$

   loop transfer function poles = poles of $F(s)$

   closed-loop transfer function poles = zeros of $1 + F(s)$

   = roots of the characteristic equation

2. The poles of $1 + F(s)$ are the same as the poles of $F(s)$, that is, the loop transfer function.

3. For a closed-loop system to be asymptotically stable, there is no restriction on the location of the poles and zeros of the loop transfer function $F(s)$, but the poles of the closed-loop transfer function or the roots of the characteristic equation must all be located in the left half of the s-plane.

"Encircled" versus "Enclosed"

It is important to distinguish between the concepts of *encircled* and *enclosed*, which are used frequently with the interpretation of the Nyquist criterion.

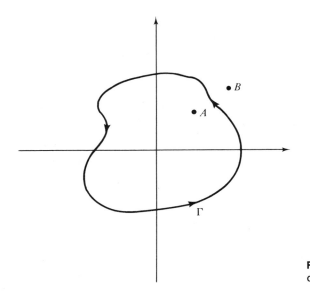

Figure 9-3 Definition of encirclement.

Encircled.   *A point is said to be encircled by a closed path if it is found inside the path.* For example, point *A* in Fig. 9-3 is encircled by the closed path Γ, since *A* is found inside the closed path. The point *B* is not encircled by the closed path Γ, since it is outside the path. In the application of the Nyquist criterion, the closed path that corresponds to the frequency-domain plot of a transfer function usually has a direction associated with it. As shown in Fig. 9-3, point *A* is said to be encircled by Γ in the counterclockwise direction.

When considering all the points inside the closed path, we can say that the region inside the closed path is encircled in the prescribed direction.

Enclosed.   *A point or region is said to be enclosed by a closed path if it is found to lie to the left of the path when the path is traversed in the prescribed direction.* For

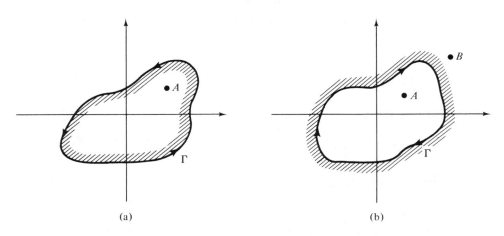

(a)   (b)

Figure 9-4   Definition of enclosed points and regions. (a) Point *A* is enclosed by Γ. (b) Point *A* is not enclosed but *B* is enclosed by the locus Γ.

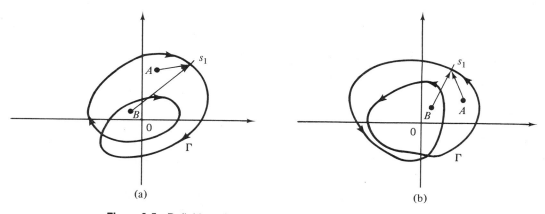

Figure 9-5 Definition of the number of encirclement and enclosure.

instance, the shaded regions shown in Fig. 9-4(a) and (b) are considered to be enclosed by the closed path Γ. In other words, point A in Fig. 9-4(a) is enclosed by Γ, but point A in Fig. 9-4(b) is not. However, in Fig. 9-4(b), point B and all the points in the region outside Γ are considered to be enclosed.

Number of Encirclement and Enclosure. When point A is encircled or enclosed by a closed path, a number N may be assigned to the number of encirclement or enclosure, as the case may be. The value of N may be determined by drawing a vector from A to any arbitrary point s_1 on the closed path Γ and then let s_1 follow the path in the prescribed direction until it returns to the starting point. The total *net* number of revolutions traversed by this vector is N. For example, point A in Fig. 9-5(a) is encircled *once* by Γ, and point B is encircled twice, all in the clockwise direction. Point A in Fig. 9-5(b) is enclosed once; point B is enclosed twice.

Principle of the Argument

The Nyquist criterion was originated as an engineering application of the well-known principle of the argument in complex variable theory. The principle is stated as follows, in a heuristic manner. Let $\Delta(s)$ be a single-valued rational function that is analytic everywhere in a specific region except at a finite number of points in the s-plane. For each point at which $\Delta(s)$ is analytic in the specified region in the s-plane, there is a corresponding point in the $\Delta(s)$-plane.

Suppose that a continuous closed path Γ_s is arbitrarily chosen in the s-plane, as shown in Fig. 9-6(a). If all the points on Γ_s are in the specified region in which $\Delta(s)$ is analytic, then curve Γ_Δ mapped by the function $\Delta(s)$ into the $\Delta(s)$-plane is also a closed one, as shown in Fig. 9-6(b). If, corresponding to point s_1 in the s-plane, point $\Delta(s_1)$ is located in the $\Delta(s)$-plane, then as the Γ_s locus is traversed starting from point s_1 in the arbitrarily chosen direction (clockwise) and then returning to s_1 after going through all the points on the Γ_s locus [as shown in Fig. 9-6(a)], the corresponding Γ_Δ locus will start from point $\Delta(s_1)$ and go through

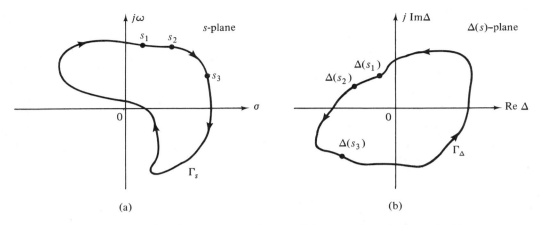

Figure 9-6 (a) Arbitrarily chosen closed path in the *s*-plane. (b) Corresponding locus Γ in the $F(s)$-plane.

points $\Delta(s_2)$ and $\Delta(s_3)$, which correspond to s_2 and s_3, respectively, and return to the starting point, $\Delta(s_1)$. The direction of traverse of Γ_Δ may be either clockwise or counterclockwise; that is, in the same direction or the opposite direction as that of Γ_s, depending on the particular function $\Delta(s)$. In Fig. 9-6(b) the direction of Γ_Δ is shown, for illustration purposes, to be counterclockwise.

It should be pointed out that, although the mapping from the *s*-plane to the $\Delta(s)$-plane is one to one for a rational function $\Delta(s)$, the reverse process is usually not a one-to-one mapping. For example, consider the function

$$\Delta(s) = \frac{K}{s(s+1)(s+2)} \tag{9-21}$$

which is analytic in the finite *s*-plane except at the points $s = 0$, -1, and -2. For each value of s in the finite *s*-plane other than the three points 0, -1, and -2, there is only one corresponding point in the $\Delta(s)$-plane. However, for each point in the $\Delta(s)$-plane, the function maps into three corresponding points in the *s*-plane. The simplest way to illustrate this is to write Eq. (9-21) as

$$s(s+1)(s+2) = \frac{K}{\Delta(s)} \tag{9-22}$$

The left side of Eq. (9-22) is a third-order equation, which has three roots when $\Delta(s)$ is chosen to be a constant.

The principle of the argument can be stated: *Let $\Delta(s)$ be a single-valued rational function that is analytic in a given region in the s-plane except at a finite number of points. Suppose that an arbitrary closed path Γ_s is chosen in the s-plane so that $\Delta(s)$ is analytic at every point on Γ_s; the corresponding $\Delta(s)$ locus mapped in the $\Delta(s)$-plane will encircle the origin as many times as the difference between the number of the zeros and the number of poles of $\Delta(s)$ that are encircled by the s-plane locus Γ_s.*

In equation form, this statement can be expressed as

$$N = Z - P \qquad (9\text{-}23)$$

where N = number of encirclement of the origin made by the $\Delta(s)$-plane locus Γ
Z = number of zeros of $\Delta(s)$ encircled by the s-plane locus Γ_s in the s-plane
P = number of poles of $\Delta(s)$ encircled by the s-plane locus Γ_s in the s-plane

In general, N can be positive ($Z > P$), zero ($Z = P$), or negative ($Z < P$). These three situations will now be described.

1. $N > 0$ ($Z > P$). If the s-plane locus encircles more zeros than poles of $F(s)$ in a certain prescribed direction (clockwise or counterclockwise), N is a positive integer. In this case the $\Delta(s)$-plane locus will encircle the origin of the $\Delta(s)$-plane N times in the same direction as that of Γ_s.
2. $N = 0$ ($Z = P$). If the s-plane locus encircles as many poles as zeros, or no poles and zeros, of $\Delta(s)$, the $\Delta(s)$-plane locus Γ_Δ will not encircle the origin of the $\Delta(s)$-plane.
3. $N < 0$ ($Z < P$). If the s-plane locus encircles more poles than zeros of $F(s)$ in a certain direction, N is a negative integer. In this case the $\Delta(s)$-plane locus, Γ_Δ, will encircle the origin N times in the opposite direction from that of Γ_s.

A convenient way of determining N with respect to the origin (or any other point) of the $\Delta(s)$ plane is to draw a line from the point in any direction to infinity; the number of net intersections of this line with the $\Delta(s)$ locus gives the magnitude of N. Figure 9-7 gives several examples of this method of determining N. It is assumed that the Γ_s locus has a counterclockwise sense.

A rigorous proof of the principle of the argument is not given here. The following illustration may be considered as a heuristic explanation of the principle. Let us consider the function $\Delta(s)$ given by

$$\Delta(s) = \frac{K(s + z_1)}{(s + p_1)(s + p_2)} \qquad (9\text{-}24)$$

where K is a positive number. The poles and zero of $\Delta(s)$ are assumed to be as shown in Fig. 9-8(a). The function $\Delta(s)$ can be written

$$\Delta(s) = |\Delta(s)| \underline{/\Delta(s)}$$

$$= \frac{K|s + z_1|}{|s + p_1|\,|s + p_2|} \left(\underline{/s + z_1} - \underline{/s + p_1} - \underline{/s + p_2} \right) \qquad (9\text{-}25)$$

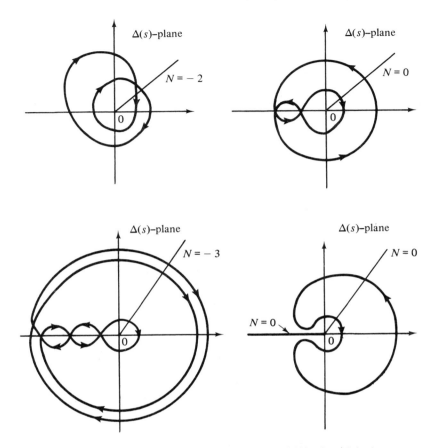

Figure 9-7 Examples of the determination of N in the $\Delta(s)$-plane.

Figure 9-8(a) shows an arbitrarily chosen trajectory Γ_s in the s-plane, with the arbitrary point s_1 on the path. The function $\Delta(s)$ evaluated at $s = s_1$ is given by

$$\Delta(s_1) = \frac{K(s_1 + z_1)}{(s_1 + p_1)(s_1 + p_2)} \tag{9-26}$$

The factor $s_1 + z_1$ can be represented graphically by the vector drawn from $-z_1$ to s_1. Similar vectors can be defined for $(s_1 + p_1)$ and $(s_1 + p_2)$. Thus $\Delta(s_1)$ is represented by the vectors drawn from the given poles and zero to the point s_1, as shown in Fig. 9-8(a). Now, if the point s_1 moves along the locus Γ_s in the prescribed counterclockwise direction until it returns to the starting point, the angles generated by the vectors drawn from the poles (and zeros if there were any) that are not encircled by Γ_s when s_1 completes one round trip are zero; whereas the vector

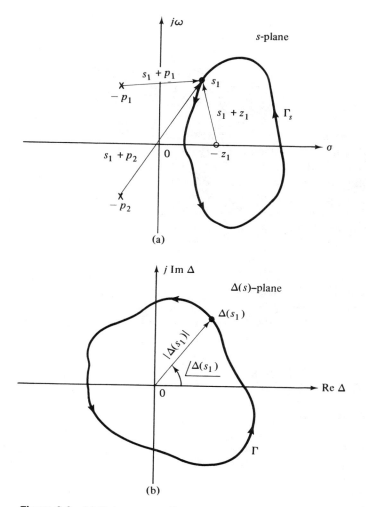

Figure 9-8 (a) Pole-zero configuration of $\Delta(s)$ in Eq. (9-24) and the s-plane trajectory Γ_s. (b) $\Delta(s)$-plane locus, Γ_Δ, which corresponds to the Γ_s, locus of (a) through the mapping of Eq. (9-24).

$(s_1 + z_1)$ drawn from the zero at $-z_1$, which is encircled by Γ_s, generates a positive angle (counterclockwise sense) of 2π rad. Then, in Eq. (9-25), the net angle or argument of $\Delta(s)$ as the point s_1 travels around Γ_s once is equal to 2π, which means that the corresponding $\Delta(s)$ plot must go around the origin 2π radians or one revolution in a counterclockwise direction, as shown in Fig. 9-8(b). This is why only the poles and zeros of $\Delta(s)$, which are inside the Γ_s path in the s-plane, would contribute to the value of N of Eq. (9-23). Since poles of $\Delta(s)$ correspond to negative phase angles and zeros correspond to positive phase angles, the value of N depends only on the difference between Z and P.

In the present case,

$$Z = 1 \qquad P = 0$$

Thus,

$$N = Z - P = 1$$

which means that the $\Delta(s)$-plane locus should encircle the origin once in the *same* direction as the s-plane locus. It should be kept in mind that Z and P refer only to the zeros and poles, respectively, of $\Delta(s)$ that are encircled by Γ_s, and not the total number of zeros and poles of $\Delta(s)$.

In general, if there are N more zeros than poles of $\Delta(s)$, which are encircled by the s-plane locus Γ_s in a prescribed direction, the net angle traversed by the $\Delta(s)$-plane locus as the s-plane locus is traversed once is equal to

$$2\pi(Z - P) = 2\pi N \qquad (9\text{-}27)$$

This equation implies that the $\Delta(s)$-plane locus will encircle the origin N times in the same direction as that of Γ_s. Conversely, if N more poles than zeros are encircled by Γ_s, in a given prescribed direction, N in Eq. (9-27) will be negative, and the $\Delta(s)$-plane locus must encircle the origin N times in the opposite direction to that of Γ_s.

A summary of all the possible outcomes of the principle of the argument is given in Table 9-1.

Nyquist Path

At this point readers may place themselves in the position of Nyquist many years ago, confronted with the problem of determining whether or not the rational function $\Delta(s) = 1 + F(s)$ has zeros in the right half of the s-plane. Apparently, Nyquist discovered that the principle of the argument could be used to solve the stability problems, if the s-plane locus, Γ_s, is taken to be one that encircles the

Table 9-1   Summary of All Possible Outcomes of the Principle of the Argument

$N = Z - P$	Sense of the s-Plane Locus	$F(s)$-Plane Locus Number of Encirclements of the Origin	$F(s)$-Plane Locus Direction of Encirclement
$N > 0$	Clockwise	N	Clockwise
	Counterclockwise		Counterclockwise
$N < 0$	Clockwise	N	Counterclockwise
	Counterclockwise		Clockwise
$N = 0$	Clockwise	0	No encirclement
	Counterclockwise		No encirclement

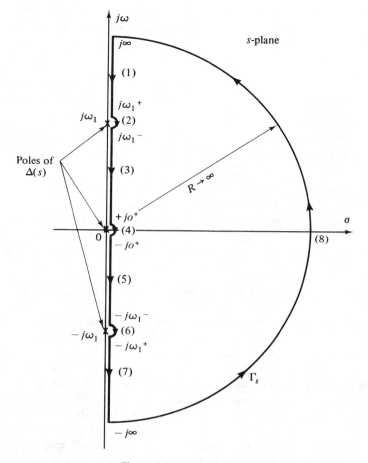

Figure 9-9 Nyquist path.

entire right half of the s-plane. Of course, as an alternative, Γ_s can be chosen to encircle the entire left-half s-plane, as the solution is a relative one. Figure 9-9 illustrates a Γ_s locus, with a counterclockwise sense, which encircles the entire right half of the s-plane. This path is often called the *Nyquist path*.

Since the Nyquist path must not pass through any singularity of $\Delta(s)$, the small semicircles shown along the $j\omega$ axis in Fig. 9-9 are used to indicate that the path should go around these singular points of $\Delta(s)$. It is apparent that if any pole or zero of $\Delta(s)$ lies inside the right half of the s-plane, it will be encircled by this Nyquist path.

For the convenience of analysis, the Nyquist path is divided into a minimum of three sections. The exact number of sections depends upon how many of those small semicircles are necessary on the imaginary axis. For the situation illustrated in Fig. 9-9, a total of eight sections needs to be defined. The order of numbering these sections is entirely arbitrary. The notations $j\omega_1^+$, $j\omega_1^-$, $+j0^+$, $-j0^+$, $-j\omega_1^+$, and $-j\omega_1^-$ are used to identify the starting and ending points of the small semicircles only.

Section 1: from $s = +j\infty$ to $+j\omega_1^+$ along the $j\omega$ axis.

Section 2: from $+j\omega_1^+$ to $+j\omega_1^-$ along the small semicircle around $s = j\omega_1$.

Section 3: from $s = j\omega_1^-$ to $+j0^+$ along the $j\omega$ axis.

Section 4: from $+j0^+$ to $-j0^+$ along the small semicircle around $s = 0$.

Section 5: from $s = -j0^+$ to $-j\omega_1^-$ along the $j\omega$ axis (mirror image of section 3).

Section 6: from $s = -j\omega_1^-$ to $-j\omega_1^+$ along the semicircle around $s = -j\omega_1$ (mirror image of section 2).

Section 7: from $s = -j\omega_1^+$ to $s = -j\infty$ along the $j\omega$ axis (mirror image of section 1).

Section 8: from $s = -j\infty$ to $s = +j\infty$ along the semicircle of infinite radius.

Nyquist Criterion and the $F(s)$ or the $G(s)H(s)$ Plot

The Nyquist criterion is a direct application of the principle of the argument when the s-plane locus is the Nyquist path. In principle, once the Nyquist path is specified, the stability of a closed-loop system can be determined by plotting the $\Delta(s) = 1 + F(s)$ locus when s takes on values along the Nyquist path, and investigating the behavior of the $\Delta(s)$ plot with respect to the origin of the $\Delta(s)$-plane. This is called the *Nyquist plot of* $\Delta(s)$. However, since $F(s)$, and in the case of a single-loop system, $G(s)H(s)$, are generally known functions, it is simpler to construct the Nyquist plot of $F(s)$ or $G(s)H(s)$, and the same conclusion on the stability of the closed-loop system can be determined from the $F(s)$ plot [$G(s)H(s)$ plot] with respect to the $(-1, j0)$ point in the $F(s)$-plane [$G(s)H(s)$-plane]. This is because the origin of the $\Delta(s)$ corresponds to the $(-1, j0)$ point of the $F(s)$-plane. In addition, if the location of the poles of $F(s)$ is not known, the stability of the open-loop system can be determined by investigating the behavior of the Nyquist plot of $F(s)$ with respect to the origin of the $F(s)$-plane. Thus, the application of the Nyquist criterion to the stability of the control systems is another example of the unique feature of control systems theory that closed-loop systems behavior is extracted from information on "open-loop" dynamics.

Let us define the origin of the $F(s)$-plane or the origin of the $\Delta(s)$-plane, as the case may be, as the *critical point*. In general, we are interested basically in two types of stability: *open-loop stability*, the stability of the open-loop systems [i.e., the stability of $F(s)$], and *closed-loop stability*, the stability of the closed-loop system [i.e., the stability of $\Delta(s)$].

It is important to note that closed-loop stability implies that $\Delta(s) = 1 + F(s)$ has zeros only in the left half of the s-plane. Open-loop stability implies that $F(s)$ has poles only in the left half of the s-plane. Again, when the closed-loop system has only one loop, $F(s) = G(s)H(s)$.

With this added dimension to the stability problem, it is necessary to define two sets of N, Z, and P, as follows:

N_0 = number of encirclements of the origin made by $F(s)$

Z_0 = number of zeros of $F(s)$ that are encircled by the Nyquist path, or in the right half of the s-plane

P_0 = number of poles of $F(s)$ that are encircled by the Nyquist path, or in the right half of the s-plane

N_{-1} = number of encirclements of the $(-1, j0)$ point made by $F(s)$

Z_{-1} = number of zeros of $1 + F(s)$ that are encircled by the Nyquist path, or in the right half of the s-plane

P_{-1} = number of poles of $F(s)$ that are encircled by the Nyquist path, or in the right half of the s-plane

Several facts become clear and should be remembered at this point:

$$P_0 = P_{-1} \qquad (9\text{-}28)$$

since $F(s)$ and $1 + F(s)$ always have the same poles. Closed-loop stability implies or requires that

$$Z_{-1} = 0 \qquad (9\text{-}29)$$

but open-loop stability requires that

$$P_0 = 0 \qquad (9\text{-}30)$$

The crux of the matter is that closed-loop stability is determined by the properties of the Nyquist plot of the open-loop transfer function.

The procedure of applying the principle of the argument for stability studies is summarized as follows:

1. Given a feedback control system that has the closed-loop transfer function such as that given in Eq. (9-11), the determinant of the closed-loop system, $\Delta(s)$, is given by Eq. (9-19). The Nyquist path is first defined according to the pole-zero properties of $F(s)$.
2. The Nyquist plot of $F(s)$ is constructed.
3. The values of N_0 and N_{-1} are determined by observing the behavior of the Nyquist plot of $G(s)H(s)$ with respect to the origin and the $(-1, j0)$ point, respectively.

4. Once N_0 and N_{-1} are determined, the value of P_0 (if it is not already known) is determined from

$$N_0 = Z_0 - P_0 \qquad (9\text{-}31)$$

if Z_0 is given. Once P_0 is determined, $P_{-1} = P_0$ [Eq. (9-28)], and Z_{-1} is determined from

$$N_{-1} = Z_{-1} - P_{-1} \qquad (9\text{-}32)$$

Since it has been established that for a stable closed-loop Z_{-1} must be zero, Eq. (9-32) gives

$$N_{-1} = -P_{-1} \qquad (9\text{-}33)$$

Therefore, the Nyquist criterion may be formally stated: For a closed-loop system to be stable, *the Nyquist plot of $F(s)$ must encircle the $(-1, j0)$ point as many times as the number of poles of $F(s)$ that are in the right half of the s-plane, and the encirclement, if any, must be made in the clockwise direction.*

A Simplified Nyquist Plot

In general, it is not always necessary to construct the Nyquist plot which corresponds to the entire Nyquist path. In fact, a majority of the practical systems have transfer functions $F(s)$ with no poles and zeros in the right-half s-plane; i.e., $Z_0 = 0$ and $P_0 = 0$. The transfer function with this property is referred to as a *minimum-phase* transfer function. It can be shown that when a transfer function $F(s)$ is of the minimum-phase type, and when as is usually the case for a control system it has more poles than zeros, when $s = j\omega$ and ω takes on values from 0 to ∞, the plot of $F(s)$ in the polar coordinates will always rotate in the clockwise direction. Putting it another way, the Nyquist plot of a minimum-phase transfer function always will have a net *counterclockwise* rotation, or will "lose phase," as the Nyquist path is traversed as shown in Fig. 9-9.

Thus, when $F(s)$ has no poles and zeros in the right half of the s-plane,

$$Z_{-1} = N_{-1} \qquad (9\text{-}34)$$

the Nyquist plot of $F(s)$ can circle the $(-1, j0)$ point only in the counterclockwise direction, since N_{-1} in Eq. (9-34) can be only zero or positive. Since counterclockwise encirclement is equivalent to enclosure, we can determine closed-loop stability by checking whether the $(-1, j0)$ critical point is *enclosed* by the $F(s)$ plot. Furthermore, if we are interested only in whether N_{-1} is zero, we need not sketch the entire Nyquist plot for $F(s)$, only the portion from the $s = j\infty$ to $s = 0$ along the imaginary axis of the s-plane.

The Nyquist criterion for the $P_{-1} = 0$ and $Z_0 = 0$ case may be stated: *If the function has no zeros and poles in the right half of the s-plane, for the closed-loop system to be stable, the Nyquist plot of $F(s)$ must not enclose the critical point $(-1, j0)$.*

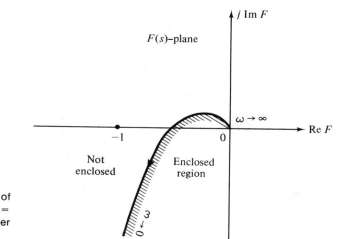

Figure 9-10 Nyquist plot of $F(s)$, which corresponds to $s = j\omega$ to $s = 0$, to indicate whether the critical point is enclosed.

Furthermore, if we are not interested in the number of roots of the characteristic equation that are in the right-half plane, but only the closed-loop stability, only the $F(s)$ plot that corresponds to the positive imaginary axis of the s-plane is necessary. Figure 9-10 illustrates how the $s = j\infty$ to $s = 0$ portion of the Nyquist plot may be used to determine whether the critical point at $(-1, j0)$ is enclosed.

9.3 APPLICATION OF THE NYQUIST CRITERION

The following examples serve to illustrate the practical application of the Nyquist criterion to the stability of control systems.

■

Example 9-1

Consider a single-loop feedback control system with the loop transfer function given by

$$F(s) = G(s)H(s) = \frac{K}{s(s + a)} \tag{9-35}$$

where K and a are positive constants. It is apparent that $G(s)H(s)$ does not have any pole in the right-half s-plane; thus, $P_0 = P_{-1} = 0$. To determine the stability of the closed-loop system, it is necessary only to sketch the Nyquist plot of $G(s)H(s)$ that corresponds to $s = j\infty$ to $s = 0$ on the Nyquist path and see if it encloses the $(-1, j0)$ point in the $G(s)H(s)$-plane. However, for the sake of illustration, we shall construct the entire $G(s)H(s)$ plot for this problem.

The Nyquist path necessary for the function of Eq. (9-35) is shown in Fig. 9-11. Since $G(s)H(s)$ has a pole at the origin, it is necessary that the Nyquist path includes a small semicircle around $s = 0$. The entire Nyquist path is divided into four sections, as shown in Fig. 9-11.

Section 2 of the Nyquist path is magnified as shown in Fig. 9-12(a). The points on this section may be represented by the phasor

$$s = \epsilon e^{j\theta} \tag{9-36}$$

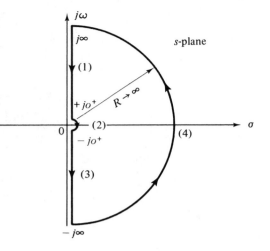

Figure 9-11 Nyquist path for the system in Example 9-1.

where $\epsilon(\epsilon \to 0)$ and θ denote the magnitude and phase of the phasor, respectively. As the Nyquist path is traversed from $+j0^+$ to $-j0^+$ along section 2, the phasor of Eq. (9-36) rotates in the clockwise direction through 180 degrees. Also, in going from $+j0^+$ to $-j0^+$, θ varies from $+90$ to -90 degrees through 0 degrees. The corresponding Nyquist plot of $G(s)H(s)$ can be determined simply by substituting Eq. (9-36) into Eq. (9-35). Thus,

$$G(s)H(s)\big|_{s=\epsilon e^{j\theta}} = \frac{K}{\epsilon e^{j\theta}(\epsilon e^{j\theta} + a)} \tag{9-37}$$

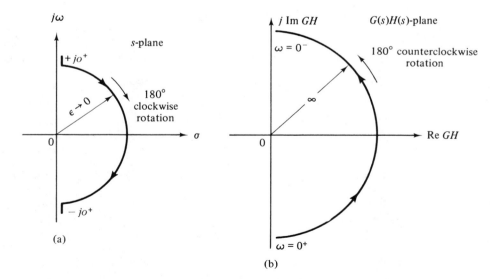

Figure 9-12 (a) Section 2 of the Nyquist path of Fig. 9-11. (b) Nyquist plot of $G(s)H(s)$ that corresponds to section 2.

Since $\epsilon \rightarrow 0$, the last expression is simplified to

$$G(s)H(s)\big|_{s=\epsilon e^{j\theta}} \cong \frac{K}{a\epsilon e^{j\theta}} = \infty e^{-j\theta} \qquad (9\text{-}38)$$

which indicates that all points on the Nyquist plot of $G(s)H(s)$ that correspond to section 2 of the Nyquist path have an infinite magnitude, and the corresponding phase is opposite that of the s-plane locus. Since the phase of the Nyquist path varies from $+90$ to -90 degrees in the clockwise direction, the minus sign in the phase relation of Eq. (9-38) indicates that the corresponding $G(s)H(s)$ plot should have a phase that varies from -90 to $+90$ degrees in the counterclockwise direction, as shown in Fig. 9-12(b).

In general, when one has acquired proficiency in the sketching of the Nyquist plots, the determination of the behavior of $G(s)H(s)$ at $s = 0$ and $s = \infty$ may be carried out by inspection. For instance, in the present problem the behavior of $G(s)H(s)$ corresponding to section 2 of the Nyquist path is determined from

$$\lim_{s \to 0} G(s)H(s) = \lim_{s \to 0} \frac{K}{s(s+a)} = \lim_{s \to 0} \frac{K}{sa} \qquad (9\text{-}39)$$

From this equation it is clear that the behavior of $G(s)H(s)$ at $s = 0$ is inversely proportional to s. As the Nyquist path is traversed by a phasor with infinitesimally small magnitude, from $+j0^+$ to $-j0^+$ through a clockwise rotation of 180 degrees, the corresponding $G(s)H(s)$ plot is traced out by a phasor with an infinite magnitude, 180 degrees in the opposite or counterclockwise direction. It can be concluded that, in general, if the limit of $G(s)H(s)$ as s approaches zero assumes the form

$$\lim_{s \to 0} G(s)H(s) = \lim_{s \to 0} Ks^{\pm n} \qquad (9\text{-}40)$$

the Nyquist plot of $G(s)H(s)$ that corresponds to section 2 of Fig. 9-12(a) is traced out by a phasor of infinitesimally small magnitude $n \times 180$ degrees in the clockwise direction if the plus sign is used, and by a phasor of infinite magnitude $n \times 180$ degrees in the counterclockwise direction if the negative sign is used.

The technique described above may also be used to determine the behavior of the $G(s)H(s)$ plot, which corresponds to the semicircle with infinite radius on the Nyquist path. The large semicircle referred to as section 4 in Fig. 9-11 is isolated, as shown in Fig. 9-13(a). The points on the semicircle may be represented by the phasor

$$s = Re^{j\phi} \qquad (9\text{-}41)$$

where $R \rightarrow \infty$. Substituting Eq. (9-41) into Eq. (9-35) yields

$$G(s)H(s)\big|_{s=Re^{j\phi}} = \frac{K}{R^2 e^{j2\phi}} = 0 e^{-j2\phi} \qquad (9\text{-}42)$$

which implies that the behavior of the $G(s)H(s)$ plot at infinite frequency is described by a phasor with infinitesimally small magnitude which rotates around the origin $2 \times 180° = 360°$ in the clockwise direction. Thus the $G(s)H(s)$ plot that corresponds to section 4 of the Nyquist path is sketched as shown in Fig. 9-13(b).

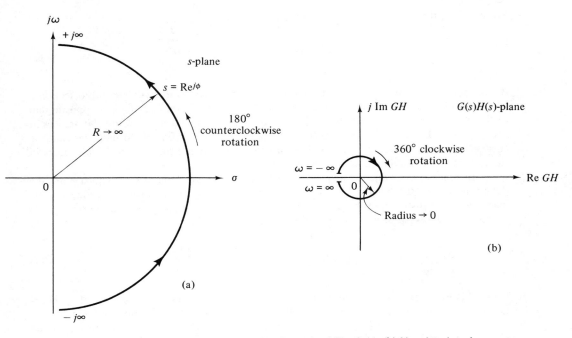

Figure 9-13 (a) Section 4 of the Nyquist path of Fig. 9-11. (b) Nyquist plot of $G(s)H(s)$ that corresponds to section 4.

Now to complete the Nyquist plot of the transfer function of Eq. (9-35) we must consider sections 1 and 3. Section 1 is usually constructed by substituting $s = j\omega$ into Eq. (9-35) and solving for the possible crossing points on the real axis of the $G(s)H(s)$-plane. Equation (9-35) becomes

$$G(j\omega)H(j\omega) = \frac{K}{j\omega(j\omega + a)} \tag{9-43}$$

which is rationalized by multiplying the numerator and denominator by the complex conjugate of the denominator. Thus,

$$G(j\omega)H(j\omega) = \frac{K(-\omega^2 - ja\omega)}{\omega^4 + a^2\omega^2} \tag{9-44}$$

The intersect of $G(j\omega)H(j\omega)$ on the real axis is determined by equating the imaginary part of $G(j\omega)H(j\omega)$ to zero. Thus the frequency at which $G(j\omega)H(j\omega)$ intersects the real axis is found from

$$\text{Im}\, G(j\omega)H(j\omega) = \frac{-Ka\omega}{\omega^4 + a^2\omega^2} = \frac{-Ka}{\omega(\omega^2 + a^2)} = 0 \tag{9-45}$$

which gives $\omega = \infty$. This means that the only intersect on the real axis in the $G(s)H(s)$-plane is at the origin with $\omega = \infty$. Since the Nyquist criterion is not concerned with the exact shape

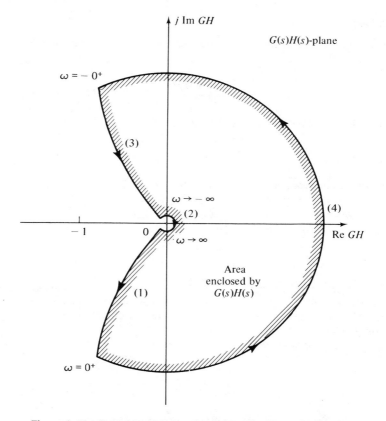

Figure 9-14 Complete Nyquist plot of $G(s)H(s) = K / [s(s + a)]$.

of the $G(s)H(s)$ locus but only the number of encirclements, it is not necessary to obtain an exact plot of the locus. The complete Nyquist plot of the function of Eq. (9-35) is now sketched in Fig. 9-14 by connecting the terminal points of the loci that correspond to sections 2 and 4, without intersecting any finite part of the real axis.

It is of interest to check all the pertinent data that can be obtained from the Nyquist plot of Fig. 9-14. First $N_0 = N_{-1} = 0$. By inspection of Eq. (9-35), $Z_0 = 0$ and $P_0 = 0$, which satisfy Eq. (9-31). Since $P_0 = P_{-1}$, Eq. (9-32) leads to

$$Z_{-1} = N_{-1} + P_{-1} = 0 \qquad (9\text{-}46)$$

Therefore, the closed-loop system is stable. This solution should have been anticipated, since for the second-order system, the characteristic equation is simply

$$s^2 + as + K = 0 \qquad (9\text{-}47)$$

whose roots will always lie in the left half of the s-plane for positive a and K.

Figure 9-14 also shows that for this problem it is necessary only to sketch the portion of $G(s)H(s)$ that corresponds to section 1 of the Nyquist path. It is apparent that the $(-1, j0)$ point will never be enclosed by the $G(s)H(s)$ plot for all positive values of K.

Example 9-2

Consider that a control system with single feedback loop has the loop transfer function

$$G(s)H(s) = \frac{K(s-1)}{s(s+1)} \tag{9-48}$$

The characteristic equation of the system is

$$s^2 + (1+K)s - K = 0 \tag{9-49}$$

which has one root in the right half of the s-plane for all positive K. The Nyquist path of Fig. 9-11 is applicable to this case.

Section 2. $s = \epsilon e^{j\theta}$:

$$\lim_{s \to 0} G(s)H(s) = \lim_{s \to 0} \frac{-K}{s} = \infty e^{-j(\theta+\pi)} \tag{9-50}$$

This means that the Nyquist plot of $G(s)H(s)$ which corresponds to section 2 of the Nyquist path is traced by a phasor with infinite magnitude. This phasor starts at an angle of $+90$ degrees and ends at -90 degrees and goes around the origin of the $G(s)H(s)$-plane counterclockwise a total of 180 degrees.

Section 4. $s = Re^{j\phi}$:

$$\lim_{s \to \infty} G(s)H(s) = \lim_{s \to \infty} \frac{K}{s} = 0 e^{-j\phi} \tag{9-51}$$

Thus the Nyquist plot of $G(s)H(s)$ corresponding to section 4 of the Nyquist path goes around the origin 180 degrees in the clockwise direction with zero magnitude.

Section 1: $s = j\omega$:

$$G(j\omega)H(j\omega) = \frac{K(j\omega - 1)}{j\omega(j\omega + 1)} = \frac{K(j\omega - 1)(-\omega^2 - j\omega)}{\omega^4 + \omega^2} = K\frac{2\omega + j(1 - \omega^2)}{\omega(\omega^2 + 1)} \tag{9-52}$$

Setting the imaginary part of $G(j\omega)H(j\omega)$ to zero, we have

$$\omega = \pm 1 \text{ rad/sec} \tag{9-53}$$

which are frequencies at which the $G(j\omega)H(j\omega)$ locus crosses the real axis. Then

$$G(j1)H(j1) = K \tag{9-54}$$

Based on the information gathered in the last three steps, the complete Nyquist plot of $G(s)H(s)$ is sketched as shown in Fig. 9-15. We can conclude that, by inspection,

$$Z_0 = 1$$
$$P_0 = P_{-1} = 0$$

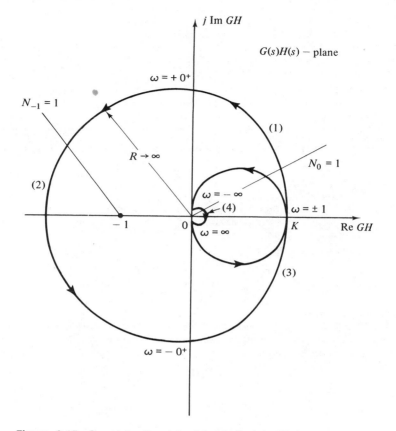

Figure 9-15 Complete Nyquist plot of $G(s)$ $H(s) = K(s-1)/$ $[s(s+1)]$.

Figure 9-15 indicates that $N_0 = 1$, which is in agreement with the Nyquist criterion. Figure 9-15 also gives $N_{-1} = 1$. Then

$$Z_{-1} = N_{-1} + P_{-1} = 1 \tag{9-55}$$

which means that the closed-loop system is unstable, since the characteristic equation has one root in the right half of the s-plane. The Nyquist plot of Fig. 9-15 further indicates that the system cannot be stabilized by changing only the value of K.

Example 9-3

Consider the control system shown in Fig. 9-16. It is desired to determine the range of K for which the system is stable. The open-loop transfer function of the system is

$$\frac{C(s)}{E(s)} = G(s) = \frac{10K(s+2)}{s^3 + 3s^2 + 10} \tag{9-56}$$

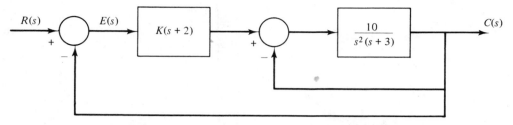

Figure 9-16 Block diagram of the control system for Example 9-3.

Since this function does not have any pole or zero on the $j\omega$ axis, the Nyquist path can consist of only three sections, as shown in Fig. 9-17. The construction of the Nyquist plot of $G(s)$ is outlined as follows:

Section 4. $s = Re^{j\phi}$:

$$\lim_{s \to \infty} G(s) = \frac{10K}{s^2} = 0e^{-j2\phi} \tag{9-57}$$

As the phasor for section 4 of the Nyquist path is traversed from -90 to $+90$ degrees through 180 degrees counterclockwise, Eq. (9-57) indicates that the corresponding Nyquist plot of $G(s)$ is traced by a phasor of practically zero length $+180$ to -180 degrees through a total of 360 degrees in the clockwise sense.

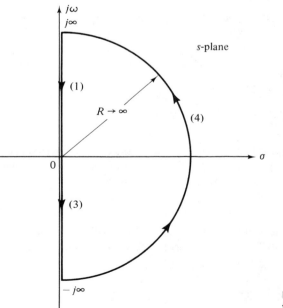

Figure 9-17 Nyquist path for the transfer function of Eq. (9-56).

Section 1. $s = j\omega$:

$$G(j\omega) = \frac{10K(j\omega + 2)}{(10 - 3\omega^2) - j\omega^3} \tag{9-58}$$

Rationalizing, Eq. (9-58) becomes

$$G(j\omega) = \frac{10K[2(10 - 3\omega^2) - \omega^4 + j\omega(10 - 3\omega^2) + j2\omega^3]}{(10 - 3\omega^2)^2 + \omega^6} \tag{9-59}$$

Setting the imaginary part of $G(j\omega)$ to zero gives

$$\omega = 0 \text{ rad/sec}$$

and

$$\omega = \pm\sqrt{10} \text{ rad/sec}$$

which correspond to the frequencies at the intersects on the real axis of the $G(s)$-plane. In this case it is necessary to determine the intersect of the $G(s)$ plot on the imaginary axis. Setting the real part of $G(j\omega)$ to zero in Eq. (9-59), we have

$$\omega^4 + 6\omega^2 - 20 = 0 \tag{9-60}$$

which gives

$$\omega = \pm 1.54 \text{ rad/sec}$$

Therefore, the intersects on the real axis of the $G(s)$-plane are at

$$G(j0) = 2K$$

and

$$G(j\sqrt{10}) = -K$$

and the intersect on the imaginary axis is

$$G(j1.54) = j5.43K$$

With the information gathered in the preceding steps, the Nyquist plot for $G(s)$ of Eq. (9-56) is sketched as shown in Fig. 9-18. The information on the imaginary axis is needed so that the direction of section 1 may be determined without actually plotting the locus point by point.

Inspection of the Nyquist plot of Fig. 9-18 reveals that $N_0 = -2$. Since Eq. (9-56) shows that $G(s)$ has no zeros inside the right half of the s-plane, $Z_0 = 0$; this means that $P_0 = 2$. Thus $P_{-1} = 2$. Now, applying the Nyquist criterion, we have

$$N_{-1} = Z_{-1} - P_{-1} = Z_{-1} - 2 \tag{9-61}$$

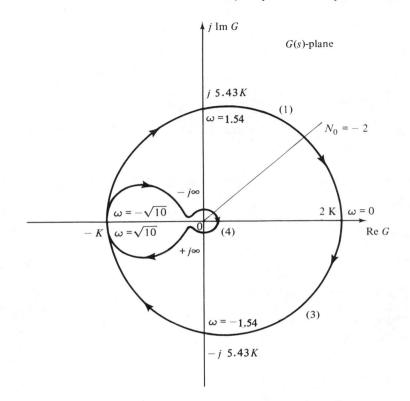

Figure 9-18   Nyquist plot of $G(s) = 10K(s + 2) / (s^3 + 3s^2 + 10)$.

Thus, for the closed-loop system to be stable, $Z_{-1} = 0$, which requires that $N_{-1} = -2$. With reference to Fig. 9-18, the stability criterion requires that the $(-1, j0)$ point must be encircled twice in the clockwise direction. In other words, the critical point should be to the right of the crossover point at $-K$. Thus, for stability,

$$K > 1 \qquad\qquad (9\text{-}62)$$

The reader can easily verify this solution by applying the Routh–Hurwitz criterion to the characteristic equation of the system.

It should be reiterated that although the Routh–Hurwitz criterion is much simpler to use in stability problems such as the one stated in this illustrative example, in general the Nyquist criterion leads to a more versatile solution, which also includes information on the relative stability of the system.

9.4 EFFECTS OF ADDITIONAL POLES AND ZEROS OF *G(s)H(s)* ON THE SHAPE OF THE NYQUIST LOCUS

Since the performance and the stability of a feedback control system are often influenced by adding and moving poles and zeros of the transfer functions, it is informative to illustrate how the Nyquist locus is affected when poles and zeros are added to a typical loop transfer function $G(s)H(s)$. This investigation will also be helpful to gain further insight on the quick sketch of the Nyquist locus of a given function.

Let us begin with a first-order transfer function

$$G(s)H(s) = \frac{K}{1 + sT_1} \qquad (9\text{-}63)$$

The Nyquist locus of $G(j\omega)H(j\omega)$ for $0 \leq \omega < \infty$ is a semicircle, as shown in Fig. 9-19.

Addition of Poles at $s = 0$. Consider that a pole at $s = 0$ is added to the transfer function of Eq. (9-63); then we have

$$G(s)H(s) = \frac{K}{s(1 + sT_1)} \qquad (9\text{-}64)$$

the effect of adding this pole is that the phase of $G(j\omega)H(j\omega)$ is reduced by -90 degrees at both zero and infinite frequencies.

In other words, the Nyquist locus of $G(j\omega)H(j\omega)$ is rotated by -90 degrees from that of Fig. 9-19 at $\omega = 0$ and $\omega = \infty$, as shown in Fig. 9-20. In addition, the magnitude of $G(j\omega)H(j\omega)$ at $\omega = 0$ becomes infinite. In general, adding a pole of multiplicity j at $s = 0$ to the transfer function of Eq. (9-63) will give the following

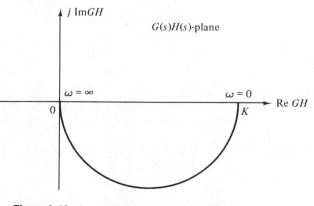

Figure 9-19   Nyquist plot of $G(s)H(s) = K/(1 + T_1s)$.

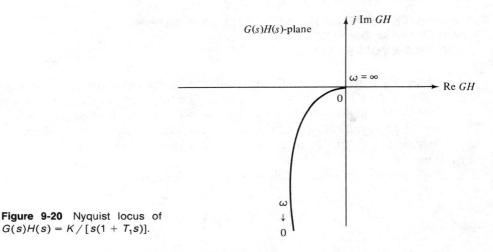

Figure 9-20 Nyquist locus of $G(s)H(s) = K/[s(1 + T_1s)]$.

properties to the Nyquist locus of $G(s)H(s)$:

$$\lim_{\omega \to \infty} \underline{/G(j\omega)H(j\omega)} = -(j + 1)\frac{\pi}{2} \tag{9-65}$$

$$\lim_{\omega \to 0} \underline{/G(j\omega)H(j\omega)} = -j\frac{\pi}{2} \tag{9-66}$$

$$\lim_{\omega \to \infty} |G(j\omega)H(j\omega)| = 0 \tag{9-67}$$

$$\lim_{\omega \to 0} |G(j\omega)H(j\omega)| = \infty \tag{9-68}$$

Figure 9-21 illustrates the Nyquist plots of

$$G(s)H(s) = \frac{K}{s^2(1 + T_1s)} \tag{9-69}$$

and

$$G(s)H(s) = \frac{K}{s^3(1 + T_1s)} \tag{9-70}$$

In view of these illustrations it is apparent that addition of poles at $s = 0$ will affect the stability adversely, and systems with a loop transfer function of more than one pole at $s = 0$ are likely to be unstable.

Addition of Finite Poles. When a pole at $s = -1/T_2$ is added to the $G(s)H(s)$ function of Eq. (9-63), we have

$$G(s)H(s) = \frac{K}{(1 + T_1s)(1 + T_2s)} \tag{9-71}$$

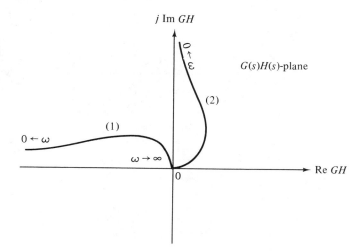

Figure 9-21 Nyquist loci for $(1)G(s)H(s) = K / [(s^2(1 + T_1s)]$. $(2)\ G(s)H(s) = K / [s^3(1 + T_1s)]$.

The Nyquist locus of $G(s)H(s)$ at $\omega = 0$ is not affected by the addition of the pole, since

$$\lim_{\omega \to 0} G(j\omega)H(j\omega) = K \qquad (9\text{-}72)$$

The Nyquist locus at $\omega = \infty$ is

$$\lim_{\omega \to \infty} G(j\omega)H(j\omega) = \lim_{\omega \to \infty} \frac{-K}{T_1T_2\omega^2} = 0\underline{/-180°} \qquad (9\text{-}73)$$

Thus, the effect of adding a pole at $s = -1/T_2$ to the transfer function of Eq. (9-63) is to shift the phase of the Nyquist locus by -90 degrees at infinite frequency, as shown in Fig. 9-22. This figure also shows the Nyquist locus of

$$G(s)H(s) = \frac{K}{(1 + T_1s)(1 + T_2s)(1 + T_3s)} \qquad (9\text{-}74)$$

These examples show the adverse effects on stability that result from the addition of poles to the loop transfer function.

Addition of Zeros. It was pointed out in Chapter 8 that the effect of the derivative control on a closed-loop control system is to make the system more stable. In terms of the Nyquist plot, this stabilization effect is easily shown, since the multiplication of the factor $(1 + T_ds)$ to the loop transfer function increases the phase of $G(s)H(s)$ by 90 degrees at $\omega = \infty$.

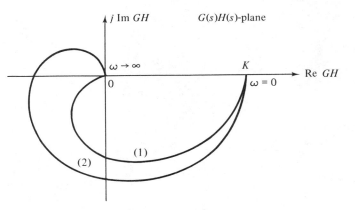

Figure 9-22 Nyquist loci for (1) $G(s)H(s) = K / [(1 + T_1s)(1 + T_2s)]$. (2) $G(s)H(s) = K / [(1 + T_1s)(1 + T_2s)(1 + T_3s)]$.

Consider that the loop transfer function of a closed-loop system is given by

$$G(s)H(s) = \frac{K}{s(1 + T_1s)(1 + T_2s)} \qquad (9\text{-}75)$$

It can be shown that the closed-loop system is stable for $0 \le K < (T_1 + T_2)/T_1T_2$. Suppose that a zero at $s = -1/T_d$ is added to the transfer function of Eq. (9-75), such as with a derivative control. Then

$$G(s)H(s) = \frac{K(1 + T_ds)}{s(1 + T_1s)(1 + T_2s)} \qquad (9\text{-}76)$$

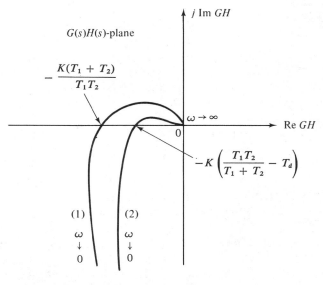

Figure 9-23 Nyquist loci for (1) $G(s)H(s) = K / [s(1 + T_1s)(1 + T_2s)]$. (2) $G(s)H(s) = K(1 + T_ds) / [s(1 + T_1s) \ s(1 + T_2s)]$.

The Nyquist loci of the two transfer functions of Eqs. (9-75) and (9-76) are sketched as shown in Fig. 9-23. The effect of the zero in Eq. (9-76) is to add 90 degrees to the phase of $G(j\omega)H(j\omega)$ at $\omega = \infty$ while not affecting the locus at $\omega = 0$. The crossover point on the real axis is moved from $-K(T_1 + T_2)/T_1T_2$ to $-K(T_1T_2 - T_1T_d - T_2T_d)/(T_1 + T_2)$, which is closer to the origin of the $G(j\omega)H(j\omega)$-plane.

9.5 STABILITY OF MULTILOOP SYSTEMS

The stability analyses conducted in the preceding sections are all centered toward systems with a single feedback loop, with the exception of the Routh–Hurwitz criterion, which apparently can be applied to systems of any configuration, as long as the characteristic equation is known. We shall now illustrate how the Nyquist criterion is applied to a linear system with multiple feedback loops.

As pointed out earlier, the stability of all linear feedback control systems regardless of the number of inputs and outputs and the number of loops can be studied by investigating the zeros of the function $\Delta(s) = 1 + F(s)$. However, for a system with multiple number of feedback loops, $F(s)$ is a sum of several rational functions, and the poles and zeros of $F(s)$ may be unknown. Thus, a systematic approach to the application of the Nyquist criterion to multiloop control systems is desirable.

Let us illustrate the procedure of applying Nyquist criterion to a multiloop control system by means of a specific example. Figure 9-24 gives the block diagram of a control system with two loops. The transfer function of each block is indicated in the diagram. In this case it is simple to derive the open-loop transfer function of the system as

$$\frac{C(s)}{E(s)} = F(s) = \frac{G_1(s)G_2(s)}{1 + G_2(s)H(s)}$$

$$= \frac{K(s + 2)}{(s + 10)[s(s + 1)(s + 2) + 5]} \tag{9-77}$$

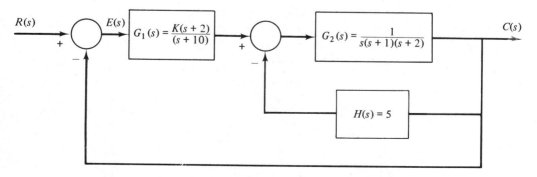

Figure 9-24 Multiloop feedback control system.

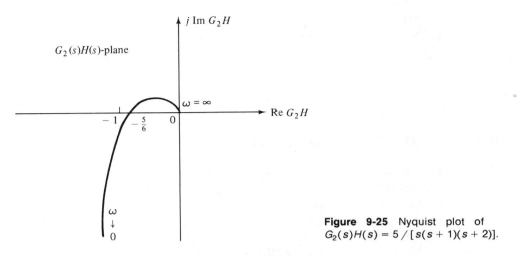

Figure 9-25 Nyquist plot of $G_2(s)H(s) = 5 / [s(s + 1)(s + 2)]$.

The stability of the overall system can be investigated by sketching the Nyquist locus of $F(s)$, except that the poles of $F(s)$ are not entirely known. To avoid the construction of the entire Nyquist locus of $F(s)$, we can attack the problem in two stages, as there are two feedback loops. First, consider only the inner loop, whose loop transfer function is $G_2(s)H(s)$. We shall first sketch the Nyquist locus of $G_2(s)H(s)$ for $0 \le \omega < \infty$. The property of the $G_2(s)H(s)$ plot with respect to the $(-1, j0)$ point gives an indication of the number of zeros of $1 + G_2(s)H(s)$ that are in the right half of the s-plane. Having found this information, we then proceed to sketch the Nyquist locus of $F(s)$ of Eq. (9-77) only for $0 \le \omega < \infty$ to determine the stability of the overall system.

Figure 9-25 shows the Nyquist locus of

$$G_2(s)H(s) = \frac{5}{s(s + 1)(s + 2)} \qquad (9\text{-}78)$$

Since the $(-1, j0)$ point is not enclosed by the locus, the inner loop is stable by itself, and the zeros of $1 + G_2(s)H(s)$ are all in the left half of the s-plane. Next, the Nyquist locus of $F(s)$ of Eq. (9-77) is sketched as shown in Fig. 9-26. Since all the poles and zeros of $F(s)$ are found to be in the left half of the s-plane, we only have to investigate the crossover point of the $F(s)$ locus with respect to the $(-1, j0)$ point to determine the requirement on K for the overall system to be stable. In this case the range of K for stability is $0 \le K < 50$.

Now, let us consider a system that is a slightly modified version of Fig. 9-24. Use the same block diagram, but with

$$G_1(s) = \frac{s + 2}{s + 1} \qquad (9\text{-}79)$$

$$G_2(s) = \frac{K}{s(s + 1)(s + 2)} \qquad (9\text{-}80)$$

$$H(s) = 5 \qquad (9\text{-}81)$$

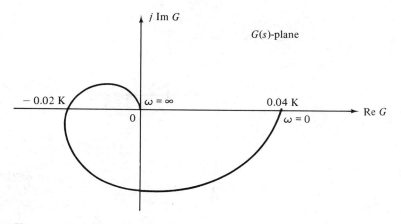

Figure 9-26 Nyquist plot of $F(s) = [K(s + 2)] / \{(s + 10)[s(s + 1)(s + 2) + 5]\}$.

In this case we cannot use the method outline above, since the unknown gain parameter is in the inner loop. However, we may still use the Nyquist criterion to solve this problem. The open-loop transfer function of the system is

$$F(s) = \frac{G_1(s)G_2(s)}{1 + G_2(s)H(s)}$$

$$= \frac{s + 2}{(s + 10)[s(s + 1)(s + 2) + 5K]} \tag{9-82}$$

Since the unknown parameter K does not appear as a gain factor of $G(s)$, it would be of no avail to sketch the Nyquist locus of $F(s)/K$. However, we can write the characteristic equation of the overall system as

$$s(s + 10)(s + 1)(s + 2) + s + 2 + 5K(s + 10) = 0 \tag{9-83}$$

In order to create an equivalent open-loop transfer function with K as a multiplying factor, we divide both sides of Eq. (9-83) by terms that do not contain K. We have

$$1 + \frac{5K(s + 10)}{s(s + 10)(s + 1)(s + 2) + s + 2} = 0 \tag{9-84}$$

Since this equation is of the form $1 + G_3(s) = 0$, the roots of the characteristic equation may be investigated by sketching the Nyquist locus of $G_3(s)$. However, the poles of $G_3(s)$ are not known, since the denominator of $G_3(s)$ is not in factored form. The zeros of the polynomial $s(s + 10)(s + 1)(s + 2) + s + 2$ may be studied

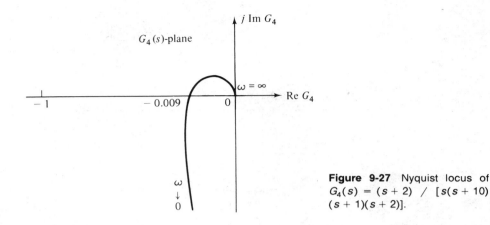

Figure 9-27 Nyquist locus of $G_4(s) = (s + 2) / [s(s + 10)(s + 1)(s + 2)]$.

by investigating the Nyquist plot of still another function $G_4(s)$, which we create as follows:

$$G_4(s) = \frac{s + 2}{s(s + 10)(s + 1)(s + 2)} \qquad (9\text{-}85)$$

Figure 9-27 shows that the Nyquist locus of $G_4(s)$ intersects the real axis to the right of the $(-1, j0)$ point. Thus, all the poles of $G_3(s)$ are in the left half of the s-plane. The Nyquist plot of $G_3(s)$ is sketched as shown in Fig. 9-28. Since the intersect of the locus on the real axis is at $-0.1K$, the range of K for stability is $0 \leq K < 10$.

In this section we have investigated the application of the Nyquist criterion to multiloop control systems. For analysis problems, the stability of the system can be investigated by applying the Nyquist criterion in a systematic fashion from the inner loop toward the outer loops. For design problems when a system parameter K is to

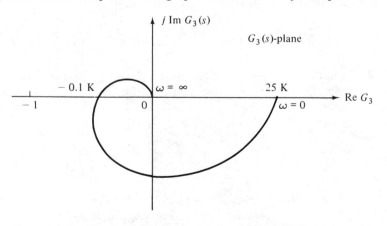

Figure 9-28 Nyquist locus of $G_3(s) = [5K(s + 10)] / [s(s + 10)(s + 1)(s + 2) + s + 2]$.

be determined for stability, it is sometimes necessary to start with the characteristic equation, which may be written in the form

$$P(s) + KQ(s) = 0 \qquad (9\text{-}86)$$

where $P(s)$ and $Q(s)$ are polynomials. The stability of the system is studied by sketching the Nyquist plot of the equivalent open-loop transfer function

$$G(s) = \frac{KQ(s)}{P(s)} \qquad (9\text{-}87)$$

Thus, we have also indicated a method of applying the Nyquist criterion to design a stable system by using the characteristic equation. The Nyquist locus represents a more convenient approach than the Routh–Hurwitz criterion, since the latter often involves the solution of an inequality equation in the high order of K, whereas in the Nyquist approach K can be determined from the intersection of the $F(s)$ locus with the real axis.

9.6 STABILITY OF LINEAR CONTROL SYSTEMS WITH TIME DELAYS

Systems with time delays and their modeling have been discussed in Section 4.9. In general, closed-loop systems with time delays in the loops will be subject to more stability problems than systems without delays. Since a pure time delay T_d is modeled by the transfer function relationship $e^{-T_d s}$, the characteristic equation of the system will no longer have constant coefficients. Therefore, the Routh–Hurwitz criterion is *not* applicable. However, we shall show in the following that the Nyquist criterion is readily applicable to a system with a pure time delay.

Let us consider that the loop transfer function of a feedback control system with pure time delay is represented by

$$G(s)H(s) = e^{-T_d s}G_1(s)H_1(s) \qquad (9\text{-}88)$$

where $G_1(s)H_1(s)$ is a rational function with constant coefficients; T_d is the pure time delay in seconds. Whether the time delay occurs in the forward path or the feedback path of the system is immaterial from the stability standpoint.

In principle, the stability of the closed-loop system can be investigated by sketching the Nyquist locus of $G(s)H(s)$ and then observing its behavior with reference to the $(-1, j0)$ point of the complex function plane.

The effect of the exponential term in Eq. (9-88) is that it rotates the phasor $G_1(j\omega)H_1(j\omega)$ at each ω by an angle of ωT_d radians in the clockwise direction. The amplitude of $G_1(j\omega)H_1(j\omega)$ is not affected by the time delay, since the magnitude of $e^{-j\omega T_d}$ is unity for all frequencies.

In control systems the magnitude of $G_1(j\omega)H_1(j\omega)$ usually approaches zero as ω approaches infinity. Thus, the Nyquist locus of the transfer function of Eq.

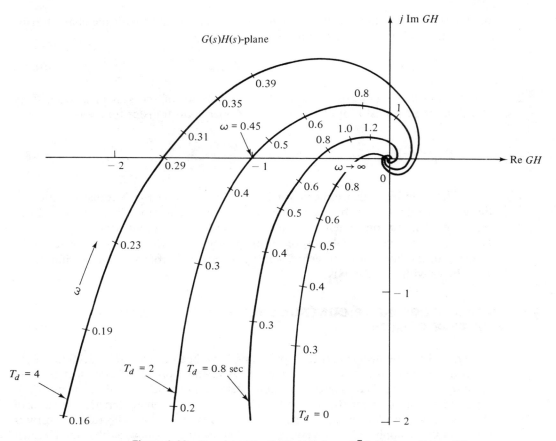

Figure 9-29   Nyquist plots of $G(s)H(s) = e^{-T_d s} / [s(s + 1)(s + 2)]$.

(9-88) will usually spiral toward the origin as ω approaches infinity, and there are infinite number of intersects on the negative real axis of the $G(s)H(s)$-plane. For the closed-loop system to be stable, all the intersects of the $G(j\omega)H(j\omega)$ locus with the real axis must occur to the right of the $(-1, j0)$ point.

Figure 9-29 shows an example of the Nyquist plot of

$$G(s)H(s) = e^{-T_d s}G_1(s)H_1(s) = \frac{e^{-T_d s}}{s(s + 1)(s + 2)} \tag{9-89}$$

for several values of T_d. It is observed from this diagram that the closed-loop system is stable when the time delay T_d is zero, but the stability condition deteriorates as T_d increases. The system is on the verge of becoming unstable when $T_d = 2$ sec. This is shown with the Nyquist plot passing through the $(-1, j0)$ point.

Unlike the rational function case, the analytical solution of the crossover points on the real axis of the $G(s)H(s)$-plane is not trivial, since the equations that govern the crossings are no longer algebraic. For instance, the loop transfer function of Eq. (9-89) may be rationalized in the usual manner by multiplying its numerator and denominator by the complex conjugate of the denominator. The result is

$$G(j\omega)H(j\omega) = \frac{(\cos \omega T_d - j \sin \omega T_d)[-3\omega^2 - j\omega(2 - \omega^2)]}{9\omega^4 + \omega^2(2 - \omega^2)^2} \qquad (9\text{-}90)$$

The condition for crossings on the real axis of the $G(s)H(s)$-plane is

$$3\omega^2 \sin \omega T_d - \omega(2 - \omega^2)\cos \omega T_d = 0$$

which is not easily solved, given T_d.

Since the term $e^{-j\omega T_d}$ always has a magnitude of 1 for all frequencies, the crossover problem is readily solved in the Bode plot domain. Since the time-delay term affects only the phase but not the magnitude of $G(j\omega)H(j\omega)$, the phase of the latter is obtained in the Bode plot by adding a negative angle of $-\omega T_d$ to the phase curve of $G_1(j\omega)H_1(j\omega)$. The frequency at which the phase curve of $G(j\omega)H(j\omega)$

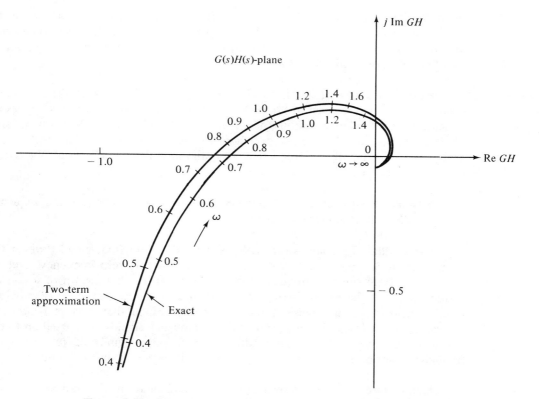

Figure 9-30 Approximation of Nyquist plot of system with time delay by truncated power-series expansion.

crosses the 180-degree axis is the place where the Nyquist locus intersects the negative real axis. In general, analysis and design problems involving pure time delays are more easily carried out graphically in the Bode diagram.

If the time delay is small, it is possible to approximate the time delay transfer function relation by a truncated power series; that is,

$$e^{-T_d s} = 1 - T_d s + \frac{T_d^2 s^2}{2!} - \frac{T_d^3 s^3}{3!} + \cdots \tag{9-91}$$

Figure 9-30 shows the Nyquist plots of the transfer function of Eq. (9-89) with $T_d = 0.8$ sec, and

$$G(s)H(s) = \frac{1 - T_d s}{s(s + 1)(s + 2)} \tag{9-92}$$

which is the result of truncating the series of Eq. (9-91) after two terms.

9.7 FREQUENCY-DOMAIN CHARACTERISTICS

In the design of a control system it is not sufficient to require that the system be stable. In addition to absolute stability, we need a set of specifications to describe the quality of the system in the frequency domain. The following frequency-domain specifications are often used in practice.

Peak Resonance M_p. The peak resonance M_p is defined as the maximum value of $M(\omega)$ that is given in Eq. (9-16). In general, the magnitude of M_p gives an indication of the relative stability of a feedback control system. Normally, a large M_p corresponds to a large peak overshoot in the step response. For most design problems it is generally accepted that an optimum value of M_p should be somewhere between 1.1 and 1.5.

Resonant Frequency ω_p. The resonant frequency ω_p is defined as the frequency at which the peak resonance M_p occurs.

Bandwidth. The bandwidth, BW, is defined as the frequency at which the magnitude of $M(j\omega)$, $M(\omega)$, drops to 70.7 percent of its zero-frequency level, or 3 dB down from the zero-frequency gain. In general, the bandwidth of a control system indicates the noise-filtering characteristics of the system. Also, bandwidth gives a measure of the transient response properties, in that a large bandwidth corresponds to a faster rise time, since higher-frequency signals are passed on to the outputs. Conversely, if the bandwidth is small, only signals of relatively low frequencies are passed, and the time response will generally be slow and sluggish.

Cutoff Rate. Often, bandwidth alone is inadequate in the indication of the characteristics of the system in distinguishing signals from noise. Sometimes it may be necessary to specify the cutoff rate of the frequency response at the high

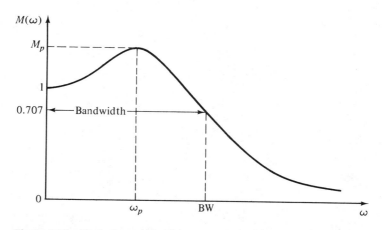

Figure 9-31 Typical magnification curve of a feedback control system.

frequencies. However, in general, a steep cutoff characteristic may be accompanied by a large M_p, which corresponds to a system with a low stability margin.

The performance criteria defined above for the frequency-domain analysis are illustrated on the closed-loop frequency response, as shown in Fig. 9-31.

There are other criteria that may be used to specify the relative stability and performance of a feedback control system. These are defined in the ensuing sections of this chapter.

9.8 M_p, ω_p, AND THE BANDWIDTH OF A SECOND-ORDER SYSTEM

For a second-order feedback control system, the peak resonance M_p, the resonant frequency ω_p, and the bandwidth are all uniquely related to the damping ratio ζ and the natural undamped frequency ω_n of the system. Consider the second-order sinusoidal steady-state transfer function of a closed-loop system,

$$M(j\omega) = \frac{C(j\omega)}{R(j\omega)} = \frac{\omega_n^2}{(j\omega)^2 + 2\zeta\omega_n(j\omega) + \omega_n^2}$$

$$= \frac{1}{1 + j2(\omega/\omega_n)\zeta - (\omega/\omega_n)^2} \tag{9-93}$$

We may simplify the last expression by letting $u = \omega/\omega_n$. Then, Eq. (9-93) becomes

$$M(ju) = \frac{1}{1 + j2u\zeta - u^2} \tag{9-94}$$

The magnitude and phase of $M(j\omega)$ are

$$|M(ju)| = M(u) = \frac{1}{\left[(1 - u^2)^2 + (2\zeta u)^2\right]^{1/2}} \tag{9-95}$$

and

$$\angle M(ju) = \phi_m(u) = -\tan^{-1}\frac{2\zeta u}{1 - u^2} \tag{9-96}$$

The resonant frequency is determined first by taking the derivative of $M(u)$ with respect to u and setting it equal to zero. Thus,

$$\frac{dM(u)}{du} = -\frac{1}{2}\left[(1 - u^2)^2 + (2\zeta u)^2\right]^{-3/2}(4u^3 - 4u + 8u\zeta^2) = 0 \tag{9-97}$$

from which

$$4u^3 - 4u + 8u\zeta^2 = 0 \tag{9-98}$$

The roots of Eq. (9-98) are

$$u_p = 0 \tag{9-99}$$

and

$$u_p = \sqrt{1 - 2\zeta^2} \tag{9-100}$$

The solution in Eq. (9-99) merely indicates that the slope of the $M(\omega)$ versus ω curve is zero at $\omega = 0$; it is not a true maximum. The solution of Eq. (9-100) gives the resonant frequency,

$$\omega_p = \omega_n\sqrt{1 - 2\zeta^2} \tag{9-101}$$

Since frequency is a real quantity, Eq. (9-101) is valid only for $1 \geq 2\zeta^2$ or $\zeta \leq 0.707$. This means simply that for all values of ζ greater than 0.707, the solution of $\omega_p = 0$ becomes the valid one, and $M_p = 1$.

Substituting Eq. (9-100) into Eq. (9-95) and simplifying, we get

$$M_p = \frac{1}{2\zeta\sqrt{1 - \zeta^2}} \quad (\zeta \leq 0.707) \tag{9-102}$$

It is important to note that M_p is a function of ζ only, whereas ω_p is a function of ζ and ω_n.

It should be noted that information on the magnitude and phase of $M(j\omega)$ of Eq. (9-93) may readily be derived from the Bode plot of Eq. (A-51), Figs. A-10 and A-12. In other words, Fig. A-12 is an exact representation of Eq. (9-96). The magnitude of $M(j\omega)$, however, may be represented in decibels versus frequency,

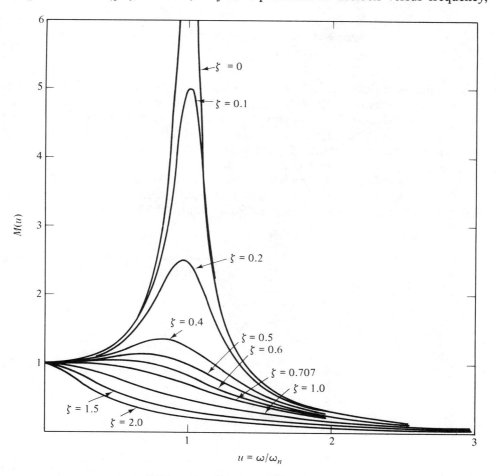

Figure 9-32 Magnification versus normalized frequency of a second-order closed-loop control system.

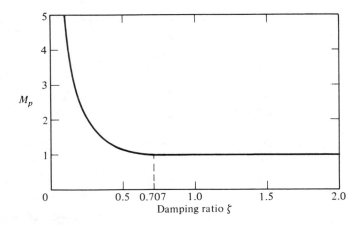

Figure 9-33 M_p-versus-damping ratio for a second-order system,
$M_p = 1 / (2\zeta\sqrt{1 - \zeta^2})$.

which is the Bode plot in Fig. A-10, or it may be plotted in absolute magnitude, $M(\omega)$, versus ω. Figure 9-32 illustrates the plots of $M(u)$ of Eq. (9-93) versus u for various values of ζ. Notice that if the frequency scale were unnormalized, the value of ω_p would increase when ζ decreases, as indicated by Eq. (9-101). When $\zeta = 0$, ω_p and ω_n become identical. Figures 9-33 and 9-34 illustrate the relationship between M_p and ζ, and $u = \omega_p/\omega_n$ and ζ, respectively.

As defined in Section 9.8, the bandwidth, BW, of a system is the frequency at which $M(\omega)$ drops to 70.7 percent of its zero-frequency value, or 3 dB down from the zero-frequency gain. Equating Eq. (9-93) to 0.707, we have

$$M(u) = \frac{1}{\left[(1 - u^2)^2 + (2\zeta u)^2\right]^{1/2}} = 0.707 \qquad (9\text{-}103)$$

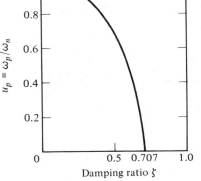

Figure 9-34 Normalized resonant frequency-versus-damping ratio for a second-order system, $u_p = \sqrt{1 - 2\zeta^2}$.

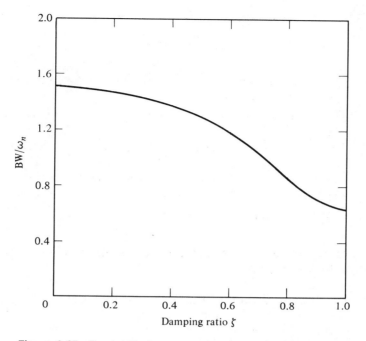

Figure 9-35 Bandwidth $/\omega_n$-versus-damping ratio for a second-order system, BW $= \omega_n [(1 - 2\zeta^2) + \sqrt{4\zeta^4 - 4\zeta^2 + 2}\,]^{1/2}$.

Thus,

$$\left[(1 - u^2)^2 + (2\zeta u)^2\right]^{1/2} = \sqrt{2} \tag{9-104}$$

This equation leads to

$$u^2 = (1 - 2\zeta^2) \pm \sqrt{4\zeta^4 - 4\zeta^2 + 2} \tag{9-105}$$

In the last expression the plus sign should be chosen, since u must be a positive real quantity for any ζ. Therefore, from Eq. (9-105), the bandwidth of the second-order system is determined as

$$\text{BW} = \omega_n \left[(1 - 2\zeta^2) + \sqrt{4\zeta^4 - 4\zeta^2 + 2}\right]^{1/2} \tag{9-106}$$

Figure 9-35 gives a plot of BW$/\omega_n$ as a function of ζ. It is of interest to note that for a fixed ω_n, as the damping ratio ζ decreases from unity, the bandwidth increases and the resonance peak M_p also increases.

For the second-order system under consideration, we easily establish some simple relationships between the time-domain response and the frequency-domain response of the system.

1. The maximum overshoot of the unit step response in the time domain depends upon ζ only [Eq. (6-96)].
2. The resonance peak of the closed-loop frequency response M_p depends upon ζ only [Eq. (9-102)].
3. The rise time increases with ζ, and the bandwidth decreases with the increase of ζ, for a fixed ω_n, Eq. (6-102), Eq. (9-106), Fig. 6-18, and Fig. 9-35. Therefore, bandwidth and rise time are inversely proportional to each other.
4. Bandwidth is directly proportional to ω_n.
5. Higher bandwidth corresponds to larger M_p.

9.9 EFFECTS OF ADDING A ZERO TO THE OPEN-LOOP TRANSFER FUNCTION

The relationships established between the time-domain and the frequency-domain responses arrived at in Section 9.8 are valid only for the second-order closed-loop transfer function of Eq. (9-93). When other transfer functions or more parameters are involved, the relationships between the time-domain and the frequency-domain responses are altered and may be more complex. It is of interest to consider the effects on the frequency-domain characteristics of a feedback control system when

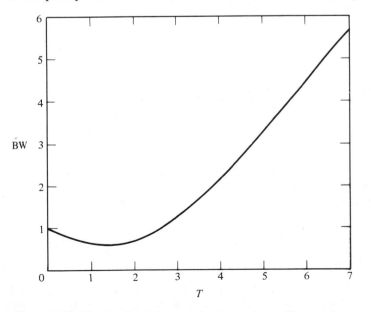

Figure 9-36 Bandwidth of a second-order system with open-loop transfer function $G(s) = (1 + Ts)/[s(s + 1.414)]$.

poles and zeros are added to the open-loop transfer function. It would be a simpler procedure to study the effects of adding poles and zeros to the closed-loop transfer function. However, it is more realistic to consider modifying the open-loop transfer function directly.

The closed-loop transfer function of Eq. (9-93) may be considered as that of a unity feedback control system with an open-loop transfer function of

$$G(s) = \frac{\omega_n^2}{s(s + 2\zeta\omega_n)} \tag{9-107}$$

Let us add a zero at $s = -1/T$ to the last transfer function so that Eq. (9-107) becomes

$$G(s) = \frac{\omega_n^2(1 + Ts)}{s(s + 2\zeta\omega_n)} \tag{9-108}$$

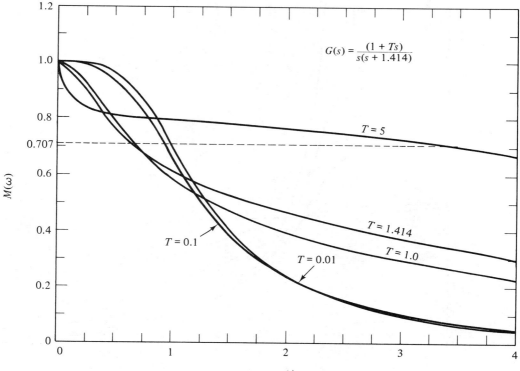

Figure 9-37 Magnification curves for a second-order system with an open-loop transfer function $G(s)$.

This corresponds to the second-order system with derivative control studied in Section 8.2. The closed-loop transfer function of the system is given by

$$M(s) = \frac{C(s)}{R(s)} = \frac{\omega_n^2(1 + Ts)}{s^2 + (2\zeta\omega_n + T\omega_n^2)s + \omega_n^2} \qquad (9\text{-}109)$$

In principle, M_p, ω_p, and BW of the system can all be derived using the same steps as illustrated in Section 9.8. However, since there are now three parameters in ζ, ω_n, and T, the exact expressions for M_p, ω_p, and BW are difficult to obtain even though the system is still of the second order. For instance, the bandwidth of the system is

$$\text{BW} = \left(-b + \frac{1}{2}\sqrt{b^2 + 4\omega_n^4}\right)^{1/2} \qquad (9\text{-}110)$$

where

$$b = 4\zeta^2\omega_n^2 + 4\zeta\omega_n^3 T - 2\omega_n^2 - \omega_n^4 T^2 \qquad (9\text{-}111)$$

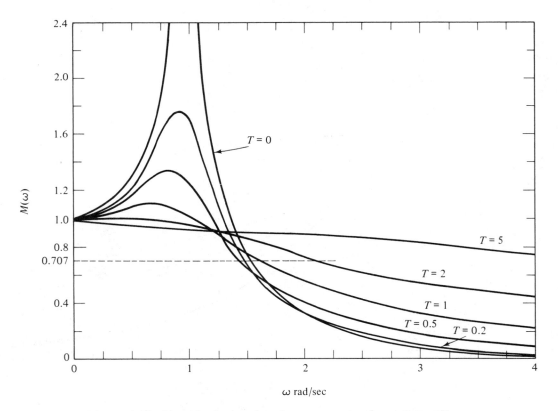

Figure 9-38 Magnification curves for a second-order system with an open-loop transfer function $G(s) = (1 + Ts)/[s(s + 0.4)]$.

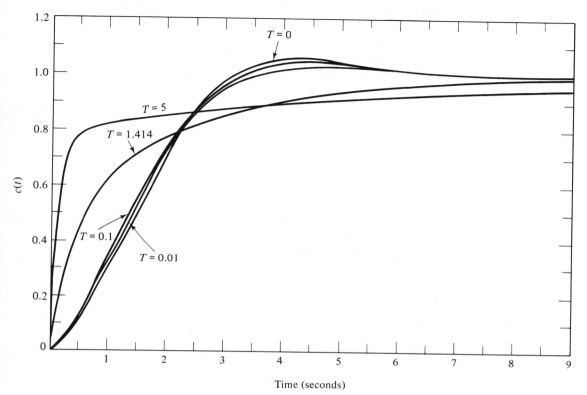

Figure 9-39 Unit step responses of a second-order system with an open-loop transfer function $G(s) = (1 + Ts) / [s(s + 1.414)]$.

It is difficult to see how each of the parameters in Eq. (9-110) affects the bandwidth. Figure 9-36 shows the relationship between BW and T for $\zeta = 0.707$ and $\omega_n = 1$. Notice that the general effect of adding a zero to the open-loop transfer function is to increase the bandwidth of the closed-loop system. However, for small values of T over a certain range the bandwidth is actually decreased. Figures 9-37 and 9-38 give the plots for $M(\omega)$ for the closed-loop system that has the $G(s)$ of Eq. (9-108) as its open-loop transfer function; $\omega_n = 1$, T is given various values, and $\zeta = 0.707$ and $\zeta = 0.2$, respectively. These curves show that for large values of T the bandwidth of the closed-loop system is increased, whereas there exists a range of smaller values of T in which the BW is decreased by the addition of the zero to $G(s)$. Figures 9-39 and 9-40 show the corresponding unit step responses of the closed-loop system. The time-domain responses indicate that a high bandwidth corresponds to a faster rise time. However, as T becomes very large, the zero of the closed-loop transfer function, which is at $s = -1/T$, moves very close to the origin, causing the system to have a large time constant. Thus Fig. 9-39 illustrates the situation that the rise time is fast but the large time constant of the zero near the origin of the s-plane causes the time response to drag out in reaching the final steady state.

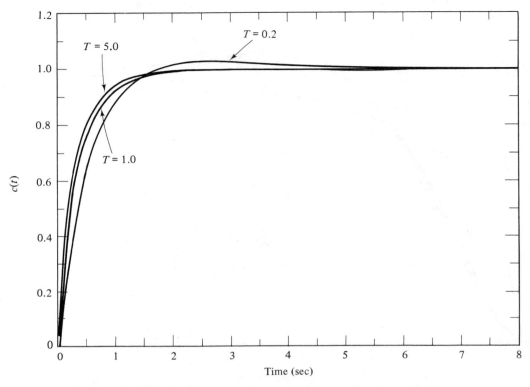

Figure 9-40   Unit step responses of a second-order system with an open-loop transfer function $G(s) = (1 + Ts) / [s(s + 0.4)]$.

9.10 EFFECTS OF ADDING A POLE TO THE OPEN-LOOP TRANSFER FUNCTION

The addition of a pole to the open-loop transfer function generally has the effect of decreasing the bandwidth of the closed-loop system. The following transfer function is arrived at by adding $(1 + Ts)$ to the denominator of Eq. (9-107):

$$G(s) = \frac{\omega_n^2}{s(s + 2\zeta\omega_n)(1 + Ts)} \tag{9-112}$$

The derivation of the bandwidth of the closed-loop system which has the $G(s)$ of Eq. (9-112) as its open-loop transfer function is quite difficult. It can be shown that the BW is the real solution of the following equation:

$$T^2\omega^6 + (1 + 4\zeta^2\omega_n^2T^2)\omega^4 + (4\zeta^2\omega_n^2 - 2\omega_n^2 - 4\zeta\omega_n^3T)\omega^2 - \omega_n^4 = 0 \tag{9-113}$$

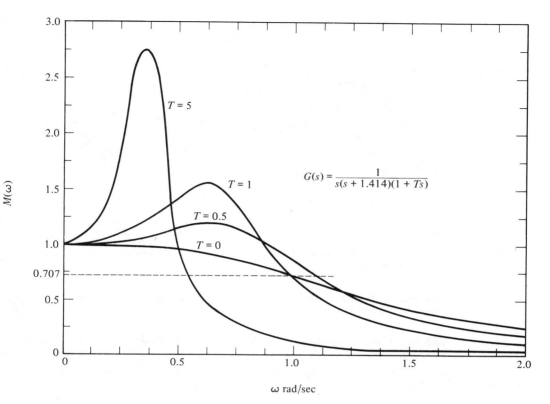

Figure 9-41 Magnification curves for a third-order system with an open-loop transfer function $G(s) = 1 / [s(s + 1.414)(1 + Ts)]$.

We may obtain a qualitative indication on the bandwidth properties by referring to Fig. 9-41, which shows the plots for $M(\omega)$ for $\omega_n = 1$, $\zeta = 0.707$, and various values of T. Since the system is now of the third order, it can be unstable for a certain set of system parameters. However, it can be easily shown by use of the Routh–Hurwitz criterion that for $\omega_n = 1$ and $\zeta = 0.707$, the system is stable for all positive values of T. The $M(\omega)$-versus-ω curves of Fig. 9-41 show that for small values of T, the bandwidth of the system is slightly increased but M_p is also increased. When T becomes large, the pole added to $G(s)$ has the effect of decreasing the bandwidth but increasing M_p. Therefore, we may conclude that, in general, the effect of adding a pole to the open-loop transfer function is to make the closed-loop system less stable. The unit step responses of Fig. 9-42 clearly show that for the larger values of T, $T = 1$ and $T = 5$, the rise time increases with the decrease of the bandwidth, and the larger values of M_p also correspond to greater peak overshoots in the step responses. However, it is important to point out that the correlation between M_p and the peak overshoot is meaningful only when the closed-loop system is stable. When the magnitude of $G(j\omega)$ equals unity, $M(\omega)$ is infinite, but if the closed-loop system is unstable with $|G(j\omega)| > 1$ at $\angle G(j\omega) = 180°$, $M(\omega)$ is finite and can assume an arbitrarily small number.

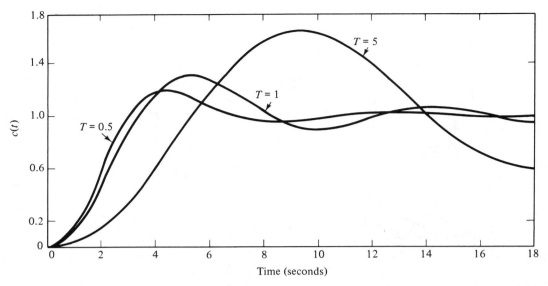

Figure 9-42   Unit step responses of a third-order system with an open-loop transfer function $G(s) = 1 / [s(s + 1.414)(1 + Ts)]$.

The objective of Sections 9.11 and 9.12 is to demonstrate the simple relationships between the bandwidth and M_p, and the characteristics of the time-domain response. The effects of adding a pole and a zero to the open-loop transfer function are discussed. However, no attempt is made to include all general cases.

9.11 RELATIVE STABILITY — GAIN MARGIN, PHASE MARGIN, AND M_p

We have demonstrated in Sections 9.8 through 9.10 the general relationship between the resonance peak M_p of the frequency response and the peak overshoot of the step response. Comparisons and correlations between frequency-domain and time-domain parameters such as these are useful in the prediction of the performance of a feedback control system. In general, we are interested not only in systems that are stable, but also in systems that have a certain degree of stability. The latter is often termed *relative stability*. In many situations we may use M_p to indicate the relative stability of a feedback control system. Another way of measuring relative stability of a closed-loop system is by means of the Nyquist plot of the loop transfer function, $G(s)H(s)$. The closeness of the $G(j\omega)H(j\omega)$ plot in the polar coordinates to the $(-1, j0)$ point gives an indication of how stable or unstable the closed-loop system is.

To demonstrate the concept of relative stability, the Nyquist plots and the corresponding step responses and frequency responses of a typical third-order system are shown in Fig. 9-43 for four different values of loop gain K. Let us consider the case shown in Fig. 9-43(a), in which the loop gain K is low, so the Nyquist plot of $G(s)H(s)$ intersects the negative real axis at a point (the phase-

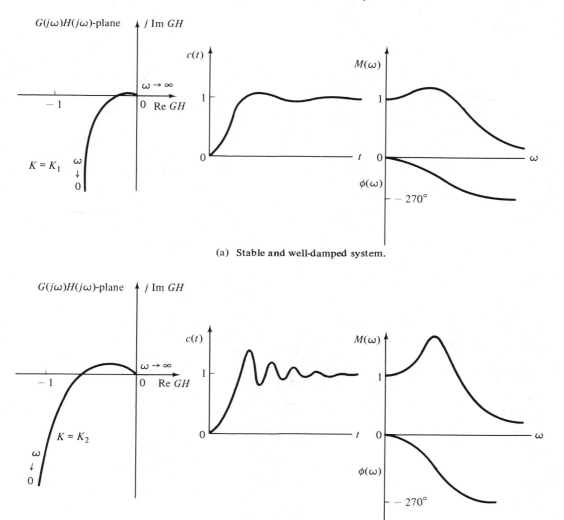

(a) Stable and well-damped system.

(b) Stable but oscillatory system.

Figure 9-43 Correlation among Nyquist plots, step responses, and frequency responses.

crossover point) quite far away from the $(-1, j0)$ point. The corresponding step response is shown to be quite well behaved, and M_p is low. As K is increased, Fig. 9-43(b) shows that the phase-crossover point is moved closer to the $(-1, j0)$ point; the system is still stable, but the step response has a higher peak overshoot, and M_p is larger. The phase curve for ϕ_m does not give as good an indication of relative stability as M_p, except that one should note the slope of the ϕ_m curve which gets steeper as the relative stability decreases. The Nyquist plot of Fig. 9-43(c) intersects the $(-1, j0)$ point, and the system is unstable with constant-amplitude oscillation, as shown by the step response; M_p becomes infinite. If K is increased still further,

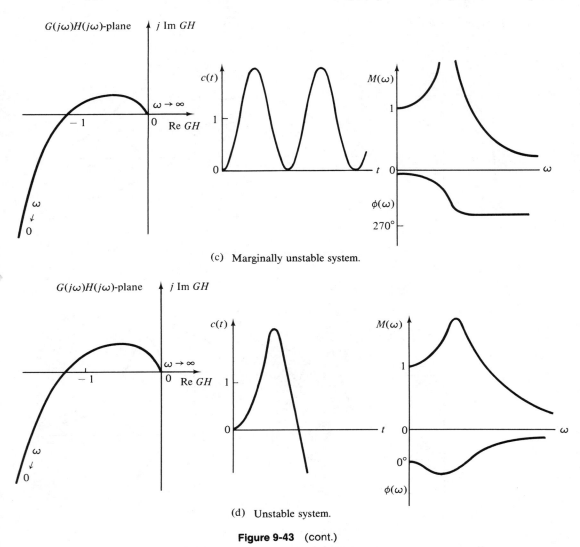

(c) Marginally unstable system.

(d) Unstable system.

Figure 9-43   (cont.)

the Nyquist plot will enclose the $(-1, j0)$ point, and the system is unstable with unbounded response, as shown in Fig. 9-43(d). In this case the magnitude curve $M(\omega)$ ceases to have any significance, and the only symptom of instability from the closed-loop frequency response is that the phase curve now has a positive slope at the resonant frequency.

Gain Margin

To give a quantitative way of measuring the relative distance between the $G(s)H(s)$ plot and the $(-1, j0)$ point, we define a quantity that is called the *gain margin*.

Specifically, the gain margin is a measure of the closeness of the phase-cross-over point to the $(-1, j0)$ point in the $G(s)H(s)$-plane. With reference to Fig. 9-44,

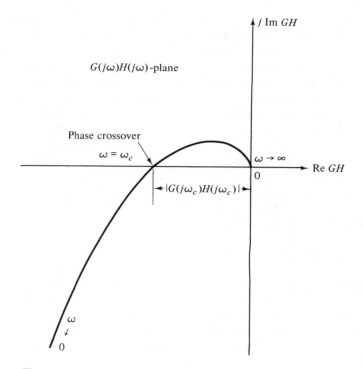

Figure 9-44 Definition of the gain margin in the polar coordinates.

the phase-crossover frequency is denoted by ω_c, and the magnitude of $G(j\omega)H(j\omega)$ at $\omega = \omega_c$ is designated by $|G(j\omega_c)H(j\omega_c)|$. Then the gain margin of the closed-loop system that has $G(s)H(s)$ as its loop transfer function is defined as

$$\text{gain margin} = \text{G.M.} = 20\log_{10}\frac{1}{|G(j\omega_c)H(j\omega_c)|} \qquad \text{dB} \qquad (9\text{-}114)$$

On the basis of this definition, it is noticed that in the $G(j\omega)H(j\omega)$ plot of Fig. 9.44, if the loop gain is increased to the extent that the plot passes through the $(-1, j0)$ point, so that $|G(j\omega_c)H(j\omega_c)|$ equals unity, the gain margin becomes 0 dB. On the other hand, if the $G(j\omega)H(j\omega)$ plot of a given system does not intersect the negative real axis, $|G(j\omega_c)H(j\omega_c)|$ equals zero, and the gain margin defined by Eq. (9-114) is infinite in decibels. Based on the foregoing evaluation, the physical significance of gain margin can be stated as follows: *Gain margin is the amount of gain in decibels that can be allowed to increase in the loop before the closed-loop system reaches instability.*

When the $G(j\omega)H(j\omega)$ plot goes through the $(-1, j0)$ point, the gain margin is 0 dB, which implies that the loop gain can no longer be increased as the system is already on the margin of instability. When the $G(j\omega)H(j\omega)$ plot does not intersect the negative real axis at any finite nonzero frequency, and the Nyquist stability criterion indicates that the $(-1, j0)$ point must not be enclosed for system stability, the gain margin is infinite in decibels; this means that, theoretically, the value of the loop gain can be increased to infinity before instability occurs.

When the $(-1, j0)$ point is to the right of the phase-crossover point, the magnitude of $G(j\omega_c)H(j\omega_c)$ is greater than unity, and the gain margin is given by Eq. (9-114) in decibels is negative. In the general case, when the above mentioned condition implies an unstable system, the gain margin is negative in decibels. It was pointed out in Section 9.2 that if $G(s)H(s)$ has poles or zeros in the right half of the s-plane, the $(-1, j0)$ point may have to be encircled by the $G(j\omega)H(j\omega)$ plot in order for the closed-loop system to be stable. Under this condition, a stable system yields a negative gain margin. In practice, we must first determine the stability of the system (i.e., stable or unstable), and then the magnitude of the gain margin is evaluated. Once the stability or instability condition is ascertained, the magnitude of the gain margin simply denotes the margin of stability or instability, and the sign of the gain margin becomes insignificant.

Phase Margin

The gain margin is merely one of the many ways of representing the relative stability of a feedback control system. In principle, a system with a large gain margin should be relatively more stable than one that has a smaller gain margin. Unfortunately, gain margin alone does not sufficiently indicate the relative stability of all systems, especially if parameters other than the loop gain are variable. For

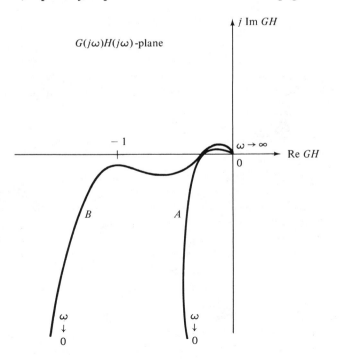

Figure 9-45 Nyquist plots showing systems with same gain margin but different amount of relative stability.

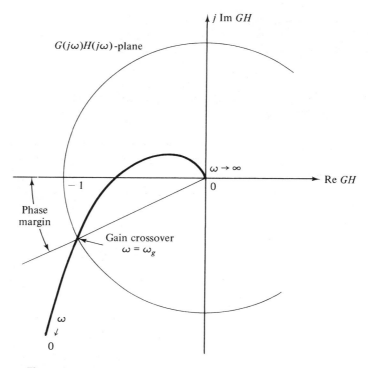

Figure 9-46   Phase margin defined in the $G(j\omega)H(j\omega)$-plane.

instance, the two systems represented by the $G(j\omega)H(j\omega)$ plots of Fig. 9-45 apparently have the same gain margin. However, locus A actually corresponds to a more stable system than locus B. The reason is that with any change in a system parameter (or parameters) other than the loop gain, it is easier for locus B to pass through or even enclose the $(-1, j0)$ point. Furthermore, system B has a much larger M_p than system A.

In order to strengthen the representation of relative stability of a feedback control system, we define the *phase margin* as a supplement to gain margin. *Phase margin is defined as the angle in degrees through which the $G(j\omega)H(j\omega)$ plot must be rotated about the origin in order that the gain-crossover point on the locus passes through the $(-1, j0)$ point.* Figure 9-46 shows the phase margin as the angle between the phasor that passes through the gain-crossover point and the negative real axis of the $G(j\omega)H(j\omega)$-plane. In contrast to the gain margin, which gives a measure of the effect of the loop gain on the stability of the closed-loop system, the phase margin indicates the effect on stability due to changes of system parameters, which theoretically alter the phase of $G(j\omega)H(j\omega)$ only.

The analytical procedure of computing the phase margin involves first the calculation of the phase of $G(j\omega)H(j\omega)$ at the gain-crossover frequency, and then subtracting 180 degrees from this phase; that is,

$$\text{phase margin} = \Phi.\text{M.} = \underline{/G(j\omega_g)H(j\omega_g)} - 180° \qquad (9\text{-}115)$$

where ω_g denotes the gain-crossover frequency. If the $(-1, j0)$ point is encircled by the $G(j\omega)H(j\omega)$ plot, the gain-crossover point would be found in the second quadrant of the $G(j\omega)H(j\omega)$-plane, and Eq. (9-115) would give a negative phase margin.

Graphic Methods of Determining Gain Margin and Phase Margin

Although the formulas for the gain margin and the phase margin are simple to understand, in practice it is more convenient to evaluate these quantities graphically from the Bode plot or the magnitude-versus-phase plot. As an illustrative example, consider that the open-loop transfer function of a control system with unity

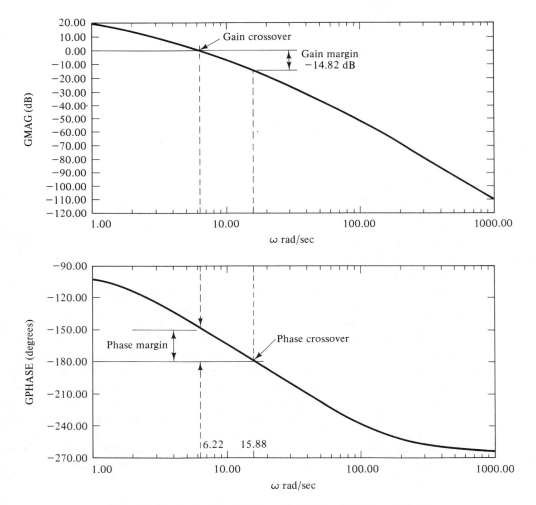

Figure 9-47   Bode plot of $G(s) = 10 / [s(1 + 0.2s)(1 + 0.02s)]$.

feedback is given by

$$G(s) = \frac{10}{s(1 + 0.02s)(1 + 0.2s)} \qquad (9\text{-}116)$$

The Bode plot of $G(j\omega)$ is shown in Fig. 9-47. Using the asymptotic approximation of $|G(j\omega)|$, the gain-crossover and the phase-crossover points are determined as shown in the figure. The phase-crossover frequency is 15.88 rad/sec, and the magnitude of $G(j\omega)$ at this frequency is about -15 dB. This means that if the loop gain of the system is increased by 15 dB, the magnitude curve will cross the 0-dB axis at the phase-crossover frequency. This condition corresponds to the Nyquist plot of $G(j\omega)$ passing through the $(-1, j0)$ point, and the system becomes

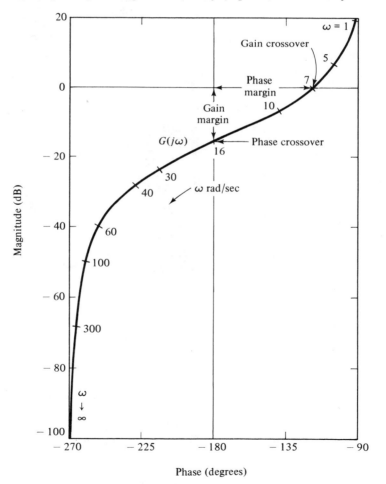

Figure 9-48 Magnitude-versus-phase plot of $G(s) = 10 / [s(1 + 0.2s)(1 + 0.02s)]$.

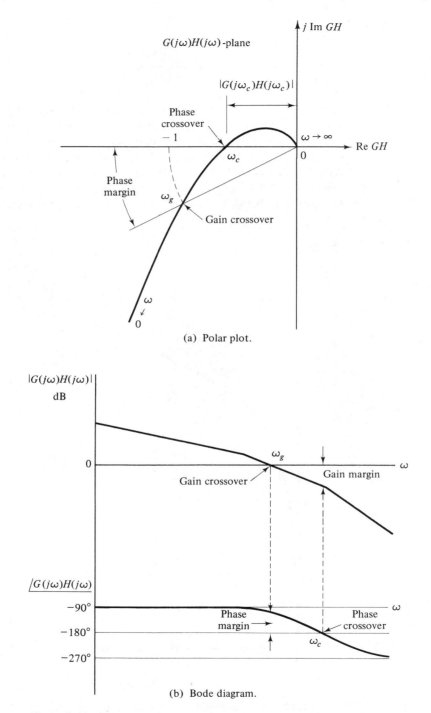

(a) Polar plot.

(b) Bode diagram.

Figure 9-49 Polar plot, Bode diagram, and magnitude-versus-phase plot, showing gain and phase margins in these domains.

marginally unstable. Therefore, from the definition of the gain margin, the gain margin of the system is 14.82 dB.

To determine the phase margin, we note that the gain-crossover frequency is at $\omega = 6.22$ rad/sec. The phase of $G(j\omega)$ at this frequency is approximately -125 degrees. The phase margin is the angle the phase curve must be shifted so that it will pass through the -180-degree axis at the gain-crossover frequency. In this case,

$$\Phi.M. = 180° - 148.28° = 31.72° \tag{9-117}$$

In general, the procedure of determining the gain margin and the phase margin from the Bode plot may be outlined as follows:

1. The gain margin is measured at the phase-crossover frequency ω_c:

$$G.M. = -|G(j\omega_c)H(j\omega_c)| \quad dB \tag{9-118}$$

2. The phase margin is measured at the gain-crossover frequency ω_g:

$$\Phi.M. = 180° + \underline{/G(j\omega_g)H(j\omega_g)} \tag{9-119}$$

The gain and phase margins are even better illustrated on the magnitude-versus-phase plot. For the transfer function of Eq. (9-116), the magnitude-versus-phase plot is shown in Fig. 9-48, which is constructed by use of the data from the Bode plot of Fig. 9-47. On the magnitude-versus-phase plot of $G(j\omega)$, the phase crossover point is where the locus intersects the -180-degree axis, and the gain

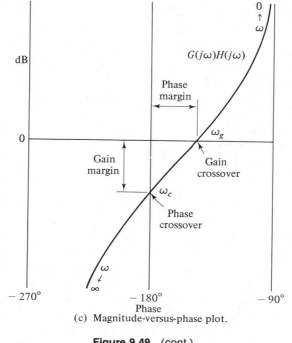

(c) Magnitude-versus-phase plot.

Figure 9.49 (cont.)

crossover is where the locus intersects the 0-dB axis. Therefore, the gain margin is simply the distance in decibels measured from the phase crossover to the critical point 0 dB and -180 degrees, and the phase margin is the horizontal distance in degrees measured from the gain crossover to the critical point.

In summarizing, the relations between the measurements of the gain margin and the phase margin in the polar plot, the Bode plot, and the magnitude-versus-phase plot are illustrated in Fig. 9-49.

9.12 RELATIVE STABILITY AS RELATED TO THE SLOPE OF THE MAGNITUDE CURVE OF THE BODE PLOT

In general, a definite relation between the relative stability of a closed-loop system and the slope of the magnitude curve of the Bode plot of $G(j\omega)H(j\omega)$ at the gain crossover can be established. For example, in Fig. 9-47, if the loop gain of the system is decreased from the nominal value, the gain crossover may be moved to the region in which the slope of the magnitude curve is only -20 dB/decade; the corresponding phase margin of the system would be increased. On the other hand, if the loop gain is increased, the relative stability of the system will deteriorate, and if the gain is increased to the extent that the gain crossover occurs in the region where the slope of the magnitude curve is -60 dB/decade, the system will definitely be unstable. The example cited above is a simple one, since the slope of the magnitude curve decreases monotonically as ω increases. Let us consider a conditionally stable system for the purpose of illustrating relative stability. Consider that a control system with unity feedback has the open-loop transfer function

$$G(s) = \frac{K(1 + 0.2s)(1 + 0.025s)}{s^3(1 + 0.01s)(1 + 0.005s)}$$

(9-120)

The Bode plot of $G(s)$ is shown in Fig. 9-50 for $K = 1$. The gain-crossover frequency is 1 rad/sec, and the phase margin is negative $(-78°)$. The closed-loop system is unstable even for a very small value of K. There are two phase crossover points: one at $\omega = 25.8$ rad/sec and the other at $\omega = 77.7$ rad/sec. The phase characteristics between these two frequencies indicate that if the gain crossover lies in this range, the system is stable. From the magnitude curve of the Bode plot, the range of K for stable operation is found to be between 69 and 85.5 dB. For values of K above and below this range, the phase lag of $G(j\omega)$ exceeds -180 degrees and the system is unstable. This system serves as a good example of the relation between relative stability and the slope of the magnitude curve at the gain crossover. As observed from Fig. 9-50, at both very low and very high frequencies, the slope of the magnitude curve is -60 dB/decade; if the gain crossover falls in either one of these two regions, the phase margin becomes negative and the system is unstable. In the two sections of the magnitude curve that have a slope of -40 dB/decade, the system is stable only if the gain crossover falls in about half of these regions, but even then the resultant phase margin is small. However, if the gain crossover falls in the region in which the magnitude curve has a slope of -20 dB/decade, the system is stable.

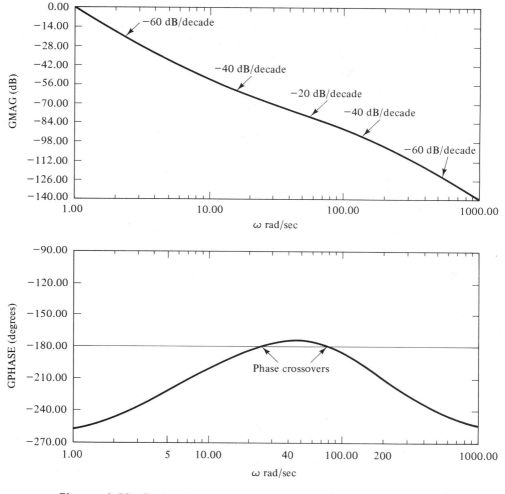

Figure 9-50 Bode plot of $G(s) = [K(1 + 0.2s)(1 + 0.025s)] / [s^3(1 + 0.01s)(1 + 0.005s)]$.

The correlation between the slope of the magnitude curve of the Bode plot at the gain crossover and the relative stability can be used in a qualitative way for design purposes.

9.13 CONSTANT M LOCI IN THE $G(j\omega)$-PLANE

In previous sections it was shown that the resonance peak M_p of the closed-loop frequency response is directly related to the maximum overshoot of the transient response. Normally, the magnification curve of $M(\omega)$ versus ω may be constructed by the Bode plot method if the closed-loop transfer function $M(s)$ is given and if its numerator and denominator are in factored form. Unfortunately, this is not the usual case, as the open-loop transfer function $G(s)$ is normally given. For the purpose of analysis we can always obtain the magnification curve by digital computation on a computer. However, our motivation is to be able to predict M_p from the plots of $G(j\omega)$, and eventually to design a system with a specified M_p.

Consider that the closed-loop transfer function of a feedback control system with unity feedback is given by

$$M(s) = \frac{C(s)}{R(s)} = \frac{G(s)}{1 + G(s)} \tag{9-121}$$

For sinusoidal steady state, $G(s) = G(j\omega)$ is written

$$G(j\omega) = \text{Re } G(j\omega) + j \text{ Im } G(j\omega)$$
$$= x + jy \tag{9-122}$$

Then

$$M(\omega) = |M(j\omega)| = \left| \frac{G(j\omega)}{1 + G(j\omega)} \right|$$

$$= \frac{\sqrt{x^2 + y^2}}{\sqrt{(1 + x)^2 + y^2}} \tag{9-123}$$

For simplicity, let $M = M(\omega)$; then Eq. (9-123) leads to

$$M\sqrt{(1 + x)^2 + y^2} = \sqrt{x^2 + y^2} \tag{9-124}$$

Squaring both sides of the last equation gives

$$M^2\left[(1 + x)^2 + y^2\right] = x^2 + y^2 \tag{9-125}$$

Rearranging this equation yields

$$(1 - M^2)x^2 + (1 - M^2)y^2 - 2M^2x = M^2 \tag{9-126}$$

This equation is conditioned by dividing through by $(1 - M^2)$ and adding the term

$[M^2/(1 - M^2)]^2$ on both sides. We have

$$x^2 + y^2 - \frac{2M^2}{1 - M^2}x + \left(\frac{M^2}{1 - M^2}\right)^2 = \frac{M^2}{1 - M^2} + \left(\frac{M^2}{1 - M^2}\right)^2 \quad \text{(9-127)}$$

which is finally simplified to

$$\left(x - \frac{M^2}{1 - M^2}\right)^2 + y^2 = \left(\frac{M}{1 - M^2}\right)^2 \quad \text{(9-128)}$$

For a given M, Eq. (9-128) represents a circle with the center at $x = M^2/(1 - M^2)$, $y = 0$. The radius of the circle is $r = |M/(1 - M)|$. Equation (9-128) is invalid for $M = 1$. For $M = 1$, Eq. (9-125) gives

$$x = -\frac{1}{2} \quad \text{(9-129)}$$

which is the equation of a straight line parallel to the $j\,\text{Im}\,G(j\omega)$ axis and passing through the $(-\frac{1}{2}, j0)$ point in the $G(j\omega)$-plane.

When M takes on different values, Eq. (9-128) describes in the $G(j\omega)$-plane a family of circles that are called the *constant M loci* or the *constant M circles*. The coordinates of the centers and the radii of the constant M loci for various values of M are given in Table 9-2 and some of the loci are shown in Fig. 9-51.

Table 9-2 Constant M Circles

M	Center $x = \dfrac{M^2}{1 - M^2}$, $y = 0$	Radius $r = \left\|\dfrac{M}{1 - M^2}\right\|$
0.3	0.01	0.33
0.5	0.33	0.67
0.7	0.96	1.37
1.0	∞	∞
1.1	-5.76	5.24
1.2	-3.27	2.73
1.3	-2.45	1.88
1.4	-2.04	1.46
1.5	-1.80	1.20
1.6	-1.64	1.03
1.7	-1.53	0.90
1.8	-1.46	0.80
1.9	-1.38	0.73
2.0	-1.33	0.67
2.5	-1.19	0.48
3.0	-1.13	0.38
4.0	-1.07	0.27
5.0	-1.04	0.21
6.0	-1.03	0.17

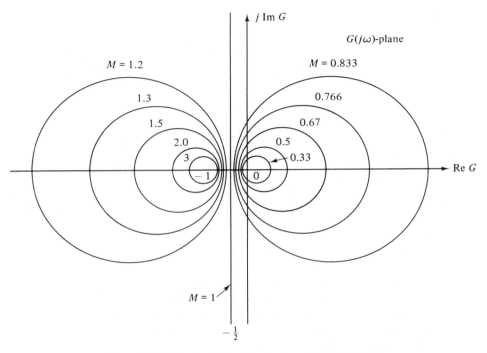

Figure 9-51 Constant M circles in the polar coordinates.

Note that when M becomes infinite, the circle degenerates into a point at the critical point, $(-1, j0)$. This agrees with the well-known fact that when the Nyquist plot of $G(j\omega)$ passes through the $(-1, j0)$ point, the system is marginally unstable and M_p is infinite. Figure 9-51 shows that the constant M loci in the $G(j\omega)$-plane are symmetrical with respect to the $M = 1$ line and the real axis. The circles to the left of the $M = 1$ locus correspond to values of M greater than 1, and those to the right of the $M = 1$ line are for M less than 1.

Graphically, the intersections of the $G(j\omega)$ plot and the constant M loci give the value of M at the frequency denoted on the $G(j\omega)$ curve. If it is desired to keep the value of M_p less than a certain value, the $G(j\omega)$ curve must not intersect the corresponding M circle at any point, and at the same time must not enclose the $(-1, j0)$ point. The constant M circle with the smallest radius that is tangent to the $G(j\omega)$ curve gives the value of M_p, and the resonant frequency ω_p is read off at the tangent point on the $G(j\omega)$ curve.

Figure 9-52(a) illustrates the Nyquist plot of $G(j\omega)$ for a unity feedback control system, together with several constant M loci. For a given loop gain $K = K_1$, the intersects between the $G(j\omega)$ curve and the constant M loci give the points on the $M(\omega)$-versus-ω curve. The peak resonance M_p is found by locating the smallest circle that is tangent to the $G(j\omega)$ plot. The resonant frequency is found at the point of tangency and is designated as ω_{p1}. If the loop gain is increased to K_2, and if the system is still stable, a constant M circle with a smaller radius that

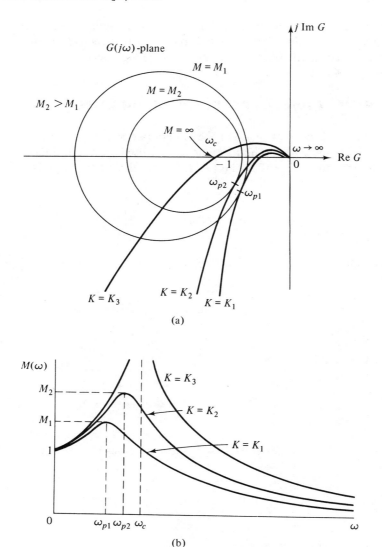

Figure 9-52 Polar plots of $G(s)$ and constant M loci showing the procedure of determining M_p and the magnification curves.

corresponds to a larger M is found tangent to the $G(j\omega)$ curve, and thus the peak resonance M_p is larger. The resonant frequency is shown to be ω_{p2}, which is closer to the phase-crossover frequency ω_c than ω_{p1}. When K is increased to K_3 so that the $G(j\omega)$ curve passes through the $(-1, j0)$ point, the system is marginally unstable, M_p is infinite, and $\omega_{p3} = \omega_c$. In all cases the bandwidth of the closed-loop system is found at the intersect of the $G(j\omega)$ curve and the $M = 0.707$ locus. For values of K

beyond K_3, the system is unstable, and the magnification curve and M_p no longer have any meaning. When enough points of intersections between the $G(j\omega)$ curve and the constant M loci are obtained, the magnification curves are plotted as shown in Fig. 9-52(b).

9.14 CONSTANT PHASE LOCI IN THE $G(j\omega)$-PLANE

The loci of constant phase of the closed-loop system may also be determined in the $G(j\omega)$-plane by a method similar to that used to secure the constant M loci. With reference to Eqs. (9-121) and (9-122), the phase of the closed-loop system is written as

$$\phi_m(\omega) = \underline{/M(j\omega)} = \tan^{-1}\left(\frac{y}{x}\right) - \tan^{-1}\left(\frac{y}{1+x}\right) \tag{9-130}$$

Taking the tangent on both sides of Eq. (9-130) and letting $\phi_m = \phi_m(\omega)$, we have

$$\tan\phi_m = \frac{y}{x^2 + x + y^2} \tag{9-131}$$

Let $N = \tan\phi_m$; then Eq. (9-131) becomes

$$x^2 + x + y^2 - \frac{y}{N} = 0 \tag{9-132}$$

Adding the term $(1/4) + (1/4N^2)$ to both sides of Eq. (9-132) yields

$$x^2 + x + \frac{1}{4} + y^2 - \frac{y}{N} + \frac{1}{4N^2} = \frac{1}{4} + \frac{1}{4N^2} \tag{9-133}$$

which is regrouped to give

$$\left(x + \frac{1}{2}\right)^2 + \left(y - \frac{1}{2N}\right)^2 = \frac{1}{4} + \frac{1}{4N^2} \tag{9-134}$$

When N assumes various values, this equation represents a family of circles with centers at $(x, y) = (-1/2, 1/2N)$. The radii are given by

$$r = \left(\frac{N^2 + 1}{4N^2}\right)^{1/2} \tag{9-135}$$

Table 9-3 Constant N Circles

$\phi_m = 180°n$ $n = 0, 1, 2 \ldots$	$N = \tan \phi_m$	Center $x = -\dfrac{1}{2},\; y = \dfrac{1}{2N}$	Radius $r\sqrt{\dfrac{N^2+1}{4N^2}}$
-90	$-\infty$	0	0.500
-60	-1.732	-0.289	0.577
-45	-1.000	-0.500	0.707
-30	-0.577	-0.866	1.000
-15	-0.268	-1.866	1.931
0	0	∞	∞
15	0.268	1.866	1.931
30	0.577	0.866	1.000
45	1.000	0.500	0.707
60	1.732	0.289	0.577
90	∞	0	0.500

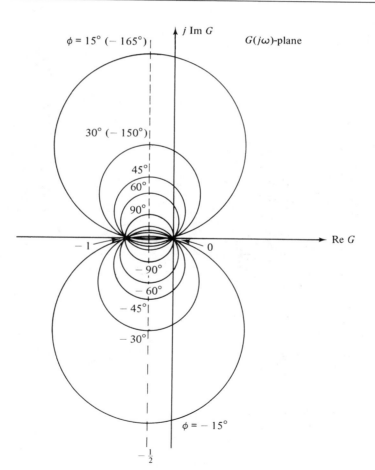

Figure 9-53 Constant N circles in the polar coordinates.

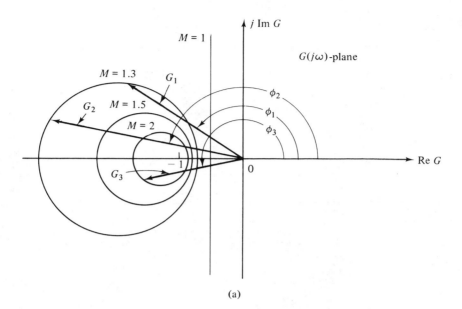

(a)

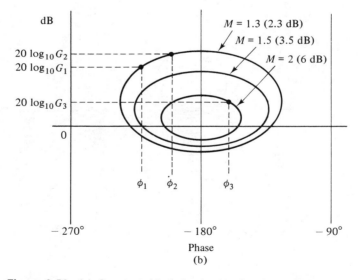

Phase
(b)

Figure 9-54 (a) Constant *M* circles in the *G(jω)*-plane. (b) Nichols chart in the magnitude-versus-phase coordinates.

The centers and the radii of the constant *N* circles for various values of *N* are tabulated in Table 9-3, and the loci are shown in Fig. 9-53.

9.15 CONSTANT *M* AND *N* LOCI IN THE MAGNITUDE-VERSUS-PHASE PLANE — THE NICHOLS CHART

In principle we need both the magnitude and the phase of the closed-loop frequency response to analyze the performance of the system. However, we have shown that the magnitude curve, which includes such information as M_p, ω_p, and BW, normally is more useful for relative stability studies.

A major disadvantage in working with the polar coordinates for the $G(j\omega)$ plot is that the curve no longer retains its original shape when a simple modification such as the change of the loop gain is made to the system. In design problems, frequently not only the loop gain must be altered, but series or feedback controllers are to be added to the original system which require the complete reconstruction of the resulting $G(j\omega)$. For design purposes it is far more convenient to work in the

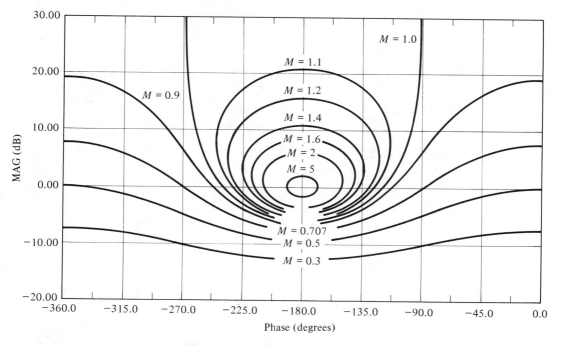

Figure 9-55   Nichols chart (for phase from −180° to 0°).

Bode diagram or the magnitude-versus-phase domain. In the Bode diagram the magnitude curve is shifted up and down without distortion when the loop gain is varied; in the magnitude-versus-phase plot, the entire $G(j\omega)$ curve is shifted up or down vertically when the gain is altered. In addition, the Bode plot can be easily modified to accommodate any modifications made to $G(j\omega)$ in the form of added poles and zeros.

The constant M and constant N loci in the polar coordinates may be transferred to the magnitude-versus-phase coordinates without difficulty. Figure 9-54 illustrates how this is done. Given a point on a constant M circle in the $G(j\omega)$-plane, the corresponding point in the magnitude-versus-phase plane may be determined by drawing a vector directly from the origin of the $G(j\omega)$-plane to the particular point on the constant M circle; the length of the vector in decibels and

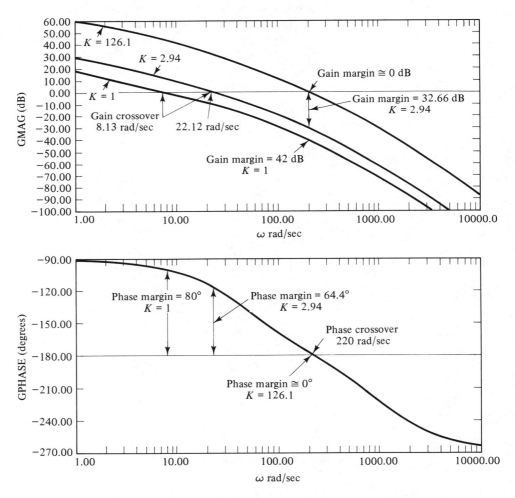

Figure 9-56 Bode plots of the open-loop transfer function of the system in Example 9-4.

the phase angle in degrees give the corresponding point in the magnitude-versus-phase plane. Figure 9-55 illustrates the process of locating three arbitrary corresponding points on the constant *M* loci in the magnitude-versus-phase plane. The critical point, $(-1, j0)$, in the $G(j\omega)$-plane corresponds to the point with 0 dB and -180 degrees in the magnitude-versus-phase plane.

Using the same procedure as described above, the constant *N* loci can also be transferred into the magnitude-versus-phase plane. These constant *M* and *N* loci in

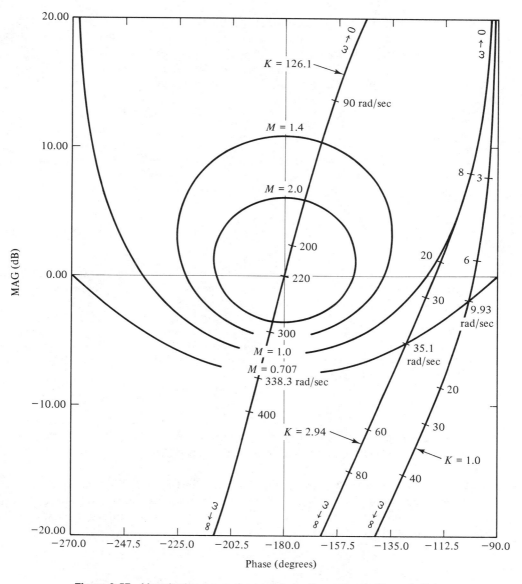

Figure 9-57 Magnitude-versus-phase plots for the system in Example 9-4.

Table 9-4

K	Peak Overshoot (%)	M_p	ω_p (rad/sec)	Gain Margin (dB)	Phase Margin (deg)
1.0	0	1.0	0	42	80
2.94	5	1.0	$\cong 9$	32	65
126.1	100	∞	220	0	0

the magnitude-versus-phase coordinates were originated by Nichols [12] and called the *Nichols chart*. A typical Nichols chart of constant-M loci is constructed in Fig. 9-55. To determine bandwidth, the $M = 0.707$ locus should be used.

The following example will illustrate the relationships among the analysis methods using the Bode plot, the magnitude-versus-phase plot, and the Nichols chart.

■

Example 9-4

Let us consider the printwheel control system discussed in Section 6.6. When the inductance of the dc motor is not neglected, the system is of the third order, and the open-loop transfer function of the system is given by Eq. (6-142), repeated here:

$$G(s) = \frac{400,000\,K}{s(s + 49)(s + 991)} \tag{9-136}$$

The Bode plot for $G(s)$ is drawn as shown in Fig. 9-56 for $K = 1$, 2.94, and 126.1. The gain margins and the phase margins of the closed-loop system can all be determined from the Bode plots of $G(j\omega)$. The data on the magnitude and phase of $G(j\omega)$ are transferred to the magnitude-versus-phase plots of Fig. 9.57. From these magnitude-versus-phase plots, together with the Nichols chart, we can find the values of the peak resonance, M_p, resonant frequency, ω_p, and the bandwidth, BW. Table 9-4 summarizes all the results of the frequency-domain analysis for the three values of K together with the time-domain results obtained in Section 6.6.

■

9.16 CLOSED-LOOP FREQUENCY RESPONSE ANALYSIS OF NONUNITY FEEDBACK SYSTEMS

The constant M and N loci and the Nichols chart analysis discussed in preceding sections are limited to closed-loop systems with unity feedback, whose transfer function is given by Eq. (9-121). When a system has nonunity feedback, the closed-loop transfer function is

$$M(s) = \frac{C(s)}{R(s)} = \frac{G(s)}{1 + G(s)H(s)} \tag{9-137}$$

The constant M loci derived earlier and the Nichols chart of Fig. 9-55 cannot be applied directly. However, we can show that with a slight modification these loci can still be applied to systems with nonunity feedback.

Let us consider the function

$$P(s) = \frac{G(s)H(s)}{1 + G(s)H(s)} \qquad (9\text{-}138)$$

Comparing Eq. (9-137) with Eq. (9-138), we have

$$P(s) = M(s)H(s) \qquad (9\text{-}139)$$

Information on the gain margin and the phase margin of the system of Eq. (9-137) can be obtained in the usual fashion by constructing the Bode plot of $G(s)H(s)$. However, the $G(j\omega)H(j\omega)$ curve and the Nichols chart together do not give the magnitude and phase plots for $M(j\omega)$, but for $P(j\omega)$. Since $M(j\omega)$ and $P(j\omega)$ are related through Eq. (9-139), once the plots for $P(\omega)$ versus ω and $\underline{/P(j\omega)}$ versus ω are obtained, the curves for $M(\omega)$ and $\phi_m(\omega)$ versus ω are determined from the following relationships:

$$M(\omega) = \frac{P(\omega)}{H(\omega)} \qquad (9\text{-}140)$$

$$\phi_m(\omega) = \underline{/P(j\omega)} - \underline{/H(j\omega)} \qquad (9\text{-}141)$$

9.17 SENSITIVITY STUDIES IN THE FREQUENCY DOMAIN [14]

The frequency-domain study of feedback control systems has an advantage in that the sensitivity of a transfer function with respect to a given parameter can be clearly interpreted. We shall show how the Nyquist plot and the Nichols chart can be utilized for analysis and design of a control system based on sensitivity considerations.

Consider that a control system with unity feedback has the transfer function

$$M(s) = \frac{C(s)}{R(s)} = \frac{G(s)}{1 + G(s)} \qquad (9\text{-}142)$$

The sensitivity of $M(s)$ with respect to $G(s)$ is defined as

$$S_G^M(s) = \frac{dM(s)/M(s)}{dG(s)/G(s)} \qquad (9\text{-}143)$$

or

$$S_G^M(s) = \frac{dM(s)}{dG(s)} \frac{G(s)}{M(s)} \qquad (9\text{-}144)$$

Substituting Eq. (9-142) into Eq. (9-144) and simplifying, we have

$$S_G^M(s) = \frac{1}{1 + G(s)} \qquad (9\text{-}145)$$

Clearly, the sensitivity function is a function of the complex variable s.

In general, it is desirable to keep the sensitivity to a small magnitude. From a design standpoint, it is possible to formulate a design criterion in the following form:

$$|S_G^M(s)| = \frac{1}{|1 + G(s)|} \leq k \qquad (9\text{-}146)$$

In the sinusoidal steady state, Eq. (9-146) is easily interpreted in the polar coordinate by a Nyquist plot. Equation (9-146) is written

$$|1 + G(j\omega)| \geq \frac{1}{k} \qquad (9\text{-}147)$$

Figure 9-58 illustrates the Nyquist plot of a stable closed-loop system. The constraint on sensitivity given in Eq. (9-147) is interpreted as the condition that the $G(j\omega)$ locus must not enter the circle with radius k. It is interesting to note that the sensitivity criterion is somewhat similar to the relative stability specifications of gain and phase margins. When the value of k is unity, the $G(j\omega)$ locus must be tangent or outside the circle with a unity radius and centered at the $(-1, j0)$ point. This corresponds to a very stringent stability requirement, since the gain margin is infinite. On the other hand, if the Nyquist plot of $G(j\omega)$ passes through the $(-1, j0)$ point, the system is unstable, and the sensitivity is infinite.

Equation (9-147) and Fig. 9-58 also indicate clearly that for low sensitivity, the magnitude of $G(j\omega)$ should be high, which reduces the stability margin. This again points to the need of compromise among the criteria in designing control systems.

Although Fig. 9-58 gives a clear interpretation of the sensitivity function in the frequency domain, in general, the Nyquist plot is awkward to use for design purposes. In this case the Nichols chart is again more convenient for the purpose of analysis and design of a feedback control system with a prescribed sensitivity. Equation (9-145) is written

$$S_G^M(j\omega) = \frac{G^{-1}(j\omega)}{1 + G^{-1}(j\omega)} \qquad (9\text{-}148)$$

which clearly indicates that the magnitude and phase of $S_G^M(j\omega)$ can be obtained by

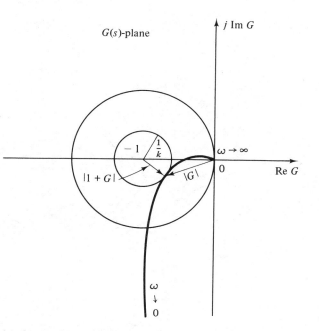

Figure 9-58 Interpretation of sensitivity criterion with the Nyquist plot.

plotting $G^{-1}(j\omega)$ in the Nichols chart and making use of the constant M loci for constant sensitivity function. Since the vertical coordinate of the Nichols chart is in decibels, the $G^{-1}(j\omega)$ curve in the magnitude-versus-phase coordinates can be easily obtained if the $G(j\omega)$ is already available, since

$$|G^{-1}(j\omega)|\,\text{dB} = -|G(j\omega)|\,\text{dB} \tag{9-149}$$

$$\underline{/G^{-1}(j\omega)} = -\underline{/G(j\omega)} \tag{9-150}$$

As an illustrative example, the function $G^{-1}(j\omega)$ for Eq. (9-136), Example 9-4 is plotted in the Nichols chart as shown in Fig. 9-59, for $K = 2.94$. The intersects of the $G^{-1}(j\omega)$ curve in the Nichols chart with the constant M loci give the magnitudes of $S_G^M(j\omega)$ at the corresponding frequencies. Figure 9-59 indicates several interesting points with regard to the sensitivity function of the feedback control system. The sensitivity function S_G^M approaches 0 dB or unity as ω approaches infinity. S_G^M becomes zero as ω approaches zero. A peak value of 1.1 dB is reached by S_G^M at $\omega = 25$ rad/sec. This means that the closed-loop transfer function is most sensitive to the change of $G(j\omega)$ at this frequency and more generally in this frequency range. This result is not difficult to comprehend, since from the Nichols chart it is observed that the stability and dynamic behavior of the closed-loop system is more directly governed by the $G(j\omega)$ curve near ω_p. Changes made to portions of $G(j\omega)$ at frequencies much higher and much lower than 25 rad/sec are not going to have a great effect on the relative stability of the system directly. When the loop gain of the system increases, the $G(j\omega)$ curve is raised in the Nichols chart domain, and the $G^{-1}(j\omega)$ curve must be lowered. If the $G(j\omega)$

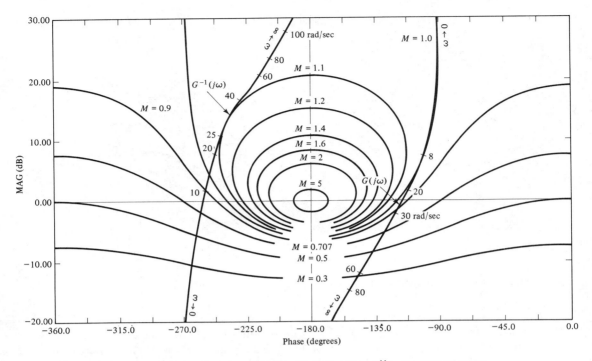

Figure 9-59   Determination of the sensitivity function S_G^M in the Nichols chart.

curve passes through the critical point at 0 dB and -180 degrees, the system becomes marginally unstable; the $G^{-1}(j\omega)$ also passes through the same point, and the sensitivity is infinite.

In this section we have simply demonstrated the use of the Nichols chart for the analysis of the sensitivity function of a closed-loop system. In a design problem, the objective may be to find a controller such that the sensitivity due to certain system parameters is small.

9.18 FREQUENCY RESPONSE OF DIGITAL CONTROL SYSTEMS

All the frequency-domain methods discussed in preceding sections can be applied directly to the analysis of digital control systems. For example, the Nyquist criterion can be applied without any modification to the stability analysis of a closed-loop digital control system.

Given any transfer function for a digital control system in the form $G^*(s)$ or $G(z)$, the frequency-domain analysis can be conducted by setting s equal to $j\omega$. Thus, for a single-loop feedback control system we can apply the Nyquist criterion to investigate the stability of the closed-loop system simply by studying the behavior of the Nyquist plot of $G^*(j\omega)$ with respect to the $(-1, j0)$ point in the $G^*(j\omega)$-plane, or of $G(e^{j\omega T})$ with respect to the $(-1, j0)$ point in the $G(e^{j\omega T})$-plane. As an

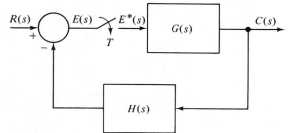

Figure 9-60 Closed-loop digital control system.

alternative, we can also define the Nyquist path to be the unit circle $|z| = 1$ in the z-plane, since its points correspond to the $j\omega$ axis in the s-plane, and plot $G(z)$ as z takes on values along this unit circle. We shall expand on the details of these methods in the following discussion.

Let us consider the digital control system with the block diagram as shown in Fig. 9-60. The closed-loop transfer function of the system is written

$$\frac{C^*(s)}{R^*(s)} = \frac{G^*(s)}{1 + GH^*(s)} \tag{9-151}$$

where

$$GH^*(s) = [G(s)H(s)]^* \tag{9-152}$$

or in z-transform notation,

$$\frac{C(z)}{R(z)} = \frac{G(z)}{1 + GH(z)} \tag{9-153}$$

where

$$GH(z) = \mathfrak{z}[G(s)H(s)] \tag{9-154}$$

Just as in the case of continuous-data systems, for absolute and relative stability analysis, we can make use of the frequency-domain plots of $GH^*(s)$ or $GH(z)$. The z-transform method of making these plots is discussed next.

Given the transfer function $GH(z)$, its frequency-domain plot is made by setting $z = e^{j\omega T}$ and then let ω vary from 0 to ∞. The plot thus obtained corresponds to the mapping of the positive half of the $j\omega$ axis in the s-plane. Alternatively, we can set z to the values along the unit circle $|z| = 1$ in the z-plane and make the corresponding plot of $GH(z)$. This brings out an interesting point: that since the unit circle repeats for every sampling frequency ω_s, when we vary ω along the $j\omega$ axis, we need only to cover the range from $\omega = 0$ to $\omega = \omega_s$. Figure 9-61 illustrates the frequency relation between the $j\omega$ axis in the s-plane and the unit circle in the z-plane.

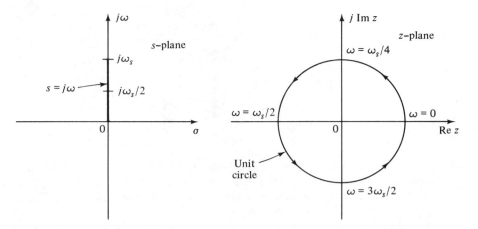

Figure 9-61 Relation between the $j\omega$ axis in the s-plane and the unit circle in the z-plane.

As an illustrative example, let

$$G(s)H(s) = \frac{1.57}{s(s+1)} \tag{9-155}$$

and the sampling frequency is 4 rad/sec. The z-transform of Eq. (9-155) is

$$GH(z) = \frac{1.243z}{(z-1)(z-0.208)} \tag{9-156}$$

As shown in Fig. 9-61, the points on the unit circle in the z-plane and the points on the $j\omega$ axis of the s-plane are related to each other as follows:

$$z = 1\underline{/90°} \qquad \omega = n\omega_s/4$$
$$z = 1\underline{/180°} \qquad \omega = 2n\omega_s/4$$
$$z = 1\underline{/270°} \qquad \omega = 3n\omega_s/4$$
$$z = 1\underline{/360°} \qquad \omega = n\omega_s$$

where n is any integer. Therefore, for Eq. (9-156),

$$GH\left(z = 1\underline{/90°}\right) = 0.863\underline{/-147°}$$
$$GH\left(z = 1\underline{/180°}\right) = 0.515\underline{/-180°}$$
$$GH\left(z = 1\underline{/270°}\right) = 0.863\underline{/147°}$$
$$GH\left(z = 1\underline{/360°}\right) = \infty\underline{/90°}$$

From these data, the plot of $GH(z)$, $z = e^{j\omega T}$, is sketched as shown in Fig. 9-62. Had the calculations of $GH(z)$ been carried out beyond $\omega = \omega_s = 4$ rad/sec, the same plot would have been obtained for $n\omega_s \le \omega < (n+1)\omega_s$, $n = \pm 1, \pm 2, \ldots$.

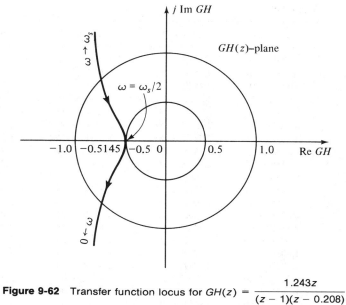

Figure 9-62 Transfer function locus for $GH(z) = \dfrac{1.243z}{(z-1)(z-0.208)}$ with $T = 1.57$ sec or $\omega_s = 4$ rad / sec.

In fact, since the frequency locus is symmetrical about the real axis, we need to compute the data only from $\omega = 0$ to $\omega_s/2$.

It should be pointed out that the $GH(z)$ plot shown in Fig. 9-62 can also be identified as $GH^*(j\omega)$, since the two functions are identical under the specific assignments of $z = e^{j\omega T}$ and $s = j\omega$.

In general, it is convenient to plot any transfer function of z by obtaining data from a digital computer.

Table 9-5 gives the values of $|GH(z)|$, $|GH(z)|$ in dB, and the phase of $GH(z)$ with $z = e^{j\omega T}$ for the transfer function in Eq. (8-156) with $T = 1.57$ sec or $\omega_s = 4$ rad/sec.

The frequency locus shown in Fig. 9-62 can be used for stability studies. The intersect of the locus on the real axis is at -0.5145. Thus, from the Nyquist criterion, the $(-1, j0)$ point is not enclosed, and the closed-loop system is stable. Clearly, if the loop gain of the system is increased by a factor of $1/0.5145$, the system would be unstable.

The frequency locus of $GH(z)$ in Fig. 9-62 is transferred to the gain-phase coordinates and superimposed on the Nichols chart. Applying the relative stability techniques in the usual fashion, we find the following quantities:

$$\text{gain margin} = 5.77 \text{ dB}$$

$$\text{phase margin} = 39 \text{ degrees}$$

$$\text{resonance peak } M_p = 4 \text{ dB or } 1.58$$

Table 9-5

OMEGA	GZMAG	GZDB	PHASE
$6.670E - 02$	$1.497E + 01$	$2.350E + 01$	$-9.457E + 01$
$1.334E - 01$	$7.453E + 00$	$1.745E + 01$	$-9.911E + 01$
$2.001E - 01$	$4.937E + 00$	$1.387E + 01$	$-1.036E + 02$
$2.668E - 01$	$3.671E + 00$	$1.129E + 01$	$-1.080E + 02$
$3.335E - 01$	$2.906E + 00$	$9.265E + 00$	$-1.122E + 02$
$4.002E - 01$	$2.392E + 00$	$7.577E + 00$	$-1.164E + 02$
$4.669E - 01$	$2.024E + 00$	$6.125E + 00$	$-1.203E + 02$
$5.336E - 01$	$1.747E + 00$	$4.846E + 00$	$-1.242E + 02$
$6.003E - 01$	$1.532E + 00$	$3.704E + 00$	$-1.279E + 02$
$6.670E - 01$	$1.360E + 00$	$2.671E + 00$	$-1.314E + 02$
$7.337E - 01$	$1.221E + 00$	$1.731E + 00$	$-1.347E + 02$
$8.004E - 01$	$1.106E + 00$	$8.716E - 01$	$-1.379E + 02$
$8.671E - 01$	$1.010E + 00$	$8.331E - 02$	$-1.410E + 02$
$9.338E - 01$	$9.289E - 01$	$-6.404E - 01$	$-1.439E + 02$
$1.001E + 00$	$8.605E - 01$	$-1.305E + 00$	$-1.468E + 02$
$1.067E + 00$	$8.022E - 01$	$-1.914E + 00$	$-1.494E + 02$
$1.134E + 00$	$7.524E - 01$	$-2.471E + 00$	$-1.520E + 02$
$1.201E + 00$	$7.097E - 01$	$-2.979E + 00$	$-1.545E + 02$
$1.267E + 00$	$6.730E - 01$	$-3.440E + 00$	$-1.569E + 02$
$1.334E + 00$	$6.416E - 01$	$-3.855E + 00$	$-1.593E + 02$
$1.401E + 00$	$6.147E - 01$	$-4.227E + 00$	$-1.615E + 02$
$1.467E + 00$	$5.918E - 01$	$-4.557E + 00$	$-1.637E + 02$
$1.534E + 00$	$5.724E - 01$	$-4.845E + 00$	$-1.659E + 02$
$1.601E + 00$	$5.563E - 01$	$-5.094E + 00$	$-1.680E + 02$
$1.668E + 00$	$5.431E - 01$	$-5.302E + 00$	$-1.700E + 02$
$1.734E + 00$	$5.326E - 01$	$-5.472E + 00$	$-1.721E + 02$
$1.801E + 00$	$5.246E - 01$	$-5.604E + 00$	$-1.741E + 02$
$1.868E + 00$	$5.189E - 01$	$-5.698E + 00$	$-1.761E + 02$
$1.934E + 00$	$5.156E - 01$	$-5.754E + 00$	$-1.780E + 02$
$2.001E + 00$	$5.145E - 01$	$-5.773E + 00$	$-1.800E + 02$

These results provide the same kind of information on the relative stability of the closed-loop digital control system as in the case of an analog system. For comparison, we evaluated the output sequence $c(kT)$ for a unit step input, with the following results: $0, 1.243, 1.1995, 0.94247, 0.96052, 1.0133, 1.0077, 0.99695, 0.9985, 1.0007, 1.0003, 0.99985, 0.99995, 1.00000, \ldots$. Thus, the unit step response has a peak overshoot of approximately 25 percent, which is in line with what $M_p = 1.58$ indicates.

REFERENCES

1. H. NYQUIST, "Regeneration Theory," *Bell System. Tech. J.*, Vol. 11, pp. 126–147, Jan. 1932.

2. C. H. HOFFMAN, "How to Check Linear Systems Stability: I. Solving the Characteristic Equation," *Control Engineering*, pp. 75–80, Aug. 1964.

3. C. H. HOFFMAN, "How to Check Linear Systems Stability: II. Locating the Roots by Algebra," *Control Engineering*, pp. 84–88, Feb. 1965.

4. C. H. HOFFMAN, "How to Check Linear Systems Stability: III. Locating the Roots Graphically," *Control Engineering*, pp. 71–78, June 1965.

5. M. R. STOJIĆ and D. D. ŠILJAK, "Generalization of Hurwitz, Nyquist, and Mikhailov Stability Criteria," *IEEE Trans. Automatic Control*, Vol. AC-10, pp. 250–254, July 1965.

6. R. W. BROCKETT and J. L. WILLEMS, "Frequency Domain Stability Criteria—Part I," *IEEE Trans. Automatic Control*, Vol. AC-10, pp. 255–261, July 1965.

7. R. W. BROCKETT and J. L. WILLEMS, "Frequency Domain Stability Criteria—Part II," *IEEE Trans. Automatic Control*, Vol. AC-10, pp. 407–413, Oct. 1965.

8. D. D. ŠILJAK, "A Note on the Generalized Nyquist Criterion," *IEEE Trans. Automatic Control*, Vol. AC-11, p. 317, Apr. 1966.

9. L. EISENBERG, "Stability of Linear Systems with Transport Lag," *IEEE Trans. Automatic Control*, Vol. AC-11, pp. 247–254, Apr. 1966.

10. T. R. NATESAN, "A Supplement to the Note on the Generalized Nyquist Criterion," *IEEE Trans. Automatic Control*, Vol. AC-12, pp. 215–216, Apr. 1967.

11. Y. CHU, "Correlation between Frequency and Transient Response of Feedback Control Systems," *AIEE Trans. Application and Industry*, Part II, Vol. 72, p. 82, 1953.

12. C. A. DESOER and W. S. CHAN, "The Feedback Interconnection of Lumped Linear Time-invariant Systems," *Journal of The Franklin Institute*, Vol. 300, Nos. 5 and 6, pp. 335–351, Nov.–Dec. 1975.

13. B. H. WILLIS and R. W. BROCKETT, "The Frequency Domain Solution of Regulator Problems," *IEEE Trans. Automatic Control*, Vol. AC-10, pp. 262–267, July 1965.

14. A. GELB, "Graphical Evaluation of the Sensitivity Function Using the Nichols Chart," *IRE Trans. Automatic Control*, Vol. AC-7, pp. 57–58, July 1962.

PROBLEMS

9.1. For the following loop transfer functions, sketch the Nyquist plots that correspond to the entire Nyquist path. In each case check the values of N, P, and Z with respect to the origin in the $G(s)H(s)$-plane. Determine the values of N, P, and Z with respect to the -1 point, and determine if the closed-loop system is stable. Specify in which case it is necessary to sketch only the Nyquist plot for $\omega = 0$ to ∞ (section 1) on the Nyquist path to investigate stability of the closed-loop system.

(a) $G(s)H(s) = \dfrac{20}{s(1 + 0.1s)(1 + 0.5s)}$

(b) $G(s)H(s) = \dfrac{10}{s^2(1 + 0.2s)(1 + 0.5s)}$

(c) $G(s)H(s) = \dfrac{100(1 + s)}{s(1 + 0.1s)(1 + 0.2s)(1 + 0.5s)}$

(d) $G(s)H(s) = \dfrac{5(s - 2)}{s(s + 1)(s - 1)}$

(e) $G(s)H(s) = \dfrac{50}{s(s + 5)(s - 1)}$

(f) $G(s)H(s) = \dfrac{3(s + 2)}{s^3 + 3s + 1}$

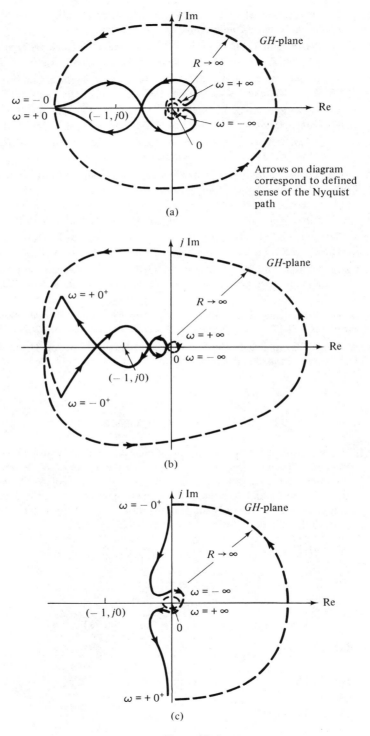

Figure 9P-4

9.2. Sketch the Nyquist plots for the following loop transfer functions. Sketch only the portion that is necessary to determine the stability of the closed-loop system. From the Nyquist plots determine the stability of the closed-loop system.

(a) $G(s)H(s) = \dfrac{100}{s(s^2 + s + 1)(s + 1)}$

(b) $G(s)H(s) = \dfrac{100}{s(s + 1)(s^2 + 2)}$

9.3. The open-loop transfer function of a unity feedback control system is

$$G(s) = \frac{K}{(s + 1)^n}$$

Determine the range of K for the closed-loop system to be stable by means of the Nyquist criterion. Sketch the necessary sections of the Nyquist plot of $G(s)$ for each of the following cases:

(a) $n = 2$
(b) $n = 3$
(c) $n = 4$

9.4. Figure 9P-4 shows the entire Nyquist plots of the loop gains $G(s)H(s)$ of some feedback control systems. It is known that in each case, the zeros of $G(s)H(s)$ are all located in the left half of the s-plane (i.e., $Z_0 = 0$). Determine the number of poles of $G(s)H(s)$ that are in the right half of the s-plane. State the stability of the open-loop systems. Determine if the closed-loop system is stable; if not, give the number of roots of the characteristic equation that are in the right half of the s-plane.

9.5. The complete Nyquist diagram of the open-loop transfer function $G(s)$ of a unity feedback control system is shown in Fig. 9P-5. Determine the open-loop stability and the closed-loop stability of the system under the following separate conditions:

(a) $G(s)$ has one zero in the right-half s-plane; the $(-1, j0)$ point is at point A.
(b) $G(s)$ has zero in the right-half s-plane; the $(-1, j0)$ point is at point B.
(c) $G(s)$ has no zeros in the right-half s-plane; the $(-1, j0)$ point is at point A.
(d) $G(s)$ has no zeros in the right-half s-plane; the $(-1, j0)$ point is at point B.

For each of the above four cases, if the open-loop or the closed-loop system (or both) is unstable under a given condition, give the number of poles of the open-loop transfer

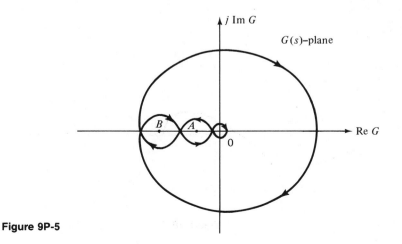

Figure 9P-5

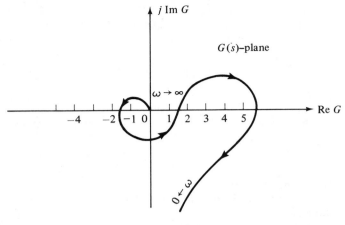

Figure 9P-6

function or the closed-loop transfer function, as the case may be, that are in the right-half s-plane.

9.6. The Nyquist plot of a unity feedback control system with *positive* feedback that corresponds to section 1 ($\omega = 0$ to $\omega = \infty$) is shown in Fig. 9P-6. It is known that the function $G(s)$ does not have any zeros or poles in the right-half s-plane. Determine the stability of the closed-loop system.

9.7. Consider a control system with unity feedback. Assume that it is only possible to measure the transfer function $1/G(s)$ experimentally. The Nyquist plot of $1/G(s)$ is sketched for $\omega = 0$ to $\omega = \infty$, as shown in Fig. 9P-7. Assuming that the function $1/G(s)$ does not have any poles or zeros in the right-half s-plane, determine the stability condition of the closed-loop system.

9.8. The open-loop transfer function of an inventory control system is

$$G(s) = \frac{4(K_P + K_D s)}{s^2}$$

and $H(s) = 1$.

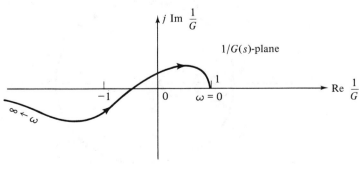

Figure 9P-7

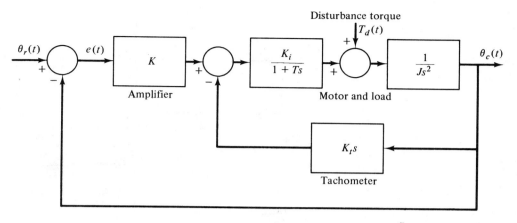

Figure 9P-10

(a) Sketch a Nyquist plot for $G(s)$ when $K_D = 0$ and K_P = variable parameter. (Only section 1, from $\omega = 0$ to $\omega = \infty$, is needed.)

(b) The parabolic error constant of the system, K_a, should be 40 sec^{-2}. Determine the value of K_P and sketch the *entire* Nyquist plot of an equivalent open-loop transfer function with K_D as a variable parameter.

9.9. The block diagram shown in Fig. 6P-27 represents the liquid-level control system described in Problem 6.27. The following system parameters are given: $K_a = 50$, $K_I = 50$, $K_b = 0.075$, $J = 0.006$, $R = 10$, $K_i = 10$, and $n = \frac{1}{100}$. The values of A, N, and K_0 are variable.

(a) For $A = 50$ and $K_0 = 50$, sketch the Nyquist plot of $G(s)$ and find the critical value of N so that the closed-loop system is asymptotically stable. Sketch only the section or sections of the Nyquist plot that are needed.

(b) Let $N = 20$, $K_0 = 50$. Sketch the Nyquist plot of an equivalent transfer function $G_{eq}(s)$ which has A as a multiplying constant. Find the critical value of A so that the closed-loop system is asymptotically stable.

(c) For $A = 50$, $N = 20$, sketch the Nyquist plot of an equivalent transfer function $G_{eq}(s)$ which has K_0 as a multiplying factor, and determine the critical value of K_0 for closed-loop stability.

9.10. The block diagram of a dc motor control system is shown in Fig. 9P-10. The fixed parameters of the system are $T = 0.1$, $J = 0.01$, and $K_i = 10$. For $K_t = 0.01$, apply the Nyquist criterion and determine the marginal value of K for the system to be asymptotically stable.

9.11. The characteristic equation of a feedback control system is

$$s^3 + 4Ks^2 + (K + 3)s + 10 = 0$$

Apply the Nyquist criterion to determine the values of K for a stable closed-loop system. Check the answer by means of the Routh–Hurwitz criterion.

9.12. The Nyquist criterion was originally devised to investigate the absolute stability of a closed-loop system. By sketching the Nyquist plot of $G(s)H(s)$ that corresponds to the Nyquist path, it is possible to tell whether the system's characteristic equation has roots in the right half of the s-plane.

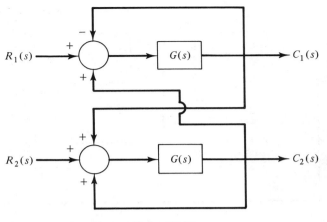

Figure 9P-13

(a) Define a new Nyquist path in the s-plane that may be used to ensure that all the complex roots of the characteristic equation have damping ratios greater than a value ζ_1.

(b) Define a new Nyquist path in the s-plane that may be used to ensure that all the characteristic equation roots are in the left half of the s-plane with real parts greater than α_1.

9.13. The block diagram of a multivariable control system is shown in Fig. 9P-13. The transfer function $G(s)$ is given by

$$G(s) = \frac{K}{(s + 1)(s + 2)}$$

and K is a positive constant. Determine the range of K so that the system is asymptotically stable.

(a) Use the Routh–Hurwitz criterion.

(b) Use the Nyquist criterion.

9.14. The system shown in Fig. 9P-14 is used to control the concentration of a chemical solution by mixing water and concentrated solution in appropriate proportions. The transfer function between the amplifier output e_a and the valve position x in inches is

$$\frac{X(s)}{E_a(s)} = \frac{K}{s^2 + 4s + 20} \qquad K = 5$$

When the sensor is viewing pure water, the amplifier output voltage e_a is zero; when it is viewing concentrated solution, $e_a = 10$ V, 0.1 in. of the valve motion changes the output concentration from zero to maximum concentration. The valve ports may be assumed to be shaped so that the output concentration varies linearly with the valve position. The output tube has a cross-sectional area of 0.1 in.^{2}, and the rate of flow is 10 in.^{3}/sec regardless of the valve position. To make sure that the sensor views a homogeneous solution, it is desirable to place it at some distance from the valve. Find the maximum distance between the valve and the sensor so that the system is stable by means of the Nyquist criterion.

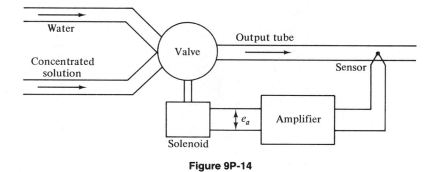

Figure 9P-14

9.15. For the system described in Problem 9.14, let K be variable and the distance between the valve and the sensor be 10 in. Find the marginal value of K so that the system is stable. Use the Nyquist criterion.

9.16. The pole-zero configuration of a closed-loop transfer function is shown in Fig. 9P-16.
  (a) Compute the bandwidth of the system.
  (b) A zero is added to the closed-loop system, as shown in Fig. 9P-16(b); how is the bandwidth affected?
  (c) Another pole is inserted on the negative real axis in Fig. 9P-16(b), but at a distance 10 times farther from the origin than the zero; how is the bandwidth affected?

9.17. The specification on a second-order control system with the closed-loop transfer function

$$\frac{C(s)}{R(s)} = \frac{\omega_n^2}{s^2 + 2\zeta\omega_n s + \omega_n^2}$$

is that the peak overshoot should not exceed 10 percent.
  (a) What are the corresponding limiting values of the damping ratio and peak resonance M_p?
  (b) Determine the corresponding values of ω_p and t_{max}.

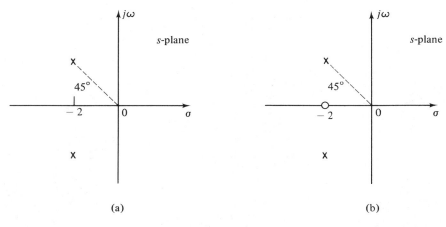

(a)   (b)

Figure 9P-16

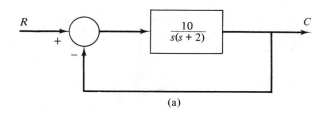

(a)

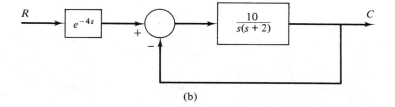

(b)

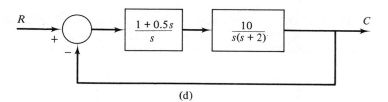

(c)

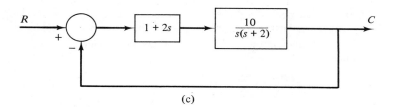

(d)

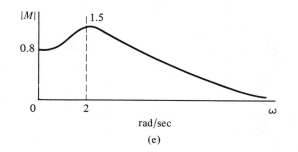

(e)

Figure 9P-18

9.18. Sketch the closed-loop frequency response $|M(j\omega)|$ as a function of frequency for the systems and responses shown in Fig. 9P-18(a)–(d).Sketch the unit step response for the system whose $|M|$-versus-ω curve is as shown in Fig. 9P-18(e). Assume that the system is of second order.

9.19. The closed-loop transfer function of a feedback control system is given by

$$M(s) = \frac{C(s)}{R(s)} = \frac{1}{(1 + 0.01s)(1 + 0.05s + 0.01s^2)}$$

(a) Plot the frequency-response curve for the closed-loop system.

(b) Determine the peak resonance peak M_p and the resonant frequency ω_p of the system.

(c) Determine the damping ratio ζ and the natural undamped frequency ω_n of the second-order system that will produce the same M_p and ω_p determined for the original system.

9.20. The open-loop transfer function of a unity feedback control system is

$$G(s) = \frac{K}{s(1 + 0.1s)(1 + s)}$$

(a) Determine the value of K so that the resonance peak M_p of the system is equal to 1.4.

(b) Determine the value of K so that the gain margin of the system is 20 dB.

(c) Determine the value of K so that the phase margin of the system is 60°.

9.21. The open-loop transfer function of a unity feedback control system is

$$G(s) = \frac{K(1 + Ts)}{s(1 + s)(1 + 0.01s)}$$

Determine the *smallest* possible value of T so that the system has an infinite gain margin.

9.22. The open-loop transfer function of a unity feedback control system is

$$G(s) = \frac{K}{s(1 + 0.1s)(1 + 0.001s)}$$

Determine the value of K if the steady-state error of the output position must be less than or equal to 0.1 percent for a ramp function input. With this value of K, what are the gain margin and the phase margin of the system? Plot $G(s)$ in the gain-phase plot and determine the resonance peak M_p and the resonant frequency ω_p.

9.23. A compensation network is added to the forward path of the system in Problem 9.22, so that now the open-loop transfer function reads

$$G(s) = \frac{K(1 + 0.0167s)}{s(1 + 0.00222s)(1 + 0.1s)(1 + 0.001s)}$$

where K is determined in Problem 9.22. Plot the gain-phase diagram of $G(s)$. Evaluate M_p, ω_p, the gain margin, the phase margin, and the bandwidth of the compensated system.

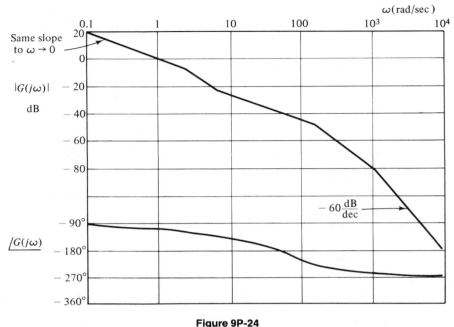

Figure 9P-24

9.24. The Bode diagram of the open-loop transfer function $G(s)$ of a unity feedback control system is shown in Fig. 9P-24.

(a) Find the gain margin and the phase margin of the system.

(b) If the open-loop transfer function is changed to $e^{-Ts}G(s)$, find the value of T so that the phase margin of the system is 45 degrees. Then find the value of T so that the gain margin is 20 dB.

(c) What is the ramp error constant of the system in part (a)? in part (b)?

9.25. The open-loop transfer function of the liquid-level control system is

$$G(s) = \frac{H(s)}{E_i(s)} = \frac{83.333N}{s(s + 12.5)(s + 1)}$$

(a) Sketch the Bode plot for $G(s)$ when $N = 1$. Find the gain crossover frequency and the phase crossover frequency on the Bode plot. Find the gain margin and the phase margin.

(b) Sketch $G(s)$ on the Nichols chart and obtain the closed-loop frequency plot of $M(j\omega)$, where $M(s)$ is the closed-loop transfer function. Find the resonance peak M_p and the resonant frequency ω_p. Find the bandwidth of the closed-loop system.

9.26. Figure 9P-26 shows the magnitude curve of the Bode plot of $G(j\omega)/K$, where $G(s)$ is the open-loop transfer function of a unity feedback control system. It is known that $G(s)$ has no poles or zeros in the right-half s-plane, and there is no pure time delay, so that the gain and phase relation of $G(s)$ are uniquely related.

(a) Find the gain margin and the phase margin of the closed-loop system for $K = 1$.

(b) Repeat part (a) for $K = 10$.

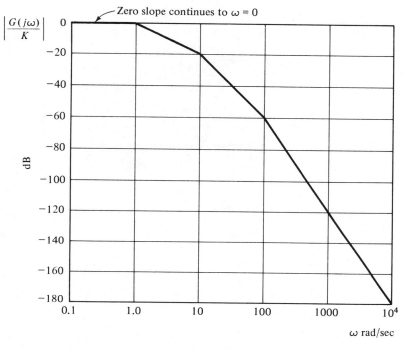

Figure 9P-26

(c) Find the steady-state error of the system when the reference input is a unit step function. Use $K = 1$.

(d) Find the resonance peak M_p and the bandwidth of the closed-loop system for $K = 10$.

9.27. Figure 9P-27 shows the block diagram of a control system which contains the robot-arm system shown in Fig. 4P-17. The constants of the system are

$$K_s = 1 \quad \text{and} \quad K_a = \text{amplifier gain}$$

For the robot-arm system,

$$K = 100 \qquad K_i = 0.4$$
$$B = 0.2 \qquad J_m = 0.2$$
$$B_L = 0.01 \qquad J_L = 0.6$$
$$B_m = 0.25$$

The objective of the control system is to control the velocity of the robot arm, ω_L.

(a) Derive the open-loop transfer function $\Omega_L(s)/E(s)$ using the coefficients given above. Assume that the disturbance torque T_L of the robot arm is zero.

(b) Find K_a if the error in the output velocity is to be less than or equal to 1 percent of the reference input velocity. Draw the Bode plot of $\Omega_L(s)/E(s)$ with the value of K_a found above.

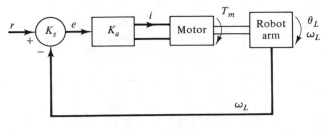

Figure 9P-27

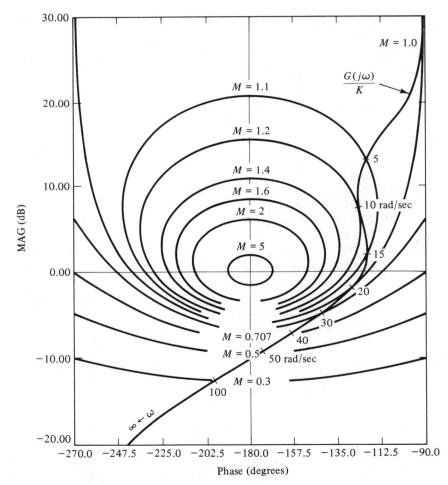

Figure 9P-28

(c) With the value of K_a found in (b), find the gain margin, gain crossover frequency, phase margin, and the phase crossover frequency.

(d) Draw the closed-loop frequency plot in $|M|$ versus ω. Find M_p, ω_p, and the bandwidth.

(e) Obtain the unit step response of $\omega_L(t)$ with the value of K_a found in (b).

9.28. The normalized open-loop transfer function of a feedback control system is shown in the Nichols chart in Fig. 9P-28. Assume that $G(s)$ is a rational function of s with constant coefficients. Find the following properties of the closed-loop system:

(a) Gain crossover frequency (rad/sec). $(K = 1)$

(b) Phase crossover frequency (rad/sec). $(K = 1)$

(c) Steady-state error when the reference input is a unit step function. $(K = 1)$

(d) Gain margin (db). $(K = 1)$

(e) Phase margin (degrees). $(K = 1)$

(f) Resonance peak M_p. $(K = 1)$

(g) Resonant frequency ω_p. $(K = 1)$

(h) Bandwidth of the open-loop system (rad/sec). $(K = 1)$

(i) Bandwidth of the closed-loop system (rad/sec). $(K = 1)$

(j) The value of K so that the gain margin is 20 db.

(k) The critical value of K so that the system will have sustained constant-amplitude oscillations. Find the frequency of oscillation in rad/sec.

9.29. For the Nichols chart of the $G(s)$ shown in Fig. 9P-28, assume that the open-loop system now has a time delay so that the transfer function becomes $G(s)e^{-Ts}$, where T is the time delay in seconds.

(a) With $K = 1$, find T so that the phase margin is $45°$.

(b) With $K = 1$, find the maximum value of T so that the system still maintains stability.

9.30. Figure 9P-30 shows the block diagram of a digital space-vehicle control system.

(a) Find the open-loop transfer function $G(z)$ and sketch it in the polar coordinates for $z = e^{j\omega T}$, $0 \leq \omega < \infty$ (Nyquist plot), for $K_P = 1,650,000$, $6,320,000$, and 10^7. Determine the stability condition of the closed-loop system under these three conditions.

(b) Find the gain margins and the phase margins of the three cases of K_P specified in part (a).

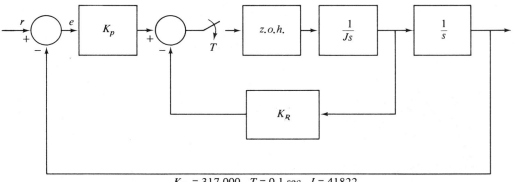

$K_R = 317{,}000$, $T = 0.1$ sec, $J = 41822$

Figure 9P-30

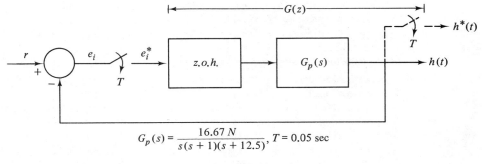

$$G_p(s) = \frac{16.67 N}{s(s + 1)(s + 12.5)}, \quad T = 0.05 \text{ sec}$$

Figure 9P-31

9.31. Figure 9P-31 shows the block diagram of the digital control version of the liquid-level control system. Construct a Nyquist plot of $G(z)/N$ with $z = e^{j\omega T}$ for $0 \le \omega < \omega_s/2$, where $\omega_s = 2\pi/T$. Determine the values of N for the closed-loop system to be stable.

chapter ten

Frequency-Domain Design of Control Systems

10.1 INTRODUCTION

In this chapter the design of control systems will be carried out in the frequency domain. The frequency-domain techniques utilizing Bode diagrams and the gain-phase plot have been discussed in Chapter 9. These methods are now applied to the design of linear control systems.

To illustrate the basic philosophy of design in the frequency domain, let us consider the following example. Let us begin by considering the transfer function of a controlled process,

$$G_p(s) = \frac{K}{s(1 + s)(1 + 0.0125s)} \tag{10-1}$$

The closed-loop system is considered to have a unity feedback. It is required that when a unit ramp input is applied to the closed-loop system, the steady-state error of the system does not exceed 1 percent of the amplitude of the input ramp, which is unity. Thus, when we use the definition of steady-state error in Chapter 6, we can find the minimum value of K needed to fulfill this error requirement:

$$\text{steady-state error} = e_{ss} = \lim_{s \to 0} \frac{1}{sG_p(s)} = \frac{1}{K} \leq 0.01 \tag{10-2}$$

Therefore, K must be greater than or equal to 100. However, applying the Routh–Hurwitz criterion to the characteristic equation of the closed-loop system, it is easy to show that the system is unstable for all values of K greater than 81. This means that some kind of compensation scheme or controller should be applied to

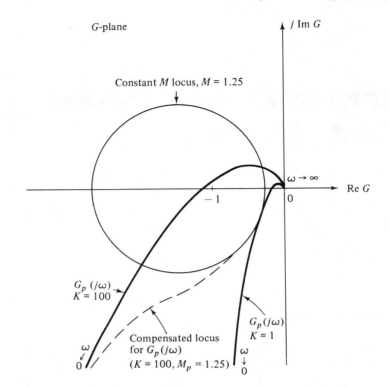

Figure 10-1   Nyquist plot for $G_p(s) = K / [s(1 + s)(1 + 0.0125s)]$.

the system so that the steady-state error and the relative stability requirement can be satisfied simultaneously. Putting it another way, the controller must be able to keep the zero-frequency gain of s times the open-loop transfer function of the compensated system effectively at 100 while maintaining a prescribed degree of relative stability. The principle of the design in the frequency domain is best illustrated by the Nyquist plot of $G_p(s)$ shown in Fig. 10-1. In practice, we prefer to use the Bode diagram for design purposes because it is simpler to construct, and the Nyquist plot is used mostly for analysis. As shown in Fig. 10-1, when $K = 100$, the Nyquist plot of $G_p(s)$ encloses the $(-1, j0)$ point, and the closed-loop system is unstable. Let us assume that we wish to realize a resonance peak of $M_p = 1.25$. This means that the Nyquist plot of $G_p(s)$ must be tangent to the constant M circle for $M = 1.25$ from below. If K is the only parameter that can be adjusted to achieve this design objective, the desired value of K is 1, as shown in Fig. 10-1. Apparently, we cannot set K to 1 since the ramp-error constant would only be 1 sec^{-1}, and the steady-state error requirement would not be satisfied.

Since the steady-state performance of the system is governed by the characteristics of the transfer function at low frequency, and the damping or the transient behavior of the system is governed by the relatively high frequency characteristics, Fig. 10-1 shows that to simultaneously satisfy the transient and the steady-state requirements, the frequency locus of $G_p(s)$ has to be reshaped so that the high-frequency portion of the locus follows the $K = 1$ trajectory and the low-frequency portion follows the $K = 100$ trajectory. The significance of this reshaping of the frequency locus is that the compensated locus shown in Fig. 10-1 will be tangent to the $M = 1.25$ circle at a relatively higher frequency, while the zero-frequency gain is maintained at 100 to satisfy the steady-state requirement.

When we inspect the loci of Fig. 10-1, we see that there are two alternative approaches in arriving at the compensated locus:

1. Starting from the $K = 100$ locus and reshaping the locus in the region near the resonant frequency ω_p, while keeping the low-frequency region of $G_p(s)$ relatively unaltered.

2. Starting from the $K = 1$ locus and reshaping the low-frequency portion of $G_p(s)$ to obtain a ramp error constant of $K_v = 100$ while keeping the locus near $\omega = \omega_p$ relatively unchanged.

In the first approach, the high-frequency portion of $G_p(s)$ is pushed in the counterclockwise direction, which means that more phase is added to the system in the positive direction in the proper frequency range. This scheme is basically referred to as *phase-lead* compensation, and controllers used for this purpose are often of the high-pass-filter type. The second approach apparently involves the shifting of the low-frequency part of the $K = 1$ trajectory in the clockwise direction, or alternatively, reducing the magnitude of $G_p(s)$ with $K = 100$ at the high-frequency range. This scheme is often referred to as phase-lag compensation, since more phase lag is introduced to the system in the low-frequency range. The type of network that is used for the phase-lag compensation is often referred to as low-pass filters.

Figures 10-2 and 10-3 further illustrate the philosophy of design in the frequency domain using the Bode diagram. In this case the relative stability of the system is more conveniently represented by the gain margin and the phase margin. In Fig. 10-2 the Bode plots of $G_p(j\omega)$ show that when $K = 100$, the gain and phase margins are both negative, and the system is unstable. When $K = 1$, the gain and phase margins are both positive, and the system has quite a comfortable safety margin. Using the first approach, the phase-lead compensation, as described earlier, we add more phase lead to $G_p(j\omega)$ so as to improve the phase margin. However, in attempting to reshape the phase curve by use of a high-pass filter, the magnitude curve of $G_p(j\omega)$ is unavoidably altered, as shown in Fig. 10-2. If the design is carried out properly, it is possible to obtain a net gain in relative stability using this approach. The Bode diagram of Fig. 10-3 serves to illustrate the principle of phase-lag compensation. If, instead of adding more positive phase to $G_p(j\omega)$ at the high-frequency range as in Fig. 10-2, we attenuate the amplitude of $G_p(j\omega)$ at the low frequency by means of a low-pass filter, a similar stabilization effect can be achieved. The Bode diagram of Fig. 10-3 shows that if the attenuation is affected at a sufficiently low frequency range, the effect on the phase shift due to the phase-lag compensation is negligible at the phase-crossover frequency. Thus, the net effect of the compensating scheme is the improvement on the relative stability of the system.

The example given above is simply for the purpose of illustrating the principle of design of control systems in the frequency domain. In general, it may not be

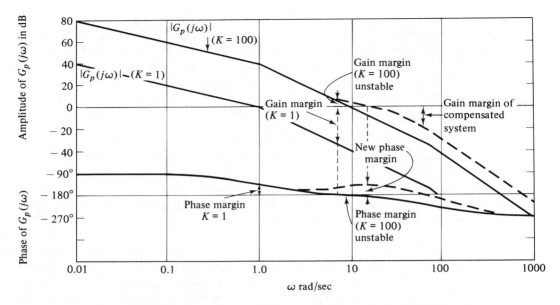

Figure 10-2 Bode plot of $G_p(s) = K / [s(1 + s)(1 + 0.0125s)]$ with phase-lead compensation.

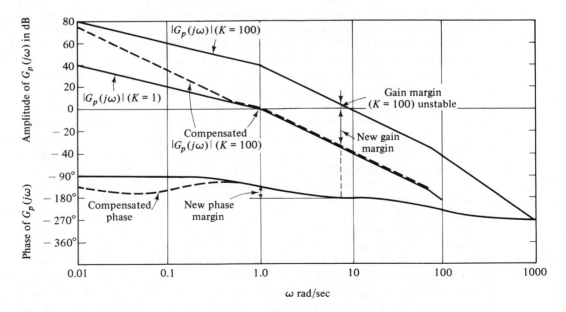

Figure 10-3 Bode plot of $G_p(s) = K / [s(1 + s)(1 + 0.0125s)]$ with phase-lag compensation.

possible to satisfy all design criteria by simply using a phase-lead or a phase-lag controller. Both the low-frequency and the high-frequency loci of the process may need to be reshaped by using a lead-lag or lag-lead controller. Or, frequently, we need to reshape the frequency locus of the controlled process only in a certain mid-frequency range so that a so-called "notch" controller such as the bridged-T networks is required.

10.2 PHASE-LEAD CONTROLLER

The phase-lead controller was used in Section 8.3 for the compensation of control systems, and the design was carried out in the time domain or the s-plane via the root loci. In this section the phase-lead controller is to be designed using frequency-domain techniques.

The transfer function of the phase-lead controller is written

$$\frac{E_2(s)}{E_1(s)} = \frac{1 + aTs}{1 + Ts} \qquad a > 1 \tag{10-3}$$

Polar Plot of the Phase-Lead Controller

The polar plot of Eq. (10-3) is shown in Fig. 10-4 for several different values of a. For any particular value of a, the angle between the tangent line drawn from the origin to the semicircle and the real axis gives the maximum phase lead ϕ_m which the network can provide. The frequency at the tangent point, ω_m, represents the frequency at which ϕ_m occurs. It is seen that, as a increases, the maximum phase lead, ϕ_m, also increases, approaching a limit of 90° as a approaches infinity. The frequency ω_m decreases with the increase in a.

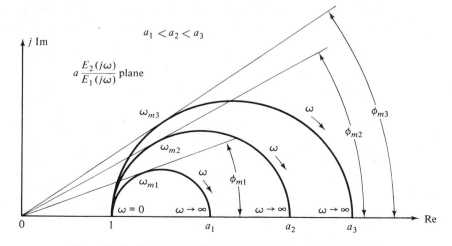

Figure 10-4   Polar plot of $E_2(s)/E_1(s) = (1 + aTs)/(1 + Ts)$.

Bode Plot of the Phase-Lead Controller

In terms of the Bode plot, the phase-lead controller has two corner frequencies: a positive corner frequency at $\omega = 1/aT$ and a negative corner frequency at $\omega = 1/T$. The Bode diagram of $E_2(j\omega)/E_1(j\omega)$ is shown in Fig. 10-5.

Analytically, ϕ_m and ω_m may be related to the parameters a and T. Since ω_m is the geometric mean of the two corner frequencies, we can write

$$\log_{10}\omega_m = \frac{1}{2}\left(\log_{10}\frac{1}{aT} + \log_{10}\frac{1}{T}\right) \tag{10-4}$$

Thus,

$$\omega_m = \frac{1}{\sqrt{a}\,T} \tag{10-5}$$

To determine the maximum phase lead, ϕ_m, we write the phase of $E_2(j\omega)/E_1(j\omega)$ as

$$\phi = \text{Arg}\left[\frac{E_2(j\omega)}{E_1(j\omega)}\right] = \tan^{-1}aT\omega - \tan^{-1}T\omega \tag{10-6}$$

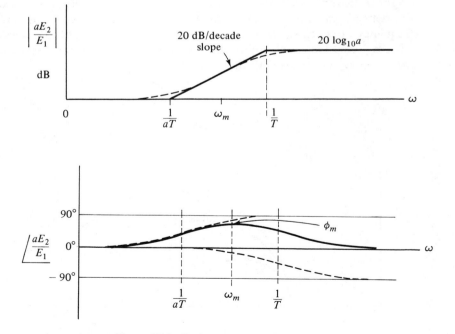

Figure 10-5   Bode plot of $(1 + Ts)$ $(a > 1)$.

from which we have

$$\tan \phi = \frac{aT\omega - T\omega}{1 + (aT\omega)(T\omega)} \tag{10-7}$$

When $\phi = \phi_m$,

$$\omega = \omega_m = \frac{1}{\sqrt{a}\,T} \tag{10-8}$$

Therefore, Eq. (10-7) gives

$$\tan \phi_m = \frac{(a-1)(1/\sqrt{a})}{1+1} = \frac{a-1}{2\sqrt{a}} \tag{10-9}$$

or

$$\sin \phi_m = \frac{a-1}{a+1} \tag{10-10}$$

This last expression is a very useful relationship in the proper selection of the value of a for compensation design.

Design of Phase-Lead Compensation by the Bode Plot Method

Design of linear control systems in the frequency domain is more preferably carried out with the aid of the Bode plot. The reason is simply because the effect of the compensation is easily obtained by adding its magnitude and phase curves, respectively, to that of the original process. The general outline of phase-lead controller design in the frequency domain is given below. It is assumed that the design specifications simply include a steady-state error and phase margin-versus-gain margin requirements.

1. The magnitude and phase-versus-frequency curves for the uncompensated process $G_p(s)$ are plotted with the gain constant K set according to the steady-state error requirement.
2. The phase margin and the gain margin of the original system are read from the Bode plot, and the additional amount of phase lead needed to provide the required degree of relative stability is determined. From the additional phase lead required, the desired value of ϕ_m is estimated accordingly, and a is calculated from Eq. (10-10).
3. Once a is determined, it is necessary only to obtain the proper value of T, and the design is in principle completed. The important step is to place the corner frequencies of the phase-lead controller, $1/aT$, and $1/T$ such that ϕ_m is located at the new gain-crossover frequency.

4. The Bode plot of the compensated system is investigated to check that all performance specifications are met; if not, a new value of ϕ_m must be chosen and the steps repeated.
5. If the specifications are all satisfied, the transfer function of the phase-lead controller is established from the values of a and T.

The following numerical example will illustrate the steps involved in the phase-lead design.

■

Example 10-1

Consider the sun-seeker control system described in Example 8-1. The block diagram of the system is shown in Fig. 8-16. The open-loop transfer function of the uncompensated system is written, from Eq. (8-30),

$$\frac{\theta_0(s)}{\alpha(s)} = \frac{2500K}{s(s + 25)} \tag{10-11}$$

The specifications of the system are given as follows:

1. The phase margin of the system should be greater than 45 degrees.
2. The steady-state error of $\alpha(t)$ due to a unit ramp function input should be less than or equal to 0.01 rad per rad/sec of the final steady-state output velocity. In other words, the steady-state error due to a ramp input should be less than or equal to 1 percent.

The following steps are carried out in the design of the phase-lead compensation:

1. Applying the final-value theorem, we have

$$\lim_{t \to \infty} \alpha(t) = \lim_{s \to 0} s\alpha(s) = \lim_{s \to 0} s\frac{\theta_r(s)}{1 + [\theta_0(s)/\alpha(s)]} \tag{10-12}$$

Since $\theta_r(s) = 1/s^2$, using Eq. (10-11), Eq. (10-12) becomes

$$\lim_{t \to \infty} \alpha(t) = \frac{0.01}{K} \tag{10-13}$$

Thus, we see that if $K = 1$, we have the steady-state error equal to 0.01. However, for this amplifier gain, the damping ratio of the closed-loop system is merely 25 percent, which corresponds to an overshoot of over 44.4 percent. Figure 10-6 shows the unit step response of the closed-loop system with $K = 1$. It is seen that the step response is quite oscillatory.
2. The Bode plot of $\theta_0(s)/\alpha(s)$ of the uncompensated system, with $K = 1$, is sketched as shown in Fig. 10-7.
3. The phase margin of the uncompensated system, read at the gain-crossover frequency, $\omega_c = 47$ rad/sec, is 28 degrees. Since the phase margin is less than the desired value of 45 degrees, more phase lead should be added to the open-loop system.

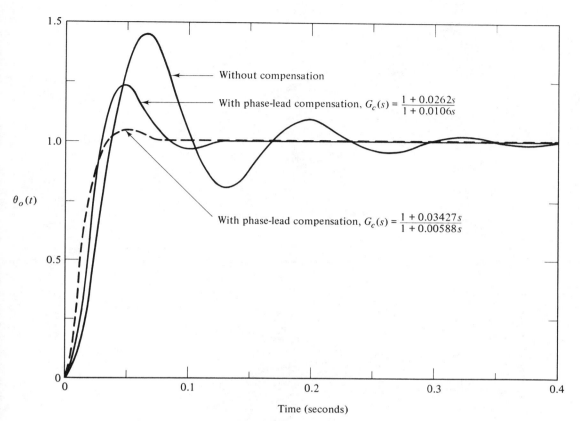

Figure 10-6   Step response of the sun-seeker system in Example 10-1.

4. Let us choose to use the phase-lead controller with the transfer function

$$\frac{U(s)}{E_0(s)} = G_c(s) = \frac{1 + aTs}{1 + Ts} \tag{10-14}$$

Since the desired phase margin is 45 degrees and the uncompensated system has a phase margin of 28 degrees, the phase-lead controller must provide the additional 17 degrees in the vicinity of the gain-crossover frequency. However, by inserting the phase-lead controller, the magnitude curve of the Bode plot is also affected in such a way that the gain-crossover frequency is shifted to a higher frequency. Although it is a simple matter to adjust the corner frequencies, $1/aT$ and $1/T$, so that the maximum phase of the network, ϕ_m, falls exactly at the new gain-crossover frequency, the original phase curve at this point is no longer 28 degrees, and could be considerably less. This represents one of the main difficulties in the phase-lead design. In fact, if the phase of the uncompensated system decreases rapidly with increasing frequency near the gain-crossover frequency, phase-lead compensation may become ineffective.

In view of the above-mentioned difficulty, when estimating the necessary amount of phase lead, it is essential to include some safety margin to account for the inevitable

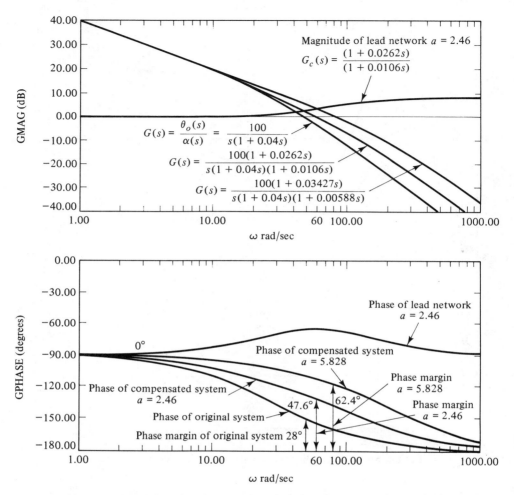

Figure 10-7 Bode plots of compensated and uncompensated systems in Example 10-1.

phase dropoff. Therefore, in the present design, instead of selecting a ϕ_m of 17 degrees, we let ϕ_m be 25 degrees. Using Eq. (10-10), we have

$$\sin \phi_m = \sin 25° = 0.422 = \frac{a-1}{a+1} \qquad (10\text{-}15)$$

from which we get

$$a = 2.46 \qquad (10\text{-}16)$$

5. To determine the proper location of the two corner frequencies, $1/aT$ and $1/T$, it is known from Eq. (10-8) that the maximum phase lead ϕ_m occurs at the geometrical mean of the corners. To achieve the maximum phase margin with the value of a

already determined, ϕ_m should occur at the new gain-crossover frequency ω_c', which is not known. Thus, the problem now is to locate the two corner frequencies so that ϕ_m occurs at ω_c'. This may be accomplished graphically as follows:

(a) The zero-frequency attenuation of the phase-lead network is calculated:

$$20 \log_{10} a = 20 \log_{10} 2.46 = 7.82 \text{ dB} \tag{10-17}$$

(b) The geometric mean ω_m of the two corner frequencies $1/aT$ and $1/T$ should be located at the frequency at which the magnitude of the uncompensated transfer function $\theta_0(j\omega)/\alpha(j\omega)$ in decibels is equal to the negative value in decibels of one half of this attenuation. This way, the magnitude plot of the compensated transfer function will pass through the 0-dB axis at $\omega = \omega_m$. Thus, ω_m should be located at the frequency where

$$\left| \frac{\theta_0(j\omega)}{\alpha(j\omega)} \right| = \frac{-7.82}{2} = -3.91 \text{ dB} \tag{10-18}$$

From Fig. 10.7, this frequency is found to be $\omega_m = 60$ rad/sec. Now using Eq. (10-8), we have

$$\frac{1}{T} = \sqrt{a}\, \omega_m = \sqrt{2.46} \times 60 = 94 \text{ rad/sec} \tag{10-19}$$

Then

$$\frac{1}{aT} = 38.2 \text{ rad/sec} \tag{10-20}$$

The parameters of the phase-lead controller are now determined. Figure 10-7 shows that the phase margin of the compensated system is 47.6 degrees. The transfer function of the phase-lead network is

$$\frac{U(s)}{E_0(s)} = \frac{1}{a} \frac{1 + aTs}{1 + Ts} = \frac{1}{2.46} \frac{1 + 0.0262s}{1 + 0.0106s} \tag{10-21}$$

Since it is assumed that the amplifier gains are increased by a factor of 2.46, the open-loop transfer function of the compensated sun-seeker system becomes

$$\frac{\theta_0(s)}{\alpha(s)} = \frac{6150(s + 38.2)}{s(s + 25)(s + 94)} \tag{10-22}$$

In Fig. 10-8 the magnitude and phase of the original and the compensated systems are plotted on the Nichols chart. These plots are obtained by taking the data directly from the Bode plot of Fig. 10-7. From the Nichols chart, the resonance peak, M_p, of the uncompensated system is found to be 2.07. The value of M_p with compensation is 1.25 or 1.9 dB. One more important point is that the resonant frequency of the system is increased from 46.67 rad/sec to approximately 53.3 rad/sec, and the bandwidth is increased from 74.3 rad/sec to 98.2 rad/sec.

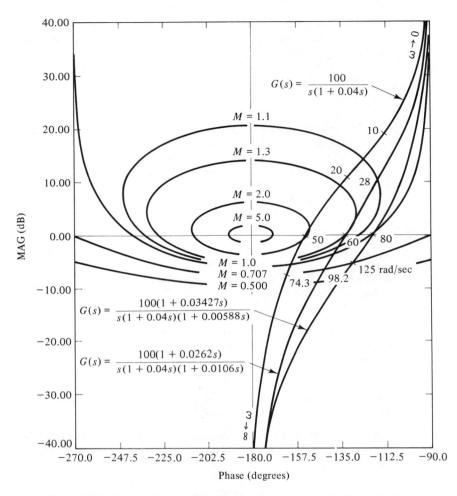

Figure 10-8 Plots of $G(s)$ in Nichols chart for the system in Example 10-1.

The unit step response of the compensated system is shown in Fig. 10-6. Note that the response of the system with the phase-lead controller is far less oscillatory than that of the original system. The overshoot is reduced from 44.4 percent to 24.5 percent, and the rise time is also reduced. The reduction of the rise time is due to the increase of the bandwidth by the phase-lead controller. On the other hand, excessive bandwidth may be objectionable in certain systems where noise and disturbance signals may be critical.

In the present design problem, we notice that a specification of 45 degrees for the phase margin yields an overshoot of 24.5 percent in the step response. To demonstrate the capability of the phase-lead controller, we select a to be 5.828. The resulting controller has the transfer function

$$G_c(s) = \frac{1}{5.828} \frac{1 + 0.03427s}{1 + 0.00588s} \tag{10-23}$$

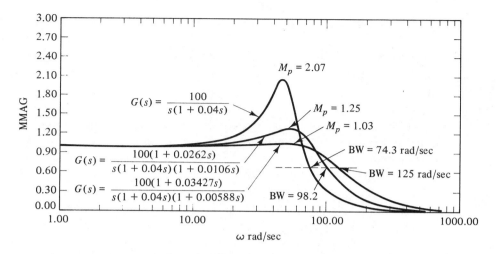

Figure 10-9 Closed-loop frequency responses of the sun-seeker system in Example 10-1.

The unit step response of the compensated system is plotted as shown in Fig. 10-6. In this case the rise time is shorter still, and the peak overshoot is reduced to 7.7 percent. The phase margin is improved to 62.4°, but the gain crossover frequency is increased to 80 rad/sec.

Using the magnitude-versus-phase plots and the Nichols chart of Fig. 10-8, the closed-loop frequency responses for the sun-seeker system, before and after compensation, are plotted as shown in Fig. 10-9.

■

The selection of $a = 5.828$ which gave rise to Eq. (10-23) was made so that the results of the present frequency-domain design can be compared with those of the time-domain design carried out in Example 8-1. Since the time-domain design and the frequency-domain design use different specifications which are very difficult to correlate, especially for high-order systems, in general, we cannot obtain unique solutions for a given design problem when these different design methods are used.

Effects and Limitations of Phase-Lead Compensation

From the results of the last illustrative example, we may summarize the general effects of phase-lead compensation on the performance of control systems as follows:

1. The phase of the open-loop transfer function in the vicinity of the gain-crossover frequency is increased. Thus the phase margin is usually improved.
2. The slope of the magnitude curve representing the magnitude of the open-loop transfer function is reduced at the gain-crossover frequency. This usually corresponds to an improvement in the relative stability of the system. In other words, the phase and gain margins are improved.

3. The bandwidths of the open-loop system and the closed-loop system are increased.
4. The overshoot of the step response is reduced.
5. The steady-state error of the system is not affected.

It was mentioned earlier that certain types of systems cannot be effectively compensated satisfactorily by phase-lead compensation. The sun-seeker system studied in Example 10-1 happens to be one for which phase-lead control is effective and practical. In general, successful application of phase-lead compensation is hinged upon the following considerations:

1. Bandwidth considerations: If the original system is unstable, the additional phase lead required to obtain a certain desirable phase margin may be excessive. This requires a relatively large value of a in Eq. (10-3), which, as a result, will give rise to a large bandwidth for the compensated system, and the transmission of noise may become objectionable. Also, if the value of a becomes too large, and if the controller is to be realized by an electric network, the values of the network elements may become disproportionate, such as a very large capacitor. Therefore, in practice, the value of a is seldom chosen greater than 15. If a larger value of a is justified, sometimes two or more phase-lead controllers are connected in cascade to achieve the large phase lead.
2. If the original system is unstable or has low stability margin, the phase plot of the open-loop transfer function has a steep negative slope near the gain-crossover frequency. In other words, the phase decreases rapidly near the gain crossover. Under this condition, phase-lead compensation usually becomes ineffective because the additional phase lead at the new gain crossover is added to a much smaller phase angle than that at the old gain crossover. The desired phase margin may be realized only by using a very large value of a. However, the resulting system may still be unsatisfactory because a portion of the phase curve may still be below the 180-degree axis, which corresponds to a conditionally stable system.

In general, the following situations may also cause the phase to change rapidly near the gain-crossover frequency:

1. The open-loop transfer function has two or more poles that are close to each other and are close to the gain-crossover frequency.
2. The open-loop transfer function has one or more pairs of complex conjugate poles near the gain-crossover frequency.

The system described in Example 8-2 will now be used to illustrate that under certain conditions phase-lead compensation may not be effective.

■

Example 10-2

Let the open-loop transfer function of a control system with unity feedback be

$$G_p(s) = \frac{K}{s(1 + 0.1s)(1 + 0.2s)} \tag{10-24}$$

It is desired that the system satisfies the following performance specifications:

1. $K_v = 100$; or the steady-state error of the system due to a unit ramp function input is 0.01 in magnitude.
2. Phase margin ≥ 40 degrees.

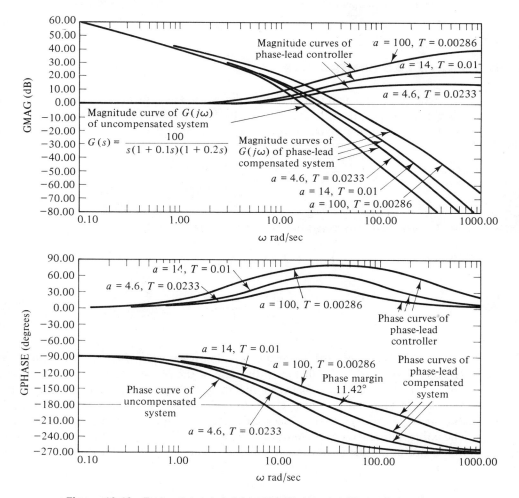

Figure 10-10 Bode plots of $G_p(s) = 100 / [s(1 + 0.1s)(1 + 0.2s)]$ and the effects of using phase-lead compensation.

From the steady-state error requirement, we set $K = 100$. The Bode plot of $G_p(s)$ when $K = 100$ is shown in Fig. 10-10. As observed from this Bode plot, the phase margin of the system is approximately -40 degrees, which means that the system is unstable. In fact, the system is unstable for all values of K greater than 15. The rapid decrease of phase at the gain-crossover frequency, $\omega_c = 17$ rad/sec, implies that the phase-lead compensation may be ineffective for this case. To illustrate the point, the phase-lead controller of Eq. (10-3) with $a = 4.6$, 14, and 100, respectively, is used to compensate the system. Figure 10-10 illustrates the effects of phase-lead compensation when the values of T are chosen according to the procedure described in Example 10-1.

It is clearly shown in Fig. 10-10 that as more phase lead is being added, the gain-crossover frequency is also being pushed to a higher value. Therefore, for this case, in which the uncompensated system is very unstable at the outset, it may be impossible to realize a phase margin of 40 degrees by the phase-lead controller of Eq. (10-3).

10.3 PHASE-LAG CONTROLLER

In Section 8.3 we demonstrated that the phase-lag controller generally has a wider range of effectiveness than the phase-lead controller. The transfer function of the phase-lag controller given in Eq. (8-51) is repeated below.

$$\frac{E_2(s)}{E_1(s)} = \frac{1 + aTs}{1 + Ts} \qquad a < 1 \tag{10-25}$$

Polar Plot of the Phase-Lag Controller

When we let $s = j\omega$, Eq. (10-25) becomes

$$\frac{E_2(j\omega)}{E_1(j\omega)} = \frac{1 + j\omega aT}{1 + j\omega T} \qquad a < 1 \tag{10-26}$$

The polar plot of this transfer function is shown in Fig. 10-11 for three values of a, $1 > a_1 > a_2 > a_3$. Just as in the case of the phase-lead network, for any value of a $(a < 1)$, the angle between the tangent line drawn from the origin to the semicircle and the real axis gives the maximum phase lag ϕ_m $(\phi_m < 0°)$ of the network. As the value of a decreases, the maximum phase lag ϕ_m becomes more negative, approaching the limit of -90 degrees as a approaches zero. As the value of a decreases, the frequency at which ϕ_m occurs, ω_m, increases; that is, Fig. 10-11, $\omega_{m3} > \omega_{m2} > \omega_{m1}$.

Bode Plot of the RC Phase-Lag Network. The Bode plot of the transfer function of Eq. (10-26) is shown in Fig. 10-12. The magnitude plot has a positive corner frequency at $\omega = 1/aT$, and a negative corner frequency at $\omega = 1/T$. Since

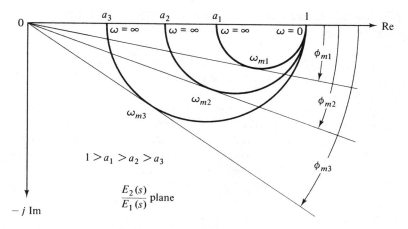

Figure 10-11 Polar plots of $E_2(s) / E_1(s) = (1 + aTs) / (1 + Ts)$ $(a < 1)$.

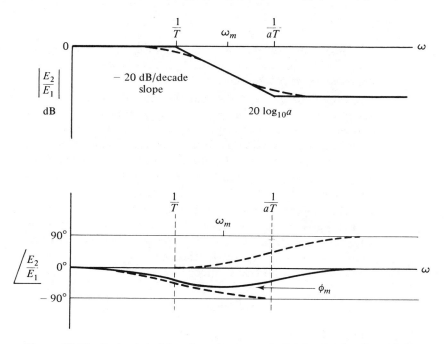

Figure 10-12 Bode plot of the phase-lag network $E_2(s) / E_1(s) = (1 + aTs)$ $/ (1 + Ts)$ $(a < 1)$.

the transfer functions of the phase-lead and phase-lag networks are identical in form except for the zero-frequency attenuation and the value of a, it can readily be shown that the maximum phase lag ϕ_m of the phase curve of Fig. 10-12 satisfies the following relation:

$$\sin \phi_m = \frac{a - 1}{a + 1} \tag{10-27}$$

Design of Phase-Lag Compensation by the Bode Plot Method

Unlike the design of phase-lead compensation, which utilizes the maximum phase lead of the network, the design of phase-lag compensation utilizes the attenuation of the network at the high frequencies. It was pointed out earlier that, for phase-lead compensation, the function of the network is to increase the phase in the vicinity of the gain-crossover frequency while keeping the magnitude curve of the Bode plot relatively unchanged near that frequency. However, usually in phase-lead design, the gain crossover frequency is increased because of the phase-lead network, and the design is essentially the finding of a compromise between the increase in bandwidth and the desired amount of relative stability (phase margin or gain margin). In phase-lag compensation, however, the objective is to move the gain-crossover frequency to a lower frequency while keeping the phase curve of the Bode plot relatively unchanged at the gain-crossover frequency.

The design procedure for phase-lag compensation using the Bode plot is outlined as follows:

1. The Bode plot of the open-loop transfer function of the uncompensated system is made. The open-loop gain of the system is set according to the steady-state error requirement.
2. The phase margin and the gain margin of the uncompensated system are determined from the Bode plot. For a certain specified phase margin, the frequency corresponding to this phase margin is found on the Bode plot. The magnitude plot must pass through the 0-dB axis at this frequency in order to realize the desired phase margin. In other words, the gain-crossover frequency of the compensated system must be located at the point where the specified phase margin is found.
3. To bring the magnitude curve down to 0 dB at the new prescribed gain-crossover frequency, ω_c', the phase-lag network must provide the amount of attenuation equal to the gain of the magnitude curve at the new gain-crossover frequency. In other words, let the open-loop transfer function of the uncompensated system be $G_p(s)$; then

$$|G_p(j\omega_c')| = -20\log_{10}a \text{ dB} \qquad a < 1 \qquad (10\text{-}28)$$

from which

$$a = 10^{-|G_p(j\omega_c')|/20} \qquad a < 1 \qquad (10\text{-}29)$$

Once the value of a is determined, it is necessary only to select the proper value of T to complete the design. Up to this point, we have assumed that although the gain-crossover frequency is altered by attenuating the gain at ω_c, the original phase curve is not affected. This is not possible, however, since any modification of the magnitude curve will being change to the phase curve, and vice versa. With reference to the phase characteristics of the phase-lag controller shown in Fig. 10-12, it is apparent that if the positive corner frequency, $1/aT$, is placed far below the new gain-crossover frequency, ω_c', the phase characteristics of the compensated system will not be appreciably affected near ω_c' by phase-lag compensation. On the other hand, the value of $1/aT$ should not be too much less than ω_c', or the bandwidth of the system will be too low, causing the system to be too sluggish. Usually, as a general guideline, it is recommended that the corner frequency $1/aT$ be placed at a frequency that is approximately 1 decade below the new gain-crossover frequency, ω_c'; that is,

$$\frac{1}{aT} = \frac{\omega_c'}{10} \qquad \text{rad/sec} \tag{10-30}$$

Therefore,

$$\frac{1}{T} = \frac{\omega_c'}{10}a \qquad \text{rad/sec} \tag{10-31}$$

5. The Bode plot of the phase-lag compensated system is investigated to see if the performance specifications are met.
6. If all the design specifications are met, the values of a and T are substituted in Eq. (10-25) to give the desired transfer function of the phase-lag controller.

◼

Example 10-3

In this example we shall design a phase-lag controller for the sun-seeker system considered in Example 10-1. The open-loop transfer function of the sun-seeker system is given by Eq. (10-11),

$$\frac{\theta_0(s)}{\alpha(s)} = \frac{2500K}{s(s + 25)} \tag{10-32}$$

The specifications of the system are repeated as follows:

1. The phase margin of the system should be greater than 45 degrees.
2. The steady-state error of $\alpha(t)$ due to a unit ramp function input should be less than or equal to 0.01 rad per rad/sec of the final steady-state output velocity. This is translated into the requirement of $K \geq 1$.

The Bode plot of $\theta_0(s)/\alpha(s)$ with $K = 1$ is shown in Fig. 10-13. As seen, the phase margin is only 28°.

For a phase-lag controller, let us choose the transfer function

$$\frac{U(s)}{E_0(s)} = G_c(s) = \frac{1 + aTs}{1 + Ts} \qquad a < 1 \tag{10-33}$$

From Fig. 10-13 it is observed that the desired 45-degree phase margin can be obtained if the

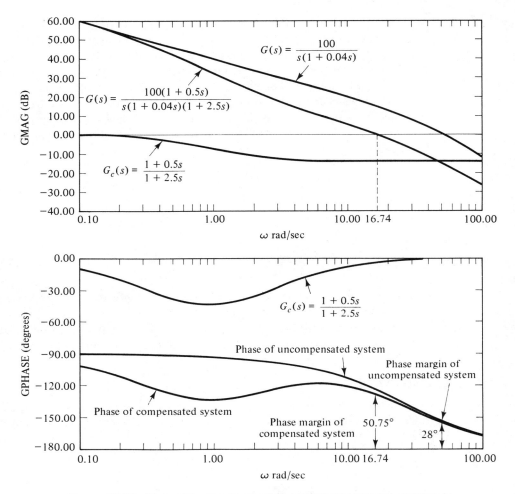

Figure 10-13 Bode plots of compensated and uncompensated systems in Example 10-3.

gain-crossover frequency ω_c' is at 25 rad/sec. This means that the phase-lag controller must reduce the magnitude of $\theta_0(j\omega)/\alpha(j\omega)$ to 0 dB at $\omega = 25$ rad/sec, while it does not appreciably affect the phase curve in the vicinity of this frequency. Since, actually, a small negative phase shift still accompanied by the phase-lag network at the new gain-crossover frequency, it is a safe measure to choose this new gain-crossover frequency somewhat less than 25 rad/sec, say 20 rad/sec.

From the magnitude plot of $\theta_0(j\omega)/\alpha(j\omega)$, the value of $|\theta_0(j\omega)/\alpha(j\omega)|$ at $\omega_c' = 20$ rad/sec is 14 dB. This means that the phase-lag controller must provide an attenuation of 14 dB at this frequency, in order to bring the magnitude curve down to 0 dB at $\omega_c' = 20$ rad/sec. Thus, using Eq. (10-29), we have

$$a = 10^{-|\theta_0(j\omega_c')/\alpha(j\omega_c')|/20}$$
$$= 10^{-0.7} = 0.2 \tag{10-34}$$

The value of a indicates the required distance between the two corner frequencies of the phase-lag controller, in order that the attenuation of 14 dB is realized.

In order that the phase lag of the controller does not appreciably affect the phase at the new gain-crossover frequency, we choose the corner frequency $1/aT$ to be at 1 decade below $\omega_c' = 20$ rad/sec. Thus,

$$\frac{1}{aT} = \frac{\omega_c'}{10} = \frac{20}{10} = 2 \text{ rad/sec} \tag{10-35}$$

which gives

$$\frac{1}{T} = 0.4 \text{ rad/sec} \tag{10-36}$$

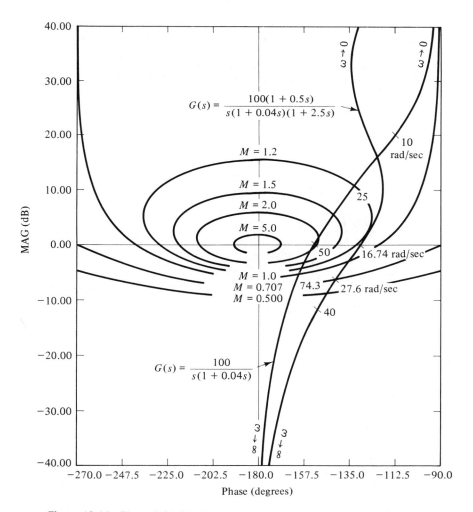

Figure 10-14 Plots of $G(s)$ in Nichols chart for the system in Example 10-3.

The transfer function of the phase-lag controller is

$$\frac{U(s)}{E_0(s)} = \frac{1 + 0.5s}{1 + 2.5s} \tag{10-37}$$

and the open-loop transfer function of the compensated system is

$$\frac{\theta_0(s)}{\alpha(s)} = \frac{500(s + 2)}{s(s + 0.4)(s + 25)} \tag{10-38}$$

The Bode plot of the open-loop transfer function of the compensated system is shown in Fig. 10-13. We see that the magnitude curve beyond $\omega = 0.4$ rad/sec is attenuated by the phase-lag controller while the low-frequency portion is not affected. In the meantime, the phase curve is not much affected by the phase-lag characteristic near the new gain-crossover frequency, which is at 16.74 rad/sec. The phase margin of the compensated system is determined from Fig. 10-13 to be 50.75 degrees.

The magnitude-versus-phase curves of the uncompensated and the compensated systems are plotted on the Nichols chart, as shown in Fig. 10-14. It is seen that the resonant

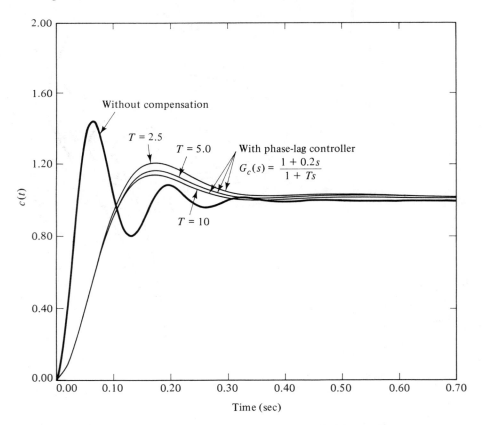

Figure 10-15 Step responses of the sun-seeker system in Example 10-3.

peak, M_p, of the compensated system is 1.2. The bandwidth of the system is reduced from 74.3 rad/sec to 27.6 rad/sec.

The unit step responses of the uncompensated and the compensated systems are shown in Fig. 10-15. The effects of the phase-lag controller are that the overshoot is reduced from 44.4 percent to 22 percent, but the rise time is increased considerably. The latter effect is apparently due to the reduction of the bandwidth by the phase-lag controller. Figure 10-15 also gives the step responses of the system when the value of T of the phase-lag controller is changed to 5 and then to 10. It is seen that larger values of T give only slight improvements on the overshoot of the step response. Earlier it was pointed out that the value of T is not critical; when $T = 5$, it is equivalent to setting $1/aT$ at 20 times below the gain-crossover frequency of $\omega'_c = 20$ rad/sec. Similarly, $T = 10$ corresponds to placing $1/aT$ at 40 times below ω'_c.

Example 10-4

Consider the system given in Example 10-2 for which the phase-lead compensation is ineffective. The open-loop transfer function of the original system and the performance specifications are repeated as follows:

$$G_p(s) = \frac{K}{s(1 + 0.1s)(1 + 0.2s)} \tag{10-39}$$

The performance specifications are:

1. $K_v = 100$ sec^{-1}.
2. Phase margin $\geq 30°$.

The phase-lag design procedure is as follows:

1. The Bode plot of $G_p(s)$ is made as shown in Fig. 10-16 for $K = 100$.
2. The phase margin at the gain-crossover frequency, $\omega_c = 16$ rad/sec, is approximately -40.3 degrees, and the closed-loop system is unstable.
3. The desired phase margin is 30 degrees; and from Fig. 10-16 this can be realized if the gain-crossover frequency is moved to approximately 4 rad/sec. This means that the phase-lag controller must reduce the magnitude of $G_p(j\omega)$ to 0 dB, while it does not affect the phase curve at this new gain crossover frequency, ω'_c. Since actually a small negative phase is still introduced by the phase-lag controller at ω'_c, it is a safe measure to choose the new gain-crossover frequency somewhat less than 4 rad/sec, say at 3.5 rad/sec. As an alternative, we may select a larger phase margin of 45 degrees.
4. From the Bode plot, the magnitude of $G_p(j\omega)$ at $\omega'_c = 3.5$ rad/sec is 30 dB, which means that the controller must introduce 30 dB of attenuation at this frequency, in order to bring down the magnitude curve of $G_p(j\omega)$ to 0 dB. Thus, from Eq. (10-29),

$$a = 10^{-|G_p(j\omega'_c)|/20} = 10^{-1.5} = 0.032 \tag{10-40}$$

This equation implies that the two corners of the phase-lag controller must be placed 1.5 decades apart, in order to produce the required 30 dB of attenuation.

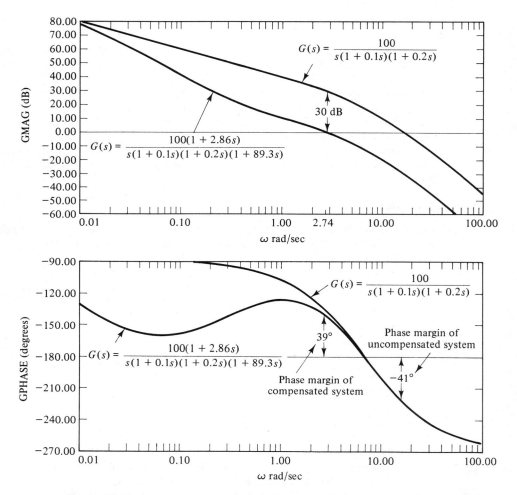

Figure 10-16 Bode plots of compensated and uncompensated systems in Example 10-4.

5. Let us place the upper corner frequency of the controller, $1/aT$, at the frequency of 1 decade below the new gain-crossover frequency, we have

$$\frac{1}{aT} = \frac{\omega_c'}{10} = \frac{3.5}{10} = 0.35 \text{ rad/sec} \tag{10-41}$$

which gives

$$T = 89.3$$

6. The Bode plot of the compensated system, with the phase-lag controller transfer function given by

$$G_c(s) = \frac{1 + aTs}{1 + Ts} = \frac{1 + 2.86s}{1 + 89.3s} \tag{10-42}$$

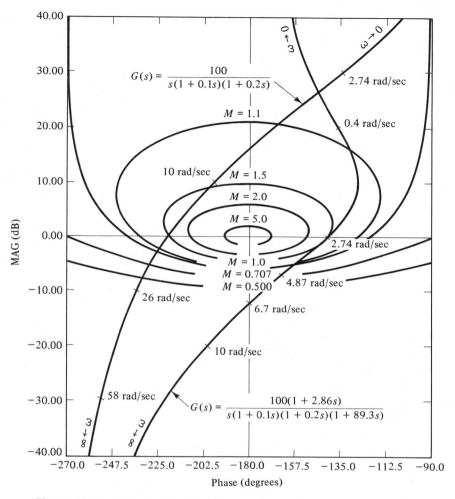

Figure 10-17 Plots of $G(s)$ of the compensated and the uncompensated systems in the Nichols chart for Example 10-4.

is sketched in Fig. 10-16. It is seen that the phase margin of the compensated system is approximately 39 degrees.

7. The open-loop transfer function of the compensated system is

$$G(s) = G_c(s)G_p(s) = \frac{100(1 + 2.86s)}{s(1 + 0.1s)(1 + 0.2s)(1 + 89.3s)} \qquad (10\text{-}43)$$

The magnitude-versus-phase curves of the compensated and the uncompensated systems are plotted on the Nichols chart, as shown in Fig. 10-17. These curves show that the

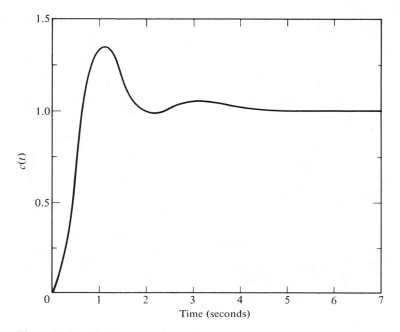

Figure 10-18 Unit step response of the system with phase-lag compensation in Example 10-4.

uncompensated system is unstable, but the compensated system has the following performance data as measured by the frequency-domain criteria:

$$\text{resonant peak } M_p = 1.49$$

$$\text{phase margin} = 39 \text{ degrees}$$

$$\text{gain margin} = 12.48 \text{ dB}$$

$$\text{bandwidth} = 4.87 \text{ rad/sec}$$

The unit step response of the system with the phase-lag controller is shown in Fig. 10-18. The peak overshoot of the system is approximately 35 percent.

■

Effects and Limitations of Phase-Lag Compensation

From the results of the preceding illustrative examples, the effects and limitations of phase-lag compensation on the performance of a control system may be summarized as follows:

1. For a given relative stability, the ramp error constant is increased.
2. The gain-crossover frequency is decreased; thus the bandwidth of the closed-loop system is decreased.

3. For a given loop gain, K, the magnitude of the open-loop transfer function is attenuated at the gain-crossover frequency, thus allowing improvement in the phase margin, gain margin, and resonance peak of the system.

4. The rise time of the system with phase-lag compensation is usually slower, since the bandwidth is usually decreased. For additional reduction in the overshoot, the rise time is further increased.

10.4 LAG – LEAD CONTROLLER

The lag–lead (or lead–lag) controller discussed in Section 8.3 can also be designed using the frequency-domain methods. When designed properly, lag–lead compensation combines the advantages of the phase-lead controller and the phase-lag controller. The following example repeats the design problem considered in Examples 10-2 and 10-4 by means of a lag–lead controller.

■

Example 10-5

In this example we design a lag–lead controller for the control system considered in Examples 10-2 and 10-4. The open-loop transfer function of the original system is repeated as

$$G_p(s) = \frac{K}{s(1 + 0.1s)(1 + 0.2s)} \qquad (10\text{-}44)$$

The performance specifications are:

1. $K_v = 100 \text{ sec}^{-1}$.
2. Phase margin ≥ 30 degrees.

These requirements have been satisfied by the phase-lag controller designed in Example 10-4. However, it is noted that the phase-lag controller yielded a step response that has a relatively large rise time. In this example we design a lag–lead controller so that the rise time is reduced.

Let the series controller be represented by the transfer function

$$G_c(s) = \frac{(1 + aT_1s)(1 + bT_2s)}{(1 + T_1s)(1 + T_2s)} \qquad (10\text{-}45)$$

For the first part we consider that the lag–lead controller is realized by the network of Fig. 8-30, so that the coefficients a and b are related through Eq. (8-75).

In general, there is no fixed procedure available for the design of the lag–lead controller. Usually, a trial-and-error procedure, using the design techniques outlined for the phase-lag and the phase-lead controllers, may provide a satisfactory design arrangement.

Let us first determine the phase-lag portion of the compensation by selecting the proper values of T_2 and b of Eq. (10-45). The Bode plot of $G_p(s)$ of Eq. (10-44) is sketched in Fig. 10-19 for $K = 100$. We arbitrarily choose to move the gain-crossover frequency of $G_p(j\omega)$ from 16 rad/sec to 5 rad/sec. Recall that in Example 10-4, the desired new gain-crossover

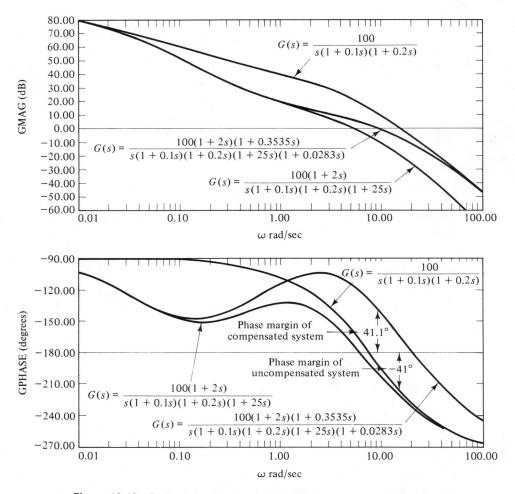

Figure 10-19 Bode plots of uncompensated system and compensated system with lag – lead controller in Example 10-5.

frequency is 3.5 rad/sec when phase-lag compensation alone is used. With the gain-crossover frequency at 6 rad/sec, the phase margin of the system should be improved to approximately 10 degrees. Using the phase-lag design technique, we notice from Fig. 10-16 that the attenuation needed to bring the magnitude of $G_p(j\omega)$ down to 0 dB at $\omega'_c = 5$ rad/sec is -22 dB. Thus, using Eq. (10-29), we have

$$b = 10^{-22/20} = 10^{-1.1} = 0.08 \qquad (10\text{-}46)$$

Placing $1/bT_2$ at 1 decade of frequency below $\omega'_c = 5$ rad/sec, we have

$$\frac{1}{bT_2} = \frac{5}{10} = 0.5 \text{ rad/sec} \qquad (10\text{-}47)$$

Thus,

$$T_2 = 25 \tag{10-48}$$

and

$$\frac{1}{T_2} = 0.04 \tag{10-49}$$

The phase-lag portion of the controller is described by

$$\frac{1 + 2s}{1 + 25s}$$

Now we turn to the phase-lead part of the controller. Since a is equal to the inverse of b, the value of a is found to be 12.5. From Eq. (10-10), the maximum phase lead that corresponds to this value of a is

$$\sin \phi_m = \frac{a - 1}{a + 1} = 0.8518 \tag{10-50}$$

or

$$\phi_m = 58.3 \text{ deg}$$

The zero-frequency attenuation of the phase-lead controller is

$$20 \log_{10} a = 20 \log_{10} 12.5 = 21.9 \text{ dB} \tag{10-51}$$

Using the procedure outlined earlier, the new gain-crossover frequency is found to be at $\omega_m = 10$ rad/sec. Then, using Eq. (10-5), we get

$$\frac{1}{T_1} = \sqrt{a}\, \omega_m = 35.35 \tag{10-52}$$

and

$$\frac{1}{aT_1} = 2.828$$

The transfer function of the lag–lead controller is determined as

$$G_c(s) = \frac{(1 + 0.3535s)(1 + 2s)}{(1 + 0.0283s)(1 + 25s)} \tag{10-53}$$

where as a usual practice, the attenuation factor in the phase-lead portion has been dropped.

Figure 10-19 shows that the phase margin of the compensated system is 41.1 degrees. The open-loop transfer function of the compensated system with $K = 100$ is

$$G(s) = G_c(s)G_p(s) = \frac{5000(s + 2.828)(s + 0.5)}{s(s + 5)(s + 10)(s + 35.35)(s + 0.04)} \tag{10-54}$$

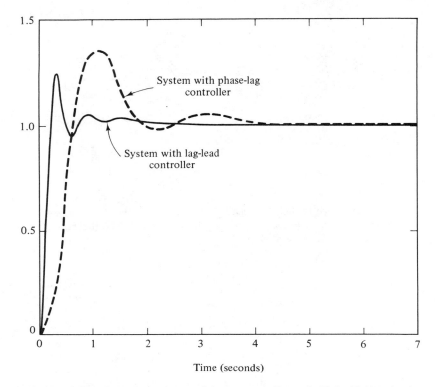

Figure 10-20   Unit step response of the system in Example 10-5 with lag – lead controller.

The unit step response of the compensated system is shown in Fig. 10-20. It is apparent that the step response of the system with the lag–lead controller is much improved over that of the phase-lead compensated system. Not only the overshoot is smaller, but the rise time is also greatly reduced. We may attribute these improvements to the addition of the phase-lead portion of the controller. It can be easily verified that the bandwidth of the closed-loop system is now approximately 16 rad/sec, which is more than three times that of the system with the phase-lag controller alone.

10.5 BRIDGED-T CONTROLLER

In Sec. 8.4 the bridged-T controllers are used for pole-zero cancellation control of control systems. The reason for this association is that in the s-domain, it is easier to grasp the controller design by canceling the undesirable poles of the process transfer function.

We can gain more perspective by investigating the properties of the bridged-T networks of Fig. 8-33 in the frequency domain. Figure 10-21 illustrates the Bode plot of the transfer function in Eq. (8-102), which is of the bridged-T network.

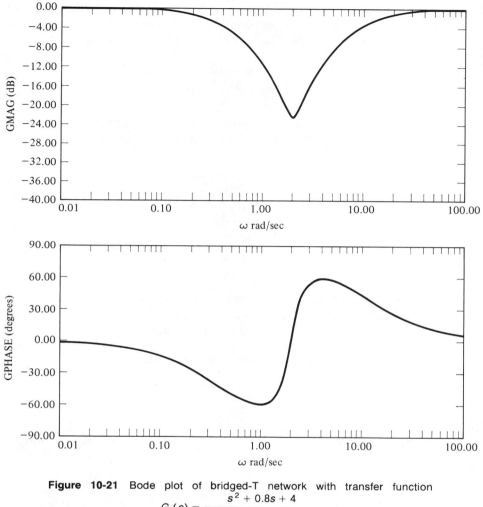

Figure 10-21 Bode plot of bridged-T network with transfer function

$$G_c(s) = \frac{s^2 + 0.8s + 4}{(s + 0.384)(s + 10.42)}.$$

Notice that the magnitude plot of the bridged-T networks typically has a "notch" at the resonant frequency ω_n. The phase plot is negative below and positive above the resonant frequency, while passing through zero degrees at the resonant frequency. The attenuation of the magnitude curve and the positive phase characteristics can be used effectively to improve the stability of a control system. Because of the "notch" characteristics in the amplitude plot, the bridged-T networks are also referred to in the industry as notch networks.

The notch networks have advantages over the phase-lag and the phase-lead networks in certain conditions, since the magnitude and phase characteristics do not affect the high- and low-frequency properties of the system. Without using the pole-zero-cancellation principle, the design of the bridged-T network for control-system compensation in the frequency domain involves the determination of the amount of attenuation required and the resonant frequency of the network.

We write the transfer function of the bridged-T network as

$$G_c(s) = \frac{E_2(s)}{E_1(s)} = \frac{s^2 + 2\zeta_z\omega_n s + \omega_n^2}{s^2 + 2\zeta_p\omega_n s + \omega_n^2} \tag{10-55}$$

where we have utilized the fact that $\omega_{nz} = \omega_{np}$. The attenuation provided by the magnitude of $G_c(j\omega)$ at the resonant frequency ω_n is simply

$$G_c = |G_c(j\omega_n)| = \frac{\zeta_z}{\zeta_p} \tag{10-56}$$

Using Eq. (8-91) or Eq. (8-95), we have

$$\zeta_z = \frac{G_c}{2(1 - G_c)} \tag{10-57}$$

Thus, given the amount of peak attenuation required, the damping ratio of the zeros of the bridged-T network is given by Eq. (10-57). The resonant frequency ω_{nz} ($= \omega_{np}$) is known, once the location of the "notch" is determined.

As an illustrative example, we can reconstruct the design problem given in Example 8-4. The transfer function of the controlled process is

$$G_p(s) = \frac{K(1 + 10s)}{s(1 + 0.2s + 0.25s^2)} \tag{8-96}$$

where $K = 1$. The transfer function of the pole-zero cancellation bridged-T controller is

$$G_c(s) = \frac{s^2 + 0.8s + 4}{(s + 0.384)(s + 10.42)} \tag{8-102}$$

Figure 10-22 shows the Bode plots of the compensated and the uncompensated transfer functions. The closed-loop frequency responses are shown in Fig. 10-23. The uncompensated system has the following characteristics in the frequency domain:

Gain margin:	infinite
Phase margin:	6.73 degrees
M_p:	8.52 (18.6 db)

The bridged-T network chosen has a resonant frequency at $\omega_n = 2$ rad/sec and a peak attenuation of approximately -22 db. The phase curve has a maximum phase

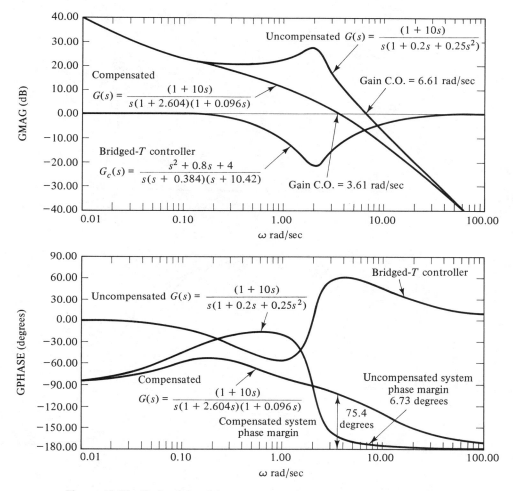

Figure 10-22 Bode plots of the compensated and uncompensated systems in Example 8-4.

angle of 60 degrees at 4 rad/sec. The combination of the gain and phase characteristics brings the following performance specifications for the compensated system:

Gain margin:	infinite
Phase margin:	75.4 degrees
M_p:	1.00

From Fig. 10-22 we see that the pole-zero cancellation compensation provides an attenuation of -22 db at 2 rad/sec, which corresponds to the peak of 22 db of the magnitude of the uncompensated transfer function at the same frequency.

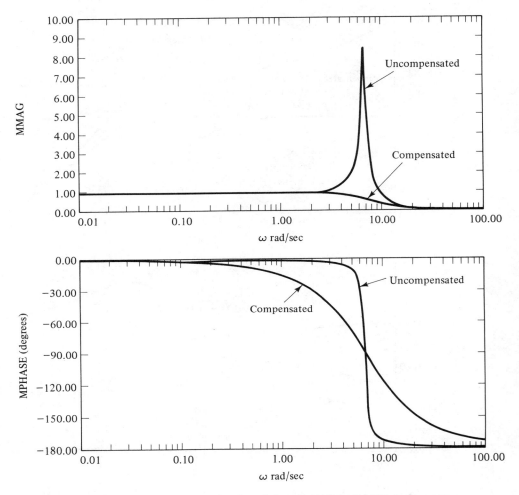

Figure 10-23 Frequency responses of the closed-loop system in Example 8-4.

In general, the frequency response of the controlled process may have to be determined experimentally without the knowledge of the transfer function. In this case, the design of the bridged-T network would have to be carried out by investigating the magnitude and phase characteristics of the process.

PROBLEMS

10.1. The process of a unity feedback control is described by the transfer function

$$G_p(s) = \frac{6}{s(s^2 + 4s + 6)}$$

Find the following quantities of the closed-loop system when a series controller with the transfer function $G_c(s)$ is applied: gain margin, phase margin, BW, and M_p.

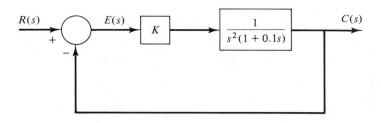

Figure 10P-3

(a) $G_c(s) = 1$

(b) $G_c(s) = \dfrac{5(s + 1)}{s + 5}$

(c) $G_c(s) = \dfrac{1 + s}{1 + 5s}$

(d) $G_c(s) = \dfrac{(s + 1)^2}{(1 + 0.2s)(1 + 5s)}$

10.2. The open-loop transfer function of a feedback control system is given by

$$G(s) = \frac{K}{s(1 + 0.2s)(1 + 0.5s)}$$

The feedback is unity. The output of the system is to follow a reference ramp input to yield a velocity of 2 rpm with a maximum steady-state error of 2 degrees.

(a) Determine the smallest value of K that will satisfy the specification given above. With this value of K, analyze the system performance by evaluating the system gain margin, phase margin, M_p, and bandwidth.

(b) A lead compensation with the transfer function $(1 + 0.4s)/(1 + 0.08s)$ is inserted in the forward path of the system. Evaluate the values of the gain margin, phase margin, M_p, and bandwidth of the compensated system. Comment on the effects of the lead compensation of the system performance.

(c) Sketch the root loci of the compensated and the uncompensated systems.

10.3. A type 2 control system is shown in Fig. 10P-3. The system must meet the following performance specifications:

(a) Parabolic error constant $K_a = 2$ sec^{-2}.

(b) The resonance peak $M_p \leq 1.5$.

Design a series phase-lead controller to satisfy these requirements. Sketch the root loci for the uncompensated and compensated systems. What are the values of the damping ratio (of the complex roots) and the bandwidth of the compensated system?

10.4. Figure 10P-4 illustrates the block diagram of a positioning control system. The system may be used for the positioning of a shaft by a command from a remote location. The parameters of the system are given as follows:

$$\tau_e = 0.01 \text{ sec}$$

$$J = 0.05 \text{ oz-in.-sec}^2$$

$$B = 0.5 \text{ oz-in.-sec}$$

$$K_m = 10 \text{ oz-in./V}$$

$$T_L = \text{disturbance torque}$$

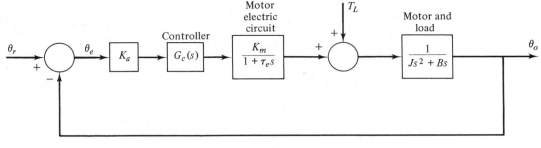

Figure 10P-4

(a) Determine the minimum value of the error sensor and amplifier gain K_a so that the steady-state error of θ_0 due to a unit step torque disturbance is less than or equal to 0.01 (1 percent).

(b) Determine the stability of the system when K_a is set at the value determined in part (a).

Design a phase-lead controller in the form

$$G_c(s) = \frac{1 + aTs}{1 + Ts} \qquad a > 1$$

so that the closed-loop system has a phase margin of approximately 30°. Determine the bandwidth of the compensated system. Plot the output response $\theta_0(t)$ when the input $\theta_r(t)$ is a unit step function. ($T_L = 0$.)

10.5. Repeat the design problem in Problem 10-4 with a phase-lag controller

$$G_c(s) = \frac{1 + aTs}{1 + Ts} \qquad a < 1$$

10.6. Human beings breathe in order to provide the means for gas exchange for the entire body. A respiratory control system is needed to ensure that the body's needs for this gas exchange are adequately met. The criterion of control is adequate ventilation, which ensures satisfactory levels of both oxygen and carbon dioxide in the arterial blood. Respiration is controlled by neural impulses that originate within the lower brain and are transmitted to the chest cavity and diaphragm to govern the rate and tidal volume. One source of the signals is the chemoreceptors located near the respiratory center, which are sensitive to carbon dioxide and oxygen concentrations. Figure 10P-6 shows the block diagram of a simplified model of the human respiratory control system. The objective is to control the effective ventilation of the lungs so that satisfactory balance of concentrations of carbon dioxide and oxygen is maintained in the blood circulated at the chemoreceptor.

(a) A normal value of the chemoreceptor gain K_f is 0.1. Determine the gain margin and the phase margin of the system.

(b) Assume that the chemoreceptor is defective so that its gain K_f is increased to 1. Design a controller in the forward path (to be inserted in front of the block representing the lungs) so that a phase margin of 45 degrees is maintained.

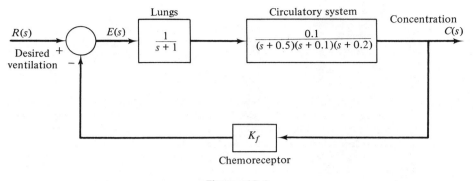

Figure 10P-6

10.7. This problem deals with the cable-reel-unwind process described in Problem 4-15 and shown in Fig. 4P-15. The inertia of the reel is

$$J_R = 18R^4 - 200 \text{ ft-lb-sec}^2$$

where R is the effective radius of the cable reel.

(a) Assume that R and J_R are constant between layers of the cable. Determine the maximum value of the amplifier gain K so that the entire unwinding process is stable from beginning until end.

(b) Let $K = 10$. Design a series controller so that the system has a phase margin of 45 degrees at the end of the unwind process ($R = 2$ ft). With this controller, what are the phase and gain margins of the system at the beginning of the unwind process? Sketch the root loci of the compensated process with $K = 10$ and indicate the variation of the roots on the loci as the unwind process progresses. It should be noted that the treatment of this problem by a transfer function is only an approximation; strictly, the process is nonlinear as well as time varying.

10.8. The inventory control system considered in Problem 8.6 has the transfer function

$$G_p(s) = \frac{4}{s^2}$$

Design a PD controller with the transfer function $G_c(s) = K_P + K_D s$ so that the peak resonance M_p is equal to 1.2, and the bandwidth BW is 1 rad/sec.

10.9. The inventory control system considered in Problem 8.6 has the transfer function

$$G_p(s) = \frac{4}{s^2}$$

Design a phase-lead controller with the transfer function

$$G_c(s) = \frac{1 + aTs}{1 + Ts} \qquad a > 1$$

so that the phase margin of the system is approximately 45 degrees.

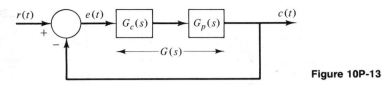

Figure 10P-13

10.10. Consider the speed-control system described in Problem 8.9. It is desired to keep the frequency variation in ω_g due to load variation T_L to within ± 0.1 percent. This sets the minimum value for KK_T. With this value of KK_T, design a series phase-lag controller with the transfer function

$$G_c(s) = \frac{1 + aTs}{1 + Ts} \qquad a < 1$$

so that the phase margin of the overall system is 45 degrees.

10.11. Consider the phase-lock-loop motor-speed-control system shown in Fig. 8P-10. Find the value of R_2 so that the phase margin of the closed-loop system is approximately 45 degrees. There are two values of R_2 that can satisfy this requirement. Sketch the Bode plots to illustrate your design. Compare the results with the root locus solution obtained in Problem 8.10.

10.12. For the computer-tape drive system described in Problem 8.14, design the bridged-T network so that its zeros cancel the complex-conjugate poles of the controlled process. Set the value of the amplifier gain K so that the ramp-error constant of the system is equal to 100. Find the gain margin and phase margin of the overall system.

10.13. Figure 10P-13 shows the block diagram of a speed regulator system. The transfer function of the controlled process is

$$G_p(s) = \frac{12{,}000(s + 333.33)}{(s + 10.052)(s^2 + 69.95s + 19{,}896.9)}$$

Design a bridged-T network controller so that the transfer function is

$$G_c(s) = \frac{s^2 + 70s + 20{,}000}{s^2 + 2\zeta_p\omega_n s + \omega_n^2} \qquad \omega_n^2 = 20{,}000$$

Determine the gain margin and the phase margin of the uncompensated and the compensated systems. Draw the Bode plots of $G(s)$. Draw the closed-loop frequency response and find the value of M_p of the compensated system. Find $c(t)$ when $r(t)$ is a unit step function.

10.14. For the liquid-level control system described in Problem 6.27, the open-loop transfer function between $e(t)$ and $h(t)$ is

$$G(s) = \frac{H(s)}{E(s)} = \frac{7.14N}{s(s + 1)(s + 5.357)}$$

where N is the number of inlets of the tank.

(a) Determine the following quantities for the closed-loop system when $N = 4$: gain margin (db), phase margin (degrees), gain-crossover frequency (rad/sec), phase-crossover frequency (rad/sec). Draw the Bode plot of $G(s)$.

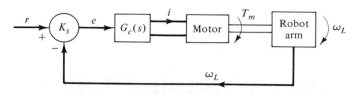

Figure 10P-15

(b) With $N = 4$, design a cascade phase-lead controller so that the phase margin of the system is 50 degrees. Use two stages of the phase-lead network if necessary. Plot the Bode plot of the compensated open-loop transfer function. Find the following quantities for the compensated system: gain margin (db), phase margin (degrees), gain-crossover frequency (rad/sec), phase-crossover frequency (rad/sec), M_p, ω_p, BW. Plot the closed-loop frequency response $|M|$ versus ω.

(c) With $N = 4$, repeat part (b), except with a phase-lag controller.

(d) Plot the single-step responses of the systems designed in parts (b) and (c). Compare the responses of the phase-lead compensated and the phase-lag compensated systems and comment on the results.

10.15. Figure 10P-15 shows the robot-arm control system originally described in Problem 9P-27. It was found in Problem 9-27 that the closed-loop system with only the amplifier with gain $K_a = 65$ has an oscillatory speed response. This is due to the high-frequency resonant mode of the arm structure.

Design a bridged-T controller with the transfer function

$$G_c(s) = \frac{K_a(s^2 + 2\zeta_z\omega_n s + \omega_n^2)}{\left(s^2 + 2\zeta_p\omega_n s + \omega_n^2\right)}$$

to cancel the lightly damped poles of $\Omega_L(s)/E(s)$. Select the value of K_a so that the relative damping ratio of the compensated system is approximately 70.7%. Assume that there is no upper limit on the value of K_a. Obtain the closed-loop frequency response of the compensated system. Determine the values of M_p, ω_p, and BW. Plot the response of $\omega_L(t)$ when the input $r(t)$ is a unit step function.

10.16. The solution of Problem 10-15 yields a system that has a very high value for K_a and a very high bandwidth for the closed-loop system. Although the response of the designed system is very fast, the high-gain and the high-bandwidth characteristics may be too high a cost for the cure of the speed ripples. Instead of using a bridged-T controller, a phase-lag controller may be attempted which will filter out the high-frequency signals in the system loop. Setting $K_a = 65$, let

$$G_c(s) = K_a\frac{1 + aTs}{1 + Ts}$$

(a) Find the values of a and T so that the compensated system has a gain margin of 30 db. Find the Bode plots of $\Omega_L(s)/E(s)$ and the closed-loop frequency response of the compensated system. Find the following quantities: gain margin, phase margin, M_p, ω_p, BW. Obtain the unit-step response of $\omega_L(t)$.

(b) Repeat part (a) by setting the gain margin at 25 db. Explain why the peak overshoot of the step response of this system is less than that of the system designed in part (a).

appendix A

Frequency-Domain Plots

Consider that $G(s)H(s)$ is the loop transfer function of a feedback control system. The sinusoidal steady-state transfer function is obtained by setting $s = j\omega$ in $G(s)H(s)$. In control systems studies, frequency-domain plots of the open-loop transfer function $G(j\omega)H(j\omega)$ are made for analyzing the performance of the closed-loop control system.

The function $G(j\omega)$ is generally a complex function of the frequency ω and can be written

$$G(j\omega) = |G(j\omega)| \underline{/G(j\omega)} \qquad \text{(A-1)}$$

where $|G(j\omega)|$ denotes the magnitude of $G(j\omega)$ and $\underline{/G(j\omega)}$ is the phase of $G(j\omega)$.

The following forms of frequency-domain plots of $G(j\omega)$ [or of $G(j\omega)H(j\omega)$] versus ω are most useful in the analysis and design of feedback control systems in the frequency domain.

1. *Polar plot:* a plot of the magnitude versus phase in the polar coordinates as ω is varied from zero to infinity.
2. *Bode plot (corner plot):* a plot of the magnitude in decibels versus ω (or $\log_{10}\omega$) in the semilog (or rectangular) coordinates.
3. *Magnitude-versus-phase plot:* a plot of the magnitude in decibels versus the phase on rectangular coordinates with ω as a variable parameter on the curve.

These various plots are described in the following sections.

A.1 POLAR PLOTS OF TRANSFER FUNCTIONS

The polar plot of a transfer function $G(s)$ is a plot of the magnitude of $G(j\omega)$ versus the phase of $G(j\omega)$ on the polar coordinates, as ω is varied from zero to infinity. From a mathematical viewpoint, the process may be regarded as a mapping of the positive half of the imaginary axis of the s-plane onto the plane of the function $G(j\omega)$. A simple example of this mapping is shown in Fig. A-1. For any frequency $\omega = \omega_1$, the magnitude and phase of $G(j\omega_1)$ are represented by a phasor that has the corresponding magnitude and phase angle in the $G(j\omega)$-plane. In measuring the phase, counterclockwise is referred to as positive, and clockwise as negative.

To illustrate the construction of the polar plot of a transfer function, consider the function

$$G(s) = \frac{1}{1 + Ts} \tag{A-2}$$

where T is a positive constant.

Putting $s = j\omega$, we have

$$G(j\omega) = \frac{1}{1 + j\omega T} \tag{A-3}$$

In terms of magnitude and phase, Eq. (A-3) is written

$$G(j\omega) = \frac{1}{\sqrt{1 + \omega^2 T^2}} \bigg/ -\tan^{-1}\omega T \tag{A-4}$$

When ω is zero, the magnitude of $G(j\omega)$ is unity, and the phase of $G(j\omega)$ is at 0 degrees. Thus, at $\omega = 0$, $G(j\omega)$ is represented by a phasor of unit length directed in the 0-degree direction. As ω increases, the magnitude of $G(j\omega)$ decreases, and the phase becomes more negative. As ω increases, the length of the phasor in the polar coordinates decreases, and the phasor rotates in the clockwise (negative) direction. When ω approaches infinity, the magnitude of $G(j\omega)$ becomes zero, and the phase reaches -90 degrees. This is often represented by a phasor with an infinitesimally small length directed along the -90-degree axis in the $G(j\omega)$-plane. By substituting other finite values of ω into Eq. (A-4), the exact plot of $G(j\omega)$ turns out to be a semicircle, as shown in Fig. A-2.

As a second illustrative example, consider the transfer function

$$G(j\omega) = \frac{1 + j\omega T_2}{1 + j\omega T_1} \tag{A-5}$$

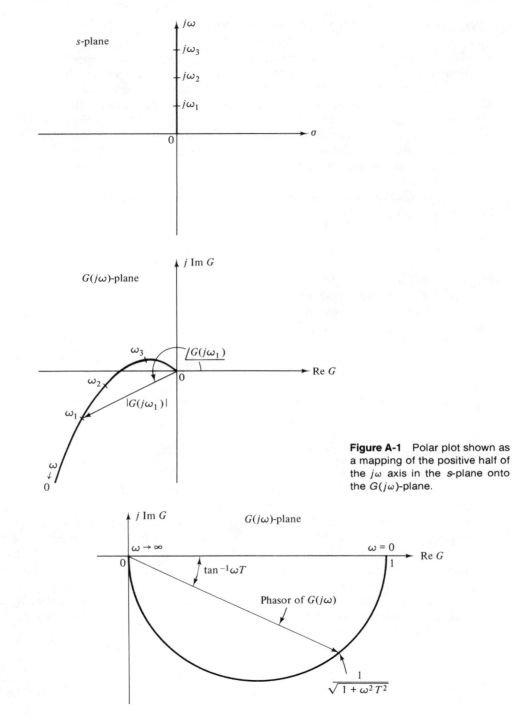

Figure A-1 Polar plot shown as a mapping of the positive half of the $j\omega$ axis in the s-plane onto the $G(j\omega)$-plane.

Figure A-2 Polar plot of $G(j\omega) = 1 / (1 + j\omega T)$.

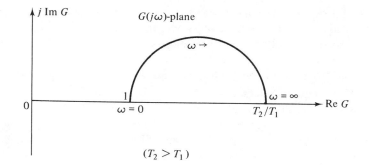

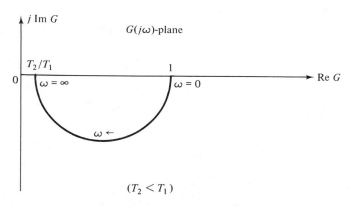

Figure A-3 Polar plots of $G(j\omega) = (1 + j\omega T_2) / (1 + j\omega T_1)$.

where T_1 and T_2 are positive constants. Equation (A-5) is written

$$G(j\omega) = \sqrt{\frac{1 + \omega^2 T_2^2}{1 + \omega^2 T_1^2}} \bigg/ \tan^{-1}\omega T_2 - \tan^{-1}\omega T_1 \qquad (A\text{-}6)$$

The polar plot of $G(j\omega)$, in this case, depends upon the relative magnitudes of T_2 and T_1. If T_2 is greater than T_1, the magnitude of $G(j\omega)$ is always greater than unity as ω is varied from zero to infinity, and the phase of $G(j\omega)$ is always positive. If T_2 is less than T_1, the magnitude of $G(j\omega)$ is always less than unity, and the phase is always negative. The polar plots of $G(j\omega)$ of Eq. (A-6) that correspond to the two above-mentioned conditions are shown in Fig. A-3.

It is apparent that the accurate plotting of the polar plot of a transfer function is generally a tedious process, especially if the transfer function is of high order. In practice, a digital computer can be used to generate the data, or even the final figure of the polar plot, for a wide class of transfer functions. However, from the analytical standpoint, it is essential that the engineer be completely familiar with the properties of the polar plot, so that the computer data may be properly interpreted. In some cases, such as in the Nyquist stability study, only the general shape of the polar plot of $G(j\omega)H(j\omega)$ is needed, and often a rough sketch of the polar plot is quite adequate for the specific objective. In general, the sketching of the polar plot is facilitated by the following information:

1. The behavior of the magnitude and the phase at $\omega = 0$ and at $\omega = \infty$.
2. The points of intersections of the polar plot with the real and imaginary axes, and the values of ω at these intersections.

The general shape of the polar plot may be determined once we have information on the two items listed above.

◼

Example A-1

Consider that it is desired to make a rough sketch of the polar plot of the transfer function

$$G(s) = \frac{10}{s(s + 1)} \tag{A-7}$$

Substituting $s = j\omega$ in Eq. (A-7), the magnitude and phase of $G(j\omega)$ at $\omega = 0$ and $\omega = \infty$ are computed as follows:

$$\lim_{\omega \to 0} |G(j\omega)| = \lim_{\omega \to 0} \frac{10}{\omega} = \infty \tag{A-8}$$

$$\lim_{\omega \to 0} \underline{/G(j\omega)} = \lim_{\omega \to 0} \underline{/\frac{10}{j\omega}} = -90° \tag{A-9}$$

$$\lim_{\omega \to \infty} |G(j\omega)| = \lim_{\omega \to \infty} \frac{10}{\omega^2} = 0 \tag{A-10}$$

$$\lim_{\omega \to \infty} \underline{/G(j\omega)} = \lim_{\omega \to \infty} \underline{/\frac{10}{(j\omega)^2}} = -180° \tag{A-11}$$

Thus, the properties of the polar plot of $G(j\omega)$ at $\omega = 0$ and $\omega = \infty$ are ascertained. Next, we determine the intersections, if any, of the polar plot with the two axes of the $G(j\omega)$-plane.

If the polar plot of $G(j\omega)$ intersects the real axis, at the point of intersection, the imaginary part of $G(j\omega)$ is zero; that is,

$$\text{Im}[G(j\omega)] = 0 \tag{A-12}$$

To express $G(j\omega)$ as

$$G(j\omega) = \text{Re}[G(j\omega)] + j\,\text{Im}[G(j\omega)] \tag{A-13}$$

we must rationalize $G(j\omega)$ by multiplying its numerator and denominator by the complex conjugate of its denominator. Therefore, $G(j\omega)$ is written

$$G(j\omega) = \frac{10(-j\omega)(-j\omega + 1)}{j\omega(j\omega + 1)(-j\omega)(-j\omega + 1)} = \frac{-10\omega^2}{\omega^4 + \omega^2} - j\frac{10\omega}{\omega^4 + \omega^2} \tag{A-14}$$

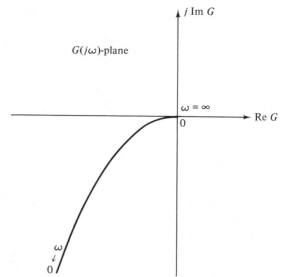

Figure A-4 Sketch of the polar plot of $G(s) = 10 / [s(s + 1)]$.

which gives

$$\text{Im}[G(j\omega)] = \frac{-10}{\omega(\omega^2 + 1)} \tag{A-15}$$

and

$$\text{Re}[G(j\omega)] = \frac{-10}{\omega^2 + 1} \tag{A-16}$$

When we set $\text{Im}[G(j\omega)]$ to zero, we get $\omega = \infty$, meaning that the $G(j\omega)$ plot intersects only with the real axis of the plane at the origin.

Similarly, the intersection of the polar plot of $G(j\omega)$ with the imaginary axis is found by setting $\text{Re}[G(j\omega)]$ of Eq. (A-16) to zero. The only real solution for ω is also $\omega = \infty$, which corresponds to the origin of the $G(j\omega)$-plane. The conclusion is that the polar plot of $G(j\omega)$ does not intersect any one of the two axes at any finite nonzero frequency. Based upon this information, as well as knowledge on the angles of $G(j\omega)$ at $\omega = 0$ and $\omega = \infty$, the polar plot of $G(j\omega)$ is easily sketched, as shown in Fig. A-4.

Example A-2
Given the transfer function

$$G(s) = \frac{10}{s(s + 1)(s + 2)} \tag{A-17}$$

it is desired to make a rough sketch of the polar plot of $G(j\omega)$. The following calculations are made for the properties of the magnitude and phase of $G(j\omega)$ at $\omega = 0$ and $\omega = \infty$:

$$\lim_{\omega \to 0} |G(j\omega)| = \lim_{\omega \to 0} \frac{5}{\omega} = \infty \tag{A-18}$$

$$\lim_{\omega \to 0} \underline{/G(j\omega)} = \lim_{\omega \to 0} \underline{/\frac{5}{j\omega}} = -90° \tag{A-19}$$

$$\lim_{\omega \to \infty} |G(j\omega)| = \lim_{\omega \to \infty} \frac{10}{\omega^3} = 0 \tag{A-20}$$

$$\lim_{\omega \to \infty} \underline{/G(j\omega)} = \lim_{\omega \to \infty} \underline{/\frac{10}{(j\omega)^3}} = -270° \tag{A-21}$$

To find the intersections of the $G(j\omega)$ curve on the real and the imaginary axes of the $G(j\omega)$-plane, we rationalize $G(j\omega)$ to give

$$G(j\omega) = \frac{10(-j\omega)(-j\omega + 1)(-j\omega + 2)}{j\omega(j\omega + 1)(j\omega + 2)(-j\omega)(-j\omega + 1)(-j\omega + 2)} \tag{A-22}$$

After simplification, Eq. (A-22) is written

$$G(j\omega) = \frac{-30\omega^2}{9\omega^4 + \omega^2(2 - \omega^2)^2} - \frac{j10\omega(2 - \omega^2)}{9\omega^4 + \omega^2(2 - \omega^2)^2} \tag{A-23}$$

We set

$$\text{Re}[G(j\omega)] = \frac{-30}{9\omega^2 + (2 - \omega^2)^2} = 0 \tag{A-24}$$

and

$$\text{Im}[G(j\omega)] = \frac{-10(2 - \omega^2)}{9\omega^3 + \omega(2 - \omega^2)^2} = 0 \tag{A-25}$$

Equation (A-24) is satisfied when

$$\omega = \infty$$

which means that the $G(j\omega)$ plot intersects the imaginary axis only at the origin. Equation (A-25) is satisfied when

$$\omega^2 = 2$$

which gives the intersection on the real axis of the $G(j\omega)$-plane when $\omega = \pm \sqrt{2}$ rad/sec. Substituting $\omega = \sqrt{2}$ into Eq. (A-23) gives the point of intersection at

$$G(j\sqrt{2}) = -\tfrac{5}{3} \tag{A-26}$$

The result of $\omega = -\sqrt{2}$ rad/sec has no physical meaning, but mathematically it simply represents a mapping point on the negative $j\omega$ axis of the s-plane. In general, if $G(s)$

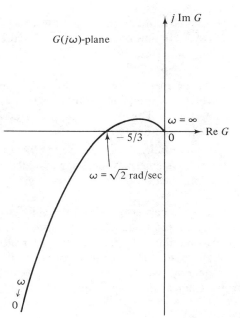

Figure A-5 Sketch of the polar plot of $G(s) = 10 / [s(s + 1)(s + 2)]$.

is a rational function of s (a quotient of two polynomials of s), the polar plot of $G(j\omega)$ for negative values of ω is the mirror image of that for positive ω, with the mirror placed on the real axis of the $G(j\omega)$-plane.

With the information collected above, it is now possible to make a sketch of the polar plot for the transfer function in Eq. (A-17), and the sketch is shown in Fig. A-5.

Although the method of obtaining the rough sketch of the polar plot of a transfer function as described above is quite straightforward, in general, for complex transfer functions that may have multiple crossings on the real and imaginary axes in the transfer function plane, the algebraic manipulation may again be quite involved. Furthermore, the polar plot is basically a tool for analysis; it is somewhat awkward for design purposes. We shall show in the next section that approximate information on the polar plot can always be obtained from the Bode plot, which is usually sketched without any computations. Thus, for more complex transfer functions, other than using the digital computer, sketches of the polar plots are preferably obtained with the help of the Bode plots.

■

A.2 BODE PLOT (CORNER PLOT) OF A TRANSFER FUNCTION

The discussions in Section A.1 show that the polar plot portrays a function $G(j\omega)$ in the polar coordinates in terms of its magnitude and phase as functions of ω. The Bode plot, on the other hand, contains two graphs, one with the magnitude of $G(j\omega)$ plotted in decibels versus log ω or ω, and the other with the phase of $G(j\omega)$ in degrees as a function of log ω or ω. The Bode plot is also known sometimes as the *corner plot* or the *logarithmic plot* of $G(j\omega)$. The name, corner plot, is used since the

Bode plot is basically an approximation method in that the magnitude of $G(j\omega)$ in decibels as a function of ω is approximated by straight-line segments.

In simple terms, the Bode plot has the following unique characteristics:

1. Since the magnitude of $G(j\omega)$ in the Bode plot is expressed in decibels, the product and division factors in $G(j\omega)$ become additions and subtractions, respectively. The phase relations are also added and subtracted from each other in a natural way.

2. The magnitude plot of the Bode plots of most $G(j\omega)$ functions encountered in control systems may be approximated by straight-line segments. This makes the construction of the Bode plot very simple.

Since the corner plot is relatively easy to construct, and usually without point-by-point plotting, it may be used to generate data necessary for the other frequency-domain plots, such as the polar plot, or the magnitude-versus-phase plot, which is discussed later in this chapter.

In general, we may represent the open-loop transfer function of a feedback control system without pure time delay by

$$G(s) = \frac{K(s + z_1)(s + z_2) \cdots (s + z_m)}{s^i(s + p_1)(s + p_2) \cdots (s + p_n)} \tag{A-27}$$

where K is a real constant and the zs and the ps may be real or complex numbers. As an alternative, the open-loop transfer function is written

$$G(s) = \frac{K(1 + T_1 s)(1 + T_2 s) \cdots (1 + T_m s)}{s^i(1 + T_a s)(1 + T_b s) \cdots (1 + T_n s)} \tag{A-28}$$

where K is a real constant, and the Ts may be real or complex numbers.

In Chapter 7, Eq. (A-27) is the preferred form for root loci construction. However, for Bode plots, the transfer function should first be written in the form of Eq. (A-28). Since practically all the terms in Eq. (A-28) are of the same form, without loss of generality, we can use the following transfer function to illustrate the construction of the Bode diagram:

$$G(s) = \frac{K(1 + T_1 s)(1 + T_2 s)}{s(1 + T_a s)(1 + j2\zeta\mu - \mu^2)} \tag{A-29}$$

where K, T_1, T_2, T_a, ζ, and μ are real coefficients. It is assumed that the second-order polynomial, $1 + 2\zeta\mu - \mu^2$, $\mu = \omega/\omega_n$, has two complex-conjugate zeros.

The magnitude of $G(j\omega)$ in decibels is obtained by multiplying the logarithm to the base 10 of $|G(j\omega)|$ by 20; we have

$$
\begin{aligned}
|G(j\omega)|_{dB} = 20\log_{10}|G(j\omega)| &= 20\log_{10}|K| + 20\log_{10}|1 + j\omega T_1| \\
&+ 20\log_{10}|1 + j\omega T_2| - 20\log_{10}|j\omega| \\
&- 20\log_{10}|1 + j\omega T_a| - 20\log_{10}|1 + j2\zeta\mu - \mu^2|
\end{aligned}
\tag{A-30}
$$

The phase of $G(j\omega)$ is written

$$
\begin{aligned}
\underline{/G(j\omega)} = \underline{/K} + \underline{/1 + j\omega T_1} + \underline{/1 + j\omega T_2} - \underline{/j\omega} \\
- \underline{/1 + j\omega T_a} - \underline{/1 + j2\zeta\mu - \mu^2}
\end{aligned}
\tag{A-31}
$$

In general, the function $G(j\omega)$ may be of higher order than that of Eq. (A-29) and have many more factored terms. However, Eqs. (A-30) and (A-31) indicate that additional terms in $G(j\omega)$ would simply produce more similar terms in the magnitude and phase expressions, so that the basic method of construction of the Bode plot would be the same. We have also indicated that, in general, $G(j\omega)$ may contain just four simple types of factors:

1. Constant factor: K
2. Poles or zeros at the origin: $(j\omega)^{\pm p}$
3. Poles or zeros not at $\omega = 0$: $(1 + j\omega T)^{\pm q}$
4. Complex poles or zeros: $(1 + j2\zeta\mu - \mu^2)^{\pm r}$

where p, q, and r are positive integers.

Equations (A-30) and (A-31) verify one of the unique characteristics of the Bode plot in that each of the four types of factors listed may be considered as a separate plot; the individual plots are then added or subtracted accordingly to yield the total magnitude in decibels and phase plot of $G(j\omega)$. The curves may be done on semilog graph paper or linear rectangular coordinate graph paper, depending on whether ω or $\log_{10}\omega$ is used as the abscissa.

We shall now investigate the sketching of the Bode plot of the different types of factors.

Constant Term, K

Since

$$
K_{dB} = 20\log_{10}K = \text{constant} \tag{A-32}
$$

and

$$
\underline{/K} = \begin{cases} 0° & K > 0 \\ 180° & K < 0 \end{cases} \tag{A-33}
$$

the Bode plot of the constant factor K is shown in Fig. A-6 in semilog coordinates.

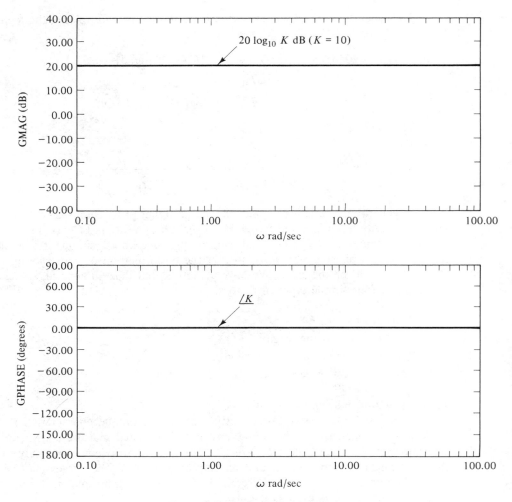

Figure A-6 Bode plot of constant K.

Poles and Zeros at the Origin, $(j\omega)^{\pm p}$

The magnitude of $(j\omega)^{\pm p}$ in decibels is given by

$$20\log_{10}\left|(j\omega)^{\pm p}\right| = \pm 20p\log_{10}\omega \text{ dB} \qquad (A\text{-}34)$$

for $\omega \geq 0$. The last expression for a given p represents the equation of a straight line in either semilog or rectangular coordinates. The slopes of these straight lines may be determined by taking the derivative of Eq. (A-34) with respect to $\log_{10}\omega$; that is,

$$\frac{d}{d\log_{10}\omega}(\pm 20p\log_{10}\omega) = \pm 20p \text{ dB} \qquad (A\text{-}35)$$

These lines all pass through the 0-dB point at $\omega = 1$. Thus a unit change in $\log_{10}\omega$ will correspond to a change of $\pm 20p$ dB in magnitude. Furthermore, a unit change in $\log_{10}\omega$ in the rectangular coordinates is equivalent to 1 decade of variation in ω, that is, from 1 to 10, 10 to 100, and so on, in the semilog coordinates. Thus the slopes of the straight lines described by Eq. (A-34) are said to be $\pm 20p$ dB/decade of frequency.

Instead of using decades, sometimes the unit *octave* is used to represent the separation of two frequencies. The frequencies ω_1 and ω_2 are separated by an octave if $\omega_2/\omega_1 = 2$. The number of decades between any two frequencies ω_1 and ω_2 is given by

$$\text{number of decades} = \frac{\log_{10}(\omega_2/\omega_1)}{\log_{10}10} = \log_{10}\left(\frac{\omega_2}{\omega_1}\right) \tag{A-36}$$

Similarly, the number of octaves between ω_2 and ω_1 is

$$\text{number of octaves} = \frac{\log_{10}(\omega_2/\omega_1)}{\log_{10}2} = \frac{1}{0.301}\log_{10}\left(\frac{\omega_2}{\omega_1}\right) \tag{A-37}$$

Thus, the relation between octaves and decades is given by

$$\text{number of octaves} = \frac{1}{0.301}\text{ decades} \tag{A-38}$$

Substituting Eq. (A-38) into Eq. (A-35), we have

$$\pm 20p \text{ dB/decade} = \pm 20p \times 0.301 \simeq \pm 6p \text{ dB/octave} \tag{A-39}$$

For a transfer function $G(j\omega)$ that has a simple pole at $s = 0$, the magnitude of $G(j\omega)$ is a straight line with a slope of -20 dB/decade, and passes through the 0-dB axis at $\omega = 1$.

The phase of $(j\omega)^{\pm p}$ is written

$$\underline{/(j\omega)^{\pm p}} = \pm p \times 90° \tag{A-40}$$

The magnitude and phase curves of the function $(j\omega)^{\pm p}$ are sketched as shown in Fig. A-7 for several values of p.

Simple Zero (1 + jωT)

Let

$$G(j\omega) = 1 + j\omega T \tag{A-41}$$

where T is a real constant. The magnitude of $G(j\omega)$ in decibels is written

$$|G(j\omega)|_{\text{dB}} = 20\log_{10}|G(j\omega)| = 20\log_{10}\sqrt{1 + \omega^2 T^2} \tag{A-42}$$

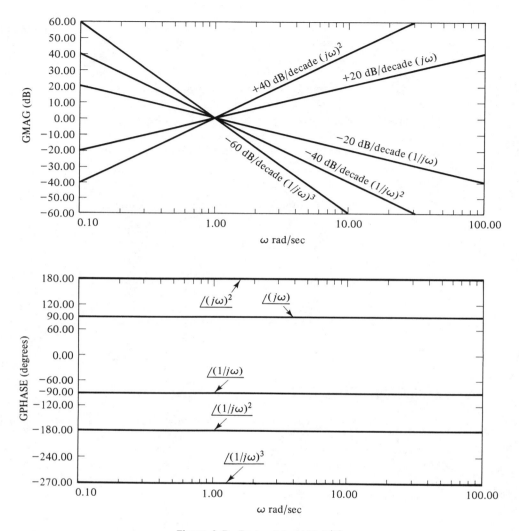

Figure A-7   Bode plots of $(j\omega)^{\pm p}$.

To obtain asymptotic approximations of the magnitude of $G(j\omega)$, we consider both very large and very small values of ω. At very low frequencies, $\omega T \ll 1$, Eq. (A-42) is approximated by

$$|G(j\omega)|_{\mathrm{dB}} = 20\log_{10}|G(j\omega)| \simeq 20\log_{10}1 = 0 \text{ dB} \qquad \text{(A-43)}$$

since $\omega^2 T^2$ is neglected when compared with 1.

At very high frequencies, $\omega T \gg 1$, we may approximate $1 + \omega^2 T^2$ by $\omega^2 T^2$; then Eq. (A-42) becomes

$$|G(j\omega)|_{\mathrm{dB}} = 20\log_{10}|G(j\omega)| \simeq 20\log_{10}\sqrt{\omega^2 T^2}$$
$$= 20\log_{10}\omega T \qquad \text{(A-44)}$$

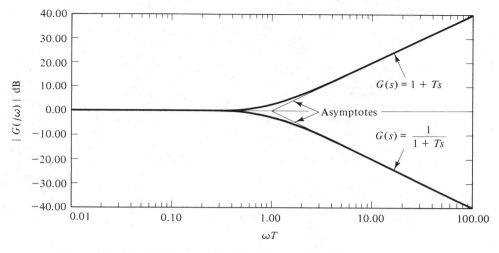

Figure A-8 Magnitude versus frequency of the Bode plots of $G(s) = 1 + Ts$ and $G(s) = 1 / (1 + Ts)$.

Equation (A-44) represents a straight line with a slope of $+20$ dB/decade of frequency. The intersect of this line with the 0-dB axis is found by equating Eq. (A-44) to zero, which gives

$$\omega = \frac{1}{T} \qquad (A\text{-}45)$$

This frequency is also the intersect of the high-frequency approximate plot and the low-frequency approximate plot which is the 0-dB line as given by Eq. (A-43). The frequency given by Eq. (A-45) is also known as the *corner frequency* of the Bode plot of the transfer function in Eq. (A-41), since the approximate magnitude plot forms the shape of a corner at that frequency, as shown in Fig. A-8. The actual magnitude curve for $G(j\omega)$ of Eq. (A-41) is a smooth curve, and deviates only slightly from the

Table A-1

ωT	$\log_{10}\omega T$	$\lvert 1 + j\omega T\rvert$	$\lvert 1 + j\omega T\rvert$ (dB)	$\underline{/\,1 + j\omega T}$
0.01	-2	1	0	0.5°
0.1	-1	1.04	0.043	5.7°
0.5	-0.3	1.12	1	26.6°
0.76	-0.12	1.26	2	37.4°
1.0	0	1.41	3	45.0°
1.31	0.117	1.65	4.3	52.7°
1.73	0.238	2.0	6.0	60.0°
2.0	0.3	2.23	7.0	63.4°
5.0	0.7	5.1	14.2	78.7°
10.0	1.0	10.4	20.3	84.3°

Table A-2

ωT	$\lvert 1 + j\omega T \rvert$ (dB)	Straight-Line Approximation of $\lvert 1 + j\omega T \rvert$ (dB)	Error (dB)
0.1 (1 decade below corner frequency)	0.043	0	0.043
0.5 (1 octave below corner frequency)	1.0	0	1
0.76	2	0	2
1.0 (at the corner frequency)	3	0	3
1.31	4.3	2.3	2
2.0 (1 octave above corner frequency)	7	6	1
10 (1 decade above corner frequency)	20.043	20	0.043

straight-line approximation. The actual values for the magnitude of the function $1 + j\omega T$ as functions of ωT are tabulated in Table A-1. Table A-2 gives a comparison of the actual values with the straight-line approximations at some significant frequencies.

The error between the actual magnitude curve and the straight-line asymptotes is symmetrical with respect to the corner frequency $1/T$. Furthermore, it is useful to remember that the error is 3 dB at the corner frequency, and 1 dB at 1 octave above $(2/T)$ and 1 octave below $(0.5/T)$ the corner frequency. At 1 decade above and below the corner frequency, the error is dropped to approximately 0.3 dB. From these facts, the procedure in obtaining the magnitude curve of the plot of the first-order factor $(1 + j\omega T)$ is outlined as follows:

1. Locate the corner frequency $\omega = 1/T$.
2. Draw the 20 dB/decade (or 6 dB/octave) line and the horizontal line at 0 dB, with the two lines intersecting at $\omega = 1/T$.
3. If necessary, the actual magnitude curve is obtained by locating the points given in Table A-1.

Usually, a smooth curve can be sketched simply by locating the 3-dB point at the corner frequency and the 1-dB points at 1 octave above and below the corner frequency.

The phase of $G(j\omega) = 1 + j\omega T$ is written as

$$\underline{/G(j\omega)} = \tan^{-1}\omega T \tag{A-46}$$

Similar to the magnitude curve, a straight-line approximation can be made for the phase curve. Since the phase of $G(j\omega)$ varies from 0° to 90° we may draw a line from 0° at 1 decade below the corner frequency to $+90°$ at 1 decade above the corner frequency. As shown in Fig. A-9, the maximum deviation of the straight-line approximation from the actual curve is less than 6°. Table A-1 gives the values of $\underline{/1 + j\omega T}$ versus ωT.

Simple Pole, $1/(1 + j\omega T)$

When

$$G(j\omega) = \frac{1}{1 + j\omega T} \tag{A-47}$$

the magnitude, $|G(j\omega)|$ in decibels is given by the negative of the right side of Eq. (A-42), and the phase $\underline{/G(j\omega)}$ is the negative of the angle in Eq. (A-46). Therefore, it is simple to extend all the analysis for the case of the simple zero to the Bode plot of Eq. (A-47). We can write

$$\omega T \ll 1 \qquad |G(j\omega)|_{dB} \simeq 0 \text{ dB} \tag{A-48}$$

$$\omega T \gg 1 \qquad |G(j\omega)|_{dB} = -20 \log_{10} \omega T \tag{A-49}$$

Thus the corner frequency of the Bode plot of Eq. (A-47) is still at $\omega = 1/T$. At high frequencies, the slope of the straight-line approximation is -20 dB/decade. The phase of $G(j\omega)$ is 0 degrees when $\omega = 0$ and -90 degrees when ω approaches infinity. The magnitude and the phase of the Bode plot of Eq. (A-47) are shown in Figs. A-8 and A-9, respectively. The data in Tables A-1 and A-2 are still useful for

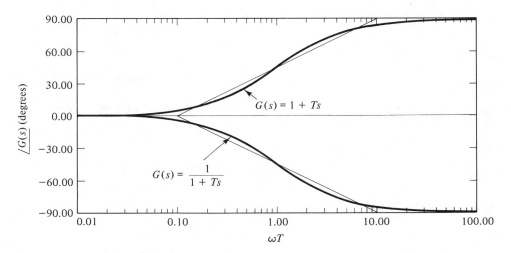

Figure A-9 Phase versus frequency of the Bode plots of $G(s) = 1 + Ts$ and $G(s) = 1/(1 + Ts)$.

the simple pole case, if appropriate sign changes are made to the numbers. For instance, at the corner frequency, the error between the straight-line approximation and the actual magnitude curve is -3 dB.

Quadratic Poles and Zeros

Now consider the second order transfer function

$$G(s) = \frac{\omega_n^2}{s^2 + 2\zeta\omega_n s + \omega_n^2}$$

$$= \frac{1}{1 + (2\zeta/\omega_n)s + (1/\omega_n^2)s^2} \tag{A-50}$$

We are interested only in the cases when $\zeta \leq 1$, since otherwise, $G(s)$ would have two unequal real poles, and the Bode plot can be determined by considering $G(s)$ as the product of two transfer functions each having a simple pole.

Letting $s = j\omega$, Eq. (A-50) becomes

$$G(j\omega) = \frac{1}{\left[1 - (\omega/\omega_n)^2\right] + j2\zeta(\omega/\omega_n)} \tag{A-51}$$

The magnitude of $G(j\omega)$ in decibels is

$$20 \log_{10}|G(j\omega)| = -20 \log_{10}\sqrt{\left[1 - \left(\frac{\omega}{\omega_n}\right)^2\right]^2 + 4\zeta^2\left(\frac{\omega}{\omega_n}\right)^2} \tag{A-52}$$

At very low frequencies, $\omega/\omega_n \ll 1$; Eq. (A-52) may be written as

$$|G(j\omega)|_{dB} = 20 \log_{10}|G(j\omega)| \cong -20 \log_{10}1 = 0 \text{ dB} \tag{A-53}$$

Thus the low-frequency asymptote of the magnitude plot of Eq. (A-50) is a straight line that lies on the 0-dB axis of the Bode plot coordinates.

At very high frequencies, $\omega/\omega_n \gg 1$; the magnitude in decibels of $G(j\omega)$ in Eq. (A-50) becomes

$$|G(j\omega)|_{dB} = 20 \log_{10}|G(j\omega)| \cong -20 \log_{10}\sqrt{\left(\frac{\omega}{\omega_n}\right)^4}$$

$$= -40 \log_{10}\left(\frac{\omega}{\omega_n}\right) \qquad \text{dB} \tag{A-54}$$

This equation represents the equation of a straight line with a slope of -40 dB/decade in the Bode plot coordinates. The intersection of the two asymptotes is

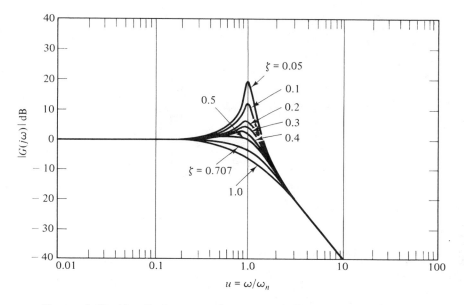

Figure A-10 Magnitude versus frequency of Bode plot of $G(s) = 1 / [1 + 2\zeta(s / \omega_n) + (s / \omega_n)^2]$.

found by equating Eq. (A-53) with Eq. (A-54), yielding

$$-40 \log_{10}\left(\frac{\omega}{\omega_n}\right) = 0 \text{ dB} \tag{A-55}$$

and from which we get

$$\omega = \omega_n \tag{A-56}$$

Thus the frequency, $\omega = \omega_n$, is considered to be the corner frequency of the second-order transfer function of Eq. (A-50), with the condition that $\zeta \leq 1$.

The actual magnitude plot of $G(j\omega)$ in this case may differ strikingly from the asymptotic lines. The reason for this is that the amplitude and phase curves of the $G(j\omega)$ of Eq. (A-50) depend not only on the corner frequency ω_n, but also on the damping ratio ζ. The actual and the asymptotic magnitude plots of $G(j\omega)$ are shown in Fig. A-10 for several values of ζ. The errors between the two sets of curves are shown in Fig. A-11 for the same set of ζs. The standard procedure of constructing the magnitude portion of the Bode plot of a second-order transfer function of the form of Eq. (A-50) is to first locate the corner frequency ω_n, then sketch the asymptotic lines; the actual curve is obtained by making corrections to the asymptotes by using either the error curves of Fig. A-11 or the curves in Fig. A-10 for the corresponding ζ.

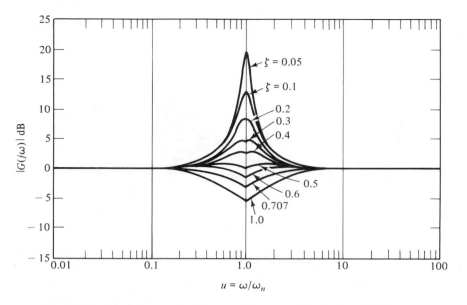

Figure A-11 Error in magnitude versus frequency of Bode plot of $G(s) = 1 /$ $[1 + 2\zeta(s / \omega_n) + (s / \omega_n)^2]$.

The phase of $G(j\omega)$ is given by

$$\angle G(j\omega) = -\tan^{-1}\left\{\frac{2\zeta\omega}{\omega_n}\ \left[1 - \left(\frac{\omega}{\omega_n}\right)^2\right]\right\} \tag{A-57}$$

and is plotted as shown in Fig. A-12 various values of ζ.

The analysis of the Bode plot of the second-order transfer function of Eq. (A-50) may be applied to a second-order transfer function with two complex zeros.

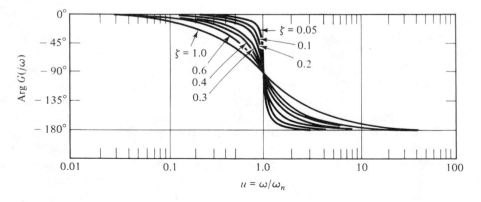

Figure A-12 Phase versus frequency of Bode plot of $G(s) = 1 / [1 + 2\zeta$ $(s / \omega_n) + (s / \omega_n)^2]$.

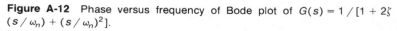

If

$$G(s) = 1 + \frac{2\zeta}{\omega_n}s + \frac{1}{\omega_n^2}s^2 \qquad \text{(A-58)}$$

the Bode plot of $G(j\omega)$ may be obtained by inverting the curves of Figs. A-10, A-11, and A-12.

■

Example A-3

As an illustrative example of the Bode plot of a transfer function, let us consider

$$G(s) = \frac{10(s + 10)}{s(s + 2)(s + 5)} \qquad \text{(A-59)}$$

The first step is to express the transfer function in the form of Eq. (A-28) and set $s = j\omega$. Thus, Eq. (A-59) becomes

$$G(j\omega) = \frac{10(1 + j0.1\omega)}{j\omega(1 + j0.5\omega)(1 + j0.2\omega)} \qquad \text{(A-60)}$$

This equation shows that $G(j\omega)$ has corner frequencies at $\omega = 10$, 2, and 5 rad/sec. The pole at the origin gives a magnitude curve that is a straight line with a slope of -20 dB/decade and passing through the $\omega = 1$ rad/sec point on the ω axis at 0 dB. The total Bode plots of the magnitude and phase of $G(j\omega)$ are obtained by adding the component curves together, point by point, as shown in Fig. A-13. The actual magnitude curve may be

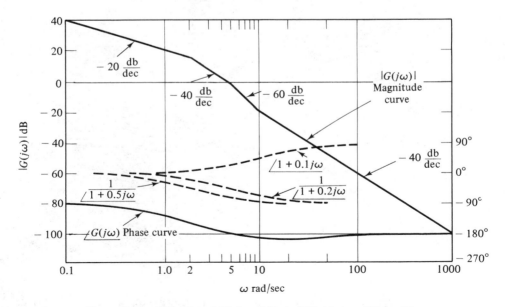

Figure A-13 Bode plot of $G(s) = [10(s + 10)] / [s(s + 2)(s + 5)]$.

obtained by considering the errors of the asymptotic curves at the significant frequencies. However, in practice, the accuracy of the asymptotic lines is deemed adequate for transfer functions with only real poles and zeros.

∎

A.3 MAGNITUDE-VERSUS-PHASE PLOT

The magnitude-versus-phase diagram is a plot of the magnitude of the transfer function in decibels versus its phase in degrees, with ω as a parameter on the curve. One of the most important applications of this plot is that it can be superposed on

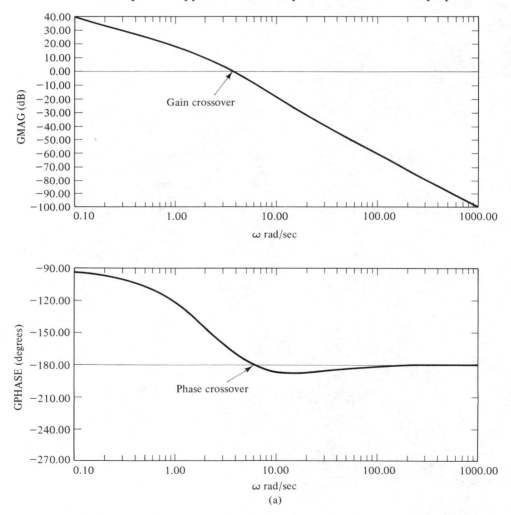

Figure A-14 $G(s) = [10(s + 10)] / [s(s + 2)(s + 5)]$. (a) Polar plot. (b) Bode diagram. (c) Magnitude-versus-phase plot.

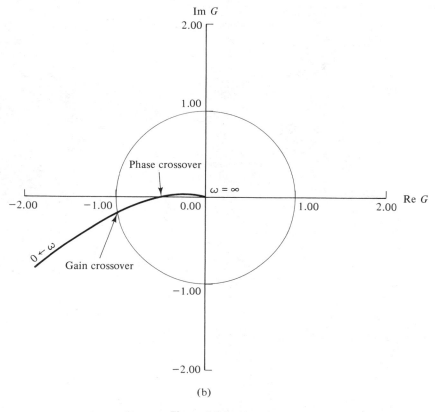

(b)

Figure A-14   (cont.)

the Nichols chart (see Chapter 9) to give information on the relative stability and the frequency response of the closed-loop system. When the gain factor K of the transfer function is varied, the plot is simply raised or lowered vertically according to the value of K in decibels. However, the unique property of adding the individual plots for cascaded terms in the Bode plot does not carry over to this case. Thus, the amount of work involved in obtaining the magnitude-versus-phase plot is equivalent to that of the polar plot, unless a digital computer is used to generate the data. Usually, the magnitude-versus-phase plots are obtained by first making the Bode plot, and then transferring the data to the decibel-versus-phase coordinates.

As an illustrative example, the Bode plot, the polar plot, and the magnitude-versus-phase plot of the function

$$G(s) = \frac{10(s + 10)}{s(s + 2)(s + 5)} \tag{A-61}$$

are sketched as shown in Fig. A-14. The Bode plot shown in Fig. A-14(a) is apparently the easiest one to sketch. The others are obtained by transferring the data from the Bode plot to the proper coordinates.

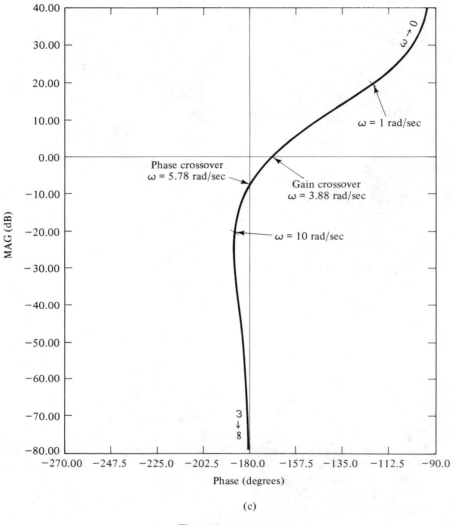

(c)

Figure A-14 (cont.)

The relationships among these three plots are easily established by comparing the curves in Fig. A-14 without the need of detailed explanation. However, for the purpose of analysis and design, it is convenient to define the following terms:

Gain-Crossover Frequency. This is the frequency at which the magnitude of the transfer function $G(j\omega)$ is unity. In logarithm scale, this corresponds to 0 dB. The following interpretations are made with respect to the three types of plots:

Polar Plot. The gain-crossover point (or points) is where $|G(j\omega)| = 1$ [Fig. A-14(a)].

Bode Plot. The gain-crossover point (or points) is where the magnitude curve of $G(j\omega)$ crosses the 0-dB axis [Fig. A-14(b)].

Magnitude-Versus-Phase Plot. The gain-crossover point (or points) is where the $G(j\omega)$ plot crosses the 0-dB axis [Fig. A-14(c)].

Phase-Crossover Frequency. This is the frequency at which the phase of $G(j\omega)$ is 180°.

Polar Plot. The phase-crossover point (or points) is where the phase of $G(j\omega)$ is 180 degrees, or where the $G(j\omega)$ plot crosses the negative real axis [Fig. A-14(a)].

Bode Plot. The phase-crossover point (or points) is where the phase curve crosses the 180° axis [Fig. A-14(b)].

Magnitude-Versus-Phase Curve. The phase-crossover point (or points) is where the $G(j\omega)$ plot intersects the 180° axis [Fig. A-14(c)].

appendix B

Laplace Transform Table

Laplace Transform, $F(s)$	Time Function, $f(t)$
$\dfrac{1}{s}$	$u(t)$ (unit step function)
$\dfrac{1}{s^2}$	t
$\dfrac{n!}{s^{n+1}}$	t^n (n = positive integer)
$\dfrac{1}{s+a}$	e^{-at}
$\dfrac{1}{(s+a)(s+b)}$	$\dfrac{e^{-at}-e^{-bt}}{b-a}$
$\dfrac{\omega_n^2}{s^2+2\zeta\omega_n s+\omega_n^2}$	$\dfrac{\omega_n}{\sqrt{1-\zeta^2}}e^{-\zeta\omega_n t}\sin\omega_n\sqrt{1-\zeta^2}\,t$
$\dfrac{1}{(1+sT)^n}$	$\dfrac{1}{T^n(n-1)!}t^{n-1}e^{-t/T}$
$\dfrac{\omega_n^2}{(1+Ts)(s^2+2\zeta\omega_n s+\omega_n^2)}$	$\dfrac{T\omega_n^2 e^{-t/T}}{1-2\zeta T\omega_n+T^2\omega_n^2}+\dfrac{\omega_n e^{-\zeta\omega_n t}\sin(\omega_n\sqrt{1-\zeta^2}\,t-\phi)}{\sqrt{(1-\zeta^2)(1-2\zeta T\omega_n-T^2\omega_n^2)}},$ where $\phi=\tan^{-1}\dfrac{T\omega_n\sqrt{1-\zeta^2}}{1-T\zeta\omega_n}$
$\dfrac{\omega_n}{s^2+\omega_n^2}$	$\sin\omega_n t$

Laplace Transform, $F(s)$	Time Function, $f(t)$
$\dfrac{\omega_n}{(1 + Ts)(s^2 + \omega_n^2)}$	$\dfrac{T\omega_n}{1 + T^2\omega_n^2}e^{-t/T} + \dfrac{1}{\sqrt{1 + T^2\omega_n^2}}\sin(\omega_n t - \phi)$ where $\phi = \tan^{-1}\omega_n T$
$\dfrac{\omega_n^2}{s(s^2 + 2\zeta\omega_n s + \omega_n^2)}$	$1 + \dfrac{1}{\sqrt{1 - \zeta^2}}e^{-\zeta\omega_n t}\sin(\omega_n\sqrt{1 - \zeta^2}\,t - \phi)$ where $\phi = \tan^{-1}\dfrac{\sqrt{1 - \zeta^2}}{-\zeta}$
$\dfrac{\omega_n^2}{s(s^2 + \omega_n^2)}$	$1 - \cos\omega_n t$
$\dfrac{1}{s(1 + Ts)}$	$1 - e^{-t/T}$
$\dfrac{1}{s(1 + Ts)^2}$	$1 - \dfrac{t + T}{T}e^{-t/T}$
$\dfrac{\omega_n^2}{s(1 + Ts)(s^2 + 2\zeta\omega_n s + \omega_n^2)}$	$1 - \dfrac{T^2\omega_n^2}{1 - 2T\zeta\omega_n + T^2\omega_n^2}e^{-t/T}$ $+ \dfrac{e^{-\zeta\omega_n t}\sin(\omega_n\sqrt{1 - \zeta^2}\,t - \phi)}{\sqrt{(1 - \zeta^2)(1 - 2\zeta T\omega_n + T^2\omega_n^2)}}$ where $\phi = \tan^{-1}\dfrac{\sqrt{1 - \zeta^2}}{-\zeta} + \tan^{-1}\dfrac{T\omega_n\sqrt{1 - \zeta^2}}{1 - T\zeta\omega_n}$
$\dfrac{\omega_n^2}{s^2(s^2 + 2\zeta\omega_n s + \omega_n^2)}$	$t - \dfrac{2\zeta}{\omega_n} + \dfrac{1}{\omega_n\sqrt{1 - \zeta^2}}e^{-\zeta\omega_n t}\sin(\omega_n\sqrt{1 - \zeta^2}\,t - \phi)$ where $\phi = 2\tan^{-1}\dfrac{\sqrt{1 - \zeta^2}}{-\zeta}$
$\dfrac{\omega_n^2}{s^2(1 + Ts)(s^2 + 2\zeta\omega_n s + \omega_n^2)}$	$t - T - \dfrac{2\zeta}{\omega_n} + \dfrac{T^3\omega_n^2}{1 - 2\zeta\omega_n T + T^2\omega_n^2}e^{-t/T}$ $+ \dfrac{e^{-\zeta\omega_n t}\sin(\omega_n\sqrt{1 - \zeta^2}\,t - \phi)}{\omega_n\sqrt{(1 - \zeta^2)(1 - 2\zeta\omega_n T + T^2\omega_n^2)}}$ where $\phi = 2\tan^{-1}\dfrac{\sqrt{1 - \zeta^2}}{-\zeta} + \tan^{-1}\dfrac{T\omega_n\sqrt{1 - \zeta^2}}{1 - T\omega_n\zeta}$
$\dfrac{1}{s^2(1 + Ts)^2}$	$t - 2T + (t + 2T)e^{-t/T}$

Laplace Transform, $F(s)$	Time Function, $f(t)$
$\dfrac{\omega_n^2(1+as)}{s^2+2\zeta\omega_n s+\omega_n^2}$	$\omega_n\sqrt{\dfrac{1-2a\zeta\omega_n+a^2\omega_n^2}{1-\zeta^2}}\,e^{-\zeta\omega_n t}\sin(\omega_n\sqrt{1-\zeta^2}\,t+\phi)$ where $\phi=\tan^{-1}\dfrac{a\omega_n\sqrt{1-\zeta^2}}{1-a\zeta\omega_n}$
$\dfrac{\omega_n^2(1+as)}{s^2+\omega_n^2}$	$\omega_n\sqrt{1+a^2\omega_n^2}\,\sin(\omega_n t+\phi)$ where $\phi=\tan^{-1}a\omega_n$
$\dfrac{\omega_n^2(1+as)}{(1+Ts)(s^2+2\zeta\omega_n s+\omega_n^2)}$	$\dfrac{\omega_n}{\sqrt{1-\zeta^2}}\sqrt{\dfrac{1-2a\zeta\omega_n+a^2\omega_n^2}{1-2T\zeta\omega_n+T^2\omega_n^2}}\,e^{-\zeta\omega_n t}$ $\times\sin(\omega_n\sqrt{1-\zeta^2}\,t+\phi)+\dfrac{(T-a)\omega_n^2}{1-2T\zeta\omega_n+T^2\omega_n^2}e^{-t/T}$ where $\phi=\tan^{-1}\dfrac{a\omega_n\sqrt{1-\zeta^2}}{1-a\zeta\omega_n}-\tan^{-1}\dfrac{T\omega_n\sqrt{1-\zeta^2}}{1-T\zeta\omega_n}$
$\dfrac{\omega_n^2(1+as)}{(1+Ts)(s^2+\omega_n^2)}$	$\dfrac{\omega_n^2(T-a)}{1+T^2\omega_n^2}e^{-t/T}+\dfrac{\omega_n\sqrt{1+a^2\omega_n^2}}{\sqrt{1+T^2\omega_n^2}}\sin(\omega_n t+\phi)$ where $\phi=\tan^{-1}a\omega_n-\tan^{-1}\omega_n T$
$\dfrac{\omega_n^2(1+as)}{s(s^2+2\zeta\omega_n s+\omega_n^2)}$	$1+\dfrac{1}{\sqrt{1-\zeta^2}}\sqrt{1-2a\zeta\omega_n+a^2\omega_n^2}\,e^{-\zeta\omega_n t}$ $\times\sin(\omega_n\sqrt{1-\zeta^2}\,t+\phi)$ where $\phi=\tan^{-1}\dfrac{a\omega_n\sqrt{1-\zeta^2}}{1-a\zeta\omega_n}-\tan^{-1}\dfrac{\sqrt{1-\zeta^2}}{-\zeta}$
$\dfrac{\omega_n^2(1+as)}{s(1+Ts)(s^2+\omega_n^2)}$	$1+\dfrac{T\omega_n^2(a-T)}{1+T^2\omega_n^2}e^{-t/T}-\sqrt{\dfrac{1+a^2\omega_n^2}{1+T^2\omega_n^2}}\cos(\omega_n t+\phi)$ where $\phi=\tan^{-1}a\omega_n-\tan^{-1}\omega_n T$
$\dfrac{\omega_n^2(1+as)}{s(1+Ts)(s^2+2\zeta\omega_n s+\omega_n^2)}$	$1+\sqrt{\dfrac{1-2\zeta a\omega_n+a^2\omega_n^2}{(1-\zeta^2)(1-2T\zeta\omega_n+T^2\omega_n^2)}}\,e^{-\zeta\omega_n t}$ $\times\sin(\omega_n\sqrt{1-\zeta^2}\,t+\phi)+\dfrac{\omega_n^2 T(a-T)}{1-2T\zeta\omega_n+T^2\omega_n^2}e^{-t/T}$ $\phi=\tan^{-1}[a\omega_n\sqrt{1-\zeta^2}/(1-a\zeta\omega_n)]$ $-\tan^{-1}\dfrac{T\omega_n\sqrt{1-\zeta^2}}{1-T\zeta\omega_n}-\tan^{-1}\dfrac{\sqrt{1-\zeta^2}}{-\zeta}$

Laplace Transform, $F(s)$	Time Function, $f(t)$
$\dfrac{1 + as}{s^2(1 + Ts)}$	$t + (a - T)(1 - e^{-t/T})$
$\dfrac{s\omega_n^2}{s^2 + 2\zeta\omega_n s + \omega_n^2}$	$\dfrac{\omega_n^2}{\sqrt{1 - \zeta^2}} e^{-\zeta\omega_n t}\sin(\omega_n\sqrt{1 - \zeta^2}\, t + \phi)$ where $\phi = \tan^{-1}\dfrac{\sqrt{1 - \zeta^2}}{-\zeta}$
$\dfrac{s}{s^2 + \omega_n^2}$	$\cos\omega_n t$
$\dfrac{s}{(s^2 + \omega_n^2)^2}$	$\dfrac{1}{2\omega_n} t \sin\omega_n t$
$\dfrac{s}{(s^2 + \omega_{n1}^2)(s^2 + \omega_{n2}^2)}$	$\dfrac{1}{\omega_{n2}^2 - \omega_{n1}^2}(\cos\omega_{n1}t - \cos\omega_{n2}t)$
$\dfrac{s}{(1 + Ts)(s^2 + \omega_n^2)}$	$\dfrac{-1}{(1 + T^2\omega_n^2)} e^{-t/T} + \dfrac{1}{\sqrt{1 + T^2\omega_n^2}}\cos(\omega_n t - \phi)$ where $\phi = \tan^{-1}\omega_n T$
$\dfrac{1 + as + bs^2}{s^2(1 + T_1 s)(1 + T_2 s)}$	$t + (a - T_1 - T_2) + \dfrac{b - aT_1 + T_1^2}{T_1 - T_2} e^{-t/T}$ $- \dfrac{b - aT_2 + T_2^2}{T_1 - T_2} e^{-t/T_2}$
$\dfrac{\omega_n^2(1 + as + bs^2)}{s(s^2 + 2\zeta\omega_n s + \omega_n^2)}$	$1 + \sqrt{\dfrac{\left(1 - a\zeta\omega_n - b\omega_n^2 + 2b\zeta^2\omega_n^2\right)^2 + \omega_n^2(1 - \zeta^2)\left(a - 2b\zeta\omega_n\right)^2}{1 - \zeta^2}}$ $\times e^{-\zeta\omega_n t}\sin(\omega_n\sqrt{1 - \zeta^2}\, t + \phi)$ where $\phi = \tan^{-1}\dfrac{\omega_n\sqrt{1 - \zeta^2}\,(a - 2b\zeta\omega_n)}{b\omega_n(2\zeta^2 - 1) + 1 - a\zeta\omega_n} - \tan^{-1}\dfrac{\sqrt{1 - \zeta^2}}{-\zeta}$
$\dfrac{s^2}{(s^2 + \omega_n^2)^2}$	$\dfrac{1}{2\omega_n}(\sin\omega_n t + \omega_n t \cos\omega_n t)$

Index

Tear out this card and fill in all necessary information. Then enclose this card with your check or money order *only* in an envelope and mail to:

Book Distribution Center
PRENTICE-HALL, Inc.
Route 59 at Brook Hill Drive
West Nyack, New York 10995

AUTOMATIC CONTROL SYSTEMS, 5/E—B. C. Kuo

Please send the item(s) checked below. PAYMENT ENCLOSED (Check or money order *only*). The Publisher will pay all shipping and handling charges.

___ Please send Computer Disks (2) and Software Manual (05495-7)—$50.00

___ Please send Software Manual *only* (05502-0)—$8.95

NAME_____

DEPT._____

SCHOOL_____

CITY_____STATE_____ZIP_____

NOTE: PROFESSIONAL/REFERENCE BOOKS ARE TAX-DEDUCTIBLE.
Prices subject to change without notice.

Dept. 1

D-ACSK-TB(2)

Tear out this card and fill in all necessary information. Then enclose this card with your check or money order *only* in an envelope and mail to:

Book Distribution Center
PRENTICE-HALL, Inc.
Route 59 at Brook Hill Drive
West Nyack, New York 10995

AUTOMATIC CONTROL SYSTEMS, 5/E—B. C. Kuo

Please send the item(s) checked below. PAYMENT ENCLOSED (Check or money order *only*). The Publisher will pay all shipping and handling charges.

___ Please send Computer Disks (2) and Software Manual (05495-7)—$50.00

___ Please send Software Manual *only* (05502-0)—$8.95

NAME_____

DEPT._____

SCHOOL_____

CITY_____STATE_____ZIP_____

NOTE: PROFESSIONAL/REFERENCE BOOKS ARE TAX-DEDUCTIBLE.
Prices subject to change without notice.

Dept. 1

D-ACSK-TB(2)